Differentiable and Complex Dynamics of Several Variables

Mathematics and Its Applications

Managing Editor:

M. HAZEWINKEL

Centre for Mathematics and Computer Science, Amsterdam, The Netherlands

Volume 483

Differentiable and Complex Dynamics of Several Variables

by

Pei-Chu Hu

Shandong University,
Jinan, China

and

Chung-Chun Yang

The Hong Kong University of Science and Technology,
Kowloon, Hong Kong

KLUWER ACADEMIC PUBLISHERS
DORDRECHT / BOSTON / LONDON

A C.I.P. Catalogue record for this book is available from the Library of Congress.

ISBN 978-90-481-5246-9

Published by Kluwer Academic Publishers,
P.O. Box 17, 3300 AA Dordrecht, The Netherlands.

Sold and distributed in North, Central and South America
by Kluwer Academic Publishers,
101 Philip Drive, Norwell, MA 02061, U.S.A.

In all other countries, sold and distributed
by Kluwer Academic Publishers,
P.O. Box 322, 3300 AH Dordrecht, The Netherlands.

Printed on acid-free paper

Printed in the Netherlands.

Contents

Preface

The development of dynamics theory began with the work of Isaac Newton. In his theory the most basic law of classical mechanics is $f = ma$, which describes the motion in $\mathbb{R}^n$ of a point of mass m under the action of a force f by giving the acceleration a. If the position of the point is taken to be a point $x \in \mathbb{R}^n$, and if the force f is supposed to be a function of x only, Newton's Law is a description in terms of a second-order ordinary differential equation:

$$m\frac{d^2 x}{dt^2} = f(x).$$

It makes sense to reduce the equations to first order by defining the velocity as an extra independent variable by $v = \dot{x} = \frac{dx}{dt} \in \mathbb{R}^n$. Then

$$\dot{x} = v, \quad m\dot{v} = f(x).$$

L. Euler, J. L. Lagrange and others studied mechanics by means of an analytical method called analytical dynamics. Whenever the force f is represented by a gradient vector field $f = -\nabla U$ of the potential energy U, and denotes the difference of the kinetic energy and the potential energy by

$$L(x, v) = \frac{1}{2}m\langle v, v \rangle - U(x),$$

the Newton equation of motion is reduced to the Euler-Lagrange equation

$$\frac{d}{dt}\left(\frac{\partial L}{\partial v}\right) = \frac{\partial L}{\partial x}.$$

If the momenta $y = \frac{\partial L}{\partial v}$ are used as the variables, the Euler-Lagrange equation can be written as

$$\dot{y} = \frac{\partial L}{\partial x}.$$

Further, W. R. Hamilton introduced the Hamiltonian function $H = \langle y, v \rangle - L$, and transformed the Euler-Lagrange equation into the system of Hamiltonian equations

$$\dot{x} = \frac{\partial H}{\partial y}, \quad \dot{y} = -\frac{\partial H}{\partial x}.$$

Thus above types of motion can always be described by a system of first-order ordinary differential equation:

$$\dot{x} = X(x).$$

In the late nineteenth century, H. Poincaré created a new branch of mathematics on the qualitative theory of differential equations. The point of his idea was to relate the geometry of the phase space with the analysis. As a result of Poincaré's qualitative approach, one of the focuses of the dynamics theory has shifted away from the existence of solutions to the study of the properties(such as stability, perturbation, and bifurcation, etc.) of the solutions of the above system of equations. Roughly, the solutions are a family $\{f^t\}_{t \in \mathbb{R}}$ of self-mappings on the phase space satisfying

$$f^0 = id, \quad f^t \circ f^s = f^{t+s}.$$

Our starting point is to study Fatou-Julia type theory of the families.

Ergodic theory originates from the so-called ergodic hypothesis posed by L. Boltzmann and J. Gibbs as the foundation of statistical mechanics. Many mathematicians try to find a rigorous proof of the hypothesis. As an attempts to prove this, the recurrence theorem of H. Poincaré and C. Carathéodory and the ergodic theorem of G. D. Birkhoff and J. von Neumann were obtained. Particularly, G. D. Birkhoff discussed many dynamical phenomena (such as invariant) in the context of transformation groups acting on general metric spaces. Since then, many mathematicians have considered dynamics theory in the environment of geometry. Now the theory is closely related to probability, geometry, and so on. All these approaches are studied on real manifolds utilizing the continuous or discrete properties of the dynamics.

Another development in the dynamics is the iteration of rational functions of one complex variable, which has its origin in long mémoires by Fatou and Julia, based on the Koebe-Poincaré Uniformization Theorem, Montel's Normality Criterion and earlier works on functional equations due to Böttcher, Poincaré and Schröder at the turn of the century. The so-called Fatou or Julia set is defined by considering the normality of sequence of iterates of an arbitrary non-linear rational function. Lately, the theory is generalized to complex Euclidean spaces $\mathbb{C}^m$ and complex projective spaces $\mathbb{P}^m$ by the theory of holomorphic mappings.

One of the major purposes of this book is to make a survey on the Fatou-Julia type of theory in complex manifolds and to provide some advanced account of dynamical system in the framework of geometry and analysis. Our main contribution here is to introduce a uniform approach to tackle Fatou-Julia type theory for both real and complex manifolds. In order to do so, we have introduced some notations and basic results of the ergodic theory and differentiable dynamics for real manifolds, and many problems connected with ergodic theory and differentiable dynamics have been posed for further studies. More specifically we have in*· ·duced some new concepts of invariant set, in terms of the integration. Accordingly, some relationships between the Julia or Fatou set and the hyperbolic sets have been established. Also criteria for the Julia sets of continuous mappings on compact manifolds to be non-empty have been derived. Moreover, we have studied the existence and quantitative measure of the fixed points of holomorphic mappings on $\mathbb{C}^m$ and the relationship between the topological and analytic descriptions of a repulsive or attractive cycle.

The book contains seven chapters: 1. Fatou-Julia type theory; 2. Ergodic theorems and invariant sets; 3. Hyperbolicity in differentiable dynamics; 4. Some topics in dynamics; 5. Hyperbolicity in complex dynamics; 6. Iteration theory on $\mathbb{P}^m$; 7. Complex dynamics in $\mathbb{C}^m$; and contains two appendices: A. Foundations of differentiable dynamics; B. Foundations of complex dynamics. In appendices A and B, we introduce basic notations, terminologies and facts used in differentiable and complex dynamics. Each chapter is self-contained and the book is appended by a comprehensive and up-dated bibliography. It is hoped that the book will provide some new research directions with unified approaches as well as challenging problems in studying dynamics of higher dimensional spaces for the readers.

We wish to thank Hong Kong's University Grant Center and China's National Natural Science Foundation for their financial support, which enabled us to do research jointly for past years and complete the writing of the book at the Hong Kong University of Science & Technology–a scenic campus with excellent library and computer facilities. Last but not least, we are indebted to our wives, Jin and Chwang-Chia, for their patience and support throughout the task.

Pei-Chu Hu & Chung-Chun Yang
Hong Kong, 1997

Chapter 1

Fatou-Julia type theory

For a continuous mapping $f : M \longrightarrow M$ on a topological space M, we will give a series of dichotomies of the space M by using different properties which are satisfied by the cascade $\{f^n\}$ generated by f. Roughly, given a point $x \in M$, if there exists a neighborhood U of x such that $\{f^n\}$ is of a property (P) on U, we write $x \in F_{(P)}(f)$. Obviously, $F_{(P)}(f)$ is open. Set $J_{(P)}(f) = M - F_{(P)}(f)$. In many cases, $F_{(P)}(f)$ and $J_{(P)}(f)$ are invariant sets on M. We will discuss these sets for some property (P).

In this chapter, we first dichotomize manifolds by using the method of Fatou and Julia, that is normality or equicontinuity, to obtain the Fatou sets and Julia sets which are the foundation of complex dynamics, and then derive criteria for the Julia sets of continuous mappings on compact manifolds to be non-empty.

1.1 Hyperbolic fixed points

Let $\mathbf{A} : \mathbb{R}^m \longrightarrow \mathbb{R}^m$ be a linear mapping. The set of eigenvalues of $\mathbf{A}$ is said to be the *spectrum* of $\mathbf{A}$, which is denoted by $\mathrm{sp}(\mathbf{A})$. Define

$$r(\mathbf{A}) = \max_{\lambda \in \mathrm{sp}(\mathbf{A})} |\lambda|,$$

and call it the *spectral radius* of $\mathbf{A}$. Given a norm $\| \cdot \|$ on $\mathbb{R}^m$, the norm of $\mathbf{A}$ is induced by

$$\|\mathbf{A}\| := \sup_{\|v\|=1} \|\mathbf{A}v\|.$$

Clearly $\|\mathbf{A}\| \geq r(\mathbf{A})$, and if $\| \cdot \|$ is the Euclidean norm, one has $\|\mathbf{A}\| = r(\mathbf{A})$ whenever $\mathbf{A}$ is diagonal. If we denote the set of norms in $\mathbb{R}^m$ by $\mathcal{N}(\mathbb{R}^m)$, then

$$r(\mathbf{A}) = \inf_{\| \cdot \| \in \mathcal{N}(\mathbb{R}^m)} \|\mathbf{A}\|,$$

that is, for every $\varepsilon > 0$ there exists a norm $\| \cdot \|$ in $\mathbb{R}^m$ such that

$$\|\mathbf{A}\| < r(\mathbf{A}) + \varepsilon,$$

see [134]. For a fixed norm $\| \cdot \|$ in $\mathbb{R}^m$, one can conclude that for every $\varepsilon > 0$ there exists $a = a_\varepsilon$ such that

$$r(\mathbf{A})^n \leq \|\mathbf{A}^n\| < a(r(\mathbf{A}) + \varepsilon)^n,$$

which implies the *spectral radius formula*

$$\|\mathbf{A}^n\|^{1/n} \to r(\mathbf{A}) \quad (n \to \infty). \tag{1.1}$$

A linear mapping $\mathbf{A} : \mathbb{R}^m \longrightarrow \mathbb{R}^m$ is said to be *hyperbolic* (resp., *parabolic*) if it has no eigenvalues of modulus one (resp., if at least one eigenvalue has norm one), and is said to be a *contraction* (*expansion*) if all its eigenvalues have modulus less than (greater than) one. The main properties of such mappings are summarized in the following theorem (cf. [134] or see Lemma B.4 and Corollary B.1):

Theorem 1.1 *If* $\mathbf{A} : \mathbb{R}^m \longrightarrow \mathbb{R}^m$ *is a hyperbolic linear mapping, then there are subspaces* E^s *and* E^u *of* $\mathbb{R}^m$ *such that*
  1) $\mathbb{R}^m = E^s \oplus E^u$;
  2) $\mathbf{A}(E^s) \subset E^s$, $\mathbf{A}(E^u) = E^u$;
  3) $\mathbf{A}|_{E^s}$ *is a contraction,* $\mathbf{A}|_{E^u}$ *is an expansion.*

The subspaces E^s and E^u of Theorem 1.1 are easily identified as the eigenspaces of eigenvalues with modulus less than 1 and greater than 1, and are usually called the *stable* and *unstable eigenspaces* of $\mathbf{A}$, respectively. Indeed, the direct sum of the identification E^s and E^u gives the whole of $\mathbb{R}^m$, since there is no eigenvalue that is not associated with either E^s or E^u. Of course, contractions ($E^s = \mathbb{R}^m$) and expansions ($E^u = \mathbb{R}^m$) are hyperbolic. When $E^s, E^u \neq \mathbb{R}^m$ the mapping is said to be '*saddle-type*'.

Given any linear mapping $\mathbf{A} : \mathbb{R}^m \longrightarrow \mathbb{R}^m$, there is a corresponding linear flow on $\mathbb{R}^m$, given by the exponential matrix $\exp(\mathbf{A}t)$. It is easy to show that $\exp(\mathbf{A}t)x$ has a vector field $\mathbf{A}x$ and thus $\exp(\mathbf{A}t)x_0$ is the solution of the linear differential equation

$$\dot{x} = \mathbf{A}x \tag{1.2}$$

passing through x_0 at $t = 0$. The linear flow $\exp(\mathbf{A}t)x$ is said to be *hyperbolic* if $\mathbf{A}$ has no eigenvalues with zero real part, and is said to be a *contraction* (*expansion*) if all the eigenvalues of $\mathbf{A}$ have negative (positive) real parts. If $\exp(\mathbf{A}t)x$ is hyperbolic then $\mathbf{A}$ must be non-singular and so $\mathbf{A}x = 0$ has only the trivial solution $x = 0$. It follows that the origin is the only fixed point of the flow. The exponential matrix $\exp(\mathbf{A}t)$ represents a hyperbolic linear automorphism for each $t \neq 0$. An eigenvector of $\mathbf{A}$ with eigenvalue λ is also an eigenvector of $\exp(\mathbf{A}t)$ with eigenvalue $\exp(\lambda t)$. Thus $\mathrm{Re}(\lambda_i) \neq 0$ for each i implies $|\exp(\lambda_i t)| = \exp(\mathrm{Re}(\lambda_i)t) \neq 1$, for all $t \neq 0$. It follows that Theorem 1.1 applies and $\mathbb{R}^m$ can be decomposed into stable (E^s) and unstable (E^u) subspaces associated with the flow. Clearly, $E^s(E^u)$ is the direct sum of eigenspaces associated with eigenvalues of $\mathbf{A}$ for which $\mathrm{Re}(\lambda_i) < 0(> 0)$. Thus $\exp(\mathbf{A}t)|_{E^s}$ is a contraction and $\exp(\mathbf{A}t)|_{E^u}$ is an expansion for $t > 0$ (cf. [134]).

A fixed point p of a differentiable mapping f on a manifold M is said to be *hyperbolic* (resp., *parabolic*) if the differential $(df)_p$ of f at p is a hyperbolic (resp., parabolic) linear mapping on the tangent space $T(M)_p$. In particular, p is said to be *attracting* if the spectrum of $(df)_p$ is contained in the unit disc $\{z \in \mathbb{C} \mid |z| < 1\}$; *repelling* if the spectrum of $(df)_p$ is contained in $\{z \in \mathbb{C} \mid |z| > 1\}$, and *saddle-type* if $(df)_p$ is saddle-type.

A equilibrium point p of a semiflow $\{f^t\}_{t \in \mathbb{R}_+}$ is said to be *hyperbolic* (resp., *parabolic, attracting, repelling,* and *saddle-type*) if it is hyperbolic (resp., parabolic, attracting, repelling, and saddle-type) as a fixed point of the time one mapping f^1. If $\{f^t\}$ is the flow obtained by integration of a C^1 vector field X, the equilibrium points of $\{f^t\}$ are the points where X vanishes (*critical points* of X). At an equilibrium point p, one has $(df^t)_p = \exp((dX)_p t)$. Therefore, p is hyperbolic (resp., attracting, repelling) iff the spectrum of $(dX)_p$ is disjoint from the imaginary axis (resp., to its left, to its right).

Theorem 1.2 *A fixed point p of a differentiable mapping f on a manifold M is hyperbolic iff the tangent space $T(M)_p$ decomposes into a direct sum $T(M)_p = E^s_p \oplus E^u_p$ of two subspaces, namely, a stable (shrinking) space E^s_p and an unstable (expanding) space E^u_p, with the following properties: for $\xi \in E^s_p, \eta \in E^u_p, n \geq 0$*

$$(df)_p E^s_p \subset E^s_p, \quad (df)_p E^u_p = E^u_p, \tag{1.3}$$

$$\|(df^n)_p \xi\| \leq a e^{-cn} \|\xi\|, \quad \|(df^n)_p \eta\| \geq \frac{1}{a} e^{cn} \|\eta\|, \tag{1.4}$$

where a, c are positive constants that are independent of ξ, η, n.

Proof. If p is a hyperbolic fixed point of f, by the definition and Theorem 1.1, the tangent space $T(M)_p$ decomposes into a direct sum $T(M)_p = E^s_p \oplus E^u_p$ with (1.3) such that $(df)_p|_{E^s_p}$ is a contraction, and $(df)_p|_{E^u_p}$ is an expansion. By the chain rule we get the identity

$$(df^n)_p = (df)_{f^{n-1}(p)} \cdot (df)_{f^{n-2}(p)} \cdots (df)_{f(p)} \cdot (df)_p = \mathbf{A}^n,$$

where $\mathbf{A} = (df)_p$. Then for $\xi \in E^s_p$

$$\begin{aligned} \|(df^n)_p \xi\| &= \|(\mathbf{A}|_{E^s_p})^n \xi\| \leq \|(\mathbf{A}|_{E^s_p})^n\| \|\xi\| \\ &< a(r(\mathbf{A}|_{E^s_p}) + \varepsilon)^n \|\xi\|, \end{aligned}$$

where the norm of the tangent vector is taken with respect to a fixed Riemannian metric on M. Choose $\varepsilon, c > 0$ such that $r(\mathbf{A}|_{E^s_p}) + \varepsilon < e^{-c} < 1$. Then $\|(df^n)_p \xi\| \leq a e^{-cn} \|\xi\|$. Note that $\{(df)_p|_{E^u_p}\}^{-1}$ is a contraction so that for $\eta \in E^u_p$, we have $\|\{(df)_p|_{E^u_p}\}^{-n} \eta\| \leq a e^{-cn} \|\eta\|$, or $\|(df^n)_p \eta\| \geq \frac{1}{a} e^{cn} \|\eta\|$.

Conversely, we easily see $\|\{(df)_p|_{E^s_p}\}^n\| \to 0$ and $\|\{(df)_p|_{E^u_p}\}^{-n}\| \to 0$, that is,

$$r((df)_p|_{E^s_p}) < 1, \quad r(\{(df)_p|_{E^u_p}\}^{-1}) < 1.$$

Hence $(df)_p$ has no eigenvalues of modulus one. □

Remark [cf. [134]]. The condition (1.4) is equivalent to that there exist $\lambda < 1 < \gamma$ such that

$$\|(df)_p|_{E^s_p}\| \leq \lambda, \quad \|\{(df)_p|_{E^u_p}\}^{-1}\| \leq \gamma^{-1}. \tag{1.5}$$

Suppose that Λ is a periodic trajectory of period k for the differentiable mapping f ; $\Lambda = \{x_0, ..., x_{k-1}\}$ with $f(x_i) = x_{i+1}$ (mod k). We say that Λ is *hyperbolic* (resp., *attracting, repelling*) if x_0 is hyperbolic (resp., attracting, repelling) as a fixed point of f^k. For $0 < i < k$, let $A = (df^i)_{x_0}$, $B = (df^{k-i})_{x_i}$. If $(df^k)_{x_0} = BA$ is hyperbolic (resp., attracting, repelling), $(df^k)_{x_i} = AB$ is also hyperbolic (resp., attracting, repelling) because, if $z \neq 0$, the existence of $(I - z^{-1}BA)^{-1} = C$ implies the existence of $(I - z^{-1}AB)^{-1} = I + z^{-1}ACB$. Thus the definition don't depend on the choice of $x_0 \in \Lambda$. The stable and unstable spaces $E^s_{x_i}, E^u_{x_i}$ for f^k at x_i are also called stable and unstable spaces for f at the periodic point x_i, which are characterized by the following result:

Theorem 1.3 *A periodic trajectory Λ of period k of a differentiable mapping f on a manifold M is hyperbolic iff for each $x_i \in \Lambda$, the tangent space $T(M)_{x_i}$ decomposes into a direct sum $T(M)_{x_i} = E^s_{x_i} \oplus E^u_{x_i}$ of the stable space $E^s_{x_i}$ and unstable space $E^u_{x_i}$ of f with the following properties: for $\xi \in E^s_{x_i}, \eta \in E^u_{x_i}, n \geq 0$*

$$(df)_{x_i} E^s_{x_i} \subset E^s_{f(x_i)}, \quad (df)_{x_i} E^u_{x_i} = E^u_{f(x_i)}, \tag{1.6}$$

$$\|(df^n)_{x_i}\xi\| \leq ae^{-cn}\|\xi\|, \quad \|(df^n)_{x_i}\eta\| \geq \frac{1}{a}e^{cn}\|\eta\|, \tag{1.7}$$

where a, c are positive constants that are independent of ξ, η, n.

Thus the periodic trajectory Λ is attracting (resp., repelling) if $E^u_{x_i} = \{0\}$ (resp., $E^s_{x_i} = \{0\}$) for $i = 0, ..., k-1$. If Λ is attracting (resp., repelling), each $x_i \in \Lambda$ is said to be a *sink* (resp., *source*). S. Smale suggested in [236] that most diffeomorphisms on S^2 with the C^r topology ought to have only finitely many sinks. However, S. E. Newhouse [186] showed that the case is not such by proving that on any compact C^∞ manifold M of dimension greater than one, there is a residual subset $\mathcal{R}$ of an open set $\mathcal{U}$ in $\text{Diff}^r(M, M)$, $r \geq 2$, such that every element of $\mathcal{R}$ has infinitely many sinks, where, by definition, $\mathcal{R}$ contains a countable intersection of dense open subsets of $\mathcal{U}$.

1.2 Attractive and repulsive fixed points

Take a topological dynamical system $\{f^t\}_{t \in \kappa}$ defined on M, where $\kappa = \mathbb{R}, \mathbb{R}_+, \mathbb{Z}$, or $\mathbb{Z}_+$. If $A = \{p\}$ is asymptotically stable, by definition, there exists a neighborhood U of p such that

$$\lim_{t \to +\infty} f^t(x) = p \quad \text{for all} \quad x \in U,$$

which implies that p is a fixed point of the DS $\{f^t\}$. For this case, the fixed point p is also called *(topological) attractive* (see [255]), or an *attractor*. Obviously, the basin of attraction satisfies

$$\text{Att}(p) = O(U) = \bigcup_{-\infty}^{\infty} f^t(U).$$

If M is a metric space with a distance function d, the basin of attraction is closely related to a stable set of the DS. Take a point $x \in M$ and a subset A of M. The *stable set* $W^s(x)$ of the point $x \in M$ for the DS $\{f^t\}_{t \in \kappa}$ is defined as

$$W^s(x) = \{y \in M \mid d(f^t(y), f^t(x)) \to 0 \text{ as } t \to +\infty\},$$

and we denote

$$W^s(A) = \bigcup_{x \in A} W^s(x).$$

For the case $\kappa = \mathbb{R}$ or $\mathbb{Z}$, we can define the *unstable set* $W^u(x)$ of the point $x \in M$ by

$$W^u(x) = \{y \in M \mid d(f^{-t}(y), f^{-t}(x)) \to 0 \text{ as } t \to +\infty\},$$

and also write

$$W^u(A) = \bigcup_{x \in A} W^u(x).$$

Further if we assume that d induces the topology of M, then we have

$$\mathrm{Att}(A) = \{x \in M \mid d(f^t(x), A) \to 0 \ \text{ as } t \to \infty\},$$

here as usually $d(x, A) = \inf_{y \in A} d(x, y)$. In particular, if p is a (topological) attractive fixed point, then

$$\mathrm{Att}(p) = W^s(p).$$

It is easy to prove that $W^s(\mathrm{Per}(f))$ is invariant with

$$\mathrm{Per}(f) \subset W^s(\mathrm{Per}(f)) \subset \mathrm{Att}(\mathrm{Per}(f)).$$

Points in $W^s(\mathrm{Per}(f))$ usually are called *asymptotically periodic*. Obviously, $x \notin W^s(\mathrm{Per}(f))$ iff

$$\limsup_{t \to +\infty} d(f^t(x), f^t(p)) > 0,$$

holds for each periodic point $p \in \mathrm{Per}(f)$.

If A is the set in Proposition A.9, then A is asymptotically stable. The zone of attraction of A is just $\mathrm{Att}(A)$ which is an open set. Clearly it contains the whole of $\mathrm{Att}(L)$ with $L \subset A$ and the whole of $W^s(x)$ with $x \in A$, but it may not be filled out by them.

Definition 1.1 *A set A is called (topological) repulsive or a repeller*
1) if for every neighborhood V of A, $V - [A]$ is non-empty;
2) if there exists a neighborhood U of A such that for each $x \in U - [A]$ there is an $t_0 > 0$ with $f^t(x) \notin U$ for all $t \geq t_0$.

Remarks. If $A = \{p\}$ is a fixed point of the DS $\{f^t\}$, Targonski[255] called the fixed point repulsive if the condition 2) is replaced by the following weak condition:
2)' if there exists a neighborhood U of p such that for each $x \in U - [p]$ there is an $t > 0$ with $f^t(x) \notin U$.

Theorem 1.4 *If one point of a k-cycle ($k \geq 1$) of a continuous self-mapping f of a topological space M is a (topological) attractive (repulsive) fixed point of f^k, then every point of the cycle is a (topological) attractive (repulsive) fixed point of f^k. Hence the cycle is asymptotically stable (topological repulsive).*

Proof. Let $A = \{x_0, ..., x_{k-1}\}$ be the cycle and let x_0 be the point which is a (topological) attractive fixed point of f^k by definition. Then $x_{k-l} = f^{k-l}(x_0)(0 < l < k)$. By definition, there exists a neighborhood U_0 of x_0 such that

$$\lim_{n \to \infty} f^{kn}(x) = x_0 \quad \text{for all} \quad x \in U_0.$$

Consider now any $y \in f^{-l}(U_0)$; for some $x \in U_0$, $f^l(y) = x$. Then

$$\begin{aligned}
\lim_{n \to \infty} f^{kn}(y) &= \lim_{n \to \infty} f^{kn-l}(f^l(y)) = \lim_{n \to \infty} f^{kn-l}(x) \\
&= \lim_{n \to \infty} f^{k(n-l)+k-l}(x) = f^{k-l}(\lim_{n \to \infty} f^{k(n-l)}(x)) \\
&= f^{k-l}(x_0) = x_{k-l}.
\end{aligned}$$

Thus x_{k-l} has a neighborhood $U_{0l} := f^{-l}(U_0)$ satisfying the definition of x_{k-l} as a (topological) attractive fixed point of f^k.

If $x \in U_0$, by the construction of U_{0l}, there exists $y_l \in U_{0l}$ such that $f^l(y_l) = x$. Note that

$$\{f^n(x)\} = \{f^{mk}(x)\} \cup \{f^{mk+1}(x)\} \cup \cdots \cup \{f^{mk+k-1}(x)\},$$

and that $f^{mk+k-l}(x) = f^{(m+1)k}(y_l)$. Then we have $L^+(x) \subset A$. By permutation, we can find a neighborhood U_l of x_l such that $L^+(x) \subset A$ if $x \in U_l$. Thus $U = U_0 \cup \cdots \cup U_{k-1}$ is a neighborhood of A with the property in Definition A.15.

Assume now that x_0 is a (topological) repulsive fixed point of f^k. Let U_0 be a neighborhood of x_0 with properties 1) and 2) in Definition 1.1. Consider the neighborhood $U_{0l} := f^{-l}(U_0)$ of x_{k-l}. Because of property 1) in Definition 1.1, there exists a sequence $\{y_n\}, y_n \in U_0 - [x_0]_{f^k}$, convergent to x_0. As f^{k-l} is continuous, $f^{k-l}(y_n) \to f^{k-l}(x_0) = x_{k-l}$ for $n \to \infty$. Since $f^{k-l}(y_n) \notin [x_{k-l}]_{f^k}$ condition 1) in Definition 1.1 is fulfilled for x_{k-l}. Let y be an element of $U_{0l} - [x_{k-l}]_{f^k}$. Then $f^l(y) \in U_0 - [x_0]_{f^k}$ and so there exists an $s_0 \in \mathbb{N}$ such that $f^{ks}(f^l(y)) \notin U_0$ for all $s \geq s_0$. Hence $f^{ks}(y) \notin f^{-l}(U_0) = U_{0l}$ for all $s \geq s_0$. So U_{0l} has property 2) in Definition 1.1.

If $x \in U_0 - [A]$, by the construction of U_{0l}, there exists $y_l \in U_{0l} - [A]$ such that $f^l(y_l) = x$. Thus $f^{mk}(y_l) \notin U_{0l}$ for all $m \geq m_{0l}(x)$, i.e., $f^{mk+l}(x) \notin U_{0l}$ for all $m \geq m_{0l}(x), 0 \leq l < k$. Then $f^n(x) \notin \cup_{l=0}^{k-1} U_{0l}$ for all $n \geq m_0(x) = \max_{0 \leq l < k} \{m_{0l}(x) + 1\}$, where $U_{00} = U_0$. By permutation, we can find a neighborhood U_{jl} of x_l such that if $x \in U_{jj} - [A]$, $f^n(x) \notin \cup_{l=0}^{k-1} U_{jl}$ for all $n \geq m_j(x)$. Obviously the set $U = \cap_{j=0}^{k-1} \{\cup_{l=0}^{k-1} U_{jl}\}$ is a neighborhood of A with the properties in Definition 1.1.   □

Theorem 1.5 *If a k-cycle ($k \geq 1$) of a continuous self-mapping f on a topological space M is (topological) repulsive, then every point of the cycle is a (topological) repulsive fixed point of f^k.*

Proof. Let $A = \{x_0, ..., x_{k-1}\}$ be the cycle. By Theorem 1.4, it is sufficient to prove that one point in A, say, x_0 is (topological) repulsive. Since A is (topological) repulsive, $[x_0] = [A]$, and a neighborhood of A also is a neighborhood of x_0, then x_0 satisfies the condition 2) in Definition 1.1.

Assume, to the contrary, that x_0 is not (topological) repulsive, i.e., x_0 does not satisfy the condition 1) in Definition 1.1. Then there exists a neighborhood U_0 of x_0 such that

$U_0 - [x_0] = \emptyset$, i.e., $U_0 \subseteq [x_0]$. Then Theorem A.3 implies $[U_0] = [x_0]$. Note that $U_l := f^{-l}(U_0)$ are neighborhood of x_{k-l} for $l = 1, 2, ..., k$. Thus we obtain a neighborhood $U = \cup_{l=1}^{k} U_l$ of A. Obviously, we have $U \subset [U_0] = [x_0] = [A]$, i.e., $U - [A] = \emptyset$. This is a contradiction. Hence x_0 is (topological) repulsive.

$\square$

Definition 1.2 *A k-cycle* $(k \geq 1)$ *of a continuous self-mapping* f *on a topological space* M *is said to be (topological) attractive if one point of the cycle is a (topological) attractive fixed point of* f^k.

Theorem 1.4 and Theorem 1.5 show that a k-cycle $(k \geq 1)$ of f is (topological) repulsive if and only if one point of the cycle is a (topological) repulsive fixed point of f^k. Also a (topological) attractive cycle is asymptotically stable, but if a cycle is asymptotically stable, is the cycle (topological) attractive?

Theorem 1.6 *Suppose that* M *is a complex manifold,* $f : M \longrightarrow M$ *a holomorphic mapping,* $p \in M$, $f(p) = p$, *and that the eigenvalues* λ_i *of* $A = f'(p)$ *satisfy* $|\lambda_1| \geq |\lambda_2| \geq ... \geq |\lambda_m|$ $(m = \dim M)$. *Then* p *is (topological) attractive if* $|\lambda_1| < 1$, *or (topological) repulsive if* $|\lambda_m| > 1$.

Proof. Assume $|\lambda_1| < 1$. Pick constants β_1, β_2, β so that $|\lambda_1| < \beta_1 < \beta_2 < \beta < 1$. The spectral radius formula gives an k so that $\|A^n\| < \beta_1^n$ for all $n \geq k$. Taking a local coordinate system ϕ at p with $\phi(p) = 0$ and approximating $\phi \circ f^k \circ \phi^{-1}$ by A^k in a coordinate neighborhood of 0 shows that there is an $r > 0$ so that for our fixed k, $z \in \mathbb{C}^m(r)$ implies

$$\|\phi \circ f^k \circ \phi^{-1}(z)\| \leq \beta_2^k \|z\|. \tag{1.8}$$

Put $c = \sup\{\|\phi \circ f^j \circ \phi^{-1}(z)\| / \|z\| \mid 0 \leq j < k, \ 0 < \|z\| < r\}$. If $n = qk + j$, $q = 1, 2, 3, ..., \ 0 \leq j < k$, and if $\|z\| < r$, then iterations of (1.8) yield

$$\begin{aligned} \|\phi \circ f^n \circ \phi^{-1}(z)\| &= \|\phi \circ f^j \circ \phi^{-1}(\phi \circ f^{qk} \circ \phi^{-1}(z))\| \\ &\leq c\|\phi \circ f^{qk} \circ \phi^{-1}(z)\| \leq c\beta_2^{qk}\|z\|. \end{aligned}$$

Thus, for all sufficiently large $n \geq n_0$ (where $n_0 \geq k$ depends only on k) we have

$$\|\phi \circ f^n \circ \phi^{-1}(z)\| < \beta^n \|z\| \text{ for all } z \in \mathbb{C}^m(r). \tag{1.9}$$

which means $\phi^{-1}(\mathbb{C}^m(r)) \subset \text{Att}(p)$ so that p is (topological) attractive.

Assume $|\lambda_m| > 1$. By the implicit function theorem, the inverse f^{-1} exists in a neighborhood of p. Note that $B = (f^{-1})'(p)$ has the eigenvalues $1/\lambda_i$. According to the proof above, there are $r > 0, n_0 > 0, 0 < \beta < 1$ such that

$$\|\phi \circ f^{-n} \circ \phi^{-1}(z)\| < \beta^n \|z\| \text{ for all } z \in \mathbb{C}^m(r), \ n \geq n_0.$$

Take $U = \phi^{-1}(\mathbb{C}^m(r))$. For any $\xi \in U - [p]$, there is a positive integer $s \geq n_0$ such that $\beta^s r \leq \|\phi(\xi)\| < r$. Then

$$f^n(\xi) \notin U \text{ for all } n \geq s.$$

In fact, if $\xi_n = f^n(\xi) \in U$ for some $n \geq s$, then $z_n = \phi(\xi_n) \in \mathbb{C}^m(r)$, so that

$$\|\phi(\xi)\| = \|\phi \circ f^{-n} \circ \phi^{-1}(z_n)\| < \beta^n \|z_n\| < \beta^s r,$$

which is a contradiction! Thus p is a (topological) repulsive fixed point.   □

According to the proof of Theorem 1.6, we, in fact, obtain the following general sufficient condition for the existence of a (topological) attractive or (topological) repulsive fixed point.

Theorem 1.7 *Suppose that $f : M \longrightarrow M$ is a C^1 mapping on a m-dimensional manifold M such that $f(p) = p$ for a point $p \in M$, and such that the eigenvalues λ_i of $(df)_p$ satisfy $|\lambda_1| \geq |\lambda_2| \geq ... \geq |\lambda_m|$. Then p is (topological) attractive if $|\lambda_1| < 1$, or (topological) repulsive if $|\lambda_m| > 1$.*

Theorem 1.8 *Suppose that M is a complex manifold, $f : M \longrightarrow M$ a holomorphic mapping, $p \in M$, $f(p) = p$, and that the eigenvalues λ_i of $A = f'(p)$ satisfy $|\lambda_1| \geq |\lambda_2| \geq ... \geq |\lambda_m|$ ($m = \dim M$). Assume that there exists a neighborhood V of p such that $f^n(V)$ is contained in a coordinate neighborhood of p for each $n \in \mathbb{Z}_+$. If p is (topological) attractive, then $|\lambda_1| < 1$.*

Proof. By assumptions, we can choose a neighborhood V of p and a chart $(U; z)$ of M with $z(p) = 0$, where $z = (z_1, ..., z_m)$, such that $f^n(V) \subset U$ for each $n \in \mathbb{Z}_+$. Take a small positive number r such that the polydisc $\Delta_r^m \subset V$. The Cauchy inequality yields

$$\left\| \frac{\partial f^n}{\partial z_i}(0) \right\| \leq \frac{\sqrt{m}}{r} \max_{\zeta \in \Gamma} \|f^n(\zeta)\| \to 0$$

as $n \to \infty$, where Γ is the skeleton of Δ_r^m. Note that

$$A^n = (f^n)'(p) = \left({}^t\left(\frac{\partial f^n}{\partial z_1}(0) \right), ..., {}^t\left(\frac{\partial f^n}{\partial z_m}(0) \right) \right).$$

Then we obtain $A^n \to 0$ as $n \to \infty$. Therefore $|\lambda_1| < 1$.   □

Remark. Now consider an k-cycle

$$f : x_0 \mapsto x_1 \mapsto \cdots \mapsto x_{k-1} \mapsto x_k = x_0$$

on a Riemann surface M. If the complex manifold M is $\mathbb{C}$, then the derivative

$$\lambda = (f^k)'(x_i) = f'(x_1) \cdot f'(x_2) \cdots f'(x_k)$$

is a well defined complex number called the *multiplier* or the eigenvalue of this cycle. More generally, for self-mappings of an arbitrary Riemann surface the multiplier of a cycle can be defined using a local coordinate chart around any point of the cycle. By definition, the cycle is either *attracting* or *repelling* or *indifferent* ($= neutral$) according as its multiplier satisfies $|\lambda| < 1$ or $|\lambda| > 1$ or $|\lambda| = 1$. The cycle is called *superattracting* if $\lambda = 0$.

Conjecture 1.1 *Suppose that M is a complex manifold, $f : M \longrightarrow M$ a holomorphic mapping, $p \in M$, $f(p) = p$, and that the eigenvalues λ_i of $A = f'(p)$ satisfy $|\lambda_1| \geq |\lambda_2| \geq ... \geq |\lambda_m|$ ($m = \dim M$). If p is (topological) repulsive, then $|\lambda_m| > 1$.*

1.3 Normality

The convention in force throughout this section is that all manifolds are *locally compact connected spaces* and all objects defined on them (differential forms, metrics, etc) are C^∞ unless stated to the contrary. It is known that all locally compact connected spaces are second countable. A second countable topology space satisfies the first countability axiom, the Lindelöf property and is separable, i.e., it has a dense subset consisting of countable points.

Let M be a smooth manifold. It is well-known that such manifolds are metrizable. A customary and useful device is to metrize these by imposing on them a Riemannian metric g, from which one derives a distance function $d_g(\,,\,)$ which converts the manifold into a metric space. In the sequel, the symbol d_g will denote the distance function reduced by some metric g. At any rate, once a distance function d is chosen on M, the following results are classic:

Lemma 1.1 (Hopf-Rinow) *A connected Riemannian manifold is a complete metric space iff every bounded infinite subset has limit points.*

Lemma 1.2 *Let M be a smooth manifold. Then the following conditions are equivalent:*
1) M is compact;
2) Every open covering of M contains a finite subcovering of M (Heine-Borel property);
3) M is complete and total bounded, that is, for any $\varepsilon > 0$, there exists a finite open covering of M by ε-balls;
4) Every infinite sequence has limit points (Bolzano-Weierstrass property).

Let N be another smooth manifold with a distance function d'. The following result is well known:

Lemma 1.3 ([135], p. 229) *A sequence $\{f_n\} \subseteq C(M, N)$ converges to $f \in C(M, N)$ in the sense of the compact-open topology iff f_n converges to f uniformly on compact sets of M.*

One of the principal results is that the uniform limit of continuous functions is continuous. Since M is second countable, there exists a countable open cover $\{U_i\}_{i=1}^s$ of M such that each U_i is relatively compact, where $s < \infty$ if M is compact. Set

$$E_k = \bigcup_{1 \le i \le k} \overline{U_i}.$$

Then E_k is compact, $E_k \subset E_{k+1}$, and $\cup_{k=1}^s E_k = M$. Obviously, every compact subset E of M is contained in an E_k. Now we define a distance function in $C(M, N)$. To do so we first replace d' by the distance function

$$\delta(w, w') = \frac{d'(w, w')}{1 + d'(w, w')},$$

which also satisfies the triangle inequality and has the advantage of being bounded. Next, for $f_1, f_2 \in C(M, N)$ we set

$$\delta_k(f_1, f_2) = \sup_{x \in E_k} \delta(f_1(x), f_2(x)),$$

which may be described as the distance between f_1 and f_2 on E_k. Finally, we adopt the definition

$$\rho(f_1, f_2) = \sum_{k=1}^{s} \frac{\delta_k(f_1, f_2)}{2^k}. \tag{1.10}$$

It is trivial to verify that $\rho(f_1, f_2)$ is finite and satisfies all conditions for a distance function.

Lemma 1.4 *Convergence of sequences in $C(M, N)$ with respect to the distance function ρ mean precisely the same as uniform convergence on compact sets.*

Proof. Suppose first that $f_n \to f$ in the sense of the ρ-distance. Then for large n we have $\rho(f_n, f) < \varepsilon$ and consequently, by the definition of ρ, $\delta_k(f_n, f) < 2^k \varepsilon$. But this implies that $f_n \to f$ uniformly on E_k, first with respect to the δ-metric, and also with respect to the d'-metric. Since every compact E is contained in an E_k it follows that the convergence is uniform on E.

Conversely, suppose that f_n converges uniformly to f on every compact set. Then $\delta_k(f_n, f) \to 0$ for every k, and because the series $\sum \delta_k(f_n, f) 2^{-k}$ has a convergent majorant with terms independent of n it follows readily that $\rho(f_n, f) \to 0$. $\square$

By Lemma 1.3 and Lemma 1.4, the metric ρ induces the compact-open topology in $C(M, N)$. If M is compact, we have defined a distance function

$$\rho_0(f_1, f_2) = \max_{x \in M} \{ d'(f_1(x), f_2(x)) \}.$$

It is easy to see that ρ and ρ_0 only differ by a bounded factor. Thus the C^0-topology of $C(M, N)$ coincides with the compact-open topology.

If N is complete, it follows easily that $C(M, N)$ is complete as a metric space with the distance ρ. Since $C(M, N)$ is second countable, Lemma 1.2 implies that compactness is equivalent to sequential compactness. Applying the fact to a set $\mathcal{F} \subset C(M, N)$, equipped with the distance ρ, we conclude that $\mathcal{F}$ is compact iff every sequence of $\mathcal{F}$ contains a convergent subsequence such that the limit mapping is in $\mathcal{F}$. Note that the fact that every sequence of $\mathcal{F}$ contains a convergent subsequence implies same fact in $\overline{\mathcal{F}}$. We have the following property:

Lemma 1.5 *A family $\mathcal{F}$ is relatively compact iff every sequence of $\mathcal{F}$ contains a subsequence which converges uniformly on every compact subset of M.*

Lemma 1.6 *If N is complete, then a family $\mathcal{F} \subset C(M, N)$ is relatively compact iff $\mathcal{F}$ is total bounded.*

Proof. If $\overline{\mathcal{F}}$ is compact, then $\overline{\mathcal{F}}$ is total bounded by Lemma 1.2, and hence $\mathcal{F}$ is total bounded. Conversely, if $\mathcal{F}$ is total bounded, so does $\overline{\mathcal{F}}$. Since N is complete, then $\overline{\mathcal{F}}$ is complete. Thus $\overline{\mathcal{F}}$ is compact by Lemma 1.2. $\square$

Lemma 1.7 *A family $\mathcal{F}$ is total bounded iff for each compact subset E of M and for every $\varepsilon > 0$, there exist $f_1, ..., f_n \in \mathcal{F}$ such that each $f \in \mathcal{F}$ satisfies $d'(f, f_j) < \varepsilon$ on E for some f_j.*

Proof. Suppose first that $\mathcal{F}$ is total bounded. Since every compact E is contained in an E_k, thus for a fixed such k and for every $\varepsilon > 0$, we can choose $\varepsilon_0 > 0$ sufficiently small such that $2^k \varepsilon_0 < 1$ and $2^k \varepsilon_0 / (1 - 2^k \varepsilon_0) < \varepsilon$. Since $\mathcal{F}$ is total bounded, there exist $f_1, ..., f_n \in \mathcal{F}$ such that each $f \in \mathcal{F}$ satisfies $\rho(f, f_j) < \varepsilon_0$ for some f_j. Consequently, by the definition of ρ, $\delta_k(f, f_j) < 2^k \varepsilon_0$, or $\delta(f, f_j) < 2^k \varepsilon_0$ on E_k. Hence $d'(f, f_j) < 2^k \varepsilon_0 / (1 - 2^k \varepsilon_0) < \varepsilon$ on E.

Conversely, choosing k_0 with $2^{-k_0} < \varepsilon/2$, by the assumption, there exist $f_1, ..., f_n \in \mathcal{F}$ such that each $f \in \mathcal{F}$ satisfies $\delta(f, f_j) \leq d'(f, f_j) < \varepsilon/(2k_0)$ on E_{k_0} for some f_j, and consequently,

$$\delta_k(f, f_j) < \begin{cases} \frac{\varepsilon}{2k_0} & : \quad k \leq k_0 \\ 1 & : \quad k > k_0. \end{cases}$$

Therefore

$$\rho(f, f_j) < k_0 \frac{\varepsilon}{2k_0} + \sum_{k=k_0+1}^{\infty} 2^{-k} = \frac{\varepsilon}{2} + 2^{-k_0} < \varepsilon.$$

□

On (N, d'), there is a natural function defined by

$$\hbar(x) = d'(x, O) \tag{1.11}$$

for any fixed point $O \in N$ and for all $x \in N$. Obviously, $\hbar(x)$ is continuous.

In this context, a sequence $\{f_n\} \subseteq C(M, N)$ is called *compactly divergent* on M iff given any compact K in M and compact K' in N, there exists an n_0 such that $f_n(K) \cap K' = \emptyset$ for all $n \geq n_0$. Obviously, if N is compact, this case don't happen. If N is complete, this fact is equivalent to that for any compact K in M,

$$\lim_{n \to \infty} \min_{x \in K} \hbar(f_n(x)) = \infty$$

i.e., $\{f_n\}$ diverges uniformly on K.

Definition 1.3 ([272]) *A subset $\mathcal{F}$ of $C(M, N)$ is called normal, or a normal family, on M iff every sequence of $\mathcal{F}$ contains a subsequence which is either relatively compact in $C(M, N)$ or compactly divergent.*

Lemma 1.5 implies that a subset $\mathcal{F}$ of $C(M, N)$ is normal on M iff every sequence of $\mathcal{F}$ contains a subsequence which is either convergent uniformly on compact sets or compactly divergent. Hence if N is compact, $\mathcal{F}$ is normal on M iff $\mathcal{F}$ is relatively compact in $C(M, N)$. Obviously, if U is a domain in M and if $\mathcal{F}$ is normal on M, then $\mathcal{F}$ is normal on U.

Definition 1.4 *A subset $\mathcal{F}$ of $C(M, N)$ is called normal at $x_0 \in M$ if there exists a neighborhood U of x_0 in M such that $\mathcal{F}|_U = \{f|_U \mid f \in \mathcal{F}\} \subset C(U, N)$ is normal on U.*

Let the distance functions of M and N be d and d' respectively. A sequence $\{f_n\}$ of mappings from the metric space (M, d) into the metric space (N, d') *converges locally uniformly* on M to some mapping f if each point x_0 of M has a disc

$$M(x_0; r) = \{x | d(x, x_0) < r\} \subset M$$

(also called an r-ball about x_0) on which f_n converges uniformly to f. In these circumstances, the convergence is then uniform on each compact subset of M. Conversely, since M is a locally compact connected metric space, then f_n converges locally uniformly on M if it is uniform on each compact subset of M. Hence convergence uniformly on compact sets in the definition of normality can be replaced by locally uniform convergence.

We say that a sequence $\{f_n\} \subset C(M, N)$ is *locally compactly divergent* on M if each point x_0 of M has a disc $M(x_0; r) \subset M$ such that $\{f_n\}$ is compactly divergent on $M(x_0; r)$. Obviously, $\{f_n\}$ is compactly divergent on M iff $\{f_n\}$ is locally compactly divergent on M. Hence we have

Lemma 1.8 *A subset $\mathcal{F}$ of $C(M, N)$ is normal on M iff every sequence $\{f_n\}$ of $\mathcal{F}$ contains a subsequence $\{f_{n_k}\}$ which satisfies either of the following conditions:*
1) $\{f_{n_k}\}$ converges locally uniformly on M.
2) $\{f_{n_k}\}$ is locally compactly divergent on M.

Lemma 1.9 *Let a subset $\mathcal{F}$ of $C(M, N)$ be normal at each point of M. Take points $x_0, x' \in M$. If a sequence $\{f_n\}$ of $\mathcal{F}$ converges uniformly (resp., diverges compactly) on a disc $M(x_0; r)$, there exists a subsequence of $\{f_n\}$ which converges uniformly (resp., diverges compactly) on a disc $M(x'; r)$.*

Proof. We use a family of geodesic segments $\{\overline{x_j x_{j+1}} | j = 0, 1, ..., n; x_{n+1} = x'\}$ in M to connect x_0 and x'. Since the family $\mathcal{F}$ is normal at each point of $\overline{x_0 x_1}$, and since $\overline{x_0 x_1}$ can be covered by finite discs, there exists a positive number R such that for each point ζ on $\overline{x_0 x_1}$, $M(\zeta; R) \subset M$ and the family $\mathcal{F}$ is normal in $M(\zeta; R)$. Take a number δ with

$$0 < \delta < \min(r, R),$$

Thus we can choose points $\zeta_0, ..., \zeta_m$ on $\overline{x_0 x_1}$ such that

$$\zeta_0 = x_0, \ \zeta_m = x_1, \ d(\zeta_{j-1}, \zeta_j) < \delta (j = 1, 2, ..., m).$$

Since the family $\mathcal{F}$ is normal in $M(\zeta_1; R)$, there exists a subsequence $\{f_{n_{1k}}\}(k = 1, 2, ...)$ of $\{f_n\}$ which satisfies one of the conditions in Lemma 1.8 on $M(\zeta_1; R)$. Due to $\zeta_1 \in M(x_0; r)$ and by assumption, $\{f_{n_{1k}}\}$ converges (resp., diverges compactly) at ζ_1 so that it converges locally uniformly (resp., diverges locally compactly) on $M(\zeta_1; R)$. By induction, there exists a subsequence $\{f_{n_{mk}}\}(k = 1, 2, ...)$ of $\{f_n\}$ which converges locally uniformly (resp., diverges locally compactly) on $M(\zeta_m; R)$ so that it converges uniformly (resp., diverges compactly) on a disc $M(x_1; r_1)$ of x_1. Finally, again by induction there exists a subsequence of $\{f_n\}$ which converges uniformly (resp., diverges compactly) on a disc $M(x'; r')$ of the point x'. $\square$

Theorem 1.9 *A family $\mathcal{F}$ in $C(M,N)$ is normal on M iff the family $\mathcal{F}$ is normal at each point of M.*

Proof. Evidently if $\mathcal{F}$ is normal in M, then $\mathcal{F}$ is normal at each point of M. Conversely, first noting that there exists a dense sequence $\{x_j | j = 1, ...\}$ in M, by assumption there exists a disc $M(x_j; r_j) \subset M$ such that $\mathcal{F}$ is normal on $M(x_j; r_j)$. Let R_j be the supremum of r_j with the property above, and set

$$D_j = M(x_j; R_j/2) \quad \text{if} \quad R_j < \infty, \quad \text{or} \quad M(x_j; 1) \quad \text{if} \quad R_j = \infty.$$

Then $D_j \subset M$ and $\mathcal{F}$ is normal on D_j. Take a sequence $\{f_n\}$ in $\mathcal{F}$. Then we can choose a subsequence $\{f_{n_{1k}}\}(k = 1, 2, ...)$ of $\{f_n\}$ which satisfies one of conditions in Lemma 1.8 on D_1, a subsequence $\{f_{n_{2k}}\}(k = 1, 2, ...)$ of $\{f_{n_{1k}}\}$ which satisfies one of conditions in Lemma 1.8 on D_2, and so on. The diagonal sequence $\{f_{n_{kk}}\}$ is a subsequence of $\{f_n\}$ which satisfies one of the conditions in Lemma 1.8 on each D_j. Next we prove that it satisfies one of the conditions in Lemma 1.8 on M. We distinguish two cases:

1) $\{f_{n_{kk}}\}$ converges locally uniformly on D_1, and does so on each D_j by Lemma 1.9. Take a point x_0 in M. By assumption $\mathcal{F}$ is normal on some disc $M(x_0; \rho) \subset M(0 < \rho < 1)$. Take x_j with $d(x_j, x_0) < \rho/4$. Obviously, $M(x_j; \rho/2) \subset M(x_0; \rho)$ so that $\mathcal{F}$ is normal on $M(x_j; \rho/2)$. Hence if $R_j < \infty$, then $\rho/2 \leq R_j$ so that $M(x_j; \rho/4) \subset D_j$, and is so if $R_j = \infty$. Therefore $\{f_{n_{kk}}\}$ converges locally uniformly on $M(x_j; \rho/4)$ and hence converges uniformly on some disc of x_0.

2) $\{f_{n_{kk}}\}$ is locally compactly divergent on D_1, and is so on each D_j by Lemma 1.9. By using the method above, we can prove that each point x_0 in X has a disc $M(x_0; r) \subset M$ such that $\{f_{n_{kk}}\}$ is compactly divergent on $M(x_0; r)$. □

Taking the collection $\{U_\alpha\}$ to be the class of all open subsets of M on which $\mathcal{F}$ is normal, this leads to the following general principle.

Theorem 1.10 *Let $\mathcal{F}$ be a family in $C(M,N)$. Then there is a maximal open subset $F(\mathcal{F})$ of M on which $\mathcal{F}$ is normal. In particular, if $f \in C(M,M)$, then there is a maximal open subset $F(f)$ of M on which the family of iterates $\{f^n\}$ is normal.*

The sets $F(\mathcal{F})$ and $F(f)$ in Theorem 1.10 is usually called *Fatou sets* of $\mathcal{F}$ and f respectively. *Julia sets* of $\mathcal{F}$ and f are defined respectively by

$$J(\mathcal{F}) = M - F(\mathcal{F}), \quad J(f) = M - F(f).$$

If $\mathcal{F}$ is finite, we define $J(\mathcal{F}) = \emptyset$. By the definition, we have the following fact:

Lemma 1.10 *Let $\mathcal{F}_1, \mathcal{F}_2$ be families in $C(M,N)$ and set $\mathcal{F} = \mathcal{F}_1 \cup \mathcal{F}_2$. Then*

$$F(\mathcal{F}) = F(\mathcal{F}_1) \cap F(\mathcal{F}_2), \quad J(\mathcal{F}) = J(\mathcal{F}_1) \cup J(\mathcal{F}_2).$$

The following result is basic:

Theorem 1.11 *If $f \in C(M,M)$ is an open mapping of a smooth manifold M into itself, then $F = F(f)$ and $J = J(f)$ are backward invariant, that is,*

$$f^{-1}(F) = F, \quad f^{-1}(J) = J. \tag{1.12}$$

Proof. Let $D \subseteq M$ be any domain and let D' be any component of $f^{-1}(D)$. Since F and J divide M, then the assertion follows from the trivial identity

$$f^n|_{D'} = f^{n-1}|_D \circ f|_{D'}$$

and by distinguishing two cases:

(a) $D \subseteq F$. Then $\{f^n\}$ is normal in D', i.e., $D' \subseteq F$. Since D and D' are arbitrary, this shows that $f^{-1}(F) \subseteq F$.

(b) $D \cap J \neq \emptyset$. This means that the sequence $\{f^n\}$ is not normal in D', and hence $D' \cap J \neq \emptyset$. If we let D shrink to some point $x_0 \in J$ then $f^{-1}(x_0) \subseteq J$ and so $f^{-1}(J) \subseteq J$ because x_0 is arbitrary. $\square$

Remark. By Corollary A.3, (1.12) implies

$$f(F) = F \cap f(M), \quad f(J) = J \cap f(M). \tag{1.13}$$

Theorem 1.12 *For each positive integer $p \geq 2$,*

$$F(f) \subseteq F(f^p), \quad J(f^p) \subseteq J(f). \tag{1.14}$$

Furthermore, if M is compact, then

$$F(f) = F(f^p), \quad J(f^p) = J(f). \tag{1.15}$$

Proof. It suffices to prove the assertion for, say, the Fatou sets. Since the family $\{f^{pn}\}$ is contained in the family $\{f^n\}$, we thus obtain (1.14). Assume that M is compact. Given any domain $D \subset M$ we set

$$\mathcal{F} = \{f^n|_D \mid n \geq 0\}, \quad \mathcal{F}_j = \{f^j \circ f^{pn}|_D \mid n \geq 0\}.$$

Then obviously,

$$\mathcal{F} = \mathcal{F}_0 \cup ... \cup \mathcal{F}_{p-1},$$

and since f^j is uniformly continuous on M, $\mathcal{F}$ is normal iff $\mathcal{F}_0$ is normal. $\square$

Theorem 1.13 *The Julia set $J(f)$ contains all (topological) repulsive fixed points.*

Proof. Assume that a fixed point ξ of f is (topological) repulsive. By definition, there exists a neighborhood U of ξ such that for every $x_j \in U - [\xi](j = 1, 2, ...)$ there is an $n_j \in \mathbb{Z}(0, \infty)$ with $f^n(x_j) \notin U$ for all $n \geq n_j$. Take the sequence $\{x_j\} \subset U - [\xi]$ such that $x_j \to \xi$ as $j \to \infty$. Assume $\xi \in F(f)$. Then we can find a neighborhood $U' \subset U$ of ξ and subsequence $\{f^{j_k}\}$ of $\{f^n\}$ which uniformly converges to ϕ on compact sets in U'. Obviously, $\phi(\xi) = \xi$ and ϕ is continuous on U'. Take $r > 0$ so that $M(\xi; r) \subset U'$. Take some j with $x_j \in M(\xi; r)$ and $d(\phi(x_j), \xi) < r/2$. Then there is k_0 such that $d(f^{j_k}(x_j), \phi(x_j)) < r/2$ for all $k \geq k_0$ so that

$$d(f^{j_k}(x_j), \xi) \leq d(f^{j_k}(x_j), \phi(x_j)) + d(\phi(x_j), \xi) < r$$

which implies $f^{j_k}(x_j) \in U$ for all $k \geq k_0$, but $f^{j_k}(x_j) \notin U$ if $j_k \geq n_j$. Contradiction! □

Thus if M is compact, Theorem 1.13 and Theorem 1.12 shows that the Julia set $J(f)$ contains the closure of its set of (topological) repulsive cycles since $J(f)$ is closed. We suggest the following questions:

Problem 1.1 *Does the Fatou set $F(f)$ contains all (topological) attractive fixed points and basins of attraction of (topological) attractive fixed points?*

We will discuss it for mappings on complex manifolds.

Conjecture 1.2 *If f is a continuous self-mapping on a compact smooth manifold M, then the set of (topological) attractive cycles is compact, and the Julia set $J(f)$ is equal to the closure of its set of (topological) repulsive cycles.*

Conjecture 1.3 *Suppose that U is some open set intersecting the Julia set. Then $f^n(U \cap J(f)) = J(f)$ for every sufficiently large n.*

Definition 1.5 *A continuous self-mapping on a manifold M is called exceptional if its Julia set contains either exceptional points or isolated points.*

We know that any rational mapping of degree at least two on the Riemannian sphere is not exceptional. Generally, non exceptional mappings maybe are dense.

Conjecture 1.4 *Let f be a continuous self-mapping on a compact smooth manifold M. Then $J(f) = M$ iff there is some x whose forward orbit $O^+(x)$ is dense in M.*

For rational functions on the Riemannian sphere, these problems are well-known, and, in Conjecture 1.3, it expresses a property of Julia sets which may be called *self-similarity*. For results related to this section, we refer the reader to Hu [118].

Let M be a smooth manifold. Take a non-empty set κ and consider a family $\mathcal{F} = \{f_t \mid t \in \kappa\} \subset C(M,M)$. Let $(\kappa^{\mathbb{N}}, \sigma)$ be the one-sided shift on κ. For each $t \in \kappa^{\mathbb{N}}$, the families $\mathbf{R}_t(\mathcal{F})$ and $\mathbf{L}_t(\mathcal{F})$ are well-defined (see § A.2). Here we discuss the following open set

$$F_\kappa = \left(\bigcap_{t \in \kappa^{\mathbb{N}}} F(\mathbf{L}_t(\mathcal{F})) \right)^o.$$

We also write $J_\kappa = M - F_\kappa$. Obviously, we have

$$F_\kappa \subset F(f^n_t), \quad J(f^n_t) \subset J_\kappa, \ t \in \kappa^{\mathbb{N}}, \ n \geq 1.$$

In particular,

$$F_\kappa \subset F(f_t), \quad J(f_t) \subset J_\kappa, \ t \in \kappa.$$

Problem 1.2 *When the following relation holds*

$$J_\kappa = \overline{\bigcup_{t \in \kappa^{\mathbb{N}}} \bigcup_{n=1}^{\infty} J(f^n_t)}?$$

Ren and his students [209] studied the properties of the Julia sets J_κ for analytic functions on $\mathbb{C}$, and confirmed this problem if $\mathcal{F}$ is a finite family of entire or meromorphic or rational functions on $\mathbb{C}$. For this case, they also proved the following results.

Theorem 1.14 *If $f_t \in C(M, M)$ is an open mapping of the smooth manifold M into itself for some $t \in \kappa$, then*

$$f_t(F_\kappa) \subset F_\kappa, \quad f_t^{-1}(J_\kappa) \subset J_\kappa.$$

Proof. Take $x \in F_\kappa$. Then there exists a neighborhood U of x with $U \subset F(\mathbf{L}_s(\mathcal{F}))$ $(s \in \kappa^{\mathbb{N}})$, and hence $U \subset F(\mathbf{L}_{(t,s)}(\mathcal{F}))$ for each $s \in \kappa^{\mathbb{N}}$, that is, the family $\{f_{n_s} \circ f_t\}$ is normal on U. Therefore $\mathbf{L}_s(\mathcal{F})$ is normal on the open set $f_t(U)$, and hence $f_t(x) \in F_\kappa$. The second part of the theorem follows easily from the first part.   □

Corollary 1.1 *If $f_t \in C(M, M)$ is an open mapping of the smooth manifold M into itself for each $t \in \kappa$, and if κ is finite, then*

$$J_\kappa = \bigcup_{t \in \kappa} f_t^{-1}(J_\kappa).$$

Proof of the corollary can be completed according to the proof of Theorem 9.5 in Ren's book [209].

Example 1.1 *Let S^1 be the unit circle in the complex plane $\mathbb{C}$*

$$S^1 = \{z \in \mathbb{C} \mid |z| = 1\} = \{e^{2\pi\theta i} \mid \theta \in \mathbb{R}\}.$$

Let R_α denote the rotation by angle $2\pi\alpha$, i.e.,

$$R_\alpha(z) = e^{2\pi\alpha i} z.$$

Then one has $R_\alpha^n(z) = e^{2\pi n\alpha i} z$. Obviously, the family $\mathcal{F} = \{R_\alpha^n\}$ is normal on S^1 since S^1 is compact. Hence $J(R_\alpha) = \emptyset$. If α is rational, writing $\alpha = \frac{q}{p}$ where p, q are relatively prime integers, then $R_\alpha^p = id$.

For the phase space of the m-dimensional torus

$$\mathbb{T}^m = S^1 \times \cdots \times S^1 \ (m \ times) = \mathbb{R}^m / \mathbb{Z}^m,$$

we can similarly define a translation

$$R_\alpha(z_1, ..., z_m) = (e^{2\pi\alpha_1 i} z_1, ..., e^{2\pi\alpha_m i} z_m), \quad \alpha = (\alpha_1, ..., \alpha_m)$$

with $J(R_\alpha) = \emptyset$.

Example 1.2 *Fix an integer $k \geq 2$ and consider the noninvertible mapping $E_k : S^1 \longrightarrow S^1$ defined by $E_k(z) = z^k$, $|z| = 1$. Then $z = 1$ is a repulsive fixed point of E_k, and hence $J(E_k) \neq \emptyset$.*

1.4   Equicontinuity

Definition 1.6 *A family $\mathcal{F}$ of mappings of a metric space (M, d) into a metric space (N, d') is called equicontinuous or an equicontinuous family at $z_0 \in M$ if and only if for every positive ε there exists a positive δ such that for all z in M, and for all f in $\mathcal{F}$,*

$$d(z, z_0) < \delta \Longrightarrow d'(f(z), f(z_0)) < \varepsilon.$$

The family $\mathcal{F}$ is said equicontinuous on M iff $\mathcal{F}$ is equicontinuous at each point of M.

Obviously, the family $\mathcal{F}$ is equicontinuous on M iff for any compact subset K of M and for every positive ε, there exists a positive $\delta = \delta(K, \varepsilon)$ such that for all z, z' in K, and for all f in $\mathcal{F}$,

$$d(z, z') < \delta \Longrightarrow d'(f(z), f(z')) < \varepsilon.$$

The latter will be refered to *strict equicontinuity* of the family on K. If the family $\mathcal{F}$ is equicontinuous on each open subsets U_α of M, then it is automatically equicontinuous on the union $\cup U_\alpha$. Taking the collection $\{U_\alpha\}$ to be the class of all open subsets of M on which $\mathcal{F}$ is equicontinuous , this leads to the following general principle.

Theorem 1.15 *Let $\mathcal{F}$ be any family of mappings, which map (M, d) into (N, d'). Then there is a maximal open subset $F_{equ}(\mathcal{F})$ of M on which $\mathcal{F}$ is equicontinuous. In particular, if f maps a metric space (M, d) into itself, then there is a maximal open subset $F_{equ}(f) = F_{equ}(f, d)$ of M on which the family of iterates $\{f^n\}$ is equicontinuous.*

Define

$$J_{equ}(\mathcal{F}) = M - F_{equ}(\mathcal{F}), \quad J_{equ}(f) = J_{equ}(f, d) = M - F_{equ}(f, d).$$

We can study these sets after the fashion of Fatou sets and Julia sets. In particular, it is easy to prove that the sets $F_{equ}(f)$ and $J_{equ}(f)$ are backward invariant if f is an open mapping.

Theorem 1.16 *Let $f : M \longrightarrow M$ be a distance decreasing mapping i.e., we have*

$$d(f(x), f(y)) \leq d(x, y) \tag{1.16}$$

for all $x, y \in M$. Then $J_{equ}(f) = \emptyset$.

Proof. By iterating (1.16), one sees that for any positive integer n

$$d(f^n(x), f^n(y)) \leq d(x, y). \tag{1.17}$$

Thus the family of iterates $\{f^n\}$ is equicontinuous on M, i.e., $F_{equ}(f) = M$. $\square$

Corollary 1.2 *Let f be a contracting self-mapping of a metric space (M, d). Then*

$$J_{equ}(f) = \emptyset.$$

Corollary 1.3 $J_{equ}(f) = \emptyset$ *for* $f \in \mathrm{Iso}^k(M)$.

According to Lemma A.2, if f is contracting on a complete metric space (M, d), then f has a unique fixed point p. In this case, obviously p is an attractor because any disc $U = M(p; \varepsilon)$ with $0 < \varepsilon < 1$ is absorbing such that

$$\bigcap_{n \geq 0} f^n(U) = \{p\},$$

and we have

$$Att(p) = W^s(p) = F_{equ}(f) = M.$$

Theorem 1.17 *Let* p *be an equilibrium point of a cascade* $\{f^n\}_{n \in \mathbb{Z}_+}$ *generated by* $f \in C(M, M)$. *Then* p *is future stable if and only if* $p \in F_{equ}(f)$.

Proof. Let p be future stable. By definition, for every positive ε, there exisits a neighborhood V of p such that $f^n(V) \subset U = M(p; \varepsilon)$ for $n \geq 0$. Choose a positive δ with $M(p; \delta) \subset V$. If $z \in M(p; \delta)$, then $f^n(z) \in U$ for all $n \geq 0$, that is,

$$d(f^n(z), f^n(p)) = d(f^n(z), p) < \varepsilon,$$

for all $n \geq 0$. Hence $\{f^n\}_{n \in \mathbb{Z}_+}$ is equicontinuous at p, and hence $p \in F_{equ}(f)$.

Conversely, if $p \in F_{equ}(f)$, then for any neighborhood U of p, by choosing a positive ε with $M(p; \varepsilon) \subset U$, there exists a positive δ such that $O^+(M(p; \delta)) \subset M(p; \varepsilon) \subset U$. Hence p is future stable.   □

Remark. Obviously, the theorem holds for a semiflow $\mathcal{F}$, that is, if p is an equilibrium point of a semiflow $\mathcal{F}$, then p is future stable if and only if $p \in F_{equ}(\mathcal{F})$.

Definition 1.7 *A mapping* $f \in C(M, M)$ *(resp.,* $f \in \mathrm{Hom}(M, M)$*) is called expansive if there exists a constant* $\lambda > 0$ *such that if* $d(f^n(x), f^n(y)) < \lambda$ *for all* $n \in \mathbb{Z}_+$ *(resp., all* $n \in \mathbb{Z}$*), then* $x = y$.

The maximal number λ_0 satisfying this property is usually called the *expansivity constant* for the dynamical system. If M is compact, the property of being expansive does not depend on the choice of metrics defining the given topology, and hence is an invariant of topological conjugacy. Also it is easy to prove that expanding mappings are expansive.

Theorem 1.18 *Let* f *be a expansive self-mapping of a metric space* (M, d). *Then*

$$J_{equ}(f) = M.$$

Proof. Assume, to the contrary, that $F_{equ}(f) \neq \emptyset$. Let $\lambda_0 > 0$ be the expansivity constant. Take $z_0 \in F_{equ}(f)$. Then there exists a positive δ such that for all z in M, and for all n in $\mathbb{Z}_+$,

$$d(z, z_0) < \delta \Longrightarrow d(f^n(z), f^n(z_0)) < \lambda_0.$$

Since f is expansive, then $M(z_0; \delta) = \{z_0\}$, i.e., $M(z_0; \delta) \cap F_{equ}(f) = \{z_0\}$ is a closed set. This is a contradiction.   □

Example 1.3 *Here we consier the mappins R_α and E_k defined in last section. Let d be the chord distance on S^1. Then $d(R_\alpha(z), R_\alpha(w)) = d(z, w)$ for $z, w \in S^1$. Hence $J(R_\alpha) = J_{equ}(R_\alpha) = \emptyset$. It is easy to prove that the mapping E_k ($k \geq 2$) is expanding, and hence is expansive. Thus we have $J(E_k) = J_{equ}(E_k) = S^1$.*

Theorem 1.19 *Assume that $f \in \mathrm{Hom}(M, M)$. If a fixed point p of f is of $W^u(p) - \{p\} \neq \emptyset$, then $p \in J_{equ}(f)$.*

Proof. Assume, to the contrary, that $p \in F_{equ}(f)$. Hence for every positive ε there exists a positive δ such that for all z in M, and for all integer $n \geq 0$,

$$d(z, p) < \delta \implies d(f^n(z), p) < \varepsilon.$$

Take $y \in W^u(p) - \{p\}$. Then $d(f^{-n}(y), p) \to 0$ as $n \to \infty$, and hence $d(f^{-n}(y), p) < \delta$ for $n \geq n_0$. Thus we obtain

$$d(y, p) = d(f^n(f^{-n}(y)), p) < \varepsilon$$

which implies $y = p$ since ε is arbitrary. This is a contradiction. $\qquad\square$

By adopting the argument used by Beardon in his proof of Theorem 4.2.9 in [31], the following general result can be obtained:

Theorem 1.20 ([126]) *If $f, g \in C(M, M)$ are open with $f \circ g = g \circ f$ and satisfying some Lipschitz condition*

$$d(f(x), f(y)) \leq \lambda d(x, y), \quad d(g(x), g(y)) \leq \lambda d(x, y),$$

on M, then $f^n(F_{equ}(g)) \subset F_{equ}(g)$ and $g^n(F_{equ}(f)) \subset F_{equ}(f)$ for all $n \in \mathbb{Z}_+$.

Proof. For any set E, we denote the diameter of E computed using the metric d by $\mathrm{diam}[E]$. Now take $x \in F_{equ}(f)$. By the equicontinuity of $\{f^n\}$ at x, given any positive ε, there is a positive δ such that for all n,

$$\mathrm{diam}[f^n(M(x; \delta))] < \varepsilon/\lambda.$$

As f and g commute we deduce that

$$\begin{aligned}
\mathrm{diam}[f^n \circ g(M(x; \delta))] &= \mathrm{diam}[g \circ f^n(M(x; \delta))] \\
&\leq \lambda \mathrm{diam}[f^n(M(x; \delta))] < \varepsilon.
\end{aligned}$$

It follows that $\{f^n\}$ is equicontinuous at $g(x)$, so, in particular, $g(x) \in F_{equ}(f)$. This proves that g, and hence each g^n, maps $F_{equ}(f)$ into itself. We conclude that $g^n : F_{equ}(f) \longrightarrow F_{equ}(f)$, and so, by symmetry, $f^n : F_{equ}(g) \longrightarrow F_{equ}(g)$. $\qquad\square$

Note that the rational function case is contained implicitly in the proof of Theorem 4.2.9 of [31]. For more information on this topic, see [24] and [32]. Finally we weaken normality and equicontinuity as follows:

Definition 1.8 *Let $\mathcal{F}$ be a family in $C(M, N)$. For $0 \leq k < m = \dim M$, a point $p \in M$ belongs to the Fatou set $F(\mathcal{F}; k)$ (resp., $F_{equ}(\mathcal{F}; k)$) if there exists a neighborhood U of p such that for every $x \in U$ there exists a submanifold C_x of codimension k such that $x \in C_x$, and $\mathcal{F}|_{C_x}$ is normal (resp., equicontinuous) on C_x. Set*

$$J(\mathcal{F}; k) = M - F(\mathcal{F}; k) \quad (resp., \ J_{equ}(\mathcal{F}; k) = M - F_{equ}(\mathcal{F}; k)).$$

In particular, if $\mathcal{F} = \{f^n\} \subset C(M, M)$, write

$$F(f; k) = F(\mathcal{F}; k) \quad (resp., \ F_{equ}(f; k) = F_{equ}(\mathcal{F}; k)),$$

and

$$J(f; k) = M - F(f; k) \quad (resp., \ J_{equ}(f; k) = M - F_{equ}(f; k)).$$

Similarly, we can define $F_{uc}(\mathcal{F}; k), F_{dc}(\mathcal{F}; k), F_{uc}(f; k)$ and $F_{dc}(f; k)$. Note that

$$F(\mathcal{F}) = F(\mathcal{F}; 0) \subset F(\mathcal{F}; 1) \subset \cdots \subset F(\mathcal{F}; m - 1),$$

and consequently,

$$J(\mathcal{F}; m - 1) \subset J(\mathcal{F}; m - 2) \subset \cdots \subset J(\mathcal{F}; 1) \subset J(\mathcal{F}; 0) = J(\mathcal{F}).$$

The sets $F_{equ}(\mathcal{F}; k)$ or $J_{equ}(\mathcal{F}; k)$ satisfy similar relations.

For results related to this section, we refer the reader to Hu [118].

1.5   Lipschitz constants and Julia sets

In this section, we will compare normality and equicontinuity, and then derive a criterion for Julia sets of holomorphic mappings on compact complex manifolds to be non-empty. To do so, the following classic results will be needed.

Lemma 1.11 (Arzela-Ascoli theorem) *A family $\mathcal{F} \subset C(M, N)$ is relatively compact if and only if*

1) $\mathcal{F}$ is equicontinuous on M;

2) $\mathcal{F}(z) = \{ f(z) | f \in \mathcal{F} \}$ is relatively compact in N for every $z \in M$.

Lemma 1.12 ([272]) *Let $\mathcal{F} \subset C(M, N)$, where M, N are connected locally compact spaces. Then*

1) If $\mathcal{F}$ is compact, then $\mathcal{F}$ is normal on M.

2) If $\mathcal{F}$ is normal on M, then its closure is locally compact.

3) If $\mathcal{F}$ is equicontinuous on M and if each bounded subset of N is relatively compact, then $\mathcal{F}$ is normal on M.

Proof. The case 1) is obvious. For the proof of 2), consider any $f \in \mathcal{F}$ and take any point $z \in M$. Let U be any relatively compact open neighborhood of $f(z)$. The neighborhood $\mathcal{W}(\{z\}, U)$ of f is then relatively compact, for the following reason: no sequence in $\mathcal{W}(\{z\}, U)$ can be compactly divergent, and so normality and Lemma 1.5 imply that every sequence in $\mathcal{W}(\{z\}, U)$ is relatively compact. Consequently, the closure of $\mathcal{W}(\{z\}, U)$ in $C(M, N)$ furnishes a compact neighborhood of f. The proof of 3) is broken into three steps:

Claim 1. If $\mathcal{G}$ is a sequence in $\mathcal{F}$ and $\mathcal{G}(K) = \{f(z) | z \in K,\ f \in \mathcal{G}\}$ is relatively compact for every compact $K \subseteq M$, then $\mathcal{G}$ is relatively compact in $C(M, N)$.

Since $\mathcal{F}$ is by assumption equicontinuous, conditions 1)-3) of Lemma 1.11 are fulfilled for the closure $\overline{\mathcal{G}}$ and $\overline{\mathcal{G}}$ is compact.

Claim 2. If $\mathcal{G} = \{f_n\}$ is a sequence in $\mathcal{F}$ and if for one compact set K_0 in M and one compact set K_1 in N, $f_n(K_0) \cap K_1 \neq \emptyset$, for all $f_n \in \mathcal{G}$, then $\mathcal{G}(K)$ is a bounded subset of N for all compact $K \subseteq M$.

Since M is connected, it will be sufficient to prove this assertion for all connected compact K containing K_0. Thus, if we take such a K, we know that $f_n(K) \cap K_1 \neq \emptyset$. Choose an arbitrary $\varepsilon > 0$, then associate with each $z \in K$ a neighborhood U such that the diameter of $f_n(U)$ is less than ε for all $f_n \in \mathcal{G}$ since $\mathcal{F}$ is equicontinuous. Then K admits a finite covering by such neighborhoods as $U_1, ..., U_l$. It is clear that the diameter of the connected $f_n(K)$ cannot exceed $2l\varepsilon$ for all n. Let the diameter of K_1 be η ($\eta < \infty$ because K_1 compact) and let w_0 be a fixed point in K_1. Because $f_n(K) \cap K_1 \neq \emptyset$, $f_n(K)$ is necessarily contained in the $(\eta + 2l\varepsilon)$-ball about w_0. Since n is arbitrary, $\mathcal{G}(K)$ is then a subset of this ball as well.

Claim 3. Let $\mathcal{G}$ be a sequence in $\mathcal{F}$. Unless $\mathcal{G}$ is compactly divergent, $\mathcal{G}$ contains a relatively compact subsequence.

The Claim 3 clearly proves 3). If we assume that $\mathcal{G}$ is not compactly divergent, then there is some compact K_0 in M and some compact K_1 in N such that for infinitely many f_n of $\mathcal{G}$, $f_n(K_0) \cap K_1 \neq \emptyset$. By Claim 2, $\{f_n\}$ carries each compact K of M into a bounded set in N, and the latter must then be relatively compact by hypothesis on N. Now Claim 1 implies that $\{f_n\}$ is a relatively compact subset of $C(M, N)$. □

It is obvious that the converse of 1) and 2) is false, and that normality does not imply equicontinuity. The condition on N in 3) implies, but is stronger than, the completeness of N; however, they coincide for Riemannian manifolds.

Definition 1.9 *A subset $\mathcal{F}$ of $C(M, N)$ is called uc-normal on M iff $\mathcal{F}$ is relatively compact in $C(M, N)$.*

Lemma 1.5 implies that a subset $\mathcal{F}$ of $C(M, N)$ is uc-normal on M iff every sequence of $\mathcal{F}$ contains a subsequence which is uniformly convergent on compact sets, where "uc-normality" means the normality in the sense of uniform convergence excluding compact divergence from the previous definition of normality. Hence if N is compact, $\mathcal{F}$ is uc-normal on M iff $\mathcal{F}$ is normal on M. Obviously, if U is a domain in M and if $\mathcal{F}$ is uc-normal on M, then $\mathcal{F}$ is uc-normal on U.

Definition 1.10 *A subset $\mathcal{F}$ of $C(M, N)$ is called uc-normal at $x_0 \in M$ if there exists a neighborhood U of x_0 in M such that $\mathcal{F}|_U = \{f|_U \mid f \in \mathcal{F}\} \subset C(U, N)$ is uc-normal on U.*

Theorem 1.21 *A family $\mathcal{F}$ in $C(M,N)$ is uc-normal on M iff the family $\mathcal{F}$ is uc-normal at each point of M.*

The proof can be completed after the fashion of Theorem 1.9. Taking the collection $\{U_\alpha\}$ to be the class of all open subsets of M on which $\mathcal{F}$ is uc-normal, this leads to the following general principle.

Theorem 1.22 *Let $\mathcal{F}$ be a family in $C(M,N)$. Then there is a maximal open subset $F_{uc}(\mathcal{F})$ of M on which $\mathcal{F}$ is uc-normal. In particular, if $f \in C(M,M)$, then there is a maximal open subset $F_{uc}(f)$ of M on which the family of iterates $\{f^n\}$ is uc-normal.*

Thus we obtain a decomposition

$$F(\mathcal{F}) = F_{uc}(\mathcal{F}) \cup F_{dc}(\mathcal{F}), \quad F(f) = F_{uc}(f) \cup F_{dc}(f)$$

such that $x \in F_{dc}(\mathcal{F})$ (resp. $F_{dc}(f)$) iff $\mathcal{F}$ (resp. $\{f^n\}$) is normal at x and there exists a sequence of $\mathcal{F}$ (resp. $\{f^n\}$) which is compactly divergent in a neighborhood of x. If U is a component of $F(\mathcal{F})$, by Lemma 1.9 we have either $U \subset F_{uc}(\mathcal{F})$ or $U \subset F_{dc}(\mathcal{F})$, i.e.,

$$F_{uc}(\mathcal{F}) \cap F_{dc}(\mathcal{F}) = \emptyset.$$

If N is compact, then $F_{dc}(\mathcal{F}) = \emptyset$ and $F_{uc}(\mathcal{F}) = F(\mathcal{F})$. Similarly, we can prove the following result:

Theorem 1.23 *If $f \in C(M,M)$ is an open mapping of a smooth manifold M into itself, then $F_{uc}(f)$ and $F_{dc}(f)$ are backward invariant.*

For $\mathcal{F} \subset C(M,N)$, define

$$\mathcal{K}(\mathcal{F}) = \{z \in M \mid \mathcal{F}(z) \text{ is relatively compact in } N \}. \tag{1.18}$$

Then we easily obtain

$$\mathcal{K}(\mathcal{F}) \cap F_{dc}(\mathcal{F}) = \emptyset.$$

Lemma 1.11 implies that

$$F_{uc}(\mathcal{F}) \subset F_{equ}(\mathcal{F}) \cap \mathcal{K}(\mathcal{F}), \quad (F_{equ}(\mathcal{F}) \cap \mathcal{K}(\mathcal{F}))^\circ \subset F_{uc}(\mathcal{F}).$$

If $F_{equ}(\mathcal{F}) \cap \mathcal{K}(\mathcal{F})$ is open, then we have

$$F_{uc}(\mathcal{F}) = F_{equ}(\mathcal{F}) \cap \mathcal{K}(\mathcal{F}).$$

If N is compact, then $\mathcal{K}(\mathcal{F}) = M$, and hence

$$F(\mathcal{F}) = F_{uc}(\mathcal{F}) = F_{equ}(\mathcal{F}),$$

that is, equicontinuity is equivalent to normality. In particular, we have

Theorem 1.24 *If M is compact, and if $f \in C(M,M)$, then*

$$F_{equ}(f) = F(f) = F_{uc}(f), \quad J_{equ}(f) = J(f).$$

Corollary 1.4 *Assume that M is compact, and that $f \in \mathrm{Hom}(M, M)$. If a fixed point p of f is of $W^u(p) - \{p\} \neq \emptyset$, then $p \in J(f)$.*

If $f \in C(M, M)$, we abbreviate

$$\mathcal{K}(f) = \mathcal{K}(\mathcal{F}) \tag{1.19}$$

for $\mathcal{F} = \{f^n\}_{n \in \mathbb{Z}_+}$. It is easy to prove that

$$f(\mathcal{K}(f)) \subset \mathcal{K}(f), \quad f^{-1}(\mathcal{K}(f)) \subset \mathcal{K}(f),$$

that is $\mathcal{K}(f)$ is backward invariant. Generally, we don't know whether $\mathcal{K}(f)^o \subset F(f)$. We will discuss the set $\mathcal{K}(f)$ in Chapter 7.

Given a family $\mathcal{F}$ of continuous mappings from a metric space (M, d) into a metric space (N, d'), define the *Lipschitz constant* $\mathrm{Lip}(f, K)$ of f for a subset $K \subset M$ as follows

$$\mathrm{Lip}(f, K) = \sup \left\{ \frac{d'(f(z), f(w))}{d(z, w)} \mid z, w \in K, \ z \neq w \right\},$$

and set

$$\mathrm{Lip}_{\mathcal{F}}(K) = \sup_{f \in \mathcal{F}} \mathrm{Lip}(f, K).$$

If f is a C^1-mapping, and if d and d' are the distances induced by Riemannian metrics, respectively, then $\mathrm{Lip}(f, K)$ is finite for any compact subset K of M. Indeed, let $K \subset M$ be compact, and take points $z_n, w_n \in K$ with $z_n \neq w_n$ such that

$$\lim_{n \to \infty} \frac{d'(f(z_n), f(w_n))}{d(z_n, w_n)} = \mathrm{Lip}(f, K).$$

Since K is compact, without loss of generality we may suppose that $z_n \to z_0$, $w_n \to w_0$ for some $z_0, w_0 \in K$, as $n \to \infty$. If $w_0 \neq z_0$, then by passing to the limit we get that

$$\mathrm{Lip}(f, K) = \frac{d'(f(z_0), f(w_0))}{d(z_0, w_0)} < +\infty.$$

Suppose that $w_0 = z_0$. Since M is locally compact, and f is continuous, we can choose relatively compact neighborhood U of z_0 such that $\overline{U}$ and $f(\overline{U})$ are contained in a local coordinate neighborhoods of z_0 and $f(z_0)$, respectively. Since $f \in C^1$, it follows that

$$f(x') - f(x'') = (df)_{z_0}(x' - x'') + \alpha(x', x'')\|x' - x''\|,$$

where $\alpha(x', x'') \to 0$ as $x', x'' \to z_0$. From this, and noting that d, d' are equivalent to the Euclidean distance on $\overline{U}$ and $f(\overline{U})$, respectively,

$$\frac{d'(f(z_n), f(w_n))}{d(z_n, w_n)} \leq c \frac{\|f(z_n) - f(w_n)\|}{\|z_n - w_n\|} \leq c \left(\|(df)_{z_0}\| + |\alpha(z_n, w_n)| \right),$$

for a constant c and for all large n, and we get by passing to the limit that

$$\mathrm{Lip}(f, K) \leq c \|(df)_{z_0}\| < \infty,$$

which is what was to be proved. If $\mathrm{Lip}_{\mathcal{F}}(K)$ is finite for any compact subset K of M, noting that

$$d'(f(z), f(w)) \leq \mathrm{Lip}_{\mathcal{F}}(K)d(z, w), \quad z, w \in K,$$

the family $\mathcal{F}$ is strict equicontinuous on K, and hence it is equicontinuous on M. Further, if $\mathcal{F}(z)$ is relatively compact in N for every $z \in M$, then $\mathcal{F}$ is uc-normal on M. Conversely, we have the following result:

Theorem 1.25 *Assume that* (M, d_g) *and* (N, d_h) *are complex manifolds and take* $\mathcal{F} \subset \mathrm{Hol}(M, N)$. *If the family* $\mathcal{F}$ *is uc-normal on* M, *then* $\mathrm{Lip}_{\mathcal{F}}(K)$ *is finite for any compact subset* K *of* M.

Proof. Assume, to the contrary, that there exists a compact subset $K \subset M$ such that $\mathrm{Lip}_{\mathcal{F}}(K) = +\infty$. Then there are a sequence $\{f_n\} \subset \mathcal{F}$ and points $z_n, w_n \in K$ with $z_n \neq w_n$ such that

$$\lim_{n \to \infty} \frac{d_h(f_n(z_n), f_n(w_n))}{d_g(z_n, w_n)} = +\infty.$$

Since K is compact and by the uc-normal family hypothesis, without loss of generality we may suppose that the sequence $\{f_n\}$ converges uniformly to a holomorphic mapping $f \in \mathrm{Hol}(M, N)$ on any compact subset of M, and that $z_n \to z_0$, $w_n \to w_0$ for some $z_0, w_0 \in K$, as $n \to \infty$. If $w_0 \neq z_0$, then

$$\lim_{n \to \infty} \frac{d_h(f_n(z_n), f_n(w_n))}{d_g(z_n, w_n)} = \frac{d_h(f(z_0), f(w_0))}{d_g(z_0, w_0)} < +\infty,$$

contradicting our assumption. Thus we have $w_0 = z_0$. Note that M is locally compact. We can choose relatively compact neighborhood U of z_0 such that $\overline{U}$ is contained in a local coordinate neighborhood. Thus by using Weierstrass theorem, we obtain that the norms

$$\|f_n'(z)\| = \sup\{\|f_n'(z)Z\|_h \mid Z \in \mathbf{T}(M)_z, \|Z\|_g = 1\}$$

are bounded uniformly on U by a constant C since the sequence $\{f_n\}$ converges uniformly to f on U. By the geometric theory, we can choose a positive number ε with $M_{d_g}(z_0; \varepsilon) \subset U$ such that arbitrary two points of $M_{d_g}(z_0; \varepsilon)$ can be joined by unique geodesic in $M_{d_g}(z_0; \varepsilon)$. Take $n_1 \in \mathbb{Z}^+$ such that $z_n, w_n \in M_{d_g}(z_0; \varepsilon)$ for all $n \geq n_1$. Let $\gamma_n : [0, 1] \longrightarrow M$ with $\gamma_n(0) = z_n, \gamma_n(1) = w_n$ be the geodesic joining z_n and w_n in $M_{d_g}(z_0; \varepsilon)$. Then λ_n uniquely defined by $\lambda_n = f_n \circ \gamma_n$ is piecewise differentiable curves joining $f_n(z_n)$ and $f_n(w_n)$ and

$$
\begin{aligned}
d_h(f_n(z_n), f_n(w_n)) &\leq \mathrm{Length}(\lambda_n) \\
&= \int_0^1 \|\dot\lambda_n(t)\| dt = \int_0^1 \|f_n'(\gamma_n(t))\dot\gamma_n(t)\| dt \\
&\leq \int_0^1 \|f_n'(\gamma_n(t))\| \cdot \|\dot\gamma_n(t)\| dt \\
&\leq C\mathrm{Length}(\gamma_n) = Cd_g(z_n, w_n),
\end{aligned}
$$

for all $n \geq n_1$, which implies

$$\lim_{n \to \infty} \frac{d_h(f_n(z_n), f_n(w_n))}{d_g(z_n, w_n)} \leq C,$$

contradicting our assumption. □

Let $f : M \longrightarrow M$ be a continuous mapping of a compact metric space M with a distance function d. Assume that f satisfies some Lipschitz condition

$$d(f(x), f(y)) \leq \lambda d(x, y).$$

For any $n \geq 1$, define a quantity by

$$\mathrm{Lip}_{f;n} = \mathrm{Lip}_{f;n}(M) = \max_{0 \leq i \leq n-1} \mathrm{Lip}(f^i, M), \tag{1.20}$$

and abbreviate

$$\mathrm{Lip}_f = \mathrm{Lip}_f(M) = \mathrm{Lip}_{\mathcal{F}}(M)$$

for $\mathcal{F} = \{f^n\}_{n \in \mathbb{Z}_+}$. Then we have $1 \leq \mathrm{Lip}_{f;n} \leq \max\{1, \lambda^{n-1}\}$. We call a finite set $K \subset M$ a *Lipschitz (n, ε)-covering set* if

$$\bigcup_{x \in K} M_d \left(x; \frac{\varepsilon}{\mathrm{Lip}_{f;n}} \right) = M,$$

where $M_d(x; \varepsilon) = \{y \in M \mid d(x, y) < \varepsilon\}$. Set

$$n_{Lip}(f, d; n, \varepsilon) = \min\{\#K \mid K \text{ is a Lipschitz } (n, \varepsilon)\text{-covering set }\},$$

and define

$$h_{Lip}(f, d) = \limsup_{\varepsilon \to 0} \limsup_{n \to \infty} \frac{1}{n} \log n_{Lip}(f, d; n, \varepsilon) \geq 0. \tag{1.21}$$

Obviously, we have $h_{Lip}(f, d) = 0$ if $\mathrm{Lip}_f < +\infty$.

Definition 1.11 *Let (M, d) be a complete metric space and let K be a compact subset of M. Let $b(\varepsilon)$ denote the minimum cardinality of a covering of K by ε-balls. The ball dimension (or fractal dimension) of K is defined by*

$$D(K) = \limsup_{\varepsilon \to 0} \frac{\log b(\varepsilon)}{|\log \varepsilon|}.$$

If d is the Euclidean metric on $\mathbb{R}^m$ and if K', K are compact subsets of $\mathbb{R}^m$, then $K' \subset K$ implies

$$0 \leq D(K') \leq D(K) \leq m,$$

see [26]. It is well known that for a compact differentiable manifold M the ball dimension $D(M)$ is the topological dimension (see [134]). For a C^1 mapping $f : M \longrightarrow M$ of a compact Riemannian manifold M, it is easy to prove that there exists a positive constant γ such that for any Riemannian metric g on M there exists another constant $a = a(f, g)$ such that for every $x, y \in M$ and every $n \in \mathbb{Z}^+$, one has (cf. [134], p.127)

$$d_g(f^n(x), f^n(y)) \leq a \gamma^n d_g(x, y).$$

Note that

$$n_{Lip}(f, d_g; n, \varepsilon) \leq b \left(\frac{\varepsilon}{a \max\{1, \gamma^{n-1}\}} \right).$$

We can prove that

$$h_{Lip}(f, d_g) \leq \dim M \log^+ \gamma,$$

where $\log^+ \gamma = \max\{0, \log \gamma\}$.

Theorem 1.26 *Let (M, d_g) be a compact complex manifold and take $f \in \mathrm{Hol}(M, M)$. If $h_{Lip}(f, d_g) > 0$, then $J(f) \neq \emptyset$.*

Proof. Assume $J(f) = \emptyset$. By Theorem 1.25, we have $L = \mathrm{Lip}_f(M) < +\infty$. Fix $\varepsilon > 0$. Since M is compact, we can find a finite set K_ε such that

$$\bigcup_{z \in K_\varepsilon} M_{d_g}\left(z; \frac{\varepsilon}{L}\right) = M.$$

Note that

$$M_{d_g}\left(z; \frac{\varepsilon}{L}\right) \subset M_{d_g}\left(z; \frac{\varepsilon}{\mathrm{Lip}_{f;n}}\right)$$

holds for $z \in K_\varepsilon$, $n \in \mathbb{Z}^+$ since $\mathrm{Lip}_{f;n} \leq L$. We have

$$n_{Lip}(f, d_g; n, \varepsilon) \leq \#K_\varepsilon$$

which yields $h_{Lip}(f, d_g) = 0$. This is a contradiction. $\qquad\square$

1.6 Degree of mappings and Julia sets

Let M and N be orientable Riemannian manifolds of dimension m. Then one has

$$H_m(M, \mathbb{Z}) \simeq \mathbb{Z}.$$

Take $f \in C(M, N)$ and fix $a \in N$. Assume that $f^{-1}(a)$ is finite. Let $f^{-1}(a) = \{b_1, ..., b_s\}$ and around each b_i choose an open ball neighborhood U_i so that $f^{-1}(a) \cap U_i = \{b_i\}$. Let U be an open ball neighborhood of a so that $f(U_i) \subset U$. If ξ_i, ξ are respectively the generators of $H_{m-1}(U_i - \{b_i\})$ and $H_{m-1}(U - \{a\})$ and are coherent with the orientations of M and N, then

$$f_*(\xi_i) = n_i \xi \tag{1.22}$$

for an integer n_i [273]. The *mapping degree* $\mu_f(b_i)$ of f at b_i is defined as follows:

$$\mu_f(b_i) = n_i, \quad i = 1, ..., s, \tag{1.23}$$

and set $\mu_f(x) = 0$ if $x \in M - \{b_1, ..., b_s\}$. For a subset $D \subset M$, write

$$n_f(D, a) = \sum_{x \in D \cap f^{-1}(a)} \mu_f(x).$$

Wu given the following results:

Lemma 1.13 (Wu [274]) *Suppose $f : D \longrightarrow N$ is a C^∞ mapping from a compact oriented manifold D with boundary ∂D into another oriented manifold N of the same dimension. Suppose Ω is an integrable form of top degree in N. Then*

$$\int_D f^*\Omega = \int_N n_f(D, a)\Omega.$$

Theorem 1.27 (Wu [273]) *Let M and N be orientable Riemannian manifolds of dimension m. Assume that N is compact and that $f \in C^\infty(M, N)$. Let D be a compact domain of M and let Ω be a C^∞ positive normalized volume element on N. Then for every $a \in N$, there exists an integrable $(m - 1)$-form λ_a on N such that*

1) if $f^{-1}(a)$ is finite and disjoint from ∂D, then

$$\int_D f^*\Omega = n_f(D, a) + \int_{\partial D} f^*\lambda_a,$$

2) λ_a is C^∞ in $N - \{a\}$ with

$$d\lambda_a = \Omega.$$

Now assume that M and N are compact. Theorem 1.27 implies that

$$\int_M f^*\Omega = n_f(M, a) \tag{1.24}$$

is an integer if $f^{-1}(a)$ is finite, so that we can define the *degree* of $f \in C^\infty(M, N)$ by

$$\deg(f) = \int_M f^*\Omega.$$

Recall a particular case which illustrates the notion of a regular value for a mapping between manifolds. A point $x \in M$ is called a *regular point* for f if $(df)_x$ is invertible. A point $a \in N$ is called a *regular value* of f if $f^{-1}(a)$ consists of regular points, or otherwise a singular value. The set of regular values is obviously open. By Sard's Theorem A.1 it has full measure and is hence also dense. If a is a regular value of f, then the compactness of M and the Implicit Function Theorem imply that $f^{-1}(a)$ is finite. Thus the degree of a smooth mapping is well defined if f is of maximal rank m. Obviously, $\deg(f) = 0$ if $\text{rank}(f) < m$. Note that if a is a regular value of f, for each $x \in f^{-1}(a)$, $\mu_f(x) = \pm 1$ according to whether $(df)_x$ preserves or reverses orientation.

Lemma 1.14 ([134], p.312) *Let $a \in N$ be a regular value of $f \in C^1(M, N)$ and Ω a positive normalized volume element on N. Then*

$$n_f(M, a) = \int_M f^*\Omega.$$

More generally, if Ψ is a continuous m-form on N with $\int_N \Psi \neq 0$, then

$$n_f(M, a) = \int_M f^*\Psi \Big/ \int_N \Psi.$$

So we can now make the following definition. If $f \in C^1(M, N)$, then the *degree* of f is defined by

$$\deg(f) = n_f(M, a)$$

for any regular value $a \in N$. Clearly, the degree

$$\deg : C^1(M, N) \longrightarrow \mathbb{Z}$$

is continuous in the C^1 topology, and is hence locally constant. We have in particular that the degree is invariant under homotopies consisting of C^1 mappings. However, any two mappings $f_0, f_1 \in C^1(M, N)$ that are sufficiently C^0-close are homotopic via C^1 mappings (see [134], p.313). Hence the degree of a C^1 mapping is homotopy invariant. This, in fact, allows us to define the degree for $f \in C(M, N)$ by smooth approximation:

$$\deg(f) = \deg(g),$$

where $g \in C^1(M, N)$ is sufficiently close to f. Evidently the degree of a continuous mapping is also homotopy invariant.

Remark. Assume that M and N are compact oriented smooth m-dimensional manifolds and $f \in C(M, N)$. Then f induces a homomorphism

$$f_* : H_r(M, \mathbb{Z}) \longrightarrow H_r(N, \mathbb{Z}) \quad (0 \leq r \leq m)$$

on homology. If ξ and η are respectively the generators of $H_m(M, \mathbb{Z})$ and $H_m(N, \mathbb{Z})$ coherent with the orientations of M and N, then

$$f_*(\xi) = k\eta \tag{1.25}$$

for an integer k. The degree of f is defined by

$$\deg(f) = k. \tag{1.26}$$

Using the duality between homology and cohomology and the description of de Rham cohomology, one easily establishes the equivalence of this definition with the volume form definition of degree.

We exhibit the following basic facts:

Lemma 1.15 *If $f \in C(M, N), g \in C(N, K)$, then*

$$\deg(g \circ f) = \deg(g)\deg(f).$$

Proof. We may assume that $f \in C^1(M, N), g \in C^1(N, K)$. Then Lemma 1.14 implies

$$\deg(g \circ f) = \int_M f^* g^* \Psi / \int_K \Psi = \deg(f) \int_N g^* \Psi / \int_K \Psi = \deg(f) \deg(g),$$

for a continuous m-form Ψ on K with $\int_K \Psi \neq 0$. $\square$

Lemma 1.16 *Let W be an $(m+1)$-dimensional oriented manifold with boundary ∂W with the induced orientation from W and $f : W \longrightarrow M$ a continuous mapping. Then $\deg(f|_{\partial W}) = 0$.*

Proof. If Ω is a volume form on M, setting $g = f|_{\partial W}$, then by the Stokes theorem,

$$\deg(g) = \int_{\partial W} g^*\Omega = \int_W d(g^*\Omega) = \int_W g^* d\Omega = 0.$$

$\square$

Here one uses the degree to give a definition of an index of an isolated fixed point. Since the definition is local, we work on $\mathbb{R}^m$. Let $U \subset \mathbb{R}^m$ be open and $f : U \longrightarrow \mathbb{R}^m$ a continuous mapping. Fix $p \in U$ and assume that $f(x) \neq x$ for all $x \in U - \{p\}$. Let $V \subset \mathbb{R}^m$ be a homeomorphic image of a ball with the natural orientation and $p \in V \subset \overline{V} \subset U$. Define the mapping

$$\nu_{f,V} : \partial V \longrightarrow S^{m-1}$$

into the unit sphere S^{m-1} in $\mathbb{R}^m$ by

$$\nu_{f,V}(x) = \frac{f(x) - x}{\|f(x) - x\|}.$$

Then the *index* of p for f is defined by

$$\iota_f(p) = \deg(\nu_{f,V}).$$

We have to show that the definition is independent of the choice of V. In fact, it suffices to show that $\deg(\nu_{f,V}) = \deg(\nu_{f,B})$ for a ball $B \subset V$ containing p. Given such B note that the mapping $w : W = \overline{V} - B \longrightarrow S^{m-1}$ defined by $w(x) = \frac{f(x)-x}{\|f(x)-x\|}$ is continuous and that by Lemma 1.16 $\deg(w|_{\partial W}) = 0$. Now ∂W is the union of ∂V and ∂B, but the latter with negative orientation. Thus

$$\deg(\nu_{f,V}) - \deg(\nu_{f,B}) = \deg(w|_{\partial W}) = 0.$$

Evidently Lemma 1.16 implies $\iota_f(p) = 0$ if p is not a fixed point.

The index of an isolated fixed point of a mapping on a manifold is defined via a coordinate chart around the point. The discussion above shows that the definition is independent of the chart chosen.

The indices of fixed points under the iterates of f can be controlled by the following theorem of Shub and Sullivan (see [229], [63] or [134]):

Theorem 1.28 *Let $U \subset \mathbb{R}^m$ be open, $0 \in U$, and $f : U \longrightarrow \mathbb{R}^m$ differentiable. If 0 is an isolated fixed point of f^n for all $n \in \mathbb{N}$, then $\{\iota_{f^n}(0) \mid n \in \mathbb{N}\}$ is bounded.*

The Shub-Sullivan Theorem 1.28 implies the following result:

Corollary 1.5 *Let M be a compact manifold and $f : M \longrightarrow M$ differentiable such that the Lefschetz number $\mathrm{Lef}(f^n)$ of f^n satisfies*

$$\limsup_{n \to \infty} \mathrm{Lef}(f^n) = \infty.$$

Then f has infinitely many periodic points.

Example 1.4 *We consider a continuous mapping $f : S^m \longrightarrow S^m$. For the sphere S^m, the Betti numbers are $b_0 = b_m = 1$ and $b_r = 0$ for $0 < r < m$. Thus $f_*|_{H_0(S^m,\mathbb{Z})}$ and $f_*|_{H_m(S^m,\mathbb{Z})}$ are the only nontrivial homomorphisms induced by f, and are given by the 1×1 matrices 1 and $\deg(f)$, respectively. Thus*

$$\mathrm{Lef}(f^n) = \sum_{r=0}^{m}(-1)^r \mathrm{tr}(f^{n*}|_{H^r_{DR}(S^m,\mathbb{R})}) = \sum_{r=0}^{m}(-1)^r \mathrm{tr}(f^n_*|_{H_r(S^m,\mathbb{Z})}) = 1 + (-1)^m \deg(f^n).$$

Therefore if $f \in C^1(S^m, S^m)$ with $|\deg(f)| \geq 2$, the mapping f has infinitely many periodic orbits.

The definition of the index of an isolated fixed point coincides with that in §A.3 if p is nondegenerate (see [134], p.319).

Theorem 1.29 *Suppose $f : U \longrightarrow \mathbb{R}^m$ is differentiable at $0 \in U \subset \mathbb{R}^m$ and 0 is a nondegenerate fixed point for f. Then $\iota_f(0) = \mathrm{sign}\det((df)_0 - id)$.*

Proof. Let $A = (df)_0$. The nondegeneracy assumption means that $A - id$ is invertible, whence there exists $\delta > 0$ such that

$$\|Ax - x\| > \delta\|x\|$$

for $x \neq 0$. On the other hand, there exists $\varepsilon > 0$ such that

$$\|f(x) - x - (Ax - x)\| = \|f(x) - Ax\| < \delta\|x\|$$

on the sphere $\mathbb{R}^m\langle 0; \varepsilon\rangle$ of radius ε centered at 0. Consequently $\nu_{f,\varepsilon}$ and $\nu_{A,\varepsilon}$ are never antipodal, where

$$\nu_{f,\varepsilon} = \nu_{f,V}, \quad \partial V = \mathbb{R}^m\langle 0; \varepsilon\rangle.$$

Hence a smooth homotopy

$$\nu_t(x) = \frac{t\nu_{f,\varepsilon} + (1-t)\nu_{A,\varepsilon}}{\|t\nu_{f,\varepsilon} + (1-t)\nu_{A,\varepsilon}\|}$$

is well defined. Therefore

$$\iota_f(0) = \deg(\nu_{f,\varepsilon}) = \deg(\nu_{A,\varepsilon}).$$

The mapping $\nu_{A,\varepsilon}$ is invertible: If $\nu_{A,\varepsilon}(x) = \nu_{A,\varepsilon}(y)$, then

$$Ax - x = \lambda(Ay - y) \text{ and } A(x - \lambda y) = x - \lambda y,$$

so $x = \lambda y$ since 0 is the only fixed point for A. But $\|x\| = \|y\|$, so $|\lambda| = 1$ and $x = \pm y$. Since

$$\nu_{A,\varepsilon}(-x) = -\nu_{A,\varepsilon}(x),$$

we have $x = y$. Thus $\nu_{A,\varepsilon}$ is a homeomorphism, and hence, $\deg(\nu_{A,\varepsilon}) = 1$ if $A - id$ preserves orientation, that is, $\mathrm{sign}\det(A - id) = 1$; $\deg(\nu_{A,\varepsilon}) = -1$ otherwise. $\square$

Theorem 1.30 ([121]) *Assume that M is a compact oriented smooth m-dimensional manifold and $f \in C(M, M)$. If $|\deg(f)| \geq 2$, then $J(f) \neq \emptyset$.*

Proof. Assume that $J(f) = \emptyset$, i.e., $\{f^n\}$ is normal on M. Let h be the limit function of some subsequence $\{f^{n_j}\}$ of $\{f^n\}$. Then $h \in C(M, M)$. Since f^{n_j} converges uniformly on M to h, there exists $j_0 \in \mathbb{Z}^+$ such that for $j \geq j_0$, f^{n_j} and h are homotopic via homotopies along the shortest geodesics connecting $f^{n_j}(x)$ and $h(x)$ with the length parameter. Hence

$$\deg(f^{n_j}) = \{\deg(f)\}^{n_j} = \deg(h) \quad \text{if } j \geq j_0.$$

This is impossible since $|\deg(f)| \geq 2$ and since $n_j \to \infty$ as $j \to \infty$. $\qquad \square$

Thus by Theorem 1.16, Theorem 1.24 and Theorem 1.30, we obtain the following fact.

Corollary 1.6 *If $f : M \longrightarrow M$ is a distance decreasing continuous mapping on a compact oriented smooth manifold M, then $|\deg(f)| \leq 1$.*

Conjecture 1.5 *Assume that M is a compact oriented smooth m-dimensional manifold and $f \in C(M, M)$ with $|\deg(f)| \geq 2$. If f is not exceptional, then $J(f)$ is perfect, and for arbitrary $x \in J(f)$, the backward orbit $O^-(x)$ is dense in $J(f)$.*

Define $\mathrm{Res} J(f)$ to be the *residual Julia set* of f, which is the set of those points in $J(f)$ not lying in the boundary of $F(f)$. Then $\mathrm{Res} J(f)$ has some fundamental properties similar to those of $J(f)$. Here we extend the Makienko's conjecture (cf. [75]) for rational functions on the Riemannian sphere to the following general case:

Conjecture 1.6 *Assume that M is a compact oriented smooth m-dimensional manifold and $f \in C(M, M)$ satisfying*
1) $|\deg(f)| \geq 2$, and
2) $F(f)$ has no components which are backward invariant,
then $\mathrm{Res} J(f) \neq \emptyset$.

S. Morosawa [177] proved the following:

Theorem 1.31 *Let f be a rational function of degree at least two. Then*
1) if $\mathrm{Res} J(f)$ is not empty, $\mathrm{Res} J(f)$ is a completely invariant subset of $J(f)$ and dense in $J(f)$, moreover, it contains uncountably many points;
2) if $J(f)$ is disconnected and if there exists no completely invariant component of $F(f)$, $\mathrm{Res} J(f)$ is not empty.

Remark. J. Y. Qiao [202], [203] also studied the Makienko's conjecture and obtained independently the result 2) in Theorem 1.31.

1.7   Topological entropies and Julia sets

We begin our discussion of entropies with the discrete-time case. Let $f : M \longrightarrow M$ be a continuous mapping of a compact metric space M with a distance function d. For any $n \geq 1$ define a new metric by

$$d_{f;n}(x,y) = \max_{0 \leq i \leq n-1} d(f^i(x), f^i(y)). \tag{1.27}$$

For any $n \geq 1, \varepsilon > 0$, we call a finite set $K \subset M$ a (n, ε)-*covering set* if, for each $x \in K$, there is a positive number $\rho = \rho(x, n, \varepsilon)$ such that when $y \in M_d(x; \rho)$, one has $d_{f;n}(x,y) < \varepsilon$, and such that

$$\bigcup_{x \in K} M_d(x; \rho) = M.$$

Set

$$n(f, d; n, \varepsilon) = \min\{\#K \mid K \text{ is a } (n, \varepsilon)\text{-covering set }\}.$$

Notice that for each $n \geq 1, \varepsilon \mapsto n(f, d; n, \varepsilon)$ is monotone decreasing. We can define

$$h(f, d) = \lim_{\varepsilon \to 0} \limsup_{n \to \infty} \frac{1}{n} \log n(f, d; n, \varepsilon) \geq 0. \tag{1.28}$$

We will extend the idea of the modified definition of entropies in next chapter.

Proposition 1.1 *If d' is another metric on M which defines the same topology as d, then $h(f, d') = h(f, d)$.*

Proof [cf.[134]]. Write

$$D_\varepsilon = \{(x, y) \in M \times M \mid d(x,y) \geq \varepsilon\}.$$

This is a compact subset of $M \times M$ with the product topology. The function d' is continuous on $M \times M$ in that topology and consequently it reaches its minimum $\delta(\varepsilon)$ on D_ε. The minimum is positive; otherwise there exist points $x \neq y$ with $d'(x, y) = 0$. Hence for each $x \in M$, we have $M_{d'}(x; \delta(\varepsilon)) \subset M_d(x; \varepsilon)$. This argument extends immediately to the metrics $d'_{f;n}$ and $d_{f;n}$. Therefore for every n we obtain

$$n(f, d'; n, \delta(\varepsilon)) \geq n(f, d; n, \varepsilon),$$

so $h(f, d') \geq h(f, d)$. Interchanging the metrics d and d' one obtain $h(f, d) \geq h(f, d')$.   □

The quantity $h(f, d)$ calculated for any metric d generating the given topology in M also is denoted by $h(f)$ or $h_{equ}(f)$. Now we compare with the standard definition (cf. [134]). For any $n \geq 1, \varepsilon > 0$, we call a finite set $K \subset M$ a *standard (n, ε)-covering set* if

$$\bigcup_{x \in K} M_{d_{f;n}}(x; \varepsilon) = M,$$

where $M_{d_{f;n}}(x; \varepsilon) = \{y \in M \mid d_{f;n}(x, y) < \varepsilon\}$. Set

$$n_{st}(f, d; n, \varepsilon) = \min\{\#K \mid K \text{ is a standard } (n, \varepsilon)\text{-covering set }\}.$$

Notice that for each $n \geq 1, \varepsilon \mapsto n_{st}(f, d; n, \varepsilon)$ is monotone decreasing. One can define

$$h_{top}(f, d) = \lim_{\varepsilon \to 0} \limsup_{n \to \infty} \frac{1}{n} \log n_{st}(f, d; n, \varepsilon) \geq 0. \tag{1.29}$$

According to the proof of Proposition 1.1, one also can prove that $h_{top}(f, d)$ is independent of the metric d generating the given topology in M. The quantity is called the *topological entropy* of f and is denoted by $h_{top}(f)$. Obviously, we have

$$n(f, d; n, \varepsilon) \geq n_{st}(f, d; n, \varepsilon),$$

and hence

$$h_{equ}(f) \geq h_{top}(f).$$

If f satisfies some Lipschitz condition, we can prove that

$$h_{Lip}(f, d) \geq h_{equ}(f).$$

We don't know when $h_{equ}(f) = h_{top}(f)$? For the discussion above, see [126].

The quantity $h_{top}(f)$ has the following standard elementary property:

$$h_{top}(f^n) = |n| h_{top}(f). \tag{1.30}$$

If $N \subset M$ is a submanifold with $f(N) \subset N$, then it is easy to see from the definition that

$$h_{top}(f|_N) \leq h_{top}(f). \tag{1.31}$$

If M, N are compact topological spaces, and if $f \in C(M, M)$ and $g \in C(N, N)$, then

$$h_{top}(f \times g) = h_{top}(f) + h_{top}(g). \tag{1.32}$$

Also the topological entropy is an invariant of topological conjugacy, i.e.,

$$h_{top}(f) = h_{top}(g) \tag{1.33}$$

if f and g are topologically conjugate (see [134], [264], p. 167). The properties above also are true for $h_{equ}(f)$. There are several quantities similar to $n(f, d; n, \varepsilon)$ that can be used to define the topological entropy.

Example 1.5 *Let $n^*(f, d; n, \varepsilon)$ be the minimal number of sets whose diameter in the metric $d_{f;n}$ is less than ε and whose union covers M. Obviously, the diameter of an ε- ball is less than or equal to 2ε so every covering by ε-balls is a covering by sets of diameter $\leq 2\varepsilon$, that is,*

$$n^*(f, d; n, 2\varepsilon) \leq n_{st}(f, d; n, \varepsilon). \tag{1.34}$$

On the other hand, any set of diameter $\leq \varepsilon$ is contained in the ε-ball around each of its points so

$$n_{st}(f, d; n, \varepsilon) \leq n^*(f, d; n, \varepsilon). \tag{1.35}$$

Hence

$$h_{top}(f) = \lim_{\varepsilon \to 0} \lim_{n \to \infty} \frac{1}{n} \log n^*(f, d; n, \varepsilon) = \lim_{\varepsilon \to 0} \liminf_{n \to \infty} \frac{1}{n} \log n_{st}(f, d; n, \varepsilon) \tag{1.36}$$

(see [134]).

Example 1.6 Let $n_*(f, d; n, \varepsilon)$ be the maximal number of points in M with pairwise $d_{f;n}$-distances at least ε. We will call such a set of points (n, ε)-separated. We can obtain the inequalities

$$n_*(f, d; n, 2\varepsilon) \leq n_{st}(f, d; n, \varepsilon) \leq n_*(f, d; n, \varepsilon), \qquad (1.37)$$

and hence

$$h_{top}(f) = \lim_{\varepsilon \to 0} \limsup_{n \to \infty} \frac{1}{n} \log n_*(f, d; n, \varepsilon) = \lim_{\varepsilon \to 0} \liminf_{n \to \infty} \frac{1}{n} \log n_*(f, d; n, \varepsilon) \qquad (1.38)$$

(see [134]).

Example 1.7 Let M be a compact Hausdorff space and let $f : M \longrightarrow M$ be a homeomorphism. We deal with open covers of M. Let $\mathcal{U}$ and $\mathcal{V}$ be open covers of M. Define the join cover by

$$\mathcal{U} \vee \mathcal{V} = \{U \cap V \mid U \in \mathcal{U}, V \in \mathcal{V}, U \cap V \neq \emptyset\}.$$

For each open cover $\mathcal{U}$ of M, let

$$n(\mathcal{U}) = \min_{\mathcal{V}}\{\#\mathcal{V} \mid \mathcal{V} \text{ is a subcovers of } \mathcal{U}\}.$$

Then

$$h(f; \mathcal{U}) = \lim_{n \to \infty} \frac{1}{n} \log n(\mathcal{U} \vee f^{-1}(\mathcal{U}) \vee \cdots \vee f^{-n+1}(\mathcal{U})) \qquad (1.39)$$

exists (see [194]). Further, if M is a metric space, then (see [194])

$$h_{top}(f) = \sup_{\mathcal{U}} h(f; \mathcal{U}). \qquad (1.40)$$

The definitions of the quantity $h_{equ}(f)$ and the *topological entropy* $h_{top}(f)$ for a flow $f = \{f^t\}_{t \in \mathbb{R}}$ are completely parallel to those for the discrete-time case. The counterpart of (1.27) is the following non-decreasing family of metrics

$$d_{f;T}(x, y) = \max_{0 \leq t \leq T} d(f^t(x), f^t(y)). \qquad (1.41)$$

The only property worth special notice is the following counterpart of (1.30)

$$h_{top}(f) = h_{top}(f^1), \quad h_{equ}(f) = h_{equ}(f^1). \qquad (1.42)$$

In fact, by compactness and continuity for $\varepsilon > 0$ one can find $\rho(\varepsilon) > 0$ such that $y \in M_d(x; \rho(\varepsilon))$ implies $y \in M_{d_{f;T}}(x; \varepsilon)$. Hence

$$M_{d_{f^1;[T]}}(x; \rho(\varepsilon)) \subset M_{d_{f;T}}(x; \varepsilon).$$

On the other hand, obviously $d_{f;n} \geq d_{f^1;n}$. These remarks imply the results.

The following theorem establishes the connection between the degree and the the topological entropy of mappings.

Theorem 1.32 (Misiurewicz-Przytycki) *If M is a compact orientable smooth manifold and if $f \in C^1(M, M)$, then*

$$h_{top}(f) \geq \log |\deg(f)|.$$

A proof can be found in [134]. There are certain cases where the inequality of the theorem becomes an equality. Holomorphic mappings of the Riemann sphere into itself are classic examples. Other examples are expanding mappings of compact Riemannian manifolds into itself. Now a question is when $J(f) \neq \emptyset$ if $|\deg(f)| < 2$? Note that if $|\deg(f)| \geq 2$, Misiurewicz-Przytycki's theorem shows

$$h_{top}(f) \geq \log |\deg(f)| > 0.$$

We conjectured: Let M be a compact oriented smooth manifold and let $f \in \mathrm{Diff}^1(M, M)$ such that $h_{top}(f) > 0$. Then $J(f) \neq \emptyset$. During the International Conference on Complex Analysis and Their Applications at Yinchuan, China (Auguest, 1996), M. Shishikura suggested to us that one might be able to prove the conjecture by utilizing the definition of $h_{top}(f)$.

Theorem 1.33 *Let M be a compact smooth manifold and let $f \in C(M, M)$ such that $h_{equ}(f) > 0$. Then $J(f) \neq \emptyset$.*

Proof. Assume $J(f) = \emptyset$. By Theorem 1.24, we have

$$J_{equ}(f, d) = J(f) = \emptyset,$$

where d is the distance function induced by a Riemann metric on M, i. e., the family of iterates $\{f^n\}$ is equicontinuous on M. Hence for every positive number ε and each $z \in M$, there exists a positive number $\delta = \delta(z) < \varepsilon$ such that the inequality $d(z, z') < \delta$ implies that

$$d(f^n(z), f^n(z')) < \varepsilon$$

hold for $n \in \mathbb{Z}_+$. Note that M is compact and note that $\{M_d(z; \delta(z))\}_{z \in M}$ is an open covering of M. Then there exists a finite set $K \subset M$ such that

$$\bigcup_{z \in K} M_d(z; \delta(z)) = M.$$

We have

$$n(f, d; n, \varepsilon) \leq \#K$$

which yields $h_{equ}(f) = h(f, d) = 0$. This is a contradiction.   □

Thus by Theorem 1.16, Theorem 1.24 and Theorem 1.33, we obtain the following fact.

Corollary 1.7 *If $f : M \longrightarrow M$ is a distance decreasing continuous mapping on a compact smooth manifold M, then $h_{equ}(f) = h_{top}(f) = 0$.*

Corollary 1.8 *Let M be a compact smooth manifold and let $f = \{f^t\}_{t \in \mathbb{R}}$ be a flow such that $h_{equ}(f) > 0$. Then $J(f) \neq \emptyset$.*

Proof. By (1.42), we have $h_{equ}(f^1) = h_{equ}(f) > 0$, and hence $J(f^1) \neq \emptyset$. Note that $J(f^1) \subset J(f)$. Thus $J(f) \neq \emptyset$.   □

According to the proof, we also see that

$$0 \leq b^+ = \inf_{t \geq 0}\{t \mid J(f^t) \neq \emptyset\} \leq 1,$$

if $h_{equ}(f) > 0$. Finally, we transform Conjecture 1.6 into the following form:

Conjecture 1.7 *Assume that M is a compact smooth m-dimensional manifold and $f \in C(M, M)$ satisfying*
   1) $h_{equ}(f) > 0$, and
   2) $F(f)$ has no components which are backward invariant,
then $\mathrm{Res}J(f) \neq \emptyset$.

Here if the condition $h_{equ}(f) > 0$ is replaced by $h_{top}(f) > 0$, we have corresponding results and the conjecture. Finally, we make a note on the Misiurewicz-Przytycki's theorem. If M is a compact orientable smooth manifold and if $f \in C(M, M)$, then f induces a homomorphism

$$f_* : H_k(M, \mathbb{Z}) \longrightarrow H_k(M, \mathbb{Z}) \quad (0 \leq k \leq m)$$

on homology. Note that the vector spaces $H_k(M, \mathbb{Z})$ are merely finite-dimensional, and f_* is linear. Thus the spectral radius $r(f_*|_{H_k(M,\mathbb{Z})})$ are well defined. The following inequality can be found in [134]:

$$h_{top}(f) \geq \log r(f_*|_{H_1(M,\mathbb{Z})}).$$

If $f \in C^1(M, M)$, the Misiurewicz-Przytycki's theorem shows

$$h_{top}(f) \geq \log |\deg(f)| = \log r(f_*|_{H_m(M,\mathbb{Z})}).$$

More generally, we have the following Shub entropy conjecture:

Conjecture 1.8 *If M is a compact orientable smooth manifold and if $f \in \mathrm{Diff}^1(M, M)$, one has the inequality*

$$h_{top}(f) \geq \max_k \log r(f_*|_{H_k(M,\mathbb{Z})}).$$

A Yomdin's inequality shows that this conjecture is true for C^∞ diffeomorphisms (cf. [199]). Related to Fatou-Julia theory, we suggest the following problem:

Conjecture 1.9 *If M is a compact orientable smooth manifold and if $f \in C(M, M)$ with $\max_k r(f_*|_{H_k(M,\mathbb{Z})}) > 1$, then $J(f) \neq \emptyset$.*

Thus the conjecture is true for C^∞ diffeomorphisms. Note that for $f \in C(S^m, S^m)$, one has

$$\max_k r(f_*|_{H_k(S^m,\mathbb{Z})}) = \max\{1, |\deg(f)|\}.$$

Hence this conjecture is true for $M = S^m$.

Example 1.8 (cf. [199]) *Let $\mathbb{T}^2$ be the 2-dimensional torus and consider the diffeomorphism $f : \mathbb{T}^2 \longrightarrow \mathbb{T}^2$ by*

$$f(z_1, z_2) = (z_1^2 z_2, z_1 z_2), \quad z_1, z_2 \in S^1.$$

For any point $z \in \mathbb{T}^2$ the derivative is of the form

$$\frac{df}{dz} = \begin{pmatrix} 2 & 1 \\ 1 & 1 \end{pmatrix}.$$

This matrix has the two eigenvalues $\lambda_1 = \frac{3+\sqrt{5}}{2}$ and $\lambda_2 = \frac{3-\sqrt{5}}{2}$. One can bound the quantity $n_{st}(f, d; n, \varepsilon)$, for sufficiently large n, by λ_1^n. In particular, one can calculate $h_{top}(f) = \log \lambda_1 > 0$. Thus $J(f) \neq \emptyset$. In fact, we have $J(f) = \mathbb{T}^2$ since the diffeomorphism f is expansive (cf. [50]).

Chapter 2

Ergodic theorems and invariant sets

In this chapter, we introduce basic notations and theorems in ergodic theory and define some invariant sets which are closely related to ergodic theorems. Also we will establish relations between these invariant sets and some quantities similar to entropies.

2.1 Poincaré recurrence domains

Take measure spaces $(M, \mathcal{B}, \mu)$ and $(N, \mathcal{R}, \nu)$. Let $f : M \longrightarrow N$ be a measurable mapping. Define a measure $f_* \mu$ on $(N, \mathcal{R})$ by

$$f_* \mu(A) = \mu(f^{-1}(A)) \quad \text{for all } A \in \mathcal{R}.$$

The measurable mapping f is called *null preserving* if $f_* \mu(A) = 0$ holds for all $A \in \mathcal{R}$ with $\nu(A) = 0$, and is said to be *measure preserving* if $f_* \mu = \nu$. A measure preserving mapping f is *invertible* if it is ono-to-one and if f^{-1} is also measurable. If f is invertible measure preserving mapping, which usually is called a *measure preserving transformation*, then

$$\nu(f(B)) = f_* \mu(f(B)) = \mu(B)$$

for all $B \in \mathcal{B}$.

Generally, if $f(\mathcal{B}) \subset \mathcal{R}$, we can define a measure $f^* \nu$ on $(M, \mathcal{B})$ by

$$f^* \nu(B) = \nu(f(B)) \quad \text{for all } B \in \mathcal{B}.$$

The measurable mapping f is called *measure decreasing* if $f^* \nu \leq \mu$, i.e.,

$$f^* \nu(B) \leq \mu(B)$$

for all measurable B. Thus If f is a measure preserving transformation, then $f^* \nu = \mu$.

Remark. Obviously, we obtain a mapping

$$f_* : \Sigma(M) \longrightarrow \Sigma(N).$$

Thus we need to study the following questions:

(i) Fix $\mu \in \Sigma(M)$ and $\nu \in \Sigma(N)$. Study the properties of mappings (for example, measure preserving or decreasing mappings, and so on);

(ii) Fix a mapping f. Study the properties of measures (for example, measures μ and ν with $f_*\mu = \nu$ or $f^*\nu \leq \mu$).

Now we consider only the case $(M, \mathcal{B}, \mu) = (N, \mathcal{R}, \nu)$. If the mapping f is measure preserving, then μ also is called an f-invariant measure on M. If so, each $f^n (n \in \mathbb{Z}_+)$ is measure preserving. Thus we have a measure preserving DS $\{f^n\}_{n \in \mathbb{Z}_+}$, i.e.,

$$f_*^n \mu = \mu, \quad n \in \mathbb{Z}_+.$$

If f is a measure preserving transformation, then f^{-1} is also a measure preserving transformation. For this case we have a measure preserving DS $\{f^n\}_{n \in \mathbb{Z}}$, i.e.,

$$f_*^n \mu = \mu, \quad n \in \mathbb{Z}.$$

Let $\Sigma_f(M)$ be the set of all f-invariant Borel probability measures on M. Then $\Sigma_f(M)$ is a convex, closed, and hence compact subset of $\Sigma(M)$.

The following statement gives another reformulation of the definition (see [199]): A measurable mapping $f : M \longrightarrow M$ is measure preserving iff

$$\int_M \phi \circ f d\mu = \int_M \phi d\mu, \text{ for all } \phi \in L^1.$$

The existence of invariant measures are given by the following Krylov-Bogolubov theorem (cf. [134]): Any continuous mapping on a metrizable compact space has an invariant Borel probability measure. Now we state the classic Poincaré recurrence theorem [197].

Theorem 2.1 *Assume that f is measure preserving on a probability space $(M, \mathcal{B}, \mu)$. If $A \in \mathcal{B}$, then for almost every $x \in A$ there is a positive integer $n = n_A(x)$ such that $f^n(x) \in A$.*

Proof. Define

$$A^u = \{x \in A \mid f^n(x) \notin A, n \in \mathbb{Z}^+\}.$$

Then $f^n(A^u) \cap A = \emptyset$ for $n \geq 1$. Since $A^u \subset A$, it follows that

$$f^n(A^u) \cap A^u = \emptyset \quad \text{for } n \geq 1,$$

which gives

$$f^{n-k}(A^u) \cap f^{-k}(A^u) \subset f^{-k}(f^n(A^u) \cap A^u) = \emptyset \quad \text{for } n \geq k \geq 0,$$

i.e., $\{f^{-n}(A^u)\}_{n=0}^{\infty}$ are pairwise disjoint. Therefore,

$$1 \geq \mu\left(\bigcup_{n=0}^{\infty} f^{-n}(A^u)\right) = \sum_{n=0}^{\infty} \mu(f^{-n}(A^u)) = \sum_{n=0}^{\infty} \mu(A^u).$$

Thus we deduce that $\mu(A^u) = 0$. $\qquad\qquad \square$

Assume that f is measure preserving on a probability space $(M, \mathcal{B}, \mu)$. Take $A \in \mathcal{B}$ and for $k \geq 0$, define

$$A_k = \bigcup_{n=k}^{\infty} f^{-n}(A), \quad A_* = \bigcap_{k=0}^{\infty} A_k.$$

Then A_* is the set of all points of M which enter A infinitely often under the iteration by f. Hence the set $A^s = A \cap A_*$ consists of all points of A which enter A infinitely often under the iteration by f. If $x \in A^s$, then there is a sequence $0 < n_1 < n_2 < \cdots$ with $f^{n_i}(x) \in A$ for all i, and hence $f^{n_i}(x) \in A^s$ for each i since $f^{n_j - n_i}(f^{n_i}(x)) \in A$ for all $j > i$. Since $f^{-1}(A_k) = A_{k+1}$, we have $\mu(A_k) = \mu(A_{k+1})$ and hence $\mu(A_0) = \mu(A_k)$ for all k. Then $\mu(A_*) = \mu(A_0)$ since $A_0 \supset A_1 \supset \cdots$, and therefore

$$\mu(A^s) = \mu(A \cap A_0) = \mu(A)$$

since $A \subset A_0$. Thus Poincaré recurrence theorem can be restated as follows:

Theorem 2.2 (cf. [264]) *Assume that f is measure preserving on a probability space $(M, \mathcal{B}, \mu)$. If $A \in \mathcal{B}$, the set A^s defined by*

$$A^s = A \cap \bigcap_{k=0}^{\infty} \bigcup_{n=k}^{\infty} f^{-n}(A)$$

satisfies $\mu(A^s) = \mu(A)$ such that for each $x \in A^s$ there is a sequence $0 < n_1 < n_2 < \cdots$ with $f^{n_i}(x) \in A^s$ for all i.

Let M be a domain in $\mathbb{C}^m$ and let λ be the Lebesgue measure. Consider the space $L^2_H(M)$ of all functions which are holomorphic and square integrable with respect to λ on M. Take $f \in \text{Aut}(M)$ and define a unitary operator $T_f : L^2_H(M) \longrightarrow L^2_H(M)$ by the formula

$$T_f \phi = \phi \circ f \mathcal{J} f,$$

where $\mathcal{J} f$ denotes the complex Jacobian of f. According to Theorem 1 of [170], it follows that if $\text{Fix}(f) \neq \emptyset$, then T_f has a non-zero, normalized eigenvector $h \in L^2_H(M)$. Thus one can define a f-invariant probability measure μ on the space $(M, \mathcal{B})$ by

$$\mu(A) = \int_A |h|^2 d\lambda$$

for every Borel subset $A \subset M$ (see [169]). Since $h \neq 0$ a. e. (λ), it follows that μ-null sets coincide with the λ-null sets. By Theorem 2.2, then $\mu(A^s) = \mu(A)$, and hence $\lambda(A^s) = \lambda(A)$. Thus one has the Mazur's result:

Theorem 2.3 ([169]) *Let M be a domain in $\mathbb{C}^m$ for which $L^2_H(M)$ is not zero space and $A \subset M$ be a Borel subset with $\lambda(A) > 0$. If $f \in \text{Aut}(M)$ has a fixed point, then almost all points of A return infinitely often to A and so $\lambda(A^s) = \lambda(A)$.*

We use set terminology to restate the Poincaré recurrence theorem. Here assume that M is a topological space and suppose that f is a measurable self-mapping on a probability space $(M, \mathcal{B}, \mu)$. A domain U in M is said to be *Poincaré recurrent* if for almost every $x \in U$ there is a positive integer $n = n_U(x)$ such that $f^n(x) \in U$. Also a point $x \in M$ is called *Poincaré recurrent* if there is a neighborhood U of x such that U is a Poincaré recurrent domain. Let $F_{Poi}(f) = F_{Poi,\mu}(f)$ be the set of Poincaré recurrent points, and set

$$J_{Poi}(f) = J_{Poi,\mu}(f) = M - F_{Poi}(f).$$

Then $F_{Poi}(f)$ is open, and $J_{Poi}(f)$ is closed. If $f \in \mathrm{Hom}(M, M)$ is null preserving, it is easy to prove that $F_{Poi}(f)$ and $J_{Poi}(f)$ are backward invariant. If f is measure preserving, the Poincaré recurrence theorem states $J_{Poi}(f) = \emptyset$.

Here we give a condition which yields $J_{Poi}(f) \neq \emptyset$. Let $f : M \longrightarrow M$ be a continuous mapping of a compact space M with a metric d. For any $n \geq 1, \varepsilon > 0$, we call a finite set $K \subset M$ a *Poincaré (n, ε)-covering set* if for each $x \in K$, there exists a positive number $\rho = \rho(x, n, \varepsilon)$ such that

$$\mu(\{y \in M(x; \rho) \mid \{f(y), f^2(y), ..., f^n(y)\} \cap M(x; \rho) = \emptyset\}) \leq \varepsilon,$$

where $M(x; \rho) = M_d(x; \rho) = \{z \in M \mid d(x, z) < \rho\}$, and such that

$$\bigcup_{x \in K} M(x; \rho) = M.$$

Set

$$n_{Poi}(f, \mu, d; n, \varepsilon) = \min\{\#K \mid K \text{ is a Poincaré } (n, \varepsilon)\text{-covering set }\}.$$

Notice that for fixed $n \geq 1$, the mapping $\varepsilon \mapsto n_{Poi}(f, \mu, d; n, \varepsilon)$ is monotone decreasing. Define

$$h_{Poi}(f, \mu, d) = \lim_{\varepsilon \to 0} \limsup_{n \to \infty} \frac{1}{n} \log n_{Poi}(f, \mu, d; n, \varepsilon) \geq 0. \tag{2.1}$$

Theorem 2.4 *Let M be a compact smooth manifold and let $f \in C(M, M)$ such that $h_{Poi}(f, \mu, d) > 0$. Then $J_{Poi}(f) \neq \emptyset$.*

Proof. Assume $J_{Poi}(f) = \emptyset$. Then for each $x \in M$, there is a positive number $\rho = \rho(x)$ such that $M(x; \rho)$ is a Poincaré recurrent domain. Set

$$E_n(x) = \{y \in M(x; \rho) \mid \{f(y), f^2(y), ..., f^n(y)\} \cap M(x; \rho) = \emptyset\}, \quad U_n(x) = M(x; \rho) - E_n(x).$$

Then $U_n(x) \subset U_{n+1}(x)$ for $n \geq 1$. Since $M(x; \rho)$ is a Poincaré recurrent domain, we have

$$\mu(M(x; \rho)) = \mu\left(\bigcup_{n=1}^{\infty} U_n(x)\right) = \lim_{n \to \infty} \mu(U_n(x)),$$

which implies

$$\lim_{n \to \infty} \mu(E_n(x)) = 0.$$

Hence for every positive number ε, there exists an integer $N(x)$ such that when $n \geq N(x)$ $\mu(E_n(x)) < \varepsilon$. Note that M is compact and note that $\{M(x; \rho)\}_{x \in M}$ is an open covering of M. Then there exists a finite set $K \subset M$ such that

$$\bigcup_{x \in K} M(x; \rho) = M.$$

Take $N = \max_{x \in K} N(x)$. Then for $n \geq N$, the set K is an Poincaré (n, ε)-covering set. Hence we have

$$n_{Poi}(f, \mu, d; n, \varepsilon) \leq \#K \quad (n \geq N),$$

which yields $h_{Poi}(f, \mu, d) = 0$. This is a contradiction. □

Thus if f is measure preserving, we obtain $h_{Poi}(f, \mu, d) = 0$. Related to the results of this section, see Hu and Yang [126].

2.2 Neumann domains

A measure preserving transformation $f : M \longrightarrow M$ determines a linear transformation

$$T_f : L^p \longrightarrow L^p; \quad T_f \phi = \phi \circ f$$

for each $p, 1 \leq p \leq \infty$ such that $\|T_f \phi\|_p = \|\phi\|_p$. That is, T_f is an isometry on L^p. Note that if ϕ is integrable, then $T_f \phi$ is integrable, and

$$\int_M T_f \phi d\mu = \int_M \phi d\mu.$$

This follows for simple functions from the fact that f is measure preserving and in the general case by a limiting argument. Thus the fact above follows from the remark.

Theorem 2.5 (Maximal ergodic theorem [114], [265], [280]) *Let T be a linear operator on L^1 with $\|T\|_1 \leq 1$ such that $T \geq 0$, i.e., $\phi \geq 0$ implies $T\phi \geq 0$. Let $\phi \in L^1$ be a real function and define*

$$M_* = \left\{ x \mid \sup_{n \geq 1} \sum_{k=0}^{n-1} T^k \phi(x) > 0 \right\}.$$

Then

$$\int_{M_*} \phi d\mu \geq 0. \tag{2.2}$$

Proof. Let

$$T_n \phi(x) = \sum_{k=0}^{n-1} T^k \phi(x), \quad \phi_n(x) = \max_{1 \leq k \leq n} T_k \phi(x) \quad (n \geq 1).$$

Then $\phi = \phi_1 \leq \phi_2 \leq \cdots$ and

$$M_{*,n} = \{x \mid \phi_n(x) > 0\},$$

is an increasing sequence of sets with union M_*. Also since $T \geq 0$,

$$\begin{aligned} T_1 \phi &= \phi \leq \phi + T\phi_n^+, \\ T_{k+1} \phi &= \phi + T(T_k \phi) \leq \phi + T\phi_n^+ \quad (1 \leq k \leq n), \end{aligned}$$

where ϕ_n^+ denotes the positive part of ϕ_n, so that

$$\phi_n \leq \phi_{n+1} \leq \phi + T\phi_n^+,$$

$$\int_{M_{*,n}} \phi d\mu \geq \int_{M_{*,n}} \phi_n d\mu - \int_{M_{*,n}} T\phi_n^+ d\mu$$

$$\geq \int_M \phi_n^+ d\mu - \int_M T\phi_n^+ d\mu$$

$$= \|\phi_n^+\|_1 - \|T\phi_n^+\|_1 \geq 0.$$

The last inequality comes from the assumption that $\|T\|_1 \leq 1$. Letting $n \to \infty$, we obtain the desired result.   □

Corollary 2.1 *For each $\phi \in L^1$ and each real r we have*

$$\int_{M_{\phi_1}(r,\infty)} \phi d\mu \geq r\mu(M_{\phi_1}(r,\infty)) \tag{2.3}$$

and

$$\int_{M_{\phi_2}(r)} \phi d\mu \leq r\mu(M_{\phi_2}(r)), \tag{2.4}$$

where

$$\phi_1(x) = \sup_{n\geq 1} \frac{1}{n} T_n\phi(x), \quad \phi_2(x) = \inf_{n\geq 1} \frac{1}{n} T_n\phi(x).$$

Proof. To prove (2.3), apply (2.2) to the function $\psi = \phi - r \in L^1$ and observe that $T_n\psi(x) > 0$ iff $\frac{1}{n} T_n\phi(x) > r$. The inequality (2.4) follows by applying (2.3) to $-\phi$ and replacing r by $-r$.   □

Definition 2.1 *A linear operator T defined for functions on M is doubly stochastic if it maps L^1 into L^1 and satisfies for all $\phi \in L^1$ the following conditions:*
  1) $\phi \geq 0 \Rightarrow T\phi \geq 0$ a.e.;
  2) $\int_M T\phi d\mu = \int_M \phi d\mu$;
  3) $T1 = 1$ a.e..

For each continuous linear operator T on L^p, there is a well-defined continuous linear operator T^* on L^q, called the *adjoint* of T, where $1 \leq p < \infty$ and

$$\frac{1}{p} + \frac{1}{q} = 1 \quad (q = \infty \text{ if } p = 1).$$

They are related by

$$(T\phi, \psi) = (\phi, T^*\psi), \quad \phi \in L^p, \psi \in L^q,$$

where

$$(\phi, \psi) = \int_M \phi\overline{\psi} d\mu.$$

Lemma 2.1 ([55]) *If T is a doubly stochastic operator, then T maps L^p into L^p for each $p(1 \leq p < \infty)$ with $\|T\|_p \leq 1$ and $\|T\|_1 = \|T\|_\infty = 1$. Moreover, T^* is also doubly stochastic.*

Theorem 2.6 (Yosida mean ergodic theorem) *Given a probability space* $(M, \mathcal{B}, \mu)$. *If* T *is a doubly stochastic operator and* $\phi \in L^p$, *then there exists* $\overline{\phi} \in L^p$ *such that*

$$\lim_{n \to \infty} \left\| \frac{1}{n} \sum_{k=0}^{n-1} T^k \phi - \overline{\phi} \right\|_p = 0.$$

A proof can be found in [55]. Yosida mean ergodic theorem yields the following Neumann mean ergodic theorem [263]:

Theorem 2.7 *If* f *is measure preserving on a measure space* $(M, \mathcal{B}, \mu)$ *and if* $\phi \in L^p$, *then there exists* $\overline{\phi} \in L^p$ *such that*

$$\lim_{n \to \infty} \left\| \frac{1}{n} \sum_{k=0}^{n-1} \phi \circ f^k - \overline{\phi} \right\|_p = 0.$$

Also $\overline{\phi}$ *satisfies* $\overline{\phi} \circ f = \overline{\phi}$ *a.e., and* $\|\overline{\phi}\|_p \le \|\phi\|_p$.

Now let M be a topological space. A domain U in M is said to be *Neumann ergodic of order* p if for each $\phi \in L^p$, there exists $\overline{\phi} \in L^p$ such that

$$\lim_{n \to \infty} \left\| \frac{1}{n} \sum_{k=0}^{n-1} \phi \circ f^k - \overline{\phi} \right\|_{p,U} = 0. \tag{2.5}$$

Also a point $x \in M$ is called *Neumann ergodic of order* p if there is a neighborhood U of x such that U is a Neumann ergodic domain of order p. Let $F_{Neu}^p(f) = F_{Neu,\mu}^p(f)$ be the set of Neumann ergodic points of order p, and set

$$J_{Neu}^p(f) = J_{Neu,\mu}^p(f) = M - F_{Neu}^p(f).$$

Then $F_{Neu}^p(f)$ is open, and $J_{Neu}^p(f)$ is closed. If f is measure preserving, the Neumann ergodic theorem states $J_{Neu}^p(f) = \emptyset$. If $\mathcal{F} = \{f^t\}$ is a flow, substituting (2.5) by

$$\lim_{t \to \infty} \left\| \frac{1}{t} \int_0^t \phi \circ f^t dt - \overline{\phi} \right\|_{p,U} = 0,$$

we also can define $F_{Neu}^p(\mathcal{F})$ and $J_{Neu}^p(\mathcal{F})$.

Now we define a quantity which is closely related to the properties of Neumann ergodic points. We begin our discussion with the discrete-time case. Let $f : M \longrightarrow M$ be a continuous mapping of a compact space M with a metric d. Fix $p \in \mathbb{R}^+$ and fix $\phi \in L^p$. For any $n \ge 1, \varepsilon > 0$, we call a finite set $K \subset M$ an *Neumann* (n, ε)-*covering set of order* p *for* ϕ if for each $x \in K$, there exist a positive number $\rho = \rho^{(p,\phi)}(x, n, \varepsilon)$ and $\overline{\phi} \in L^p$ such that

$$\left\| \frac{1}{n} \sum_{k=0}^{n-1} \phi \circ f^k - \overline{\phi} \right\|_{p,M(x;\rho)} \le \varepsilon, \tag{2.6}$$

and such that

$$\bigcup_{x \in K} M(x; \rho) = M.$$

Set

$$n_{Neu}^{(p,\phi)}(f, \mu, d; n, \varepsilon) = \min\{\#K \mid K \text{ is a Neumann } (n, \varepsilon)\text{-covering set of order } p \text{ for } \phi \}.$$

Notice that for each $n \geq 1, \varepsilon \mapsto n_{Neu}^{(p,\phi)}(f, \mu, d; n, \varepsilon)$ is monotone decreasing. Define

$$h_{Neu}^{(p,\phi)}(f, \mu, d) = \lim_{\varepsilon \to 0} \limsup_{n \to \infty} \frac{1}{n} \log n_{Neu}^{(p,\phi)}(f, \mu, d; n, \varepsilon) \geq 0, \qquad (2.7)$$

$$h_{Neu}^{(p)}(f, \mu, d) = \sup_{\phi \in L^p} h_{Neu}^{(p,\phi)}(f, \mu, d). \qquad (2.8)$$

If $\mathcal{F} = \{f^t\}$ is a flow, substituting (2.6) by

$$\left\| \frac{1}{t} \int_0^t \phi \circ f^t dt - \overline{\phi} \right\|_{p, M(x; \rho)} \leq \varepsilon,$$

we also can define $h_{Neu}^{(p)}(\mathcal{F}, \mu, d)$.

Theorem 2.8 *Let M be a compact smooth manifold and let $f \in C(M, M)$ such that $h_{Neu}^{(p)}(f, \mu, d) > 0$. Then $J_{Neu}^p(f) \neq \emptyset$.*

Proof. Assume $J_{Neu}^p(f) = \emptyset$. Then for each $x \in M$, there is a positive number $\rho = \rho(x)$ such that for every $\phi \in L^p$ and for every positive number ε, there exist an integer $N(x)$ and $\overline{\phi} \in L^p$ such that when $n \geq N(x)$

$$\left\| \frac{1}{n} \sum_{k=0}^{n-1} \phi \circ f^k - \overline{\phi} \right\|_{p, U} < \varepsilon,$$

where $U = U(x) = M(x; \rho)$. Note that M is compact and note that $\{U(x)\}_{x \in M}$ is an open covering of M. Then there exists a finite set $K \subset M$ such that

$$\bigcup_{x \in K} U(x) = M.$$

Take $N = \max_{x \in K} N(x)$. Then for $n \geq N$, the set K is an Neumann (n, ε)-covering set of order p for ϕ. Hence we have

$$n_{Neu}^{(p,\phi)}(f, \mu, d; n, \varepsilon) \leq \#K \quad (n \geq N)$$

which yields $h_{Neu}^{(p,\phi)}(f, \mu, d) = 0$ for every $\phi \in L^p$, and hence $h_{Neu}^{(p)}(f, \mu, d) = 0$. This is a contradiction.   □

Thus if f is measure preserving, then $J_{Neu}^p(f) = \emptyset$. Hence we have $h_{Neu}^{(p)}(f, \mu, d) = 0$. Related to the results of this section, see Hu and Yang [126].

2.3 Birkhoff domains

First we prove the Birkhoff individual ergodic theorem:

Theorem 2.9 ([43]) *If f is measure preserving on a finite measure space $(M, \mathcal{B}, \mu)$ and $\phi \in L^1$, then there exists $\overline{\phi} \in L^1$ such that*

$$\lim_{n \to \infty} \frac{1}{n} \sum_{k=0}^{n-1} \phi \circ f^k = \overline{\phi} \ a.e..$$

Also $\overline{\phi}$ satisfies $\overline{\phi} \circ f = \overline{\phi}$ a.e., and $\|\overline{\phi}\|_1 \le \|\phi\|_1$. Furthermore, if $A \in \mathcal{B}$ with $f^{-1}(A) = A$ and with $\mu(A) < \infty$, then

$$\int_A \overline{\phi} d\mu = \int_A \phi d\mu.$$

Proof. Define

$$\overline{\phi} = \lim_{n \to \infty} \sup \frac{1}{n} \sum_{k=0}^{n-1} \phi \circ f^k, \quad \tilde{\phi} = \lim_{n \to \infty} \inf \frac{1}{n} \sum_{k=0}^{n-1} \phi \circ f^k.$$

For fixed s and r with $s < r$, let

$$N = N_{sr} = M_{\tilde{\phi}}(s) \cap M_{\overline{\phi}}(r, \infty). \tag{2.9}$$

Since

$$\overline{\phi} \circ f = \overline{\phi}, \quad \tilde{\phi} \circ f = \tilde{\phi},$$

it is clear that $f : N \longrightarrow N$. Assuming that $\mu(N) = a > 0$, we can apply Corollary 2.1 to the dynamical system $D_{sr} = (N, \mathcal{B} \cap N, \frac{1}{a}\mu, f)$. Since

$$\inf_n \frac{1}{n} \sum_{k=0}^{n-1} \phi \circ f^k = \phi_2 \le \tilde{\phi} \le \overline{\phi} \le \phi_1 = \sup_n \frac{1}{n} \sum_{k=0}^{n-1} \phi \circ f^k,$$

we have for D_{sr} that

$$N_{\phi_1}(r, \infty) = N_{\phi_2}(s) = N.$$

It follows that

$$r \le \frac{1}{a} \int_N \phi d\mu \le s,$$

which contradicts $s < r$. Thus we have $\mu(N) = 0$. Since

$$A = \{x \mid \tilde{\phi}(x) < \overline{\phi}(x)\} = \bigcup_{s < r; s, r \in \mathbb{Q}} N_{sr},$$

it follows that $\mu(A) = 0$. Thus $\overline{\phi}(x) = \tilde{\phi}(x)$ a.e., and the proof of convergence is complete.
To see that $\overline{\phi} \in L^1$, note that

$$\int_M \left| \frac{1}{n} \sum_{k=0}^{n-1} \phi \circ f^k \right| d\mu \le \frac{1}{n} \int_M \sum_{k=0}^{n-1} |\phi \circ f^k| d\mu = \int_M |\phi| d\mu.$$

By Fatou's lemma,

$$\int_M |\overline{\phi}| d\mu = \int_M \lim_{n\to\infty} \left| \frac{1}{n} \sum_{k=0}^{n-1} \phi \circ f^k \right| d\mu$$

$$\leq \liminf_{n\to\infty} \int_M \left| \frac{1}{n} \sum_{k=0}^{n-1} \phi \circ f^k \right| d\mu \leq \int_M |\phi| d\mu.$$

$\square$

Remark. One also has the general Birkhoff-Khinchin ergodic theorem as follows (cf. [66]): Given a finite measure space $(M, \mathcal{B}, \mu)$ and $\phi \in L^1$, then there exists $\overline{\phi} \in L^1$ such that for almost every $x \in M$,

(i) if f is measure preserving

$$\lim_{n\to\infty} \frac{1}{n} \sum_{k=0}^{n-1} \phi(f^k(x)) = \overline{\phi}(x);$$

(ii) if f is invertible measure preserving

$$\lim_{n\to\infty} \frac{1}{n} \sum_{k=0}^{n-1} \phi(f^k(x)) = \lim_{n\to\infty} \frac{1}{n} \sum_{k=0}^{n-1} \phi(f^{-k}(x)) = \overline{\phi}(x);$$

(iii) if $\{f^t\}$ is a measurable action of $\mathbb{R}$ and each $f^t, t \in \mathbb{R}$, preserves the measure μ

$$\lim_{t\to\infty} \frac{1}{t} \int_0^t \phi(f^t(x)) dt = \lim_{t\to\infty} \frac{1}{t} \int_0^t \phi(f^{-t}(x)) dt = \overline{\phi}(x);$$

(iv) if $\{f^t\}$ is a measurable action of $\mathbb{R}_+$ and each $f^t, t \in \mathbb{R}_+$, preserves the measure μ

$$\lim_{t\to\infty} \frac{1}{t} \int_0^t \phi(f^t(x)) dt = \overline{\phi}(x);$$

Moreover $\overline{\phi}$ satisfies $\int_M \overline{\phi} d\mu = \int_M \phi d\mu$. The function $\overline{\phi}$ is invariant, i.e., $\overline{\phi} \circ f^n = \overline{\phi}$ for $n \geq 0$ in the case (i) and for $n \in \mathbb{Z}$ in the case (ii); $\overline{\phi} \circ f^t = \overline{\phi}$ for $t \in \mathbb{R}$ in the case of a flow and for $n \in \mathbb{R}_+$ in the case of a semiflow;

Let M be a topological space again. A domain U in M is said to be *Birkhoff (ergodic)* if for each $\phi \in L^1$, there exists $\overline{\phi} \in L^1$ such that for almost every $x \in U$,

$$\lim_{n\to\infty} \frac{1}{n} \sum_{k=0}^{n-1} \phi(f^k(x)) = \overline{\phi}(x). \tag{2.10}$$

Also a point $x \in M$ is called *Birkhoff (ergodic)* if there is a neighborhood U of x such that U is a Birkhoff domain. Let $F_{Bir}(f) = F_{Bir,\mu}(f)$ be the set of Birkhoff points, and set

$$J_{Bir}(f) = J_{Bir,\mu}(f) = M - F_{Bir}(f).$$

Then $F_{Bir}(f)$ is open, and $J_{Bir}(f)$ is closed. If $f \in \text{Hom}(M, M)$ is null preserving, we can prove that $F_{Bir}(f)$ and $J_{Bir}(f)$ are backward invariant. If f is measure preserving, the Birkhoff ergodic theorem states $J_{Bir}(f) = \emptyset$. If $\mathcal{F} = \{f^t\}$ is a flow, substituting (2.10) by

$$\lim_{t \to \infty} \frac{1}{t} \int_0^t \phi(f^t(x))dt = \overline{\phi}(x),$$

we also can define $F_{Bir}(\mathcal{F})$ and $J_{Bir}(\mathcal{F})$.

Now we define a quantity which is closely related to the properties of Birkhoff ergodic points. We begin our discussion with the discrete-time case. Let $f : M \longrightarrow M$ be a continuous mapping of a compact space M with a metric d. Fix $p \in \mathbb{R}^+$ and $\phi \in L^1$. For any $n \geq 1, \varepsilon > 0, 0 < \delta < \mu(M)$, we call a finite set $K \subset M$ a *Birkhoff (n, ε, δ)-covering set for ϕ* if for each $x \in K$, there exist a positive number $\rho = \rho(x, \phi; n, \varepsilon)$ and $\overline{\phi} \in L^1$ such that

$$\mu\left(\bigcup_{x \in K} \left\{ y \in M(x; \rho) \mid \left| \frac{1}{n} \sum_{k=0}^{n-1} \phi(f^k(y)) - \overline{\phi}(y) \right| < \varepsilon \right\}\right) > \delta. \qquad (2.11)$$

Set

$$n_{Bir,\phi}(f, \mu, d; n, \varepsilon, \delta) = \min\{\#K \mid K \text{ is a Birkhoff } (n, \varepsilon, \delta)\text{-covering set for } \phi \}.$$

Notice that for fixed δ, $n \geq 1$, the mapping $\varepsilon \mapsto n_{Bir,\phi}(f, \mu, d; n, \varepsilon, \delta)$ is monotone decreasing and for fixed ε, $n \geq 1$, the mapping $\delta \mapsto n_{Bir,\phi}(f, \mu, d; n, \varepsilon, \delta)$ is monotone increasing. Define

$$h_{Bir,\phi}(f, \mu, d) = \lim_{\delta \to \mu(M)} \lim_{\varepsilon \to 0} \limsup_{n \to \infty} \frac{1}{n} \log n_{Bir,\phi}(f, \mu, d; n, \varepsilon, \delta) \geq 0, \qquad (2.12)$$

$$h_{Bir}(f, \mu, d) = \sup_{\phi \in L^1} h_{Bir,\phi}(f, \mu, d). \qquad (2.13)$$

If $\mathcal{F} = \{f^t\}$ is a flow, substituting (2.11) by

$$\mu\left(\bigcup_{x \in K} \left\{ y \in M(x; \rho) \mid \left| \frac{1}{t} \int_0^t \phi(f^t(y))dt - \overline{\phi}(y) \right| < \varepsilon \right\}\right) > \delta,$$

we also can define $h_{Bir,\phi}(\mathcal{F}, \mu, d)$ and $h_{Bir}(\mathcal{F}, \mu, d)$.

Theorem 2.10 *Let M be a compact smooth manifold and let $f \in C(M, M)$ such that $h_{Bir}(f, \mu, d) > 0$. Then $J_{Bir}(f) \neq \emptyset$.*

Proof. Assume $J_{Bir}(f) = \emptyset$. Then for each $x \in M$, there is a positive number $\rho = \rho(x)$ such that for each $\phi \in L^1$, there exists $\overline{\phi} \in L^1$ satisfying that for almost every $y \in U(x) = M(x; \rho)$,

$$\lim_{n \to \infty} \frac{1}{n} \sum_{k=0}^{n-1} \phi(f^k(y)) = \overline{\phi}(y). \qquad (2.14)$$

Note that M is compact and note that $\{U(x)\}_{x \in M}$ is an open covering of M. Then there exists a finite set $K \subset M$ such that

$$\bigcup_{x \in K} U(x) = M.$$

Set $l = \#K$ and take $\varepsilon > 0, 0 < \delta < \mu(M)$. By a basic theorem in measure theory, for $x \in K$ there is a subset $E(x) \subset U(x)$ with $\mu(E(x)) < (\mu(M) - \delta)/l$, such that $\frac{1}{n}\sum_{k=0}^{n-1} \phi \circ f^k$ converges uniformly to $\overline{\phi}$ on $U(x) - E(x)$. Hence there exists a positive integer $N(x)$ such that when $n \geq N(x)$, $y \in U(x) - E(x)$,

$$\left| \frac{1}{n}\sum_{k=0}^{n-1} \phi(f^k(y)) - \overline{\phi}(y) \right| < \varepsilon.$$

Take $N = \max_{x \in K} N(x)$. Then if $n \geq N$, we have

$$\mu\left(\bigcup_{x \in K} \left\{ y \in U(x) \mid \left| \frac{1}{n}\sum_{k=0}^{n-1} \phi(f^k(y)) - \overline{\phi}(y) \right| < \varepsilon \right\} \right)$$

$$\geq \mu\left(\bigcup_{x \in K} \{U(x) - E(x)\} \right)$$

$$= \mu(M) - \mu\left(\bigcup_{x \in K} E(x) \right) \geq \mu(M) - \sum_{x \in K} \mu(E(x))$$

$$> \mu(M) - \sum_{x \in K} \frac{\mu(M) - \delta}{l} = \delta.$$

Therefore for $n \geq N$, the set K is an Birkhoff (n, ε, δ)-covering set for ϕ. Hence we have

$$n_{Bir,\phi}(f, \mu, d; n, \varepsilon, \delta) \leq l \quad (n \geq N),$$

which yields $h_{Bir,\phi}(f, \mu, d) = 0$ for every $\phi \in L^1$, and hence $h_{Bir}(f, \mu, d) = 0$. This is a contradiction. □

Thus if f is measure preserving, then $J_{Bir}(f) = \emptyset$, and hence $h_{Bir}(f, \mu, d) = 0$. Related to the results of this section, see Hu and Yang [126].

2.4   Ergodic points

Definition 2.2 *A measure preserving mapping f in $(M, \mathcal{B}, \mu)$ is called ergodic if all backward invariant sets A have the property that $\mu(A) = 0$ or $\mu(A^c) = 0$. The measure μ also is called ergodic.*

We know the following existence theorems of ergodic measure: If M is a compact topological space, $\mathcal{B}$ the usual Borel σ-algebra, and $f \in \mathrm{Hom}(M, M)$, then there always exists at least one ergodic measure (cf. [199]). Every continuous mapping f on a metrizable compact space M has an ergodic f-invariant Borel probability measure (Katok [134], p.139). Here we exhibit some equivalent conditions of the ergodicity of f:

Theorem 2.11 ([150],[55]) *For a measure preserving mapping f of a finite measure space $(M, \mathcal{B}, \mu)$ each of the following conditions is equivalent to the ergodicity of f:*
    1) $\mu(A) > 0 \Longrightarrow \mu(O^-(A)^c) = 0;$

2) $\mu(A) > 0, \mu(B) > 0 \Longrightarrow \mu(f^{-k}(A) \cap B) > 0$ *for some* $k \geq 0$;

3) for all $A, B \in \mathcal{B}$,

$$\lim_{n\to\infty} \frac{1}{n} \sum_{k=0}^{n-1} \mu(f^{-k}(A) \cap B) = \frac{1}{\mu(M)} \mu(A)\mu(B);$$

4) there exists a family $\mathcal{G} \subset \mathcal{B}$ *such that the linear combinations of the characteristic functions* χ_E $(E \in \mathcal{G})$ *of* E *lie dense in* L^2 *and which has the property that*

$$\lim_{n\to\infty} \frac{1}{n} \sum_{k=0}^{n-1} \chi_E \circ f^k = \frac{\mu(E)}{\mu(M)} \quad a.e.;$$

5) for all $\phi \in L^1$,

$$\lim_{n\to\infty} \frac{1}{n} \sum_{k=0}^{n-1} \phi \circ f^k = \frac{1}{\mu(M)} \int_M \phi d\mu \quad a.e.;$$

6) for all $\phi, \psi \in L^2$,

$$\lim_{n\to\infty} \frac{1}{n} \sum_{k=0}^{n-1} (\phi \circ f^k, \psi) = \frac{1}{\mu(M)} (\phi, 1)(1, \psi);$$

7) $\phi \circ f = \phi$, *for some* $\phi \in L^1 \Rightarrow \phi = constant\ a.e.$ (μ).

Remark. The following Birkhoff ergodic theorem [44] holds for flows: Let M be an m-dimensional manifold. Let dV be the volume element and $d\mu = \lambda dV$, where $\lambda \geq 0$ is a continuous function on M. Assume that $\mu(M) < \infty$. If each f^t of a flow $\{f^t\}$ is an ergodic measure preserving transformation on the measure space $(M, \mathcal{B}, \mu)$ and $\phi \in L^\infty$, then

$$\lim_{t\to\infty} \frac{1}{t} \int_0^t \phi(f^t(x))dt = \frac{1}{\mu(M)} \int_M \phi d\mu \quad a.e. \ .$$

A proof can be found, say, in [137] or [256].

Thus, f is ergodic if the space M cannot be decomposed into two non trivial backward invariant sets. If an measure preserving mapping is ergodic and $0 < \mu(M) < \infty$, the limit $\overline{\phi}$ is simply given by the space average

$$\overline{\phi} = \frac{1}{\mu(M)} \int_M \phi d\mu. \tag{2.15}$$

This means that for large n the space average is very close to the time average

$$\frac{1}{n} \sum_{k=0}^{n-1} \phi \circ f^k.$$

An important *ergodic hypothesis* in statistical mechanics is the assumption that these two averages are assymptotically equal for the measure preserving transformation arising in the

Hamiltonian flow in a phase space. Birkhoff's theorem made it clear that the limit of the time average does exist and that it is equal to the space average in the ergodic case.

A domain U in a topological space M is said to be *ergodic* if for each $\phi \in L^1$,

$$\lim_{n \to \infty} \frac{1}{n} \sum_{k=0}^{n-1} \phi(f^k(x)) = \overline{\phi} \tag{2.16}$$

holds for almost every $x \in U$, where $\overline{\phi}$ is given by (2.15). Also a point $x \in M$ is called *ergodic* if there is a neighborhood U of x such that U is an ergodic domain. Let $F_{erg}(f) = F_{erg,\mu}(f)$ be the set of ergodic points, and set

$$J_{erg}(f) = J_{erg,\mu}(f) = M - F_{erg}(f).$$

Then $F_{erg}(f)$ is open, and $J_{erg}(f)$ is closed. Obviously, we have

$$J_{Bir}(f) \subset J_{erg}(f).$$

If $f \in \mathrm{Hom}(M, M)$ is null preserving, we also can prove that $F_{erg}(f)$ and $J_{erg}(f)$ are backward invariant. According to Theorem 2.11, the mapping f is ergodic if and only if $J_{erg}(f) = \emptyset$. If $\mathcal{F} = \{f^t\}$ is a flow, substituting (2.16) by

$$\lim_{t \to \infty} \frac{1}{t} \int_0^t \phi(f^t(x)) dt = \overline{\phi},$$

we also can define $F_{erg}(\mathcal{F})$ and $J_{erg}(\mathcal{F})$.

We also can define a quantity which is closely related to the properties of ergodic points. We begin our discussion with the discrete-time case. Let $f : M \longrightarrow M$ be a continuous mapping of a compact space M with a metric d. Fix $p \in \mathbb{R}^+$ and $\phi \in L^1$. For any $n \geq 1, \varepsilon > 0, 0 < \delta < \mu(M)$, we call a finite set $K \subset M$ an *ergodic (n, ε, δ)-covering set for* ϕ if for each $x \in K$, there exist a positive number $\rho = \rho(x, \phi; n, \varepsilon)$ such that

$$\mu \left(\bigcup_{x \in K} \left\{ y \in M(x; \rho) \mid \left| \frac{1}{n} \sum_{k=0}^{n-1} \phi(f^k(y)) - \overline{\phi} \right| < \varepsilon \right\} \right) > \delta. \tag{2.17}$$

Set

$$n_{erg,\phi}(f, \mu, d; n, \varepsilon, \delta) = \min\{ \#K \mid K \text{ is an ergodic } (n, \varepsilon, \delta)\text{-covering set for } \phi \}.$$

Notice that for fixed δ, $n \geq 1$, the mapping $\varepsilon \mapsto n_{erg,\phi}(f, \mu, d; n, \varepsilon, \delta)$ is monotone decreasing and for fixed ε, $n \geq 1$, the mapping $\delta \mapsto n_{erg,\phi}(f, \mu, d; n, \varepsilon, \delta)$ is monotone increasing. Define

$$h_{erg,\phi}(f, \mu, d) = \lim_{\delta \to \mu(M)} \lim_{\varepsilon \to 0} \limsup_{n \to \infty} \frac{1}{n} \log n_{erg,\phi}(f, \mu, d; n, \varepsilon, \delta) \geq 0, \tag{2.18}$$

$$h_{erg}(f, \mu, d) = \sup_{\phi \in L^1} h_{erg,\phi}(f, \mu, d). \tag{2.19}$$

By the definition, we see

$$n_{Bir,\phi}(f,\mu,d;n,\varepsilon,\delta) \leq n_{erg,\phi}(f,\mu,d;n,\varepsilon,\delta).$$

Thus we obtain

$$h_{Bir,\phi}(f,\mu,d) \leq h_{erg,\phi}(f,\mu,d),$$

for $\phi \in L^1$, and hence $h_{Bir}(f,\mu,d) \leq h_{erg}(f,\mu,d)$. If $\mathcal{F} = \{f^t\}$ is a flow, substituting (2.17) by

$$\mu\left(\bigcup_{x \in K} \left\{y \in M(x;\rho) \mid \left|\frac{1}{t}\int_0^t \phi(f^t(y))dt - \overline{\phi}\right| < \varepsilon\right\}\right) > \delta,$$

we also can define $h_{erg,\phi}(\mathcal{F},\mu,d)$ and $h_{erg}(\mathcal{F},\mu,d)$. Similarly, we can prove the following result:

Theorem 2.12 *Let M be a compact smooth manifold and let $f \in C(M,M)$ such that $h_{erg}(f,\mu,d) > 0$. Then $J_{erg}(f) \neq \emptyset$.*

Thus if f is ergodic, then $J_{erg}(f) = \emptyset$, and hence $h_{erg}(f,\mu,d) = 0$.

Example 2.1 *Here we consider the mappings E_k on S^1 defined in Example 1.2. A point on S^1 may be given by putting $z = e^{2\pi xi}$, in the additive notation, as the system of real number x considered mod 1. In this case, we may assume that $0 \leq x < 1$. Using the additive notation, the mapping E_k is given by*

$$E_k(x) = kx(mod\ 1).$$

Now the space $L^1(S^1)$ is contained the set of Lebesgue-integrable real-valued functions on $\mathbb{R}$ with period 1. Raikov[205] (see also F. Riesz[212]) proved that if $k \geq 2$, for every Lebesgue-integrable real-valued function ϕ on $\mathbb{R}$ with period 1,

$$\frac{1}{n}\sum_{j=0}^{n-1}\phi(E_k^j(x)) = \frac{1}{n}\sum_{j=0}^{n-1}\phi(k^j x) \to \int_0^1 \phi(x)dx$$

holds for almost all real x as $n \to \infty$. Thus each point on S^1 is ergodic for E_k, and hence $J_{Bir}(E_k) = J_{erg}(E_k) = \emptyset$.

Example 2.2 *Here we consider the translations defined on the torus $\mathbb{T}^m$ as in Example 1.1. A point on $\mathbb{T}^m$ may be given either in the multiplicative notation or by putting $z_k = e^{2\pi x_k i}$, in the additive notation, as the system of m real numbers $x_1, ..., x_m$ considered mod 1. In this case, we may assume that $0 \leq x_k < 1, 1 \leq k \leq m$. Using the additive notation, the translation R_α is given as follows*

$$R_\alpha(x_1,...,x_m) = (x_1 + \alpha_1(mod\ 1),...,x_m + \alpha_m(mod\ 1)).$$

It is clear that the Lebesgue measure on $\mathbb{T}^m$

$$d\mu = dx_1 \wedge \cdots \wedge dx_m,$$

is invariant with respect R_α. Also R_α is ergodic if and only if the numbers $1, \alpha_1, ..., \alpha_m$ are rationally independent. If so, the translation R_α is minimal, and is uniquely ergodic, i.e., the Lebesgue measure μ on $\mathbb{T}^m$ is the only invariant normalized Borel measure (see [66]). Thus by Theorem 2.11, each point on $\mathbb{T}^m$ is Birkhoff ergodic for R_α, that is, $J_{Bir}(R_\alpha) = J_{erg}(R_\alpha) = \emptyset$. If the numbers $1, \alpha_1, ..., \alpha_m$ are rationally dependent, then $J_{erg}(R_\alpha) \neq \emptyset$.

2.5   Regular ergodic points

A continuous mapping $f \in C(M, M)$ on a metrizable compact space M is called *uniquely ergodic* if it has only one invariant Borel probability measure. The following theorem makes clear the meaning of the notion of unique ergodicity.

Theorem 2.13 (H. Furstenberg, cf. [230], p.9) *Suppose f is a homeomorphism of the compact metric space M and μ is a f-invariant Borel probability measure. The following statements are equivalent:*

1) f is uniquely ergodic;

2) for any continuous function ϕ on M and any $x \in M$ one has

$$\lim_{n \to \infty} \frac{1}{n} \sum_{k=0}^{n-1} \phi(f^k(x)) = \int_M \phi d\mu;$$

3) for any continuous function ϕ on M the convergence $\frac{1}{n} \sum\limits_{k=0}^{n-1} \phi(f^k(x)) \to \int_M \phi d\mu$ is uniform on M.

The invariant Borel probability measure of an uniquely ergodic mapping f is ergodic (see Proposition 4.1.8 of [134]). For a Borel measure μ on a separable metrizable space M define the *support* of μ to be the following set

$$\operatorname{supp}\mu = \{x \in M \mid \mu(U) > 0 \text{ whenever } x \in U, \ U \text{ open }\}.$$

Then $\operatorname{supp}\mu$ is a closed set with $\mu(M - \operatorname{supp}\mu) = 0$ such that any set of full measure is dense in $\operatorname{supp}\mu$. If f is a measure preserving transformation, then $\operatorname{supp}\mu$ is an invariant set.

Proposition 2.1 (cf. [134]) *Let M be a complete separable metrizable space and take $f \in \operatorname{Hom}(M, M)$. Then*

1) $\operatorname{supp}\mu \subset P(f)$ for any f-invariant Borel probability measure μ.

2) If μ is ergodic, then $f|_{\operatorname{supp}\mu}$ has a dense orbit.

3) If M is compact and $f|_{\operatorname{supp}\mu}$ is uniquely ergodic, then $\operatorname{supp}\mu$ is a minimal set.

Problem 2.1 *Let M be a compact space and take $f \in \operatorname{Hom}(M, M)$. If $f|_{J(f)}$ is uniquely ergodic, is $J(f)$ a minimal set? Can we find a f-invariant Borel probability measure μ such that $\operatorname{supp}\mu \subset J(f)$?*

In this section, we fix a measure space $(M, \mathcal{B}, \mu)$ with $0 < \mu(M) < \infty$, where M is a topological space. Recall that

$$C_0(M) = \{\phi \in C(M, \mathbb{R}) \mid \operatorname{supp}(\phi) \text{ is compact }\}. \tag{2.20}$$

If $\phi \in L^1(M)$, we write

$$\overline{\phi} = \frac{1}{\mu(M)} \int_M \phi d\mu,$$

and define

$$G_{\mu,\phi}(f) = \left\{ x \in M \mid \lim_{n \to \infty} \frac{1}{n} \sum_{k=0}^{n-1} \phi \circ f^k(x) = \overline{\phi} \right\}.$$

It is easy to prove that

$$f^{-1}(G_{\mu,\phi}(f)) \subset G_{\mu,\phi}(f), \quad f(G_{\mu,\phi}(f)) \subset G_{\mu,\phi}(f). \tag{2.21}$$

Hence $G_{\mu,\phi}(f)$ is backward invariant, i.e.,

$$f^{-1}(G_{\mu,\phi}(f)) = G_{\mu,\phi}(f). \tag{2.22}$$

Definition 2.3 *Take $f \in C(M, M)$. We say that $x \in M$ is a generic point of f for μ if $x \in G_{\mu,\phi}(f)$ for every $\phi \in C_0(M)$. Denote the set of generic points for μ by $G_\mu(f)$.*

Then $G_\mu(f)$ also is backward invariant with

$$G_\mu(f) = \bigcap_{\phi \in C_0(M)} G_{\mu,\phi}(f).$$

Thus if f is a homeomorphism of a compact metric space M and if μ is uniquely f-invariant Borel probability measure, then $G_\mu(f) = M$.

Definition 2.4 *A point $x \in M$ is said to be regularly ergodic for μ if there exists a neighborhood U of x such that every $y \in U$ is a generic point for μ.*

Let $F_\mu(f)$ be the set of regular ergodic points of f on M for μ. Obviously, the set $F_\mu(f)$ is open, and the set

$$J_\mu(f) = M - F_\mu(f)$$

is closed. If M is a compact manifold and if f is a measure preserving transformation for μ, then f is uniquely ergodic if and only if $J_\mu(f) = \emptyset$.

Theorem 2.14 *For $f \in C(M, M)$, we have*

$$f^{-1}(F_\mu(f)) \subset F_\mu(f). \tag{2.23}$$

Further if f is an open mapping, then

$$f(F_\mu(f)) \subset F_\mu(f), \tag{2.24}$$

and hence $F_\mu(f)$ and $J_\mu(f)$ are backward invariant.

Proof. Take $x \in F_\mu(f)$. By definition, there exist a neighborhood U of x such that $U \subset G_\mu(f)$. Thus for any $y \in U, \tilde{y} \in f^{-1}(y)$, noting that

$$\lim_{n \to \infty} \frac{1}{n} \sum_{k=0}^{n-1} \phi \circ f^k(\tilde{y}) = \lim_{n \to \infty} \left\{ \frac{1}{n} \phi(\tilde{y}) + \frac{1}{n} \sum_{k=0}^{n-2} \phi \circ f^k(y) \right\}$$

$$= \overline{\phi},$$

i.e., $f^{-1}(U) \subset G_\mu(f)$, then (2.23) follows.

If any $y \in U, \hat{y} = f(y)$, noting that

$$\lim_{n \to \infty} \frac{1}{n} \sum_{k=0}^{n-1} \phi \circ f^k(\hat{y}) = \lim_{n \to \infty} \left\{ -\frac{1}{n}\phi(y) + \frac{n+1}{n}\frac{1}{n+1} \sum_{k=0}^{n+1-1} \phi \circ f^k(y) \right\}$$
$$= \overline{\phi},$$

then (2.24) follows. $\square$

Let $\mathcal{F} = \{f^t\}$ be a flow (or semiflow). We say that $x \in M$ is a generic point of $\mathcal{F}$ for μ if

$$\lim_{t \to \infty} \frac{1}{t} \int_0^t \phi \circ f^t(x) dt = \overline{\phi},$$

for every $\phi \in C_0(M)$. Also denote the set of generic points for μ by $G_\mu(\mathcal{F})$. Similarly, we can define regularly ergodic points of $\mathcal{F}$ for μ, and obtain the set $F_\mu(\mathcal{F})$ of regularly ergodic points of $\mathcal{F}$ on M for μ. Also set $J_\mu(\mathcal{F}) = M - F_\mu(\mathcal{F})$. It is interesting to study the number

$$\inf_{t \geq 0}\{t \mid J_\mu(f^t) \neq \emptyset\}.$$

For example, is it finite when $h_{top}(f) > 0$?

Now we define a quantity which is closely related to the properties of regular ergodic points. Let $f : M \longrightarrow M$ be a continuous mapping of a compact space M with a metric d. Define

$$h_{erg}^0(f, \mu, d) = \sup_{\phi \in C_0(M)} h_{erg,\phi}(f, \mu, d) \leq h_{erg}(f, \mu, d). \tag{2.25}$$

After the fashion of Theorem 2.10, we also can prove the following result:

Theorem 2.15 *Let M be a compact smooth manifold and let $f \in C(M, M)$ such that $h_{erg}^0(f, \mu, d) > 0$. Then $J_\mu(f) \neq \emptyset$.*

Definition 2.5 *A point $x \in M$ is said to be uniformly ergodic for μ if there exists a neighborhood U of x such that for any function $\phi \in C_0(M)$ the convergence $\frac{1}{n} \sum_{k=0}^{n-1} \phi(f^k(x)) \to \overline{\phi}$ is uniform on U.*

Let $F_{uni}(f) = F_{uni,\mu}(f)$ be the set of uniform ergodic points of f on M for μ. Obviously, the set $F_{uni}(f)$ is open, and the set

$$J_{uni}(f) = J_{uni,\mu}(f) = M - F_{uni}(f)$$

is closed with $J_\mu(f) \subset J_{uni}(f)$. We also have the following result:

Theorem 2.16 *For $f \in C(M, M)$, we have*

$$f^{-1}(F_{uni}(f)) \subset F_{uni}(f). \tag{2.26}$$

Further if f is an open mapping, then

$$f(F_{uni}(f)) \subset F_{uni}(f), \tag{2.27`}$$

and hence $F_{uni}(f)$ and $J_{uni}(f)$ are backward invariant.

Now we define a quantity which is closely related to the properties of uniform ergodic points. Let $f : M \longrightarrow M$ be a continuous mapping of a compact space M with a metric d. Fix $\phi \in C_0(M)$. For any $n \geq 1, \varepsilon > 0$, we call a finite set $K \subset M$ a (n, ε)-*covering set for* ϕ if, for each $x \in K$, there is a positive number $\rho = \rho(x, n, \varepsilon)$ such that when $y \in M_d(x; \rho)$, we have

$$\left| \frac{1}{n} \sum_{k=0}^{n-1} \phi(f^k(y)) - \overline{\phi} \right| < \varepsilon,$$

and such that

$$\bigcup_{x \in K} M_d(x; \rho) = M.$$

Set

$$n_{uni,\phi}(f, \mu, d; n, \varepsilon) = \min\{\#K \mid K \text{ is a } (n, \varepsilon)\text{-covering set for } \phi\}.$$

Notice that for each $n \geq 1, \varepsilon \mapsto n_{uni,\phi}(f, \mu, d; n, \varepsilon)$ is monotone decreasing. We can define

$$h_{uni,\phi}(f, \mu, d) = \lim_{\varepsilon \to 0} \limsup_{n \to \infty} \frac{1}{n} \log n_{uni,\phi}(f, \mu, d; n, \varepsilon) \geq 0. \qquad (2.28)$$

Define

$$h_{uni}(f, \mu, d) = \sup_{\phi \in C_0(M)} h_{uni,\phi}(f, \mu, d). \qquad (2.29)$$

Theorem 2.17 *Let M be a compact smooth manifold and let $f \in C(M, M)$ such that $h_{uni}(f, \mu, d) > 0$. Then $J_{uni}(f) \neq \emptyset$.*

The proof can be completed after the fashion of Theorem 2.10.

Problem 2.2 *Are there relations between $J(f)$ and $J_{uni}(f)$ for some measure μ ?*

Example 2.3 *We consider the Gauss transformation $f : [0, 1] \longrightarrow [0, 1]$ defined by*

$$f(x) = \begin{cases} \frac{1}{x} - \left[\frac{1}{x}\right] & : \quad x \neq 0 \\ 0 & : \quad x = 0 \end{cases}$$

The mapping f possesses the important property that it preserves the Borel measure μ in $[0, 1]$ given by

$$\mu(A) = \frac{1}{\log 2} \int_A \frac{1}{1+x} d\lambda,$$

where λ is Lebesgue measure, that is, $\mu(f^{-1}(A)) = \mu(A)$ for every Borel set $A \subset [0, 1]$. This is the so-called Gauss measure which also is uniquely ergodic (see [164]). Therefore $J_\mu(f) = J_{uni}(f) = \emptyset$.

2.6 L^p-Ergodic points

In this section, we fix a measure space $(M, \mathcal{B}, \mu)$ with $0 < \mu(M) < \infty$, where M is a topological space. For $\phi \in L^1(M)$, write

$$\overline{\phi} = \frac{1}{\mu(M)} \int_M \phi d\mu.$$

Definition 2.6 *Take $f \in C(M, M)$. We say that $x \in M$ is a L^p-ergodic point of f for μ if there exists a neighborhood U of x such that*

$$\lim_{n \to \infty} \left\| \frac{1}{n} \sum_{k=0}^{n-1} \phi \circ f^k - \overline{\phi} \right\|_{p,U} = 0 \quad (0 < p \leq \infty),$$

for every $\phi \in C_0(M)$.

Let $F_\mu^p(f)$ be the set of L^p-ergodic points of f on M for μ, and set

$$J_\mu^p(f) = M - F_\mu^p(f).$$

Then $F_\mu^p(f)$ is open, and $J_\mu^p(f)$ is closed.

Assume $0 < p < q < \infty$. Then Hölder's inequality yields

$$\left\| \frac{1}{n} \sum_{k=0}^{n-1} \phi \circ f^k - \overline{\phi} \right\|_{p,U} \leq \mu(U)^{\frac{1}{p}-\frac{1}{q}} \left\| \frac{1}{n} \sum_{k=0}^{n-1} \phi \circ f^k - \overline{\phi} \right\|_{q,U}$$

$$\leq \mu(U)^{\frac{1}{p}-\frac{1}{q}} \left\| \frac{1}{n} \sum_{k=0}^{n-1} \phi \circ f^k - \overline{\phi} \right\|_{\infty,U}.$$

Hence we have

$$F_\mu^\infty(f) \subset F_\mu^q(f) \subset F_\mu^p(f) \tag{2.30}$$
$$J_\mu^p(f) \subset J_\mu^q(f) \subset J_\mu^\infty(f). \tag{2.31}$$

Theorem 2.18 *Take $f \in \mathrm{Diff}^\infty(M, M)$. Suppose that M is compact, orientable and that f, f^{-1} are orientation preserving. Let μ be the measure induced by a volume form Ω of M. Then*

$$f^{-1}(F_\mu^p(f)) = F_\mu^p(f), \quad f^{-1}(J_\mu^p(f)) = J_\mu^p(f) \quad (0 < p \leq \infty). \tag{2.32}$$

Proof. Take $x \in F_\mu^p(f)$. Then there exists a neighborhood U of x such that

$$\lim_{n \to \infty} \left\| \frac{1}{n} \sum_{k=0}^{n-1} \phi \circ f^k - \overline{\phi} \right\|_{p,U} = 0,$$

for every $\phi \in C_0(M)$. Since f is orientation preserving and since M is compact, then there is a positive number c such that $0 \leq f^*\Omega/\Omega \leq c$. Thus if $0 < p < 1$,

$$
\begin{aligned}
c^{-1}\left\|\frac{1}{n}\sum_{k=0}^{n-1}\phi\circ f^k - \overline{\phi}\right\|_{p,f(U)}^p &\leq \left\|\frac{1}{n}\sum_{k=0}^{n-1}\phi\circ f^{k+1} - \overline{\phi}\right\|_{p,U}^p \\
&\leq \left\|\frac{1}{n}\sum_{k=0}^{n-1}\phi\circ f^k - \overline{\phi}\right\|_{p,U}^p + \frac{2}{n^p}\|\phi\|_{p,M}^p \\
&\to 0 \quad \text{as } n\to\infty,
\end{aligned}
$$

and if $p \geq 1$,

$$
\begin{aligned}
c^{-1}\left\|\frac{1}{n}\sum_{k=0}^{n-1}\phi\circ f^k - \overline{\phi}\right\|_{p,f(U)} &= \left\|\frac{1}{n}\sum_{k=0}^{n-1}\phi\circ f^{k+1} - \overline{\phi}\right\|_{p,U} \\
&\leq \left\|\frac{1}{n}\sum_{k=0}^{n-1}\phi\circ f^k - \overline{\phi}\right\|_{p,U} + \frac{2}{n}\|\phi\|_{p,M} \\
&\to 0 \quad \text{as } n\to\infty,
\end{aligned}
$$

therefore $f(x) \in F_\mu^p(f)$, i.e., $f(F_\mu^p(f)) \subset F_\mu^p(f)$.

Similarly, we can prove that $f^{-1}(F_\mu^p(f)) \subset F_\mu^p(f)$. Hence $F_\mu^p(f)$ is backward invariant. Consequently, $J_\mu^p(f)$ is backward invariant as well. $\square$

Remark. Under the assumptions in Theorem 2.18, we also can prove that $F_{Neu}^p(f)$ and $J_{Neu}^p(f)$ are backward invariant.

Remark. Let $\mathcal{F} = \{f^t\}$ be a flow (or semiflow). We say that $x \in M$ is a L^p-ergodic point of $\mathcal{F}$ for μ if there exists a neighborhood U of x such that

$$
\lim_{t\to\infty}\left\|\frac{1}{t}\int_0^t \phi\circ f^t dt - \overline{\phi}\right\|_{p,U} = 0 \quad (0 < p \leq \infty),
$$

for every $\phi \in C_0(M)$. Similarly, we can define $F_\mu^p(\mathcal{F})$ and $J_\mu^p(\mathcal{F})$, and study the number

$$
\inf_{t\geq 0}\{t \mid J_\mu^p(f^t) \neq \emptyset\}.
$$

Note that the definition of the topological entropy is related to equicontinuity so that one can find the relation in Theorem 1.33. Now we define a quantity which is closely related to the properties of L^p-ergodic points. We begin our discussion with the discrete-time case. Let $f : M \longrightarrow M$ be a continuous mapping of a compact space M with a metric d. Fix $p \in \mathbb{R}_+$ and fix $\phi \in C_0(M)$. For any $n \geq 1, \varepsilon > 0$, we call a finite set $K \subset M$ an (n,ε)-covering set of type (p,ϕ) if for each $x \in K$, there exists a positive number $\rho = \rho^{(p,\phi)}(x,n,\varepsilon)$ such that

$$
\left\|\frac{1}{n}\sum_{k=0}^{n-1}\phi\circ f^k - \overline{\phi}\right\|_{p,M(x;\rho)} \leq \varepsilon, \tag{2.33}
$$

and such that

$$\bigcup_{x \in K} M(x; \rho) = M.$$

Set

$$n^{(p,\phi)}(f, \mu, d; n, \varepsilon) = \min\{\#K \mid K \text{ is a } (n, \varepsilon)\text{-covering set of type } (p, \phi)\}.$$

Notice that for each $n \geq 1$, the mapping $\varepsilon \mapsto n^{(p,\phi)}(f, \mu, d; n, \varepsilon)$ is monotone decreasing. Define

$$h^{(p,\phi)}(f, \mu, d) = \lim_{\varepsilon \to 0} \limsup_{n \to \infty} \frac{1}{n} \log n^{(p,\phi)}(f, \mu, d; n, \varepsilon) \geq 0, \tag{2.34}$$

$$h^{(p)}(f, \mu, d) = \sup_{\phi \in C_0(M)} h^{(p,\phi)}(f, \mu, d). \tag{2.35}$$

By the definition, we obviously have

$$n^{(p,\phi)}_{Neu}(f, \mu, d; n, \varepsilon) \leq n^{(p,\phi)}(f, \mu, d; n, \varepsilon).$$

Therefore for $\phi \in C_0(M)$,

$$h^{(p,\phi)}_{Neu}(f, \mu, d) \leq h^{(p,\phi)}(f, \mu, d).$$

If $\mathcal{F} = \{f^t\}$ is a flow, substituting (2.33) by

$$\left\| \frac{1}{t} \int_0^t \phi \circ f^t dt - \overline{\phi} \right\|_{p, M(x; \rho)} \leq \varepsilon,$$

we also can define $h^{(p)}(\mathcal{F}, \mu, d)$.

Theorem 2.19 *Let M be a compact smooth manifold and let $f \in C(M, M)$ such that $h^{(p)}(f, \mu, d) > 0$. Then $J^p_\mu(f) \neq \emptyset$.*

The proof can be completed after the fashion of Theorem 2.8. Related to the results of this sections, see Hu and Yang [122], [124] and [126].

2.7 L^0-Ergodic points

In this section, we fix a measure space $(M, \mathcal{B}, \mu)$ with $0 < \mu(M) < \infty$, where M is a topological space. If $\phi \in L^1(M)$, we write

$$\overline{\phi} = \frac{1}{\mu(M)} \int_M \phi d\mu.$$

Definition 2.7 *Take $f \in C(M, M)$. We say that $x \in M$ is a L^0-ergodic point of f for μ if there exists a neighborhood U of x such that for any $\delta > 0$, and for every $\phi \in C_0(M)$, we have*

$$\lim_{n \to \infty} \mu(U_{n, \phi, \delta}) = 0,$$

where

$$U_{n, \phi, \delta} = \left\{ x \in U \mid \left| \frac{1}{n} \sum_{k=0}^{n-1} \phi \circ f^k(x) - \overline{\phi} \right| \geq \delta \right\}.$$

Let $F_\mu^0(f)$ be the set of L^0-ergodic points of f on M for μ, and set

$$J_\mu^0(f) = M - F_\mu^0(f).$$

Then $F_\mu^0(f)$ is open, and $J_\mu^0(f)$ is closed. Noting that

$$\left\| \frac{1}{n} \sum_{k=0}^{n-1} \phi \circ f^k - \overline{\phi} \right\|_{p,U} \geq \left\| \frac{1}{n} \sum_{k=0}^{n-1} \phi \circ f^k - \overline{\phi} \right\|_{p,U_{n,\phi,\delta}} \geq \delta\mu(U_{n,\phi,\delta})^{\frac{1}{p}},$$

we have

$$F_\mu^p(f) \subset F_\mu^0(f), \quad J_\mu^0(f) \subset J_\mu^p(f). \tag{2.36}$$

For $\hat{y} = f(y), \tilde{y} \in f^{-1}(y)$, noting that

$$\left| \sum_{k=0}^{n-1} \phi \circ f^k(\tilde{y}) - \overline{\phi} \right| \leq \left| \sum_{k=0}^{n-1} \phi \circ f^k(y) - \overline{\phi} \right| + \frac{2}{n} \sup_{x \in M} |\phi(x)|,$$

$$\left| \sum_{k=0}^{n-1} \phi \circ f^k(\hat{y}) - \overline{\phi} \right| \leq \left| \sum_{k=0}^{n-1} \phi \circ f^k(y) - \overline{\phi} \right| + \frac{2}{n} \sup_{x \in M} |\phi(x)|,$$

we can prove the following theorem.

Theorem 2.20 *Take $f \in \mathrm{Hom}(M, M)$. If f is measure preserving, then*

$$f^{-1}(F_\mu^0(f)) = F_\mu^0(f), \quad f^{-1}(J_\mu^0(f)) = J_\mu^0(f).$$

Remark. Let $\mathcal{F} = \{f^t\}$ be a flow (or semiflow). We say that $x \in M$ is a L^0-ergodic point of $\mathcal{F}$ for μ if there exists a neighborhood U of x such that for any $\delta > 0$, and for every $\phi \in C_0(M)$, we have

$$\lim_{t \to \infty} \mu(U_{t,\phi,\delta}) = 0,$$

where

$$U_{t,\phi,\delta} = \left\{ x \in U \mid \left| \frac{1}{t} \int_0^t \phi \circ f^t(x) - \overline{\phi} \right| \geq \delta \right\}.$$

Then we also have the notations $F_\mu^0(\mathcal{F})$ and $J_\mu^0(\mathcal{F})$.

Problem 2.3 *When does $J_\mu(f) \neq \emptyset$ or $J_\mu^p(f) \neq \emptyset$?*

Now we define a quantity which is closely related to the properties of L^0-ergodic points. We begin our discussion with the discrete-time case. Let $f : M \longrightarrow M$ be a continuous mapping of a compact space M with a metric d. Fix $\phi \in C_0(M)$. For any $n \geq 1, \varepsilon > 0, \delta > 0$, we call a finite set $K \subset M$ an (n, ε)-*covering set of type* $(0, \phi)$ *for* δ if for each $x \in K$, there exists a positive number $\rho = \rho^{(0,\phi,\delta)}(x, n, \varepsilon)$ such that

$$\mu(U_{n,\phi,\delta}) \leq \varepsilon,$$

where $U = M(x; \rho)$, and such that

$$\bigcup_{x \in K} M(x; \rho) = M.$$

Set

$$n^{(0,\phi,\delta)}(f,\mu,d;n,\varepsilon) = \min\{\#K \mid K \text{ is a } (n,\varepsilon)\text{-covering set of type } (0,\phi) \text{ for } \delta\}.$$

Notice that for fixed δ, $n \geq 1$, the mapping $\varepsilon \mapsto n^{(0,\phi,\delta)}(f,\mu,d;n,\varepsilon)$ is monotone decreasing, and for fixed ε, $n \geq 1$, the mapping $\delta \mapsto n^{(0,\phi,\delta)}(f,\mu,d;n,\varepsilon)$ also is monotone decreasing. Define

$$h^{(0,\phi)}(f,\mu,d) = \lim_{\delta \to 0} \lim_{\varepsilon \to 0} \limsup_{n \to \infty} \frac{1}{n} \log n^{(0,\phi,\delta)}(f,\mu,d;n,\varepsilon) \geq 0, \qquad (2.37)$$

$$h^{(0)}(f,\mu,d) = \sup_{\phi \in C_0(M)} h^{(0,\phi)}(f,\mu,d). \qquad (2.38)$$

If $\mathcal{F} = \{f^t\}$ is a flow, we also can define $h^{(0)}(\mathcal{F},\mu,d)$ similarly.

Theorem 2.21 *Let M be a compact smooth manifold and let $f \in C(M,M)$ such that $h^{(0)}(f,\mu,d) > 0$. Then $J_\mu^0(f) \neq \emptyset$.*

Proof. Assume $J_\mu^0(f) = \emptyset$. Then for each $x \in M$, there is a positive number $\rho = \rho(x)$ such that for every $\phi \in C_0(M)$, and for any $\delta > 0$,

$$\lim_{n \to \infty} \mu(U_{n,\phi,\delta}) = 0,$$

where $U = M(x;\rho)$, and hence for every positive number ε, there exists an integer $N(x)$ such that when $n \geq N(x)$ $\mu(U_{n,\phi,\delta}) < \varepsilon$. Note that M is compact and note that $\{U(x)\}_{x \in M}$ is an open covering of M. Then there exists a finite set $K \subset M$ such that

$$\bigcup_{x \in K} U(x) = M.$$

Take $N = \max_{x \in K} N(x)$. Then for $n \geq N$, the set K is an (n,ε)-covering set of type $(0,\phi)$ for δ. Hence we have

$$n^{(0,\phi,\delta)}(f,\mu,d;n,\varepsilon) \leq \#K \quad (n \geq N)$$

which yields $h^{(0,\phi)}(f,\mu,d) = 0$ for every $\phi \in C_0(M)$, and hence $h^{(0)}(f,\mu,d) = 0$. This is a contradiction. $\square$

Related to the results of this sections, see Hu and Yang [122], [124] and [126].

Chapter 3

Hyperbolicity in differentiable dynamics

In this chapter, we compare Julia sets with the important sets in differentiable dynamics, say, nonwandering sets, chain recurrent sets and hyperbolic sets, and conjecture some relations between Julia sets and Lyapunov exponents. Finally, we suggest problems for Julia sets related to Hausdorff measures and metrics.

3.1 Nonwandering sets and chain recurrent sets

Definition 3.1 *Let M be a topological space and take $\mathcal{F} \subset C(M, M)$. A point x of M is wandering if there exists a neighborhood U of x such that $U \cap f(U) = \emptyset$ for all $f \in \mathcal{F}$. A point x is nonwandering if the above does not hold; that is, if for all neighborhoods U of x, there is some $f \in \mathcal{F}$ such that $U \cap f(U) \neq \emptyset$. We denote by $\Omega(\mathcal{F})$ the set of nonwandering points.*

Obviously, the set $\Omega^c(\mathcal{F}) = M - \Omega(\mathcal{F})$ of wandering points of $\mathcal{F}$ is open, and hence $\Omega(\mathcal{F})$ is closed. By the definition, we see that $\text{Fix}(f) \subset \Omega(\mathcal{F})$ for all $f \in \mathcal{F}$. For a semiflow (or flow) $\{f^t\}$, the definition usually is applied to a family $\mathcal{F} = \{f^t\}_{t \geq t_0}$ for some $t_0 > 0$, and we will write

$$\Omega(f) = \Omega_{t_0}(f) = \Omega(\mathcal{F}).$$

As usual, we define

$$\Omega(f) := \Omega(\{f^n\}_{n \in \mathbb{Z}^+}),$$

for $f = f^1 \in C(M, M)$.

Proposition 3.1 $O^+(\Omega(f)) = \Omega(f)$. *Moreover, if $\kappa = \mathbb{R}$, or $\mathbb{Z}$, $\Omega(f)$ is invariant. If $\kappa = \mathbb{Z}$, a point is nonwandering for f iff it is nonwandering for f^{-1}.*

Proof. Take $x \in \Omega(f)$ and a neighborhood U of $f^s(x)$ for any $0 \leq s \in \kappa$. Then $f^{-s}(U)$ is a neighborhood of x. Therefore, there is a t such that $f^t(f^{-s}(U)) \cap f^{-s}(U) \neq \emptyset$; the images of these intersections under f^s are contained in $f^t(U) \cap U$, which are thus non-empty. Thus it follows that $f^s(\Omega(f)) \subset \Omega(f)$ so that $O^+(\Omega(f)) = \Omega(f)$.

If $\kappa = \mathbb{R}$, or $\mathbb{Z}$ and if $x \in \Omega(f)$, then for every neighborhood U of $f^{-s}(x)$, $f^s(U)$ is a neighborhood of x. Hence there are some $t > t_0$ such that $f^t(f^s(U)) \cap f^s(U) \neq \emptyset$. The f^{-s} images of these intersections are $f^t(U) \cap U$, which are non-empty, so $f^{-s}(x) \in \Omega(f)$. Thus $O^-(\Omega(f)) = \Omega(f)$. Combining the results above, we obtain that $\Omega(f)$ is invariant.

If $\kappa = \mathbb{Z}$ and if $x \in \Omega(f)$, then for every neighborhood U of x, there is some n such that $f^n(U) \cap U \neq \emptyset$. The images of these intersections under f^{-n} are contained in $U \cap f^{-n}(U)$, which are non-empty, so $x \in \Omega(f^{-1})$. $\qquad\square$

Since $\Omega(f)$ is closed and a plus Poisson stable point is clearly nonwandering, then

$$\overline{\mathrm{Per}(f)} \subset P^+(f) \subset \Omega(f).$$

Let y be a point of $L^+(x)$ and U a neighborhood of y. Then there is a $t > 0$ and there are infinitely many $s > t$ such that $f^s(x), f^t(x) \in U$. Therefore, $f^{s-t}(U) \cap U \neq \emptyset$ so y is nonwandering. Hence we obtain

$$\mathrm{Per}(f) \subset L^+(M) \subset \Omega(f) \subset \bigcap_{t \geq 0} f^t(M).$$

Thus if M is compact, then $\Omega(f) \neq \emptyset$ since $L^+(x)$ is nonempty for every $x \in M$.

If $\kappa = \mathbb{R}$, or $\mathbb{Z}$, a similar argument applying to the set of α-limit points yields

$$L^-(M) \subset \Omega(f), \quad L^+(M) \cup L^-(M) \subset \Omega(f).$$

Generally, we only obtain

$$L^-(M) \subset \bigcap_{t \geq 0} f^t(M).$$

Theorem 3.1 (Pugh, cf. [228]) *If M is a compact smooth manifold, then $\overline{\mathrm{Per}(f)} = \Omega(f)$ generically in $\mathrm{Diff}^1(M, M)$.*

A similar result holds for flows. Assume that M is a topological space and suppose that f is a continuous self-mapping on a probability space $(M, \mathcal{B}, \mu)$. If $x \in \mathrm{supp}\mu \cap \{M - \Omega(f)\}$, then x is wandering. Hence there exists a neighborhood U of x such that $U \cap f^n(U) = \emptyset$ for all $n \geq 1$. Thus we have $x \in J_{Poi}(f)$, that is,

$$\mathrm{supp}\mu \cap \{M - \Omega(f)\} \subset J_{Poi}(f),$$

and hence

$$\mathrm{supp}\mu \subset \Omega(f) \cup J_{Poi}(f).$$

In particular if f is measure preserving or $J_{Poi}(f) = \emptyset$, we have $\mathrm{supp}\mu \subset \Omega(f)$.

According to Theorem 1.13 and Conjecture 1.2, we can expect that there are some relations between $J(f)$ and $\Omega(f)$ generally. If f is a rational mapping of degree at least two on the Riemann sphere, and if U is any non-empty open set which meets $J(f)$, we know that $J(f) \subset f^n(U)$ for all sufficiently large integers n so that $J(f) \subset \Omega(f)$ (see Beardon [31], Theorem 4.2.5). We make the following conjecture:

Conjecture 3.1 *If M is a compact smooth manifold, then $J(f) \subset \Omega(f)$, and generically in $\mathrm{Diff}^1(M, M)$*

$$J(f) = \bigcap_{n=1}^{\infty} \Omega(f^n),$$

i.e., this fact holds on a countable intersection of open dense sets of $\mathrm{Diff}^1(M, M)$ with C^1-topology.

Pugh[201] proved the following "closing lemma": Let $\{f^t\}$ be a flow on a compact manifold M generated by a vector field X. If $x \in \Omega(f)$, then every C^1-neighborhood of X contains a vector field Y such that x lies on a certain closed trajectory of the flow generated by Y. An analogous statement for fields of class C^0 also is true. The question is open for fields of class C^r with $r \geq 2$.

Definition 3.2 *Let M be a topological space with a distance d compatible with the topology and take $f \in C(M, M)$. A point $x \in M$ is recurrent if*

$$\liminf_{n \to \infty} d(f^n(x), x) = 0.$$

Obviously, recurrent points are nonwandering. Pugh[201] also proved the following closing lemma: Let f be a diffeomorphism on a compact manifold M. If x is recurrent, then every C^1-neighborhood of f contains a diffeomorphism g such that x is periodic for g. The question is open for the C^r topology with $r \geq 2$. Fornaess and Sibony [88] proved the closing lemma for various classes of holomorphic self-mappings on $\mathbb{C}^m$, which answer a question in the survey by Yoccoz [278].

Definition 3.3 *Let κ be an additive semigroup and take a non-empty set $\Lambda \subset \kappa$. Suppose M is endowed with a metric d and given a positive real number ε. A mapping $x : \kappa \to M$ is said to be a (Λ, ε)-trajectory of a family $\mathcal{F} = \{f_t\}_{t \in \kappa} \subset C(M, M)$ if there exists a mapping $\tau : \kappa \to \Lambda$ such that*

$$d(x(t + \tau(t)), f_{\tau(t)}(x(t))) < \varepsilon,$$

for all $t \in \kappa$. A point $p \in M$ is chain recurrent for Λ if for any $\varepsilon > 0$, there exists an (Λ, ε)-trajectory $\{x(t)\}_{t \in \kappa}$ such that $x(t_0) = p$ for some $t_0 \in \kappa$ and such that $x(t_0 + \cdots + t_n) = p$ for some $n > 0$, where $t_i = \tau(t_0 + \cdots + t_{i-1}), i = 1, ..., n$. We denote the set of chain recurrent points by $R_\Lambda(\mathcal{F})$.

Assume that $\mathcal{F} = \{f^n\}_{n \in \mathbb{Z}_+} \subset C(M, M)$ is a cascade. Take $\Lambda = \{1\}$. A (Λ, ε)-trajectory of $\mathcal{F}$ is called a ε-trajectory of the cascade $\{f^n\}$, and a chain recurrent point for $\Lambda = \{1\}$ is called a chain recurrent point of the cascade $\{f^n\}$. For this case, a ε-trajectory of $\mathcal{F}$ is a sequence $\{x(i)\}$ on M such that

$$d(x(i + 1), f(x(i))) < \varepsilon, \quad i = 0, 1, \dots .$$

Obviously, a point $p \in M$ is *chain recurrent* if and only if for any ε, there exists $\{x_i\}_{i=1}^{n+1} \subset M$ such that $x_1 = x_{n+1} = p$, and such that

$$d(x_{i+1}, f(x_i)) < \varepsilon, \quad i = 1, ..., n.$$

We denote the set of chain recurrent points by $R(f)$.

Take $\Lambda = \{\tau\}$ for some positive integer τ. Then a (τ, ε)-trajectory of $\mathcal{F}$ is a sequence $\{x(i)\}$ on M such that

$$d(x(i + \tau), f^\tau(x(i))) < \varepsilon, \quad i = 0, 1, \dots .$$

We can prove that a point $p \in M$ is *chain recurrent* for $\Lambda = \{\tau\}$ if and only if for any ε, there exists $\{x_i\}_{i=1}^{n+1} \subset M$ such that $x_1 = x_{n+1} = p$, and such that

$$d(x_{i+1}, f^\tau(x_i)) < \varepsilon, \quad i = 1, \dots, n.$$

Hence we obtain $R_\tau(\mathcal{F}) = R(f^\tau)$.

Proposition 3.2 *Let M be a compact metric space and $f : M \longrightarrow M$ a continuous mapping. Then $R(f)$ is closed, and*

$$\Omega(f) \subset R(f).$$

Proof. Let $x \in \Omega(f)$ and ε be positive. Since M is compact, $f : M \longrightarrow M$ is uniformly continuous so that we can choose $0 < \delta < \varepsilon/2$ such that

$$d(x, y) < \delta \Longrightarrow d(f(x), f(y)) < \varepsilon/2.$$

Let U be a neighborhood of x contained in the ball $M(x; \delta)$ of radius δ about x. Since x is nonwandering, we can find some n such that $f^n(U) \cap U \neq \emptyset$. If $n = 1$, $\{x, x\}$ defines an ε-trajectory starting from x and returning to x after time $T = 1$; if $n > 1$, we can find a $y \in U$ with $f^n(y) \in U$ and thus $\{x, f(y), \dots, f^{n-1}(y), x\}$ defines an ε-trajectory starting from x and returning to x after time $T = n$.

Let $\{x^{(k)}\}$ be a sequence in $R(f)$ such that $x^{(k)} \to x'$ as $k \to \infty$. Then $d(x^{(k)}, x') < \delta$ for some $k > 0$. Since $x^{(k)} \in R(f)$, there is a $\delta/2$-trajectory defined by $\{x^{(k)}, x_1, \dots, x_n, x^{(k)}\}$. Therefore $\{x', x_1, \dots, x_n, x'\}$ defines an ε-trajectory. Since ε is arbitrary, we have $x' \in R(f)$. $\qquad \square$

An (Λ, ε)-trajectory of a flow (or semiflow) $\{f^t\}_{t \in \kappa}$ is a parametrized, possibly discontinuous, curve $x(t)$ such that

$$d(x(t + \tau(t)), f^{\tau(t)}(x(t))) < \varepsilon,$$

for some function $\tau : \kappa \to \Lambda$ and for all $t \in \kappa$. Usually we take $\Lambda = \mathbb{R}^+$. For this case, (Λ, ε)-trajectories also are called ε-trajectories, and denote the set of chain recurrent points by $R(f)$.

Lemma 3.1 *For a semiflow $\mathcal{F} = \{f^t\}_{t \in \mathbb{R}_+} \subset C(M, M)$, $p \in R(f)$ if and only if for any $\varepsilon > 0$, there exist*

$$\{x_1, \dots, x_{n+1}\} \subset M, \quad \{t_1, \dots, t_n\} \subset \mathbb{R}^+$$

such that $x_1 = x_{n+1} = p$, and such that

$$d(x_{i+1}, f^{t_i}(x_i)) < \varepsilon, \quad i = 1, \dots, n.$$

Proof. ($\Leftarrow$) Take $r \in \mathbb{R}^+$ with $0 < r < \min_i\{t_i\}$, set $t_{n+i} = t_n, x_{n+1+i} = f^{it_n}(x_{n+1})$ for $i = 1, 2, ...$, and define

$$t_0 = 0, \quad s_i = \sum_{k=0}^{i} t_k, \quad i = 0, 1, ..., n, ...,$$

and define

$$\tau(t) = \begin{cases} t_i & : \quad s_{i-1} \leq t < r + s_{i-1}, \quad i = 1, ... \\ \ell_i(t) & : \quad r + s_{i-1} \leq t < s_i, \quad i = 1, ... \end{cases} \tag{3.1}$$

where

$$\ell_i(t) = t_{i+1} + \frac{t_{i+1} - t_i}{t_i - r}(t - s_i) \tag{3.2}$$

with $\ell_i(t_0 + s_{i-1}) = t_i, \ell_i(s_i) = t_{i+1}$. Define

$$x(t) = \begin{cases} x_i & : \quad s_{i-1} \leq t < r + s_{i-1}, \quad i = 1, ... \\ f^t(x_1) & : \quad r + s_{i-1} \leq t < s_i, \quad i = 1, ... \end{cases} \tag{3.3}$$

Then we have

$$d(x(t + \tau(t)), f^{\tau(t)}(x(t))) < \varepsilon, \quad t \geq 0. \tag{3.4}$$

Obviously, we have

$$t_{i+1} = \tau(t_0 + t_1 + \cdots + t_i), \quad x_{i+1} = x(t_0 + t_1 + \cdots + t_i).$$

($\Rightarrow$) By definition, $p \in R(f)$ iff for any $\varepsilon > 0$, there exists an ε-trajectory $x(t)$ passing p at t_0 and returning to p after $T = t_0 + \cdots + t_n > 0$ for some n, where $t_i = \tau(t_0 + \cdots + t_{i-1}), i = 1, ..., n$. Define

$$x_{i+1} = x(t_0 + \cdots + t_i), \quad i = 0, 1, ..., n.$$

Then the lemma follows. $\quad\square$

It is easy to prove that the lemma also is true for a flow. We know that a number of basic generic properties of homeomorphisms were described in [192] (also see [67], [129], [130], which extend some of these results, and correct a mistake in the proof of one of them). Among these results are that, generically, a homeomorphism f of a compact smooth manifold has no periodic attractors, and that

$$\Omega(f) = \overline{Per(f)} = R(f).$$

We also make the following conjecture:

Conjecture 3.2 *If M is a compact smooth manifold, then $J(f) \subset R(f)$, and generically in $\mathrm{Hom}(M, M)$*

$$J(f) = \bigcap_{n=1}^{\infty} R(f^n).$$

It is an easy exercise to show that the chain recurrent set is invariant under the flow or diffeomorphism and is closed and hence compact since we are assuming M to be compact. The importance of chain recurrence for the description of asymptotic behavior of orbits (or trajectories) of the system is shown in the following theorem of Conley.

Theorem 3.2 ([64], [90]) *If f^t is a continuous flow on M, there exists a continuous function $\phi : M \longrightarrow \mathbb{R}$ such that*

1) $\phi(f^t(x)) < \phi(f^s(x))$ if $t > s, x \notin R(f)$;

2) $\phi(x) = \phi(y)$ for $x, y \in R(f)$, iff for $\varepsilon > 0$ there exist points

$$x_1 = x, x_2, ..., x_n = y, x_{n+1}, ..., x_{2n} = x$$

in $R(f)$ and real numbers $t_i > 0, 1 \leq i < 2n$, such that

$$d(x_{i+1}, f^{t_i}(x_i)) < \varepsilon, \quad 1 \leq i < 2n.$$

The analogue of this theorem for diffeomorphisms is also valid. A function $\phi : M \longrightarrow \mathbb{R}$ satisfying the conclusion of this theorem is called a *Lyapunov function*. Results of W. Wilson [266] show that it is possible to choose a Lyapunov function ϕ which is C^∞ and satisfies

$$\frac{d}{dt}\phi(f^t(x))|_{t=0} < 0 \quad \text{for } x \notin R(f).$$

The existence of Lyapunov functions is very important in the work of Franks [90].

3.2   Measure theoretic entropies

Here we introduce a definition of measure theoretic entropy. Let $(M, \mathcal{B}, \mu)$ be a compact probability space with a metric d and let $f \in C(M, M)$. For any $0 < \delta < 1, n \geq 1, \varepsilon > 0$, we call a finite set $K \subset M$ an (n, ε, δ)-*covering set* if for each $x \in K$, there is a positive number $\rho = \rho(x, n, \varepsilon)$ such that when $y \in M_d(x; \rho)$, one has

$$d_{f;n}(x, y) = \max_{0 \leq i \leq n-1} d(f^i(x), f^i(y)) < \varepsilon,$$

and such that

$$\mu\left(\bigcup_{x \in K} M_d(x; \rho)\right) > \delta.$$

Set

$$n(f, \mu, d; n, \varepsilon, \delta) = \min\{\#K \mid K \text{ is a } (n, \varepsilon, \delta)\text{-covering set }\}.$$

Notice that for fixed n, δ the mapping $\varepsilon \mapsto n(f, \mu, d; n, \varepsilon, \delta)$ is monotone decreasing and for fixed n, ε the mapping $\delta \mapsto n(f, \mu, d; n, \varepsilon, \delta)$ is monotone increasing. We can define

$$h_{equ}(f, \mu, d) = \lim_{\delta \to 1} \lim_{\varepsilon \to 0} \liminf_{n \to \infty} \frac{1}{n} \log n(f, \mu, d; n, \varepsilon, \delta) \geq 0.$$

According to the proof of Proposition 1.1, we also can prove $h_{equ}(f, \mu, d') = h_{equ}(f, \mu, d)$ if d' is another metric on M which defines the same topology as d. The quantity $h_{equ}(f, \mu, d)$ calculated for any metric d generating the given topology in M also is denoted by $h_{equ}(f, \mu)$ with $h_{equ}(f, \mu) \leq h_{equ}(f)$.

Here we introduce the definition of measure theoretic entropy due to Katok. For any $0 < \delta < 1, n \geq 1, \varepsilon > 0$, we call a finite set $K \subset M$ a *standard* (n, ε, δ)-*covering set* if

$$\mu \left(\bigcup_{x \in K} M_{d_{f;n}}(x; \varepsilon) \right) > \delta.$$

Set

$$n_{st}(f, \mu, d; n, \varepsilon, \delta) = \min\{\#K \mid K \text{ is a standard } (n, \varepsilon, \delta)\text{-covering set }\}.$$

Define

$$h(f, \mu, d) = \lim_{\delta \to 1} \lim_{\varepsilon \to 0} \liminf_{n \to \infty} \frac{1}{n} \log n_{st}(f, \mu, d; n, \varepsilon, \delta) \geq 0.$$

In fact, the limit

$$\lim_{\varepsilon \to 0} \liminf_{n \to \infty} \frac{1}{n} \log n_{st}(f, \mu, d; n, \varepsilon, \delta)$$

is independent of δ (see [133]). Also $h(f, \mu, d)$ is independent of the metric d generating the given topology in M, and is called the *measure theoretic entropy* of f which also is denoted by $h(f, \mu)$. Obviously, we have

$$h(f, \mu) \leq h_{equ}(f, \mu).$$

It is not difficult to show that the entropy of any iterate f^n of f is given by

$$h(f^n, \mu) = nh(f, \mu), \quad \text{for any } n \in \mathbb{Z}^+.$$

If f is invertible, then $h(f^{-1}, \mu) = h(f, \mu)$ and hence

$$h(f^n, \mu) = |n| h(f, \mu), \quad \text{for any } n \in \mathbb{Z}.$$

Measure theoretic entropy plays a very important role in the history of ergodic theory. It was originally introduced by Kolmogorov and the definition was further refined by Sinai. The following example is the standard definition.

Example 3.1 *Let f be a measure preserving mapping on a probability space $(M, \mathcal{B}, \mu)$. Let Λ be a finite or countable set of indices. A collection of measurable subsets $\alpha = \{C_i \in \mathcal{B}\}_{i \in \Lambda}$ is called a measurable partition of M if*

$$\mu(M - \bigcup_{i \in \Lambda} C_i) = 0; \quad \mu(C_i \cap C_j) = 0 \quad (i \neq j).$$

Define

$$f^{-1}(\alpha) = \{f^{-1}(C_i)\}_{i \in \Lambda}.$$

If $\beta = \{D_i \in \mathcal{B}\}_{i \in \Xi}$ is another measurable partition of M, define the joint partition by

$$\alpha \vee \beta = \{C \cap D \mid C \in \alpha, D \in \beta, \mu(C \cap D) > 0\}.$$

The entropy of a measurable partition α is given by

$$H(\mu, \alpha) = -\sum_{i \in \Lambda} \mu(C_i) \log \mu(C_i) \geq 0.$$

Here we agree that $0 \log 0 = 0$. *For countable* α *the entropy may be infinite. Obviously*

$$H(\mu, f^{-1}(\alpha)) = H(\mu, \alpha).$$

If α *is finite or countable measurable partition with finite entropy, then the limit*

$$h(f, \mu; \alpha) = \lim_{n \to \infty} \frac{1}{n} H(\mu, \alpha \vee f^{-1}(\alpha) \vee \cdots \vee f^{-n+1}(\alpha)) \geq 0$$

exists (see [50], p. 30, [134], p. 168), which is called the entropy of the mapping f *relative to the partition* α. *Finally, define the (Kolmogorov-Sinai) measure theoretic entropy of* f *with respect to* μ *(or the entropy of* μ*) by*

$$h_{KS}(f, \mu) = \sup_{\alpha} \{ h(f, \mu; \alpha) \mid \alpha \text{ is a measurable partition with } H(\mu, \alpha) < \infty \}$$

(see [134], [199], [264]). If μ_1 *and* μ_2 *are two* f*-invariant probability measures, then*

$$h_{KS}(f, t\mu_1 + (1-t)\mu_2) = t h_{KS}(f, \mu_1) + (1-t) h_{KS}(f, \mu_2) \quad (0 \leq t \leq 1)$$

(see [134], [264], p. 183). If $f : M \longrightarrow M$ *is a homeomorphism of compact metric space* (M, d), *and if* $\mu \in \Sigma_f(M)$, *then (see Katok [133])*

$$h(f, \mu) = h_{KS}(f, \mu). \tag{3.5}$$

Given two measure preserving mappings f on $(M, \mathcal{B}, \mu)$ and g on $(N, \mathcal{R}, \nu)$, we call f and g *isomorphic* if there exists an invertible measure preserving mapping $h : M \longrightarrow N$, i.e., $h_*\mu = \nu$, such that $g \circ h = h \circ f$. If so, Kolmogorov (cf. [264], p. 89) showed that

$$h_{KS}(f, \mu) = h_{KS}(g, \nu).$$

The following Variational Principle is referred to [134], [264]:

Proposition 3.3 *Let* (M, d) *be a compact metric space and let* $f \in \mathrm{Hom}(M, M)$. *Then*

$$h_{top}(f) = \sup\{ h(f, \mu) \mid \mu \in \Sigma_f(M) \}. \tag{3.6}$$

If a measure μ satisfies $h(f, \mu) = h_{top}(f)$, then μ is called a *measure of maximal entropy*. A Newhouse's theorem shows that if $f : M \longrightarrow M$ is a C^∞ diffeomorphism of a compact C^∞ manifold, then f has a measure of maximal entropy. We suspect

$$h_{equ}(f) = \sup\{ h_{equ}(f, \mu) \mid \mu \in \Sigma_f(M) \}.$$

For the rest of this section, we give an explanation of the meaning of the notion $h_{equ}(f, \mu)$.

Definition 3.4 *A family* $\mathcal{F}$ *of mappings of a metric space* (M, d) *into a metric space* (N, d') *is called equicontinuous or an equicontinuous family at* $z_0 \in M$ *for* μ *if and only if for every positive* ε *there exist a positive* ρ *and a subset* $E \subset M_d(z_0; \rho)$ *such that for all* f *in* $\mathcal{F}$,

$$z \in M_d(z_0; \rho) - E \Longrightarrow d'(f(z), f(z_0)) < \varepsilon,$$

where $\mu(E) = 0$ *if* $z_0 \notin \mathrm{supp}\mu$, *and* E *is of arbitrary small positive measure if* $z_0 \in \mathrm{supp}\mu$. *The family* $\mathcal{F}$ *is said equicontinuous on* M *for* μ *iff* $\mathcal{F}$ *is equicontinuous at each point of* M *for* μ.

We also have the following general principle.

Theorem 3.3 *Let $\mathcal{F}$ be any family of mappings, which map (M, d) into (N, d'). Then there is a maximal open subset $F_{eq,\mu}(\mathcal{F})$ of M on which $\mathcal{F}$ is equicontinuous for μ. In particular, if f maps a metric space (M, d) into itself, then there is a maximal open subset $F_{eq,\mu}(f) = F_{eq,\mu}(f, d)$ of M on which the family of iterates $\{f^n\}$ is equicontinuous for μ.*

Define

$$J_{eq,\mu}(\mathcal{F}) = M - F_{eq,\mu}(\mathcal{F}), \quad J_{eq,\mu}(f) = J_{eq,\mu}(f, d) = M - F_{eq,\mu}(f, d).$$

We can study these sets after the fashion of Fatou sets and Julia sets. For example, we can define a quantity which is closely related to the set $J_{eq,\mu}(f)$ as follows: For any $0 < \delta < 1, n \geq 1, \varepsilon > 0$, we call a finite set $K \subset M$ a *weak (n, ε, δ)-covering set* if for each $x \in K$, there is a positive number $\rho = \rho(x, n, \varepsilon)$ such that

$$\mu\left(\bigcup_{x \in K} \{y \in M_d(x; \rho) \mid d_{f;n}(x, y) < \varepsilon\}\right) > \delta.$$

Set

$$n_{weak}(f, \mu, d; n, \varepsilon, \delta) = \min\{\#K \mid K \text{ is a weak } (n, \varepsilon, \delta)\text{-covering set }\}.$$

We can define

$$h_{weak}(f, \mu, d) = \lim_{\delta \to 1} \lim_{\varepsilon \to 0} \liminf_{n \to \infty} \frac{1}{n} \log n_{weak}(f, \mu, d; n, \varepsilon, \delta) \geq 0,$$

with $h_{weak}(f, \mu, d) \leq h_{equ}(f, \mu, d)$.

Theorem 3.4 *Let M be a compact smooth manifold and let $f \in C(M, M)$ such that $h_{weak}(f, \mu, d) > 0$. Then $J_{eq,\mu}(f) \neq \emptyset$.*

Proof. Assume $J_{eq,\mu}(f) = \emptyset$. Then for $\varepsilon > 0$ and for each $x \in M$, there is a positive number $\rho = \rho(x)$ such that

$$d(f^n(x), f^n(y)) < \varepsilon, \quad n \geq 0$$

holds for every $y \in U(x) = M_d(x; \rho)$ except for a subset of arbitrary small measure. Note that M is compact and note that $\{U(x)\}_{x \in M}$ is an open covering of M. Then there exists a finite set $K \subset M$ such that

$$\bigcup_{x \in K} U(x) = M.$$

Set $l = \#K$ and take $0 < \delta < 1$. Then for $x \in K$ there is a subset $E(x) \subset U(x)$ with $\mu(E(x)) < (\mu(M) - \delta)/l$, such that

$$d(f^n(x), f^n(y)) < \varepsilon, \quad n \geq 0$$

holds on $U(x) - E(x)$. Then for any $n > 1$, we have

$$
\mu \left(\bigcup_{x \in K} \{ y \in U(x) \mid d_{f;n}(x, y) < \varepsilon \} \right)
$$

$$
\geq \mu \left(\bigcup_{x \in K} \{ U(x) - E(x) \} \right)
$$

$$
= \mu(M) - \mu \left(\bigcup_{x \in K} E(x) \right) \geq \mu(M) - \sum_{x \in K} \mu(E(x))
$$

$$
> \mu(M) - \sum_{x \in K} \frac{\mu(M) - \delta}{l} = \delta.
$$

Therefore the set K is an weak (n, ε, δ)-covering set. Hence we have

$$
n_{weak}(f, \mu, d; n, \varepsilon, \delta) \leq l \quad (n \geq 1),
$$

which yields $h_{weak}(f, \mu, d) = 0$. This is a contradiction. $\square$

3.3   Lyapunov exponents

Let f be a C^1 mapping on a m-dimensional manifold M with a Riemannian metric g. In fact, the following discussion also works on Finsler metrics of M (see § 5.3). For $x \in M$, $X \in T(M)_x$, the number (possibly infinite)

$$
\chi(x, X) = \limsup_{n \to \infty} \frac{1}{n} \log \| (df^n)_x X \|
$$

is called the (upper) Lyapunov exponent of (x, X). Here we define $\log 0 = -\infty$. Obviously, the Lyapunov exponents have the following properties:

$$
\chi(x, X) = \chi(x, aX) \ (a \in \mathbb{R} - \{0\}),
$$

$$
\chi(x, X + Y) \leq \max\{\chi(x, X), \chi(x, Y)\} \ (X, Y, \in T(M)_x),
$$

and the latter is an equality if $\chi(x, X) \neq \chi(x, Y)$. Thus for each real number χ and each $x \in M$, the set

$$
E_\chi = \{ X \in T(M)_x \mid \chi(x, X) \leq \chi \}
$$

is a linear subspace of $T(M)_x$. Furthermore for each $x \in M$, there are an integer $k(x) \leq m$, a collection of numbers $\chi_i(x)$ and linear subspaces $E_{\chi_i(x)}$ such that

$$
\chi_1(x) > \chi_2(x) > \cdots > \chi_{k(x)}(x), \tag{3.7}
$$

$$
\{0\} \subset E_{\chi_{k(x)}(x)} \subset \cdots \subset E_{\chi_2(x)} \subset E_{\chi_1(x)} = T(M)_x, \tag{3.8}
$$

and such that

$$
\chi(x, X) = \chi_i(x), \quad X \in E_i(x), \tag{3.9}
$$

where
$$E_i(x) = E_{\chi_i(x)} - E_{\chi_{i+1}(x)} \ (1 \le i < k(x)), \quad E_{k(x)}(x) = E_{\chi_{k(x)}(x)}, \cdot \tag{3.10}$$

These numbers $\chi_i(x)$ are the so called *(upper) Lyapunov exponent* at x. The number $\dim E_i(x)$ is called the *multiplicity* of the exponent $\chi_i(x)$. Define
$$E_x^s = \bigcup_{\chi_i(x)<0} E_i(x), \quad E_x^u = \bigcup_{\chi_i(x)>0} E_i(x), \quad E_x^c = \bigcup_{\chi_i(x)=0} E_i(x). \tag{3.11}$$

Then the tangent space $T(M)_x$ decomposes into a direct sum
$$T(M)_x = E_x^s \oplus E_x^u \oplus E_x^c \tag{3.12}$$

of *stable*, *unstable* and *neutral* subspaces, respectively, with
$$\chi(x, X) \begin{cases} < 0 & : \quad X \in E_x^s \\ = 0 & : \quad X \in E_x^c \\ > 0 & : \quad X \in E_x^u. \end{cases}$$

Note that for $x \in M$, $X \in T(M)_x$,
$$\begin{aligned} \chi(x, X) &= \limsup_{n \to \infty} \frac{1}{n+1} \log \|(df^{n+1})_x X\| \\ &= \limsup_{n \to \infty} \frac{1}{n} \log \|(df^n)_{f(x)} \circ (df)_x X\| = \chi(f(x), (df)_x X). \end{aligned}$$

We have
$$(df)_x E_x^s \subset E_{f(x)}^s, \quad (df)_x E_x^u \subset E_{f(x)}^u, \quad (df)_x E_x^c \subset E_{f(x)}^c.$$
If $(df)_x : T(M)_x \longrightarrow T(M)_{f(x)}$ is nondegenerate, we also have
$$\chi(f(x), Y) = \chi(x, (df)_x^{-1} Y),$$

for $Y \in T(M)_{f(x)}$. Thus in this case, we obtain
$$(df)_x E_x^s = E_{f(x)}^s, \quad (df)_x E_x^u = E_{f(x)}^u, \quad (df)_x E_x^c = E_{f(x)}^c.$$

For more details, see [134].

Example 3.2 *Let $p \in M$ be a fixed point of f and let $A = (df)_p$. Then $(df^n)_p = A^n$. If $\lambda_1, ..., \lambda_k$ are the eigenvalues of A with $|\lambda_1| > \cdots > |\lambda_k|$, then $\chi_i(p) = \log |\lambda_i|$ for $i = 1, ..., k$. Moreover in this case the limits always exist.*

Define the *stable* and *unstable index functions* $s : M \longrightarrow \mathbb{Z}_+$ and $u : M \longrightarrow \mathbb{Z}_+$, respectively, by setting
$$s(x) = \dim E_x^s, \quad u(x) = \dim E_x^u.$$
Define
$$\mathcal{K}_{Lya}(f) = \{x \in M \mid s(x) + u(x) = m\}.$$

Then $x \in \mathcal{K}_{Lya}(f)$ if and only if all the Lyapunov exponents of x are different from zero. Obviously, if f is a diffeomorphism, then $\mathcal{K}_{Lya}(f)$ is invariant.

Let $\{f^t\}$ be a C^1 semiflow (or flow) on a m-dimensional Riemannian manifold M. For $x \in M$, $X \in T(M)_x$, define the *(upper) Lyapunov exponent* (possibly infinite) of (x, X) by

$$\chi(x, X) = \limsup_{t \to +\infty} \frac{1}{t} \log \|(df^t)_x X\|.$$

Then we also have the decomposition (3.11) and (3.12). Similarly, we can define the functions $s(x)$ and $u(x)$ for $\{f^t\}$, but we have to say more about the set $\mathcal{K}_{Lya}(f)$. Let X be the vector field generating $\{f^t\}$ and let $E_x^v \subset T(M)_x$ be spanned by X_x. Note that

$$(df^t)_x X_x = X_{f^t(x)}.$$

Thus if $x \in \mathcal{K}(\mathcal{F})$ for $\mathcal{F} = \{f^t\}_{t \geq 0}$, we have

$$\chi(x, X_x) = \limsup_{t \to +\infty} \frac{1}{t} \log \|X_{f^t(x)}\| \leq 0.$$

By the remark, it is rational to define

$$\mathcal{K}_{Lya}(f) = \mathcal{K}_{Lya}(\mathcal{F}) = \{x \in M \mid E_x^c \subset E_x^v\}.$$

Let f be a C^1 mapping on a m-dimensional Riemannian manifold M with the induced distance d and define

$$h^{ss}(f) = \limsup_{n \to \infty} \frac{1}{n} \sup_{x \in M} \log \|(df^n)_x\|.$$

For any $x \in M$, we also define

$$h^s(f; x) = \limsup_{n \to \infty} \frac{1}{n} \log \|(df^n)_x\|.$$

Then for $x \in M$, $X \in T(M)_x - \{0\}$, we have

$$\chi(x, X) \leq h^s(f; x) \leq h^{ss}(f).$$

With respect to some Borel probability measure $\mu \in \Sigma(M)$, define

$$h^s(f, \mu) = \int_M h^s(f; x) d\mu(x),$$

$$h_{Lya}(f, \mu) = \int_M \sum_{\chi_i(x) > 0} m_i(x) \chi_i(x) d\mu(x),$$

where $m_i(x)$ is the multiplicity of the exponent $\chi_i(x)$. Then we have

$$h_{Lya}(f, \mu) \leq m h^s(f, \mu) \leq m h^{ss}(f).$$

Write

$$h_{Lya}(f) = \sup_{\mu \in \Sigma_f(M)} h_{Lya}(f, \mu), \quad h^s(f) = \sup_{\mu \in \Sigma_f(M)} h^s(f, \mu).$$

If M is compact, and if $f \in \text{Diff}^1(M)$, $\mu \in \Sigma_f(M)$, then one has the Pesin-Ruelle inequality [219]:

$$h(f, \mu) \leq h_{Lya}(f, \mu).$$

For this case, by variational principle one has

$$h_{top}(f) \leq h_{Lya}(f) \leq m h^s(f) \leq m h^{ss}(f).$$

We do not know when the following inequality holds

$$\sup_{x \in M} h^s(f; x) \leq h_{top}(f)?$$

Conjecture 3.3 *Let M be a compact manifold and suppose $f \in \text{Diff}^1(M, M)$. Then $J(f) \neq \emptyset$ if $h^{ss}(f) > 0$.*

This conjecture is true for complex dynamics (see Corollary 5.10). Here we transform Conjecture 1.6 into the following form:

Conjecture 3.4 *Assume that M is a compact smooth m-dimensional manifold and $f \in \text{Diff}^1(M, M)$ with $h^{ss}(f) > 0$ such that $F(f)$ has no components which are backward invariant. Then $\text{Res} J(f) \neq \emptyset$.*

If the condition $h^{ss}(f) > 0$ in these problems is replaced by $h_{Lya}(f) > 0$, its strong versions also are raised. For the following discussion, we will need the subadditive ergodic theorem:

Theorem 3.5 (cf. [199]) *Let $(M, \mathcal{B}, \mu)$ be a probability space and let f be an ergodic measure preserving mapping on M. If a sequence $\{\phi_n\}_{n=1}^{\infty} \in L^1(M)$ satisfies the subadditivity condition*

$$\phi_{n+k}(x) \leq \phi_n(x) + \phi_k(f^n(x)) \quad a.e. \ (\mu),$$

for $n, k \in \mathbb{Z}^+$, then

$$\lim_{n \to \infty} \frac{1}{n} \phi_n(x) = \inf_{n \geq 1} \left\{ \frac{1}{n} \int_M \phi_n d\mu \right\} \quad a.e. \ (\mu).$$

Assume that M is a compact Riemannian manifold and takes $\mathcal{B}$ to be the Borel σ-algebra. Assume that μ is an ergodic f-invariant Borel probability measure on M. Then each mapping $x \mapsto (df^n)_x$ is continuous, $(df^n)_x$ is bounded, and thus $\phi_n(x) = \log \|(df^n)_x\|, n \geq 1$ is integrable. By the chain rule we have the inequality

$$\|(df^{n+k})_x\| = \|(df^k)_{f^n(x)} \cdot (df^n)_x\| \leq \|(df^k)_{f^n(x)}\| \cdot \|(df^n)_x\|.$$

By taking logarithms of each side we see that the subadditivity condition holds for ϕ_n. Applying Theorem 3.5 gives that (see [199])

$$h^s(f; x) = \lim_{n \to \infty} \frac{1}{n} \log \|(df^n)_x\| = h^s \quad a.e. \ (\mu), \tag{3.13}$$

where

$$h^s = \inf_{n \geq 1} \left\{ \frac{1}{n} \int_M \log \|(df^n)_x\| d\mu \right\}. \tag{3.14}$$

Thus we obtain $h^s(f, \mu) = h^s$. If we choose an equivalent Riemannian metric for M we would get exactly the same limit h^s, since the expression $\log \|(df^n)_x\|$ will only change by a bounded quantity and thus the difference in $\frac{1}{n} \log \|(df^n)_x\|$ will disappear as $n \to \infty$. If $f \in \mathrm{Diff}^1(M, M)$, by the chain rule we get the identity

$$(df^{-n})_{f^n(x)} = (df^{-1})_{f(x)} \cdots (df^{-1})_{f^{n-1}(x)} \cdot (df^{-1})_{f^n(x)},$$

which implies

$$\|(df^{-n})_{f^n(x)}\| \leq c^n \quad (c = \sup_{x \in M} \|(df^{-1})_x\|).$$

Thus we have

$$\|(df^n)_x\| \geq \frac{1}{\|(df^{-n})_{f^n(x)}\|} \geq \frac{1}{c^n},$$

so that the sequence $\frac{1}{n} \log \|(df^n)_x\|, n \geq 1$ is uniformly bounded below, and the limit h^s must be finite.

Theorem 3.6 (Oseledec, cf. [199]) *Let $f \in \mathrm{Diff}^1(M, M)$ be an ergodic measure preserving transformation on a probability space $(M, \mathcal{B}, \mu)$. If M is a compact manifold of dimension m, then there exist real numbers $\chi_1 > \cdots > \chi_k (k \leq m)$, integers $m_1, \cdots, m_k \in \mathbb{Z}^+$ with $m_1 + \cdots + m_k = m$, and a splitting*

$$T(M)_x = E_1(x) \oplus \cdots \oplus E_k(x),$$

with $\dim E_i(x) = m_i$ and $(df)_x E_i(x) = E_i(f(x))$, such that the mappings $x \mapsto E_i(x)$ are measurable, and such that whenever $X \in E_i(x)$,

$$\chi(x, X) = \lim_{n \to \infty} \frac{1}{n} \log \|(df^n)_x X\| = \chi_i \quad a.e. \ (\mu).$$

The numbers $\chi_1, \cdots, \chi_k$ in Theorem 3.6 are called the *Lyapunov exponents* of the ergodic measure μ. A proof can be found in [199]. Now one has $h_{Lya}(f, \mu) \leq mh^s$. The Pesin-Ruelle inequality (also see [199]) is of the following form:

$$h(f, \mu) \leq h_{Lya}(f, \mu) = \sum_{\chi_i > 0} m_i \chi_i. \tag{3.15}$$

As consequence, if $f \in \mathrm{Diff}^1(M)$ with $h_{top}(f) > 0$, then there exists an ergodic f-invariant measure μ with at least one positive and one negative Lyapunov exponent. In particular, if $\dim M = 2$, the Lyapunov exponents satisfy $\chi_1 > 0 > \chi_2$ (see [134] or [199]). If we consider arbitrary Borel probability measure μ, then the preceding inequality may be strict (cf. [133]). When μ is a smooth measure, i.e., absolutely continuous with respect to volume, and $f \in \mathrm{Diff}^2(M, M)$, Pesin showed that (3.15) is actually an equality (see [199]). The following solution was conjectured by Ruelle, and eventually proved by Ledrappier and Young [157].

Theorem 3.7 *Let M be a compact manifold and let $f \in \mathrm{Diff}^2(M, M)$ be an ergodic measure preserving transformation on a probability space $(M, \mathcal{B}, \mu)$ with associated Lyapunov exponents $\chi_1, \cdots, \chi_k$. Then the following equality*

$$\sum_{\chi_i > 0} m_i \chi_i = h(f, \mu)$$

holds iff the measure μ induces a smooth measure on unstable manifolds $W^u(x)$.

An (ergodic) measure μ is said to be *hyperbolic* if no Lyapunov exponent is zero. The number of periodic points can be estimated as follows (see Katok and Mendoza [134], p.698, or Pollicott [199]):

Theorem 3.8 *Let μ be a hyperbolic ergodic measure for a $C^{1+\alpha}$-diffeomorphism f of a compact Riemannian surface M with $h(f, \mu) > 0$. Then one has*

$$\limsup_{n \to \infty} \frac{\log^+ \#\mathrm{Fix}(f^n)}{n} \geq h_{top}(f),$$

where $\log^+ x = \max\{\log x, 0\}$.

In the theorem, $f \in C^{1+\alpha}$ means that f is C^1 and furthermore the first derivative $x \mapsto (df)_x$ satisfy Hölder condition of degree α with $0 < \alpha \leq 1$, i.e., there exists a positive number $K > 0$ such that $\|(df)_y - (df)_x\| \leq K d(y, x)^\alpha$, whenever y is sufficiently close to x. If we assume that f is expansive, then $\#\mathrm{Fix}(f^n)$ is finite, and one gets the opposite inequality

$$\limsup_{n \to \infty} \frac{\log^+ \#\mathrm{Fix}(f^n)}{n} \leq h_{top}(f),$$

(see [264], p.203). We end this section by the following:

Conjecture 3.5 *Let M be a compact manifold and suppose $f \in \mathrm{Diff}^1(M, M)$. Then $J(f) \neq \emptyset$ iff there exists a ergodic f-invariant probability measure μ on M with a positive Lyapunov exponent.*

The sufficient condition holds for complex dynamics (see Corollary 5.11).

Example 3.3 *Let $\mathbb{T}^2$ be the 2-dimensional torus and consider the diffeomorphism $f : \mathbb{T}^2 \longrightarrow \mathbb{T}^2$ by*

$$f(z_1, z_2) = (z_1^2 z_2, z_1 z_2), \quad z_1, z_2 \in S^1.$$

The Lyapunov exponents are $\log \frac{3+\sqrt{5}}{2} > 0$ and $\log \frac{3-\sqrt{5}}{2} < 0$.

3.4 Hyperbolic sets

Let f be a C^1 mapping on a m-dimensional Riemannian manifold M. For $x \in M$, $X \in T(M)_x$, set (possibly infinite)

$$\chi^u(x, X) = \liminf_{n \to \infty} \frac{1}{n} \log \|(df^n)_x X\|.$$

Now one singles out a class of sets from $\mathcal{K}_{Lya}(f)$. For a non-negative real number c, let $\mathcal{K}_{Lya,c}(f)$ denote the subset of M such that $x \in \mathcal{K}_{Lya,c}(f)$ iff either $\chi(x, X) < -c$ or $\chi^u(x, X) > c$ holds for every unit vector X in the tangent space $T(M)_x$. By the definition, we have $\mathcal{K}_{Lya,c}(f) \subset \mathcal{K}_{Lya}(f)$ such that for each point $x \in \mathcal{K}_{Lya,c}(f)$,

$$\chi \left(x, \frac{\xi}{\|\xi\|} \right) < -c, \quad \chi^u \left(x, \frac{\eta}{\|\eta\|} \right) > c$$

hold for $\xi \in E_x^s - \{0\}, \eta \in E_x^u - \{0\}$. Hence there exists an integer $N(x) > 0$ such that

$$\frac{1}{n} \log \left\| (df^n)_x \frac{\xi}{\|\xi\|} \right\| < -c, \quad \frac{1}{n} \log \left\| (df^n)_x \frac{\eta}{\|\eta\|} \right\| > c$$

hold for $n > N(x)$, and hence there exists a minimal positive number $a \left(x, \frac{\xi}{\|\xi\|}, \frac{\eta}{\|\eta\|} \right) \geq 1$ such that

$$\|(df^n)_x \xi\| \leq a \left(x, \frac{\xi}{\|\xi\|}, \frac{\eta}{\|\eta\|} \right) e^{-nc} \|\xi\|, \quad \|(df^n)_x \eta\| \geq a \left(x, \frac{\xi}{\|\xi\|}, \frac{\eta}{\|\eta\|} \right)^{-1} e^{nc} \|\eta\|$$

hold for all $n \geq 0$. Thus for some compact subset Λ in $\mathcal{K}_{Lya,c}(f)$, we can expect that there is a positive constant a which is independent of x, ξ, η, n such that

$$a \left(x, \frac{\xi}{\|\xi\|}, \frac{\eta}{\|\eta\|} \right) \leq a$$

hold for $x \in \Lambda$, $\xi \in E_x^s, \eta \in E_x^u$. These are the so-called hyperbolic sets:

Definition 3.5 *By a hyperbolic set of a cascade $\{f^n\}_{n \in \mathbb{Z}_+}$ (or differentiable mapping f) on a manifold M, we mean a forward invariant compact set $\Lambda \subset M$ such that for each point $x \in \Lambda$ the tangent space $T(M)_x$ decomposes into a direct sum*

$$T(M)_x = E_x^s \oplus E_x^u \tag{3.16}$$

of the stable space E_x^s and the unstable space E_x^u, with the following properties: for $\xi \in E_x^s, \eta \in E_x^u, n \geq 0$

$$(df)_x E_x^s \subset E_{f(x)}^s, \quad (df)_x E_x^u \subset E_{f(x)}^u, \tag{3.17}$$

$$\|(df^n)_x \xi\| \leq ae^{-cn} \|\xi\|, \quad \|(df^n)_x \eta\| \geq \frac{1}{a} e^{cn} \|\eta\|, \tag{3.18}$$

where a, c are positive constants that are independent of x, ξ, η, n.

Remark 1. The condition (3.18) is independent of the metric on M, since if $\| \; \|_1$ and $\| \; \|_2$ are two equivalent norms of $T(M)_x$, there are strictly positive constants a_1' and a_2' such that $a_1'\| \; \|_2 \leq \| \; \|_1 \leq a_2'\| \; \|_2$.

Remark 2. By induction, (3.17) yields

$$(df^n)_x E_x^s \subset E_{f^n(x)}^s, \quad (df^n)_x E_x^u \subset E_{f^n(x)}^u \quad (n \in \mathbb{Z}_+). \tag{3.19}$$

If f is a diffeomorphism, noting that $f^n \circ f^{-n} = f^{-n} \circ f^n = id$ induces the identity transformation

$$(df^n)_{f^{-n}(x)} \cdot (df^{-n})_x = (df^{-n})_{f^n(x)} \cdot (df^n)_x : T(M)_x \longrightarrow T(M)_x,$$

then (3.18) and (3.19) give

$$\|(df^{-n})_x \xi\| \geq \frac{1}{a} e^{cn}\|\xi\|, \quad \|(df^{-n})_x \eta\| \leq a e^{-cn}\|\eta\|, \tag{3.20}$$

for $\xi \in E_x^s, \eta \in E_x^u, n \geq 0$. The condition (3.18) is equivalent to that there exist $\lambda < 1 < \gamma$ such that

$$\|(df)_{f^n(x)}|_{E_{f^n(x)}^s}\| \leq \lambda, \quad \|(df)_{f^n(x)}^{-1}|_{E_{f^{n+1}(x)}^u}\| \leq \gamma^{-1} \quad (n \in \mathbb{Z}_+). \tag{3.21}$$

Remark 3. It is easy to prove that dimensions of the subspaces E_x^s, E_x^u are locally constant (as functions of $x \in \Lambda$), while the subspaces themselves depend continuously on x. The unions

$$E^s = \bigcup_{x \in \Lambda} E_x^s, \quad E^u = \bigcup_{x \in \Lambda} E_x^u$$

are vector subbundles of the restriction $T(M)|_\Lambda$ of the tangent bundle of M to Λ, and

$$T(M)|_\Lambda = E^s \oplus E^u \quad \text{(Whitney sum)} .$$

These subbundles are (forward) invariant with respect to df and are called *stable* and *unstable bundles*, respectively (for Λ, f and $\{f^n\}$). Here we allow the slight but fairly obvious generalization of the notion of a vector bundle in which the fibres can have different dimensions over different parts of the base Λ which, of course, are at a positive distance from each other. If the dimension $\dim E_x^u$ is constant for a hyperbolic set Λ, then it is called the *Morse index* of Λ, denoted by u_Λ.

Remark 4. Extending the notion of a hyperbolic set to noninvertible systems presents a curious problem. Here we exhibit the definition of Ruelle [220] by considering the negative trajectory of a point x to determine the unstable part E_x^u. Assume that $\Lambda \subset M$ is a backward invariant compact set. The set of negative trajectories,

$$\hat{\Lambda} = \{\{x_n\}_{-\infty}^0 \mid f(x_n) = x_{n+1}\} \subset \prod_{n \leq 0} \Lambda,$$

is compact in the product topology. We define the tangent bundle $T(\hat{\Lambda})$ of $\hat{\Lambda}$ as the set of $(\hat{x}, \xi)$ where $\hat{x} = \{x_n\}^0_{-\infty} \in \hat{\Lambda}$ and where $\xi \in T(M)_{x_0}$ is a tangent vector. Then f lifts to a homeomorphism $\hat{f} : \hat{\Lambda} \longrightarrow \hat{\Lambda}$ by setting

$$\hat{f}(\{..., x_{-1}, x_0\}) = \{..., x_{-1}, x_0, f(x_0)\}.$$

Similarly, df lifts to a mapping $d\hat{f}$ on $T(\hat{\Lambda})$. Then Λ is called a *prehyperbolic set*, with the *hyperbolic cover* $\hat{\Lambda}$, if for each point $\hat{x} \in \hat{\Lambda}$ the tangent space $T(\hat{\Lambda})_{\hat{x}}$ decomposes into a direct sum

$$T(\hat{\Lambda})_{\hat{x}} = E^s_{\hat{x}} \oplus E^u_{\hat{x}} \qquad (3.22)$$

with the following properties: for $\xi \in E^s_{\hat{x}}, \eta \in E^u_{\hat{x}}, n \geq 0$

$$(d\hat{f})_{\hat{x}} E^s_{\hat{x}} \subset E^s_{\hat{f}(\hat{x})}, \quad (d\hat{f})_{\hat{x}} E^u_{\hat{x}} = E^u_{\hat{f}(\hat{x})}, \qquad (3.23)$$

$$\|(d\hat{f}^n)_{\hat{x}}\xi\| \leq ae^{-cn}\|\xi\|, \quad \|(d\hat{f}^n)_{\hat{x}}\eta\| \geq \frac{1}{a}e^{cn}\|\eta\|, \qquad (3.24)$$

where a, c are positive constants that are independent of $\hat{x}, \xi, \eta, n$.

An forward invariant subset of a hyperbolic set and unions of a finite number of hyperbolic sets are both hyperbolic sets.

Example 3.4 *Let $M = \mathbb{T}^2 = \mathbb{R}^2/\mathbb{Z}^2$ be the 2-torus and let $f : M \longrightarrow M$ be the diffeomorphism defined by*

$$f(x_1, x_2) = (x_1 + 2x_2(mod\ 1), x_1 + x_2(mod\ 1)).$$

At every point $x \in M$ the derivative is represented by the same matrix

$$(df)_x = \begin{pmatrix} 2 & 1 \\ 1 & 1 \end{pmatrix}.$$

Then E^s_x (resp., E^u_x) is the space of eigenvectors for the eigenvalue $\frac{3-\sqrt{5}}{2}$ (resp., $\frac{3+\sqrt{5}}{2}$). For this splitting of the unit tangent bundle we can explicitly compute the norm of the iterates of the tangent mapping using the eigenvalues to get that

$$\left\|(df)_{f^n(x)}|E^s_{f^n(x)}\right\| = \frac{3-\sqrt{5}}{2} < 1, \quad \left\|(df)^{-1}_{f^n(x)}|E^u_{f^{n+1}(x)}\right\| = \frac{3-\sqrt{5}}{2} \ (n \in \mathbb{Z}_+).$$

Then M is a hyperbolic set.

Definition 3.6 *By a hyperbolic set of a flow $\{f^t\}_{t \in \mathbb{R}}$ on a manifold M, we mean a forward invariant compact set $\Lambda \subset M$ such that for each point $x \in \Lambda$ the tangent space $T(M)_x$ decomposes into a direct sum*

$$T(M)_x = E^s_x \oplus E^u_x \oplus E^v_x \qquad (3.25)$$

of subspaces with the following properties: for $\xi \in E^s_x, \eta \in E^u_x, t \geq 0$

$$(df^t)_x E^s_x = E^s_{f^t(x)}, \quad (df^t)_x E^u_x = E^u_{f^t(x)}, \quad (df^t)_x E^v_x = E^v_{f^t(x)}, \qquad (3.26)$$

$$\|(df^t)_x\xi\| \leq ae^{-ct}\|\xi\|, \quad \|(df^t)_x\eta\| \geq \frac{1}{a}e^{ct}\|\eta\|, \qquad (3.27)$$

where a, c are positive constants that are independent of x, ξ, η, t.

Here E^v is spanned by the vector field generating f^t. For a hyperbolic set Λ of the flow $\{f^t\}$, obviously we have $\Lambda \subset \mathcal{K}_{Lya}(f)$, and

$$E_x^c = E_x^v, \quad x \in \Lambda.$$

Now we exhibit some results on hyperbolic set.

Theorem 3.9 (Shadowing lemma [51]) *Suppose $f : M \longrightarrow M$ is a diffeomorphism with hyperbolic chain recurrent set R. Given $\varepsilon > 0$ there is a $\delta > 0$ such that if $\{x_i\}_{i \in \mathbb{Z}}$ is a δ-trajectory of a cascade $\{f^n\}$, then there is a unique $x \in R$ satisfying*

$$d(x_i, f^i(x)) < \varepsilon \quad \text{for all } i.$$

Theorem 3.10 (Spectral decomposition theorem [232]) *If the nonwandering point set $\Omega(f)$ of a diffeomorphism f is hyperbolic and if the set $\mathrm{Per}(f)$ is dense in $\Omega(f)$, then $\Omega(f)$ is a finite disjoint union of compact invariant sets $\Lambda_1, ..., \Lambda_k$ and each Λ_i contains an orbit of the system which is dense in Λ_i.*

If $\Omega(f)$ satisfies the conditions in this theorem, the mapping f is usually said to satisfy Axiom A which is introduced by S. Smale in [232]. If a diffeomorphism f satisfies Axiom A, S. Smale [232] also proved the following:

$$M = \bigcup_{i=1}^{k} \mathrm{Att}(\Lambda_i).$$

Further, there is a neighborhood U of Λ_i such that

$$\bigcap_{n \in \mathbb{Z}} f^n(U) = \Lambda_i,$$

that is, Λ_i is locally maximal, and

$$\mathrm{Att}(\Lambda_i) = W^s(\Lambda_i),$$

which was first proved in [112]; a simpler proof was given by Bowen in (3.10) of [50]. The sets $\Lambda_1, ..., \Lambda_k$ are called *basic sets*.

Theorem 3.11 *A diffeomorphism $f \in \mathrm{Diff}^1(M, M)$ is C^1 strongly structurally stable iff f satisfies Axion A and the geometric strong transversality condition.*

Here a diffeomorphism f is said to satisfy the *geometric strong transversality condition* if for any $x, y \in \Omega(f)$, the manifolds $W^s(x), W^u(y)$ are transversal. J. Robbin [213] proved that the conditions are sufficient for C^2 diffeomorphisms, C. Robinson [214] did the same for C^1 diffeomorphisms. R. Mañé [163] proved the necessity.

Let Λ be a compact hyperbolic set of $f \in \mathrm{Diff}^1(M, M)$. For each $p \in \Lambda$ let us fix a base in $T(M)_p$ such that the decomposition $E_p^s \oplus E_p^u$ is identified with the standard decomposition

$\mathbb{R}^m = \mathbb{R}^l \oplus \mathbb{R}^{m-l}$, and the Riemannian metric in $T(M)_p$ becomes the Euclidean metric in $\mathbb{R}^m$. Set

$$B_\varepsilon^l = \{(x,y) \in \mathbb{R}^l \times \mathbb{R}^{m-l} \mid \|x\| < \varepsilon, \|y\| < \varepsilon\}.$$

Since Λ is a compact, we can take $\varepsilon' > 0$ such that the exponential mapping $\exp_p : B_\varepsilon^l \longrightarrow M$ is injective for every $p \in \Lambda$, $0 < \varepsilon \le \varepsilon'$, and hence is a diffeomorphism between B_ε^l and a certain neighborhood $D_\varepsilon^l(p)$ of p in M. If necessary, choose $0 < \varepsilon'' \le \varepsilon'$ such that $f(D_\varepsilon^l(p)) \subset D_{\varepsilon'}^l(f(p))$ for each $p \in \Lambda$, $0 < \varepsilon \le \varepsilon''$. Thus for $0 < \varepsilon \le \varepsilon''$, one obtains a family of diffeomorphisms

$$f_{p,\varepsilon} = \exp_{f(p)}^{-1} \circ f \circ \exp_p : B_\varepsilon^l \longrightarrow f_{p,\varepsilon}(B_\varepsilon^l) \subset B_{\varepsilon'}^l \subset \mathbb{R}^m$$

with $f_{p,\varepsilon}(0) = 0$. It is convenient to extend the mappings to the whole space $\mathbb{R}^m$ using the following fact:

Lemma 3.2 (Extension Lemma, cf. [134]) *Let U be an open bounded neighborhood of $0 \in \mathbb{R}^m$ and $f : U \longrightarrow \mathbb{R}^m$ a local diffeomorphism with $f(0) = 0$. For $\delta > 0$, there exist $\rho > 0$ and a diffeomorphism $\hat{f} : \mathbb{R}^m \longrightarrow \mathbb{R}^m$ such that $\|\hat{f} - (df)_0\|_{C^1} < \delta$ and $\hat{f} = f$ on the ρ-ball $\mathbb{R}^m(\rho) = \{x \in \mathbb{R}^m \mid \|x\| < \rho\}$.*

Note that $\exp_p$ depends smoothly on p, and its differential at the origin is the identity mapping so that $(df_{p,\varepsilon})_0 = (df)_p$. Hence for $\delta > 0$, there exist $\varepsilon_1 > 0$ and a diffeomorphism $\hat{f}_{p,\varepsilon} : \mathbb{R}^m \longrightarrow \mathbb{R}^m$ such that for each $p \in \Lambda$, $\|\hat{f}_{p,\varepsilon} - (df)_p\|_{C^1} < \delta$ and $\hat{f}_{p,\varepsilon} = f_{p,\varepsilon}$ on $\mathbb{R}^m(\varepsilon_1)$. Thus, along each orbit $O(p)$ for $p \in \Lambda$, one obtains a sequence of mappings

$$f_n = f_{n,p} = \hat{f}_{f^n(p),\varepsilon} : \mathbb{R}^m \longrightarrow \mathbb{R}^m$$

satisfying the conditions of the Hadamard-Perron Theorem (cf. [134]):

Theorem 3.12 *Let $\lambda < \gamma$, $r \ge 1$, and for each $n \in \mathbb{Z}$ let $f_n : \mathbb{R}^m \longrightarrow \mathbb{R}^m$ be a C^r diffeomorphism such that for $(x,y) \in \mathbb{R}^l \times \mathbb{R}^{m-l}$,*

$$f_n(x,y) = (A_n x + \alpha_n(x,y), B_n y + \beta_n(x,y))$$

for some linear mappings $A_n : \mathbb{R}^l \longrightarrow \mathbb{R}^l$ and $B_n : \mathbb{R}^{m-l} \longrightarrow \mathbb{R}^{m-l}$ with $\|A_n\| \le \lambda$, $\|B_n^{-1}\| \le \gamma^{-1}$ and $\alpha_n(0) = 0$, $\beta_n(0) = 0$. Then there exists $\chi_0 = \chi_0(\lambda, \gamma)$ such that for $\chi \in (0, \chi_0)$, there is a $\delta = \delta(\lambda, \gamma, \chi)$ satisfying

$$0 < \delta < \min \left(\frac{\gamma - \lambda}{\chi + \frac{1}{\chi} + 2}, \frac{\gamma - (1+\chi)^2 \lambda}{(2+\chi)(1+\chi)} \right)$$

with the following property: If $\|\alpha_n\|_{C^1} < \delta$ and $\|\beta_n\|_{C^1} < \delta$ for all $n \in \mathbb{Z}$, then there is a unique family $\{W_n^s\}_{n \in \mathbb{Z}}$ of l-dimensional C^1 manifolds

$$W_n^s = \{(x, \varphi_n^s(x)) \mid x \in \mathbb{R}^l\} = \text{graph}\varphi_n^s$$

and a unique family $\{W_n^u\}_{n \in \mathbb{Z}}$ of $(m-l)$-dimensional C^1 manifolds

$$W_n^u = \{(\varphi_n^u(y), y) \mid y \in \mathbb{R}^{m-l}\} = \text{graph}\varphi_n^u,$$

where $\varphi_n^s \in C^1(\mathbb{R}^l, \mathbb{R}^{m-l})$, $\varphi_n^u \in C^1(\mathbb{R}^{m-l}, \mathbb{R}^l)$, $\sup_{n \in \mathbb{Z}} \|d\varphi_n^v\| < \chi$ $(v = s, u)$, *and the following properties hold:*

1) $f_n(W_n^v) = W_{n+1}^v$ $(v = s, u)$;

2) $\|f_n(z)\| < \tilde{\lambda}\|z\|$ *for* $z \in W_n^s$ *and* $\|f_{n-1}^{-1}(z)\| < \tilde{\gamma}^{-1}\|z\|$ *for* $z \in W_n^u$, *where* $\tilde{\lambda} = (1+\chi)(\lambda + \delta(1+\chi))$ *and* $\tilde{\gamma} = \left(\frac{\gamma}{1+\chi} - \delta\right)$;

3) *Let* $\tilde{\lambda} < \nu < \tilde{\gamma}$. *Then* $z \in W_n^s$ *if*

$$\|f_{n+k-1} \circ \cdots \circ f_n(z)\| < c\nu^k \|z\|$$

for all $k \geq 0$ *and some* $c > 0$, *similarly* $z \in W_n^u$ *if*

$$\|f_{n-k}^{-1} \circ \cdots \circ f_{n-1}^{-1}(z)\| < c\nu^{-k}\|z\|$$

for all $k \geq 0$ *and some* $c > 0$. *Finally, in the hyperbolic case* $\lambda < 1 < \gamma$ *the families* $\{W_n^s\}_{n \in \mathbb{Z}}$ *and* $\{W_n^u\}_{n \in \mathbb{Z}}$ *consist of* C^r *manifolds.*

Since Λ is hyperbolic, we can take $\lambda < 1 < \gamma$ so that $\tilde{\lambda} < 1 < \tilde{\gamma}$ by choosing χ and δ sufficiently small. Then by 2),

$$f_n(W_n^s \cap B_\varepsilon^l) \subset W_{n+1}^s \cap B_\varepsilon^l, \quad f_{n-1}^{-1}(W_n^u \cap B_\varepsilon^l) \subset W_{n-1}^u \cap B_\varepsilon^l,$$

and hence there exists $0 < \varepsilon_0 \leq \min\{\varepsilon'', \varepsilon_1\}$ such that for $0 < \varepsilon \leq \varepsilon_0$,

$$f : \tilde{W}_{f^n(p)}^s \cap D_\varepsilon^l(f^n(p)) \longrightarrow \tilde{W}_{f^{n+1}(p)}^s \cap D_\varepsilon^l(f^{n+1}(p)),$$

$$f^{-1} : \tilde{W}_{f^n(p)}^u \cap D_\varepsilon^l(f^n(p)) \longrightarrow \tilde{W}_{f^{n-1}(p)}^u \cap D_\varepsilon^l(f^{n-1}(p)),$$

where

$$\tilde{W}_{f^n(p)}^s = \exp_{f^n(p)}(W_n^s), \quad \tilde{W}_{f^n(p)}^u = \exp_{f^n(p)}(W_n^u).$$

A reformulation of the Hadamard-Perron Theorem in terms of the original map f yields the following result which is usually called stable and unstable manifold theorem. The reader can find a proof in [196], Chap. 1 or [134]. Also see [195].

Theorem 3.13 *Assume that M is a smooth compact manifold of dimension m with a Riemannian metric d. Assume that $f \in \text{Diff}^r(M, M), r \geq 1$ and that Λ is a hyperbolic set of f. Then there exists $\varepsilon_0 > 0$ have the following properties. If $p \in \Lambda$, $\dim E_p^s = l$, then*

1) *there exist immersions b^s, b^u of class C^r:*

$$b^s : \mathbb{R}^l \longrightarrow M, \quad b^s(0) = p, \quad b^s(\mathbb{R}^l) = W^s(p),$$

$$b^u : \mathbb{R}^{m-l} \longrightarrow M, \quad b^u(0) = p, \quad b^u(\mathbb{R}^{m-l}) = W^u(p);$$

2) *for any $\varepsilon \in \mathbb{R}(0, \varepsilon_0]$ there exist $\rho > 0$ and a pair of embedded smooth discs (of class C^r) $W_\varepsilon^s(p)$ and $W_\varepsilon^u(p)$ being subsets of $W^s(p)$ and $W^u(p)$ containing p respectively, called the local stable manifold and the local unstable manifold of p respectively, such that*

2.1) $T(W_\varepsilon^s(p))_p = E_p^s, T(W_\varepsilon^u(p))_p = E_p^u$;

2.2) *for $x \in M(p; \rho) - W_\varepsilon^s(p)$ there is $n_1 > 0$ with $d(f^{n_1}(x), f^{n_1}(p)) \geq \varepsilon$;*

2.3) *for $x \in M(p; \rho) - W_\varepsilon^u(p)$ there is $n_2 < 0$ with $d(f^{n_2}(x), f^{n_2}(p)) \geq \varepsilon$;*

2.4) $f(W_\varepsilon^s(p)) \subset W_\varepsilon^s(f(p))$,　　$f^{-1}(W_\varepsilon^u(p)) \subset W_\varepsilon^u(f^{-1}(p))$;

2.5) for every $\delta > 0$ there exists $a(\delta)$ such that for $n \in \mathbb{N}$

$$d(f^n(x), f^n(p)) < a(\delta)(e^{-c} + \delta)^n d(x, p) \text{ for } x \in W_\varepsilon^s(p),$$

$$d(f^{-n}(x), f^{-n}(p)) < a(\delta)(e^{-c} + \delta)^n d(x, p) \text{ for } x \in W_\varepsilon^u(p);$$

2.6) there exist a family of neighborhoods $D_\varepsilon^l(p)$ containing the ball $M(p; \rho)$ around p of radius ρ and families $\tilde{W}_p^s$ and $\tilde{W}_p^u$ such that

$$
\begin{aligned}
W_\varepsilon^s(p) &= \{x \mid f^n(x) \in \tilde{W}_{f^n(p)}^s \cap D_\varepsilon^l(f^n(p)), \quad n = 0, 1, 2, \ldots\}, \\
W_\varepsilon^u(p) &= \{x \mid f^{-n}(x) \in \tilde{W}_{f^{-n}(p)}^u \cap D_\varepsilon^l(f^{-n}(p)), \quad n = 0, 1, 2, \ldots\}.
\end{aligned}
$$

Obviously, we have

$$
\begin{aligned}
W^s(p) &= \bigcup_{n=0}^{\infty} f^{-n}(W_\varepsilon^s(f^n(p))), \\
W^u(p) &= \bigcup_{n=0}^{\infty} f^n(W_\varepsilon^u(f^{-n}(p))).
\end{aligned}
\tag{3.28}
$$

Thus if $d(f^n(x), f^n(p)) < \varepsilon$ for $n \in \mathbb{Z}$, then $f^n(x) \in M(f^n(p); \varepsilon)$ and $x \in W_\varepsilon^s(p) \cap W_\varepsilon^u(p) = \{p\}$. Therefore one obtain

Corollary 3.1 *The restriction of a diffeomorphism to a hyperbolic set is expansive.*

3.5　Notes on hyperbolic sets

Theorem 3.14 ([125]) *Let Λ be a hyperbolic set of a C^r differentiable transformation f on a manifold M. If the Morse index $u_\Lambda = 0$, then Λ is asymptotically stable, and $\mathrm{Att}(\Lambda) \subset F_{equ}(f)$. If the Morse index $u_\Lambda = m = \dim M$, then $\Lambda \subset J_{equ}(f)$. Both cases have*

$$J_{equ}(f) \cap J_{equ}(f^{-1}) \cap \Lambda = \emptyset.$$

Proof. Assume that $u_\Lambda = 0$. By Theorem 3.13, for $p \in \Lambda$, there exist immersions b^s of class C^r:

$$b^s : \mathbb{R}^m \longrightarrow M, \quad b^s(0) = p, \quad b^s(\mathbb{R}^m) = W^s(p).$$

Also there exists $\varepsilon_0 > 0$ have the following properties: for any $\varepsilon \in \mathbb{R}(0, \varepsilon_0]$ and for every $\delta > 0$ there exists $a(\delta)$ such that for $n \in \mathbb{N}$

$$d(f^n(x), f^n(p)) < a(\delta)(e^{-c} + \delta)^n d(x, p) \text{ for } x \in W_\varepsilon^s(p) \supset M(p; \rho).$$

Take $\delta > 0$ such that $e^{-c} + \delta < 1$. We see that the family $\mathcal{F} = \{f^n\}$ is equicontinuous at p, i.e., $p \in F_{equ}(f)$. Hence $\Lambda \subset F_{equ}(f)$. Note that $F_{equ}(f)$ is invariant. Then $W_\varepsilon^s(f^n(p)) \subset F_{equ}(f)$ imply

$$W^s(p) = \bigcup_{n=0}^{\infty} f^{-n}(W_\varepsilon^s(f^n(p))) \subset F_{equ}(f).$$

Hence

$$\text{Att}(\Lambda) = W^s(\Lambda) = \bigcup_{p \in \Lambda} W^s(p) \subset F_{equ}(f).$$

Therefore Λ is asymptotically stable. Note that $W^u(p) = \{p\}$. Then the family $\{f^{-n}\}$ is not equicontinuous at p. Hence $p \in J_{equ}(f^{-1})$, i.e., $\Lambda \subset J_{equ}(f^{-1})$. For the case $u_\Lambda = m$, we can prove the conclusion similarly. □

Corollary 3.2 *Let Λ be a hyperbolic set of a C^r differentiable transformation f on a manifold M such that the Morse index satisfies $0 < u_\Lambda < \dim M$. Then*

$$\Lambda \subset J_{equ}(f) \cap J_{equ}(f^{-1}).$$

The proof follows from 2.2) and 2.3) of Theorem 3.13 directly.

Conjecture 3.6 *Let Λ be a hyperbolic set of a C^r differentiable transformation f on a manifold M. If the Morse index $u_\Lambda = \dim M$, then Λ is repulsive.*

Hyperbolic sets were introduced by Smale (see [231], [232]). Smale discovered a class of hyperbolic sets, the *Smale "horseshoes"*. Anosov[16], [17] singled out a class of DS's as follows: A dynamical system on a compact manifold M is called *Anosov* (*Anosov diffeomorphism* or *Anosov flow* respectively) if M is a hyperbolic set for the DS. The set of all Anosov diffeomorphisms is open in $\text{Diff}^1(M, M)$ (see [254]). Every Anosov dynamical system (Anosov diffeomorphism or Anosov flow) is C^1 structurally stable (The proof can be found, for example, in Mather [168]). Recall that a homeomorphism f is called *topologically transitive* if it possesses an everywhere dense trajectory. One also has the following result:

Theorem 3.15 (cf. [230]) *Assume that f is a C^2-Anosov diffeomorphism. Then we have*
1) If f is topologically transitive, then $\overline{W^s(x)} = M, \overline{W^u(x)} = M$ for any $x \in M$.
2) $\overline{\text{Per}(f)} = \Omega(f)$, and the number $\#\text{Fix}(f^n)$ is finite satisfying

$$\limsup_{n \to \infty} \frac{\log^+ \#\text{Fix}(f^n)}{n} = h_{top}(f),$$

where $\log^+ x = \max\{\log x, 0\}$.

The following theorem is a special case of Theorem 3.14:

Theorem 3.16 *Let M be a smooth compact manifold and let f be an Anosov diffeomorphism. Then $J_{equ}(f) = \emptyset$ if $u_\Lambda = 0$, or $J_{equ}(f) = M$ if $u_\Lambda = \dim M$. Both cases have $J_{equ}(f) \cap J_{equ}(f^{-1}) = \emptyset$. If $0 < u_\Lambda < \dim M$, then $J_{equ}(f) = J_{equ}(f^{-1}) = M$.*

We denote m-torus by $\mathbb{T}^m = \mathbb{R}^m / \mathbb{Z}^m$. Let $\mathbf{A} : \mathbb{T}^m \longrightarrow \mathbb{T}^m$ be an algebraic automorphism given by an integer matrix $\mathbf{A} = (a_{ij})$ and let $\lambda_1, ..., \lambda_m$ be the eigenvalues of $\mathbf{A}$. Then the algebraic automorphism is Anosov if and only if $\mathbf{A}$ is hyperbolic, i.e., $|\lambda_i| \neq 1$ for all $i \in \mathbb{Z}[1, m]$ (cf. [17]). Moreover any Anosov diffeomorphism f on the torus is topologically conjugate to some hyperbolic automorphism $\mathbf{A}$. Then $J(f) = \emptyset$ if $\mathbf{A}$ is contraction, or $J(f) = \mathbb{T}^m$ if $\mathbf{A}$ is expansion, otherwise,

$$J(f) = J(f^{-1}) = \mathbb{T}^m.$$

We know the following fact: If Λ is a locally maximal hyperbolic set of a diffeomorphism f, then Λ is an attractor if and only if $W^u(x) \subset \Lambda$ for any $x \in \Lambda$ (see [193]). Thus the hyperbolic attractor Λ is related to the unstable manifold and the Julia set $J(f)$ by Theorem 3.14 and Corollary 3.2.

According to the definition by Li and Yorke [159], a continuous mapping $f \in C(M, M)$ is *(Li-Yorke) chaotic* if there are an uncountable set $S \subset M - W^s(\text{Per}(f))$ with the following properties: For every $x, y \in S$ with $x \neq y$,

$$\limsup_{n \to \infty} d(f^n(x), f^n(y)) > 0, \quad \liminf_{n \to \infty} d(f^n(x), f^n(y)) = 0.$$

Such a set S is called a *scrambled set*. Obviously, if $f \in \text{Hom}(M, M)$ and if S is a scrambled set, then $f(S)$ and $f^{-1}(S)$ all are scrambled sets since $M - W^s(\text{Per}(f))$ is invariant. Setting

$$W_+^s(x) = \{y \in M \mid \liminf_{n \to \infty} d(f^n(y), f^n(x)) = 0\},$$

if x is a point in a scrambled set S, then we see

$$S - \{x\} \subset \{W_+^s(x) - W^s(x)\} \cap \{M - W^s(\text{Per}(f))\}.$$

Li and Yorke [159] proved that if M is an interval, then f is chaotic if there is a periodic point with period 3. Smital [238], [239] and Misiurewicz [175] showed that there exists a continuous mapping of the interval $[0, 1]$ onto itself for which there exists a scrambled set of Lebesgue measure 1. In the general multidimensional case, one suggests the following sufficient condition for the existence of chaos:

Theorem 3.17 ([166]) *Let $f : \mathbb{R}^m \longrightarrow \mathbb{R}^m$ be a differentiable mapping with a fixed point p, and assume that for some $\varepsilon > 0$ and a positive integer n, there is another point $q \in \mathbb{R}^m(p; \varepsilon)$ such that $f^n(q) = p$, $\det\{(df^n)_q\} \neq 0$, and $\|f(x) - f(y)\| > \|x - y\|$ for any two distinct points $x, y \in \mathbb{R}^m(p; \varepsilon)$. Then f is chaotic.*

It is known that there exist continuous Li-Yorke chaotic mappings with zero topological entropy [176], but we still suspect that $J(f) \neq \emptyset$ if f is chaotic. The following questions are natural:

Problem 3.1 *Are the mappings with hyperbolic Julia sets dense in $C^r(M, M)$?*

Problem 3.2 ([232]) *Are mappings satisfying Axiom A dense in $C^r(M, M)$?*

R. Abraham and Smale [7] showed that Axiom A is not dense in $\text{Diff}^1(\mathbb{T}^2 \times S^2, \mathbb{T}^2 \times S^2)$. S. E. Newhouse [185] showed that there is an open set $\mathcal{U}$ in $\text{Diff}^2(S^2, S^2)$ such that if $f \in \mathcal{U}$, then f does not satisfy Axiom A and f is not C^2 structurally stable. If M is an interval, Problem 3.2 is open (see [235]). Thus we may think that Problem 3.1 also is not true for the case $\dim M > 1$. Here we give a weak version of hyperbolic sets (see [126]).

Definition 3.7 *Assume that M is orientable and let Ω be a volume element on M. Take $f \in C^1(M, M)$ and define a function*

$$v_{f,\Omega} = f^*(\Omega)/\Omega.$$

Then f is said to be the volume contracting (resp., volume expanding) on a subset Λ of M if there exists $0 \le \lambda < 1$ (resp., $\gamma > 1$) such that $|v_{f,\Omega}| \le \lambda$ (resp., $|v_{f,\Omega}| \ge \gamma$) on Λ.

Note that

$$v_{f^n,\Omega} = (v_{f,\Omega} \circ f^{n-1}) \cdot (v_{f,\Omega} \circ f^{n-2}) \cdots (v_{f,\Omega} \circ f) \cdot v_{f,\Omega}.$$

Thus if Λ is plus invariant and if f contracts or expands the volume Ω on Λ, then $|v_{f^n,\Omega}| \le \lambda^n$ or $\ge \gamma^n$.

Definition 3.8 *Assume that M is orientable and let Ω be a volume element on M. Take $f \in C^1(M, M)$. A forward invariant compact subset Λ of M is said to be volume contracting hyperbolic (resp., volume expanding hyperbolic) if there exist $a > 0$ and $0 \le \lambda < 1$ (resp., $b > 0$ and $\gamma > 1$) such that $|v_{f^n,\Omega}| \le a\lambda^n$ (resp., $|v_{f^n,\Omega}| \ge b\gamma^n$) on Λ, and is called volume hyperbolic if each forward invariant component of Λ is either volume contracting or expanding hyperbolic.*

Note that the definition don't depend on the choice of the volume elements, but so does the former. Let Λ be a forward invariant compact subset of M. If f is volume contracting (resp., expanding) on Λ, then Λ is volume contracting (resp., expanding) hyperbolic. Conversely, it is easy to prove that there is an positive integer n_0 such that f^n is volume contracting (resp., expanding) on Λ for each $n \ge n_0$.

Here we can define functions by

$$h_\Omega^s(f; x) = \limsup_{n\to\infty} \frac{1}{n} \log |v_{f^n,\Omega}(x)| = \limsup_{n\to\infty} \frac{1}{n} \sum_{k=0}^{n-1} \log |v_{f,\Omega}(f^k(x))|,$$

$$h_\Omega^u(f; x) = \liminf_{n\to\infty} \frac{1}{n} \log |v_{f^n,\Omega}(x)| = \liminf_{n\to\infty} \frac{1}{n} \sum_{k=0}^{n-1} \log |v_{f,\Omega}(f^k(x))|.$$

If Λ is volume contracting (or expanding) hyperbolic, then $h_\Omega^s(f; x) \le \log \lambda < 0$ (or $h_\Omega^u(f; x) \ge \log \gamma > 0$) on Λ, that is, $h_\Omega^s(f; x)$ is bounded above by a negative constant (resp., $h_\Omega^u(f; x)$ is bounded low by a positive constant) on Λ. If M is compact, these functions don't depend on the choice of the volume elements. So we also write

$$h_{vol}^s(f; x) = h_\Omega^s(f; x), \quad h_{vol}^u(f; x) = h_\Omega^u(f; x).$$

Further, define

$$h_{vol}^{ss}(f) = h_\Omega^{ss}(f) = \limsup_{n\to\infty} \frac{1}{n} \sup_{x\in M} \log |v_{f^n,\Omega}(x)|,$$

$$h_{vol}^s(f, \mu) = \int_M h_{vol}^s(f; x) d\mu(x),$$

for some $\mu \in \Sigma(M)$ with $h_{vol}^s(f, \mu) \le h_{vol}^{ss}(f)$, and set

$$h_{vol}^s(f) = \sup_{\mu \in \Sigma_f(M)} h_{vol}^s(f, \mu).$$

These notions can be extend to the case of flows similarly.

Assume that M is a compact manifold and takes $\mathcal{B}$ to be the Borel σ-algebra. Assume that μ is an ergodic f-invariant Borel probability measure on M. Then each function $v_{f^n,\Omega}$ is continuous, bounded, and thus $\phi_n(x) = \log |v_{f^n,\Omega}(x)|, n \ge 1$ is integrable such that the subadditivity condition

$$\phi_{n+k}(x) = \phi_n(x) + \phi_k(f^n(x))$$

holds. Applying Theorem 3.5 gives that

$$h_{vol}^s(f; x) = h_{vol}^u(f; x) = \lim_{n \to \infty} \frac{1}{n} \log |v_{f^n,\Omega}(x)| = h' \quad \text{a.e. } (\mu), \tag{3.29}$$

where

$$h' = \inf_{n \ge 1} \{ \frac{1}{n} \int_M \log |v_{f^n,\Omega}(x)| d\mu \}. \tag{3.30}$$

Thus we obtain $h_{vol}^s(f, \mu) = h' \le h_{vol}^{ss}(f)$. If we choose another volume form for M we would get exactly the same limit h', since the expression $\log |v_{f^n,\Omega}(x)|$ will only change by a bounded quantity and thus the difference in $\frac{1}{n} \log |v_{f^n,\Omega}(x)|$ will disappear as $n \to \infty$. If $f \in \text{Diff}^1(M, M)$, we get

$$|v_{f^{-n},\Omega}(x)| \le c^n \quad (c = \sup_{x \in M} |v_{f^{-1},\Omega}(x)|).$$

Thus we have

$$|v_{f^n,\Omega}(x)| = \frac{1}{|v_{f^{-n},\Omega}(x)|} \ge \frac{1}{c^n},$$

so that the sequence $\frac{1}{n} \log |v_{f^n,\Omega}(x)|, n \ge 1$ is uniformly bounded below, and the limit h' must be finite.

Conjecture 3.7 *Let M be a compact orientable manifold and suppose $f \in \text{Diff}^1(M, M)$. Then $J(f) \ne \emptyset$ if $h_{vol}^{ss}(f) > 0$.*

For the case of complex dynamics, this conjecture is true (see § 5.7). Here we transform Conjecture 1.6 into the following form:

Conjecture 3.8 *Assume that M is a compact orientable smooth m-dimensional manifold and $f \in \text{Diff}^1(M, M)$ with $h_{vol}^{ss}(f) > 0$ such that $F(f)$ has no components which are backward invariant. Then $\text{Res} J(f) \ne \emptyset$.*

If the condition $h_{vol}^{ss}(f) > 0$ in these problems is replaced by $h_{vol}^s(f) > 0$, its strong versions also are raised.

Let N be another orientable manifold of same dimension with M and let Ψ be a volume element on N. Take $f \in C^1(M, N)$ and define a function

$$v_{f,\Psi,\Omega} = f^*(\Psi)/\Omega.$$

Assume that M is compact and let d be a distance function on N. For $f, g \in C^1(M, N)$, define

$$\rho_1(f, g) = \max_{x \in M}\{d(f(x), g(x)) + |v_{f, \Psi, \Omega}(x) - v_{g, \Psi, \Omega}(x)|\}.$$

It is easy to see that ρ_1 is a metric on $C^1(M, N)$. The topology generated by the metric ρ_1 is called the *volume topology* of $C^1(M, N)$.

Definition 3.9 *A $f \in C^1(M, M)$ is volume stable if there exists a neighborhood $\mathcal{U}$ of f in the volume topology such that every mapping $g \in \mathcal{U}$ is topologically conjugate to f, and is strongly volume stable if it is volume stable and in addition for any $g \in \mathcal{U}$ one can choose a conjugating homeomorphism h_g in such a way that both h_g and h_g^{-1} uniformly converge to the identity as g converges to f in the volume topology.*

Note that if Λ is hyperbolic such that the Morse index of Λ is equal to the dimension of the base space (resp., $u_\Lambda = 0$), it is easy to prove that Λ is volume expanding (resp., contracting) hyperbolic. We suggest the following question:

Conjecture 3.9 *The set of C^r self-mappings $(r \geq 1)$ on M with volume hyperbolic Julia sets is dense in $C^r(M, M)$.*

If we replace Axiom A by the following
Axiom B:
i) $\Omega(f)$ is volume hyperbolic;
ii) the set $\mathrm{Per}(f)$ is dense in $\Omega(f)$,
we also suggest the following questions:

Problem 3.3 *Is the set of C^r self-mappings $(r \geq 1)$ satisfying Axiom B on M dense in $\mathrm{Diff}^r(M, M)$?*

Problem 3.4 *Is a diffeomorphism $f \in \mathrm{Diff}^1(M, M)$ strongly volume stable iff f satisfies Axiom B and the geometric strong transversality condition?*

3.6 Mappings with bounded distortion

Let V and V' be real vector spaces of dimensions m and m' with the inner products κ and κ', respectively. Let $\mathbf{A} : V \longrightarrow V'$ be a linear mapping, set $\mathrm{Im}(\mathbf{A}) = \mathbf{A}(V)$, $\mathrm{Ker}(\mathbf{A}) = \mathbf{A}^{-1}(0)$, and define

$$\|\mathbf{A}\| = \sup_{\|x\|_\kappa = 1} \|\mathbf{A}x\|_{\kappa'},$$

where

$$\|x\|_\kappa = \sqrt{\kappa(x, x)}.$$

There exists a unique linear mapping $\mathbf{A}^* : V' \longrightarrow V$ such that

$$\kappa'(\mathbf{A}x, x') = \kappa(x, \mathbf{A}^*x')$$

for any $x \in V$ and $x' \in V'$. This mapping $\mathbf{A}^*$ is called the *adjoint* of $\mathbf{A}$. Take a basis $e = (e_1, ..., e_m)$ in V and a basis $e' = (e'_1, ..., e'_{m'})$ in V'. Set

$$\kappa_{ij} = \kappa(e_i, e_j), \quad \kappa'_{ij} = \kappa'(e'_i, e'_j),$$

define

$$\mathbf{A}e_i = \sum_{j=1}^{m'} a_{ij} e'_j, \quad \mathbf{A}^* e'_i = \sum_{j=1}^{m} a^*_{ij} e_j,$$

and write $A = (a_{ij})$, $A^* = (a^*_{ij})$. Then

$$\kappa \, {}^t A^* = A \kappa',$$

where we also denote the matrixes (κ_{ij}) and (κ'_{ij}) by using κ and κ', respectively.

If $\mathbf{A} \neq 0$, then the subspaces $\text{Ker}(\mathbf{A})$ and $\text{Im}(\mathbf{A}^*)$ are completely orthogonal, and their direct sum concides with V. Further, there exist orthonormal systems of vectors $v_1, ..., v_l$ in V and $v'_1, ..., v'_l$ in V' and numbers $\lambda_i > 0$ $(i = 1, ..., l)$ such that $(v_1, ..., v_l)$ is a basis in $\text{Im}(\mathbf{A}^*)$, $(v'_1, ..., v'_l)$ is a basis in $\text{Im}(\mathbf{A})$, and

$$\mathbf{A}v_i = \lambda_i v'_i, \quad \mathbf{A}^* v'_i = \lambda_i v_i, \quad i = 1, ..., l. \tag{3.31}$$

Thus we have

$$\mathbf{A}^* \mathbf{A} v_i = \lambda_i^2 v_i, \quad \mathbf{A} \mathbf{A}^* v'_i = \lambda_i^2 v'_i, \quad i = 1, ..., l,$$

and all the remaining eigenvalues of the mappings $\mathbf{A}^* \mathbf{A}$ and $\mathbf{A} \mathbf{A}^*$ are equal to zero. The quantities $\lambda_1, ..., \lambda_l$ are called the *principal dilation coefficients* or *singular numbers* of $\mathbf{A}$. We will assume that

$$0 < \lambda_1 \leq \lambda_2 \leq \cdots \leq \lambda_l.$$

The vectors $v_1, ..., v_l$ are called *principal vectors* of $\mathbf{A}$. It is easy to prove that

$$\|\mathbf{A}\| = \lambda_l.$$

For more details, see [211].

Now assume that $\dim V = \dim W$. If $\det(A) \neq 0$, then $l = m$ and

$$\lambda_1 \lambda_2 \cdots \lambda_m = \sqrt{\det(AA^*)} = |\det(A)| \sqrt{\frac{\det(\kappa')}{\det(\kappa)}}$$

is independent of the choice of the bases e and e', and so does

$$K(\kappa, \kappa'; \mathbf{A}) = \frac{\lambda_m^m}{\lambda_1 \cdots \lambda_m} \geq 1.$$

Also define $K(\kappa, \kappa'; \mathbf{A}) = 1$ if $A = 0$, and $K(\kappa, \kappa'; \mathbf{A}) = \infty$ if $A \neq 0$ but $\det(A) = 0$. Thus if $V = V'$ with $\kappa = \kappa'$ and if $\det(A) \neq 0$, then

$$K(\mathbf{A}) = K(\kappa, \kappa; \mathbf{A}) = \frac{\|\mathbf{A}\|^m}{|\det(A)|}.$$

Let M and N be smooth manifolds of same dimension m with Riemannian metrics g and h, respectively, and take $f \in C^1(M, N)$. Define the *distortion coefficient* $K_{g,h}(f, S)$ of f in a subset $S \subset M$ by

$$K_{g,h}(f, S) = \sup_{x \in S} K(g(x), h(f(x)); (df)_x),$$

and abbreviate $K_{g,h}(f) = K_{g,h}(f, M)$. Let M and N be orientable and let Ω and Ψ be the associated volume elements of the metrics g and h, respectively. Define

$$v_{f,\Psi,\Omega} = f^*(\Psi)/\Omega.$$

Take compatible local coordinate systems $(U; u_i)$ and $(V; v_i)$ of M and N, respectively, such that $U \subset\subset M$, $f(U) \subset V$. Then

$$\Omega|_U = \sqrt{\det(g)}\,du_1 \wedge \cdots \wedge du_m, \quad \Psi|_V = \sqrt{\det(h)}\,dv_1 \wedge \cdots \wedge dv_m,$$

and hence

$$v_{f,\Psi,\Omega} = \det\left(\frac{df}{du}\right)\sqrt{\frac{\det(h \circ f)}{\det(g)}}.$$

If $v_{f,\Psi,\Omega}(x) \neq 0$, then we have

$$|v_{f,\Psi,\Omega}(x)| = \lambda_1(x)\lambda_2(x)\cdots\lambda_m(x),$$

where $\lambda_1(x), ..., \lambda_m(x)$ are the principal dilation coefficients of $(df)_x$. Therefore

$$K(g(x), h(f(x)); (df)_x) = \frac{\|(df)_x\|^m}{|v_{f,\Psi,\Omega}(x)|} \geq 1.$$

Note that $\overline{U}$ and $f(\overline{U})$ are compact. It follows from continuity that there exist positive constants a_0, a_1, b_0 and b_1 such that

$$a_0^2\|\xi\|^2 \leq \sum g_{ij}(x)\xi_i\xi_j \leq a_1^2\|\xi\|^2,$$

while

$$b_0^2\|\xi\|^2 \leq \sum h_{ij}(f(x))\xi_i\xi_j \leq a_1^2\|\xi\|^2$$

for $x \in U$, which implies

$$a_0^m \leq \sqrt{\det(g)} \leq a_1^m, \quad b_0^m \leq \sqrt{\det(h \circ f)} \leq b_1^m,$$

where $\|\xi\|$ is taked with respect to the Euclidean norm. This allows us to conclude that

$$\left(\frac{b_0}{a_1}\right)^m \left|\det\left(\frac{df}{du}\right)\right| \leq |v_{f,\Psi,\Omega}| \leq \left(\frac{b_1}{a_0}\right)^m \left|\det\left(\frac{df}{du}\right)\right|.$$

Set

$$\|(df)_x\|_E = \sup_{\|\xi\|=1} \left\|\frac{df}{du}(x)\xi\right\|.$$

It is easy to prove that

$$\frac{b_0}{a_1}\|(df)_x\|_E \leq \|(df)_x\| \leq \frac{b_1}{a_0}\|(df)_x\|_E.$$

Thus one obtains

$$\left(\frac{a_0 b_0}{a_1 b_1}\right)^m K((df)_x) \leq K(g(x), h(f(x)); (df)_x) \leq \left(\frac{a_1 b_1}{a_0 b_0}\right)^m K((df)_x).$$

Definition 3.10 *A mapping $f \in C^1(M,N)$ is said to be quasiconformal if f is a homeomorphism such that $v_{f,\Psi,\Omega}$ does not change sign in M, i.e., either $v_{f,\Psi,\Omega}(x) \geq 0$ in M or $v_{f,\Psi,\Omega}(x) \leq 0$ in M and such that $K_{g,h}(f) < \infty$, and to be conformal if f is quasiconformal with $K_{g,h}(f) = 1$.*

Now assume that $(M,g) = (N,h)$ and set

$$v_{f,\Omega} = v_{f,\Omega,\Omega}, \quad K_g(f) = K_{g,g}(f).$$

We always have $|v_{f^n,\Omega}(x)| \leq \|(df^n)_x\|^m$ for $x \in M$, and hence

$$h_\Omega^s(f;x) \leq mh^s(f;x).$$

Thus if $\Lambda \subset M$ is volume expanding hyperbolic, then $h^s(f;x) > 0$ for $x \in \Lambda$. If the distortion coefficient $K_g(f)$ is finite, then

$$\|(df^n)_x\|^m \leq K_g(f)^n |v_{f^n,\Omega}(x)|,$$

and hence

$$K_g(f^n) \leq K_g(f)^n.$$

Thus we have

$$\begin{aligned} h^s(f;x) &\leq \frac{1}{m}\left\{h_\Omega^s(f;x) + \limsup_{n\to\infty}\frac{1}{n}\log K_g(f^n)\right\} \\ &\leq \frac{1}{m}\{h_\Omega^s(f;x) + \log K_g(f)\}. \end{aligned}$$

In particular, if f is conformal $h_\Omega^s(f;x) = mh^s(f;x)$ for each $x \in M$.

Lemma 3.3 ([211]) *Suppose that $M \subset \mathbb{R}^m$ is an open set and take $f \in C^1(M,\mathbb{R}^m)$. Assume that the Jacobian $\det(df)_x$ of f does not change sign in M, and that*

$$K((df)_x) = \frac{\|(df)_x\|^m}{|\det(df)_x|} \leq K, \quad x \in M.$$

Assume that

$$\int_M \|(df)_x\|^m dx \leq I < \infty.$$

If $U \subset\subset M$ is relatively compact in M, then for any $x,y \in U$

$$\|f(x) - f(y)\| \leq \lambda\|x - y\|^{1/K},$$

where the constant λ depends only on U, the distance from U to the boundary of M, the constants I and K.

Corollary 3.3 *Suppose that $M \subset \mathbb{R}^m$ is an open set and take $f \in C^1(M, M)$. Assume that the Jacobian $\det(df)_x$ of f does not change sign in M, and that*

$$K((df^n)_x) = \frac{\|(df^n)_x\|^m}{|\det(df^n)_x|} \leq K, \quad x \in M, \ n \in \mathbb{Z}^+.$$

Assume that

$$\int_M \|(df^n)_x\|^m dx \leq I < \infty, \quad n \in \mathbb{Z}^+.$$

Then $J_{equ}(f) = \emptyset$.

3.7 Hausdorff measure and metric

Let (M, d) be a metric space and let α be a non-negative real number. Given $\varepsilon > 0$, for a subset A of M we set

$$\mu_{\alpha,\varepsilon}(A) = \inf\left\{\sum_i (\delta(A_i))^\alpha \ | \quad A \subset \bigcup_i A_i, \quad \delta(A_i) < \varepsilon\right\},$$

where $\delta(A_i)$ denotes the diameter of A_i, and the infimum is taken over all possible coverings of A. Then the α-dimensional *Hausdorff measure* $\mu_\alpha(A)$ of A is defined as follows.

$$\mu_\alpha(A) = \sup_{\varepsilon > 0} \mu_{\alpha,\varepsilon}(A).$$

If $\alpha = 0$, obviously μ_0 is the counting measure:

$$\mu_0(A) = \begin{cases} 0 & : \quad \text{if } A \text{ is empty,} \\ n & : \quad \text{if } A \text{ is a finite set of } n \text{ points,} \\ \infty & : \quad \text{if } A \text{ is an infinite set.} \end{cases}$$

If $\alpha < \beta$, then $\mu_\alpha(A) \geq \mu_\beta(A)$; in fact $\alpha < \beta$ and $\mu_\alpha(A) < \infty$ imply $\mu_\beta(A) = 0$ which follows from that

$$\mu_{\beta,\varepsilon}(A) \leq \sum_i (\delta(A_i))^\beta \leq \varepsilon^{\beta-\alpha} \sum_i (\delta(A_i))^\alpha \leq \varepsilon^{\beta-\alpha}(\mu_{\alpha,\varepsilon}(A) + 1),$$

as $\varepsilon \to 0$ since $\delta(A_i) < \varepsilon$, where we choose the covering of A such that $\sum_i (\delta(A_i))^\alpha \leq \mu_{\alpha,\varepsilon}(A) + 1$. Thus $\alpha < \beta$ and $\mu_\beta(A) > 0$ imply $\mu_\alpha(A) = \infty$. Obviously, we have

$$\mu_\alpha(A) \leq \mu_\alpha(B) \text{ if } A \subset B.$$

If A is compact, then $\mu_\alpha(A) = 0$ iff for each $\varepsilon > 0$, there exists a finite covering $\{A_1, ..., A_k\}$ of A such that

$$\sum_{i=1}^k (\delta(A_i))^\alpha < \varepsilon,$$

see Hurewicz and Wallman [128], Chapter VII.

The *Hausdorff dimension* of A is defined to be the supremum of all real numbers α such that $\mu_\alpha(A) > 0$. We denote the number by $\dim_H A$. Then we see that

$$
\begin{aligned}
\dim_H A &= \sup\{\alpha \mid \mu_\alpha(A) > 0\} = \sup\{\alpha \mid \mu_\alpha(A) = \infty\} \\
&= \inf\{\beta \mid \mu_\beta(A) = 0\} = \inf\{\beta \mid \mu_\beta(A) < \infty\}.
\end{aligned}
$$

Hence $\mu_\alpha(A) = 0$ or ∞ if $\alpha > \dim_H A$ or $\alpha < \dim_H A$, respectively, and $\alpha = \dim_H A$ if $0 < \mu_\alpha(A) < \infty$. Also the Hausdorff dimension has the property of monotonicity:

$$\dim_H A \leq \dim_H B \text{ if } A \subset B.$$

If M is a space of dimension m $(0 \leq m < \infty)$, then $\mu_m(M) > 0$ (see Hurewicz and Wallman [128], Chapter VII). Thus we have

$$\dim M \leq \dim_H M.$$

If A is a compact subset of $\mathbb{R}^m$ with the Euclidean metric, then the ball dimension $D(A)$ of A satisfies

$$0 \leq \dim_H A \leq D(A) \leq m,$$

see [26].

Let $\overline{\mathcal{B}}$ be the space of closed subsets of a metric space (M, d). As usual, write

$$d(A, B) = \inf_{x \in A} \inf_{y \in B} d(x, y)$$

for any subsets A and B of M. We introduce the following metric in $\overline{\mathcal{B}}$:

Definition 3.11 *The Hausdorff metric is defined on $\overline{\mathcal{B}}$ by setting*

$$d_H(A, B) = \sup_{x \in A} d(x, B) + \sup_{y \in B} d(y, A)$$

for any two closed sets $A, B \in \overline{\mathcal{B}}$.

It is easy to check that this is a distance function on $\overline{\mathcal{B}}$, i.e.,

$$
\begin{aligned}
d_H(A, B) &= 0 \Leftrightarrow A = B, d_H(A, B) = d_H(B, A), \\
d_H(A, B) &\leq d_H(A, C) + d_H(C, B)
\end{aligned}
$$

for $A, B, C \in \overline{\mathcal{B}}$. We refer to a limit with respect to the topology induced by the Hausdorff metric as a *Hausdorff limit* . If M is compact, the Hausdorff metric d_H defines a compact topology(see [134]).

Let $\mathcal{I}_f(M)$ be the collection of all closed invariant subsets of a dynamical system $\{f^t\}_{t \in \kappa}$ defined on M. Note that any homeomorphism of a compact metric space M induces natural homeomorphism of $\overline{\mathcal{B}}$. Thus if $\kappa = \mathbb{Z}$ or $\mathbb{R}$, then the dynamical system $\{f^t\}_{t \in \kappa}$ induces a DS on $\overline{\mathcal{B}}$ such that $\mathcal{I}_f(M)$ is just the set of fixed points of the induced DS. Hence $\mathcal{I}_f(M)$ is a closed set with respect to the topology induced by d_H.

Fix $f \in C(M, M)$ and let $\mathcal{F} = \{f_t \mid t \in \kappa\} \subset C(M, M)$ be a neighborhood of f in the compact-open topology. For any $t \in \kappa^{\mathbb{N}}$, the sequence $\mathcal{F}_t = \{f_{t_j}\}_{j=1}^{\infty} \subset \mathcal{F}$ will be called a *random perturbation* of f on $\mathcal{F}$. The family

$$\mathbf{L}_t(\mathcal{F}) = \{f_{nt} = f_{t_n} \circ f_{t_{n-1}} \circ \cdots \circ f_{t_1} \mid n = 1, 2, \ldots\}$$

is said to be a *random perturbation* of the DS $\{f^n\}$ on $\mathcal{F}$. If $f_{t_j} = g$ for all $j \geq 1$, we obtain a constant perturbation $\mathcal{F}_t = \{g\}$ of f with $\mathbf{L}_t(\mathcal{F}) = \{g^n\}$. An interesting question is to compare $J(\mathbf{L}_t(\mathcal{F}))$ and $J(f)$ when $\mathcal{F}_t \to f$, that is, $f_{t_j} \to f$ for all $j \geq 1$.

Conjecture 3.10 *If M is compact and if f is topological stable, there exists a neighborhood $\mathcal{F}$ of f in the C^0 topology such that every mapping $g \in \mathcal{F}$ is topologically conjugate to f. Then for any random perturbation $\mathcal{F}_t$ of f on $\mathcal{F}$, $d_H(J(\mathbf{L}_t(\mathcal{F})), J(f)) \to 0$ as $\mathcal{F}_t \to f$.*

Obviously, if $g \in \mathcal{F}$, there is $h \in \text{Hom}(M, M)$ such that $J(g) = h(J(f))$. Thus we have

$$d_H(J(g), J(f)) = d_H(h(J(f)), J(f)) \to 0$$

as $h \to id$, i.e., $g \to f$. Hence the conjecture is true for this special case. Under the condition of the conjecture, can we have $\dim_H J(\mathbf{L}_t(\mathcal{F})) \to \dim_H J(f)$?

Conjecture 3.11 *Let M be a compact metric space and let $\mathcal{F}_t$ be a random perturbation of f on $C(M, M)$. Then $d_H(J(\mathbf{L}_t(\mathcal{F})), J(f)) \to 0$ as $\mathcal{F}_t \to f$ if and only if each component of $F(f)$ is a basin of attraction.*

If M is the Riemann sphere, this conjecture is true (see [72], [204], [277]).

Given a dynamical system $\mathcal{F} = \{f^t\}_{t \in \kappa}$ defined on M. If $\kappa = \mathbb{Z}_+$, then

$$J(\mathcal{F}) = J(f) \quad (f = f^1).$$

If $\kappa = \mathbb{Z}$, then $\mathcal{F} = \{f^n\}_{n \in \mathbb{Z}_+} \cup \{(f^{-1})^n\}_{n \in \mathbb{Z}_+}$. Hence

$$J(\mathcal{F}) = J(f) \cup J(f^{-1}).$$

Generally we have

$$F(\mathcal{F}) \subset F(f^t), \quad J(f^t) \subset J(\mathcal{F}), \quad t \in \kappa.$$

Note that $J(f^0) = \emptyset$. We can define

$$b^+ = \inf_{t \geq 0}\{t \mid J(f^t) \neq \emptyset\}.$$

Problem 3.5 *If $f \in \text{Hom}(M, M)$, when $J(f) \cap J(f^{-1}) = \emptyset$? When $b^+ < \infty$?*

Take a positive number δ and set

$$T = T_{t,\delta} = (t - \delta, t + \delta) \cap \kappa.$$

For a point $s = (s_1, s_2, \ldots) \in T^{\mathbb{N}}$, then $\mathcal{F}_s = \{f^{s_n} \mid n = 1, 2, \ldots\}$ is a random perturbation of f^t on T. It is interesting to find some relation between $J(f^t)$ and $J(\mathbf{L}_s(\mathcal{F}))$. For example, define

$$p(t) = \limsup_{\delta \to 0} \sup_{s \in T^{\mathbb{N}}} d_H(J(f^t), J(\mathbf{L}_s(\mathcal{F}))).$$

Problem 3.6 *Can we have* $p(t) = 0$ *a.e.? What can we say about the function* $q(t) = d_H(J(f^t), J(\mathcal{F}))$ *?*

Let M be compact and let X be the vector field generating the flow $\mathcal{F} = \{f^t\}_{t \in \mathbb{R}}$. Let Ω be an open set in $\mathbb{R}^l$ which contains the origin. Let X_ε be a vector field depending on a parameter $\varepsilon \in \Omega$ such that $X_\varepsilon \to X_0 = X$ as $\varepsilon \to 0$ in the C^1-topology of vector fields. Let $\mathcal{F}_\varepsilon = \{f_\varepsilon^t\}_{t \in \mathbb{R}}$ be the flow generated by X_ε.

In perturbation theory, it is usual to take M as a smooth fiber bundle $\pi : M \longrightarrow B$. A vector field X on the bundle M is said to be *vertical* if it is tangent to every fiber. The functions on the base B of the fibering π determine first integrals of the equation $\dot{x} = X(x)$ on M. The vertical vector field X is said to be *unperturbed*. A *perturbed field* is defined to be a field $X_\varepsilon = X + \varepsilon X_1$ close to X which determines the following perturbed differential equation:

$$\dot{x} = X(x) + \varepsilon X_1(x).$$

Problem 3.7 *When we have* $d_H(J(\mathcal{F}_\varepsilon), J(\mathcal{F})) \to 0$ *and* $\dim_H J(\mathcal{F}_\varepsilon) \to \dim_H J(\mathcal{F})$ *as* $\varepsilon \to 0$? *We conjecture that it is true if* $\{f^t\}_{t \in \mathbb{R}}$ *is structurally stable.*

For a dynamical system $\mathcal{F} = \{f^t\}_{t \in \kappa}$ defined on M, define

$$J(\mathcal{F} : x) = J(f : x) = \{y \in J(\mathcal{F}) \mid x \notin \overline{O^+(y)}, \ x \in J(\mathcal{F})\}.$$

If the DS $\mathcal{F} = \{f^n\}_{n \in \mathbb{Z}_+}$ is given by an rational mapping f of degree at least 2 on the Riemannian sphere $\mathbb{P}^1$ such that $J(f)$ is hyperbolic with top Morse index (i.e., $J(f)$ is volume expanding hyperbolic), Hill and Velani [109] noticed that $J(\mathcal{F} : x)$ is a null set with respect to Hausdorff measure μ_α, where $\alpha = \dim_H J(f)$, and Abercrombie and Nair [6] proved

$$\dim_H J(f : x) = \dim_H J(f)$$

for each $x \in J(f)$. Generally, can we extend these results to higher dimensional spaces?

Let Λ be an invariant subset of a C^1-diffeomorphism f of a compact Riemannian manifold M. We denote by $\Sigma_f^*(\Lambda)$ the set of all f-invariant Borel probability ergodic measures on Λ. Consider the ordered Lyapunov exponent $\chi_1 \geq \cdots \geq \chi_m$ ($m = \dim M$) of some $\mu \in \Sigma_f^*(\Lambda)$ and define the *Lyapunov dimension* :

$$\dim_L \mu = \begin{cases} 0 & : \quad \text{if } \chi_1 < 0 \\ \sup_{0 \leq \alpha \leq m} \{\alpha \mid \sum_{i=1}^{[\alpha]} \chi_i + (\alpha - [\alpha])\chi_{[\alpha]+1} \geq 0\} & : \quad \text{if } \chi_1 \geq 0. \end{cases}$$

The following estimate of Hausdorff dimension obtained by Ledrappier [156] is very interesting.

$$\dim_H \Lambda \leq \sup_{\mu \in \Sigma_f^*(\Lambda)} \dim_L \mu. \tag{3.32}$$

Theorem 3.18 (cf. [52], [193]) *If* Λ *is a hyperbolic attractor of a diffeomorphism* f *and* $f|_\Lambda$ *is topologically transitive, then there exist a neighborhood* U *of* Λ *and a measure* μ *on* Λ

such that for any $\phi \in C(U, \mathbb{R})$ and for almost every point x (with respect to the Riemannian volume)

$$\lim_{n \to \infty} \frac{1}{n} \sum_{k=0}^{n-1} \phi \circ f^k(x) = \overline{\phi}.$$

Such μ are called *Sinai-Ruelle-Bowen measures*, which is invariant under f. Here we exhibit two related results:

Theorem 3.19 ([165]) Let Λ be a locally maximal hyperbolic set of a $C^{1+\alpha}$-diffeomorphism f of a compact Riemannian manifold M. For $\mu \in \Sigma_f^*(\Lambda)$, consider a set G_μ which consists of $x \in \Lambda$ satisfying

$$\lim_{n \to \infty} \frac{1}{n} \sum_{k=0}^{n-1} \phi \circ f^k(x) = \int_\Lambda \phi d\mu,$$

for every $\phi \in C(\Lambda, \mathbb{R})$. Suppose that $\dim M = 2$ and that $f|_\Lambda$ is topologically transitive. Then for any $\mu \in \Sigma_f^*(\Lambda)$ and for any $p \in \Lambda$

$$\dim_H G_\mu = \dim_H(G_\mu \cap W_\varepsilon^s(p)) + \dim_H(G_\mu \cap W_\varepsilon^u(p)),$$

$$\dim_H(G_\mu \cap W_\varepsilon^s(p)) = \frac{h(f, \mu)}{\chi_1},$$

$$\dim_H(G_\mu \cap W_\varepsilon^u(p)) = \frac{h(f, \mu)}{|\chi_2|},$$

where the Lyapunov exponents satisfy $\chi_1 > 0 > \chi_2$.

A similar formula is proved by Lai-Sang Young [281].

Theorem 3.20 ([165]) Let Λ be a locally maximal hyperbolic set of a $C^{1+\alpha}$-diffeomorphism f of a compact Riemannian manifold M. Suppose that $\dim M = 2$ and that $f|_\Lambda$ is topologically transitive. Then for any $p \in \Lambda$

$$\dim_H \Lambda = \dim_H(\Lambda \cap W_\varepsilon^s(p)) + \dim_H(\Lambda \cap W_\varepsilon^u(p)),$$

$$\dim_H(\Lambda \cap W_\varepsilon^s(p)) = \sup_{\mu \in \Sigma_f^*(\Lambda)} \dim_H(G_\mu \cap W_\varepsilon^s(p)),$$

$$\dim_H(\Lambda \cap W_\varepsilon^u(p)) = \sup_{\mu \in \Sigma_f^*(\Lambda)} \dim_H(G_\mu \cap W_\varepsilon^u(p)).$$

Further let Λ be a hyperbolic attractor. Then $W_\varepsilon^u(p) \subset \Lambda$ for any $p \in \Lambda$; thus

$$\dim_H(\Lambda \cap W_\varepsilon^u(p)) = 1,$$

and consequently

$$\dim_H \Lambda = 1 + \sup_{\mu \in \Sigma_f^*(\Lambda)} \dim_H(G_\mu \cap W_\varepsilon^s(p)).$$

If μ is a Sinai-Ruelle-Bowen measure on Λ, then

$$\dim_H(G_\mu \cap W_\varepsilon^s(p)) = \frac{\chi_1}{|\chi_2|},$$

which implies

$$\dim_H G_\mu = 1 + \frac{\chi_1}{|\chi_2|}.$$

Chapter 4

Some topics in dynamics

In this chapter, we will discuss some topics which are related to the Fatou-Julia type theory, say, Hamiltonian systems, linearization, L_p-normality, and so on.

4.1 Properties of Hamiltonian systems

Let (M, φ) be a Hamiltonian manifold of dimension $2m$, and let $H : M \longrightarrow \mathbb{R}$ be a smooth function. Then the vector field X_H defined by

$$\varphi(X_H, \ ) = X_H \llcorner \varphi = dH \tag{4.1}$$

is called the *Hamiltonian vector field* associated with H or the *symplectic gradient* of H. The flow $\{f^t\}$ defined by the *Hamiltonian equations*

$$\dot{x} = X_H(x) \tag{4.2}$$

is called the *Hamiltonian flow* of H, and for simplicity, H is said to be a *Hamiltonian*.

In terms of a local coordinate system $\{x_k, y_k\}$, the usual Hamiltonian equations are

$$\dot{x}_k = \frac{\partial H}{\partial y_k}, \quad \dot{y}_k = -\frac{\partial H}{\partial x_k}, \quad k = 1, ..., m. \tag{4.3}$$

Let us show that (4.1) and (4.2) are indeed a formulation of (4.3). To see this, we need to check that

$$X_H = \sum_{k=1}^{m} \frac{\partial H}{\partial y_k} \frac{\partial}{\partial x_k} - \sum_{k=1}^{m} \frac{\partial H}{\partial x_k} \frac{\partial}{\partial y_k}$$

satisfies (4.1) in Darboux coordinates $\{x_k, y_k\}$. Write

$$X_H = \sum_{k=1}^{m} \xi_k \frac{\partial}{\partial x_k} + \sum_{k=1}^{m} \eta_k \frac{\partial}{\partial y_k}.$$

Then

$$X_H \llcorner \varphi \ = \ X_H \llcorner (\sum_{k=1}^{m} dx_k \wedge dy_k)$$

99

$$
\begin{aligned}
&= \sum_{k=1}^{m} \langle X_H, dx_k \rangle dy_k - \sum_{k=1}^{m} \langle X_H, dy_k \rangle dx_k \\
&= \sum_{k=1}^{m} \xi_k dy_k - \sum_{k=1}^{m} \eta_k dx_k = dH \\
&= \sum_{k=1}^{m} \frac{\partial H}{\partial x_k} dx_k + \sum_{k=1}^{m} \frac{\partial H}{\partial y_k} dy_k,
\end{aligned}
$$

that is,

$$
\xi_k = \frac{\partial H}{\partial y_k}, \quad \eta_k = -\frac{\partial H}{\partial x_k}.
$$

It is well known that the form of the equations (4.3) is preserved under the group of canonical coordinate transformations, that is, under coordinate transformations

$$
x = x(\xi, \eta), \quad y = y(\xi, \eta)
$$

where x, y are vector functions of $\xi = (\xi_1, ..., \xi_m)$, $\eta = (\eta_1, ..., \eta_m)$ for which the identity

$$
\sum_{k=1}^{m} dx_k \wedge dy_k = \sum_{k=1}^{m} d\xi_k \wedge d\eta_k
$$

holds.

Remark. Let X be a vector field on a manifold M and let ω be an r-form on M. The notation $X \angle \omega$ also is called the *contraction* of ω with X. It satisfy

$$
\begin{aligned}
X \angle (\omega \wedge \eta) &= (X \angle \omega) \wedge \eta + (-1)^r \omega \wedge (X \angle \eta), && (4.4) \\
L_X \omega &= X \angle d\omega + d(X \angle \omega). && (4.5)
\end{aligned}
$$

Let $\alpha, \beta : M \longrightarrow \mathbb{R}$ be smooth functions. Then the *Poisson bracket* of α and β is defined by

$$
\{\alpha, \beta\} = \langle X_\alpha \wedge X_\beta, \varphi \rangle, \tag{4.6}
$$

which is expressed by

$$
\{\alpha, \beta\} = \sum_{k=1}^{m} \left(\frac{\partial \alpha}{\partial x_k} \frac{\partial \beta}{\partial y_k} - \frac{\partial \alpha}{\partial y_k} \frac{\partial \beta}{\partial x_k} \right)
$$

in Darboux coordinates $\{x_k, y_k\}$. By the definition, we have

$$
\{\alpha, \beta\} = -\{\beta, \alpha\},
$$

$$
\{\alpha, \beta\} = \langle X_\beta, X_\alpha \angle \varphi \rangle = \langle X_\beta, d\alpha \rangle = X_\beta \alpha = L_{X_\beta} \alpha.
$$

The functions α and β are said to be *in involution* if their Poisson bracket vanishes. If $\mathcal{F} = \{f^t\}$ is the Hamiltonian flow for H, a smooth function α on M is said to be an *integral* of $\mathcal{F}$ iff

$$
\frac{d\alpha(f^t)}{dt} = 0.
$$

Note that

$$\frac{d\alpha(f^t)}{dt} = f^{t*}X_H\alpha = f^{t*}\{\alpha, H\}.$$

Then α is an integral of $\mathcal{F}$ iff $\{\alpha, H\} = 0$. The set $\mathcal{H}$ of integrals of $\mathcal{F}$ forms a vector space on $\mathbb{R}$.

Theorem 4.1 *Let (M, φ) be a Hamiltonian manifold of dimension $2m$, $H : M \longrightarrow \mathbb{R}$ a smooth function, and $\mathcal{F} = \{f^t\}$ the Hamiltonian flow for H. Then*
  1) f^t is symplectic and hence volume preserving for all $t \in \mathbb{R}$ (Liouville);
  2) $H \circ f^t$ does not depend on t;
  3) α is an integral of $\mathcal{F}$ if $H \circ g^t$ does not depend on t, where $\{g^t\}$ is the one-parameter family of symplectic transformations generated by α (Noether);
  4) $C^\infty(M, \mathbb{R})$ with the Poisson bracket is a Lie algebra;
  5) $X_{\{\alpha,\beta\}} = -[X_\alpha, X_\beta]$, and $\{\alpha, \beta\}$ is an integral of $\mathcal{F}$ if α and β are (Poisson).

Proof. 1) follows from

$$\begin{aligned}\frac{d}{dt}f^{t*}\varphi &= f^{t*}L_{X_H}\varphi = f^{t*}(d(X_H\angle\varphi) + X_H\angle d\varphi) \\ &= f^{t*}(d(X_H\angle\varphi)) = f^{t*}(ddH) = 0,\end{aligned}$$

and 2) follows from

$$\frac{dH(f^t)}{dt} = f^{t*}X_H H = f^{t*}\{H, H\} = 0.$$

The hypothesis in 3) is that H is an integral for the flow $\{g^t\}$ of α, that is, $\{\alpha, H\} = 0$, and so conversely α is an integral for the flow $\{f^t\}$ of H. It is trivial to prove that the Poisson bracket satisfies the Jacobi identity. 5) follows from 4) easily. $\square$

Theorem 4.2 (Liouville-Arnold Theorem [134]) *Suppose M is a Hamiltonian manifold of dimension $2m$, $H_1, ..., H_m \in C^\infty(M, \mathbb{R})$, $\{H_i, H_j\} = 0 (i, j = 1, ..., m)$, and $z \in \mathbb{R}^m$ is such that the differentials dH_i are pointwise linearly independent on*

$$M_z = \{x \in M \mid H_i(x) = z_i, i = 1, ..., m\}.$$

Then
  1) M_z is a smooth Lagrangian submanifold invariant under the Hamiltonian flows $\{f_i^t\}$ of H_i;
  2) M_z is diffeomorphic to the m-torus $\mathbb{T}^m$ if M_z is compact and connected;
  3) $f_i^t|_{M_z}$ are conjugate to a linear flow via this diffeomorphism.

Let μ be a Borel measure on M. Take $\alpha \in \mathcal{H} \cap C_0(M)$. Then $\alpha \circ f^t = \alpha$ for all $t \in \mathbb{R}$. Thus we have

$$G_{\mu,\alpha}(\mathcal{F}) = \left\{x \in M \mid \lim_{t\to\infty}\frac{1}{t}\int_0^t \alpha(f^t(x))dt = \overline{\alpha}\right\} = \{x \in M \mid \alpha(x) = \overline{\alpha}\},$$

and hence

$$G_{\mu,\alpha}(\mathcal{F})^c = M - G_{\mu,\alpha}(\mathcal{F}) = \{x \in M \mid \alpha(x) \neq \overline{\alpha}\},$$

where

$$\overline{\alpha} = \frac{1}{\mu(M)} \int_M \alpha d\mu.$$

If $x \in \operatorname{supp}\mu \cap G_{\mu,\alpha}(\mathcal{F})^c$, w.l.o.g, letting $\alpha(x) > \overline{\alpha}$, by continuity of α, there is a neighborhood U_0 of x such that $\alpha(y) > \overline{\alpha}$ for all $y \in U_0$. Thus for any neighborhood U of x with $U \subset U_0$, we have $U \cap G_{\mu,\alpha}(\mathcal{F}) = \emptyset$. Note that $x \in \operatorname{supp}\mu$, and hence $\mu(U) > 0$. Therefore $x \notin F_{erg}(\mathcal{F}) \cup F_\mu(\mathcal{F})$, i.e., $x \in J_{erg}(\mathcal{F}) \cap J_\mu(\mathcal{F})$. Thus we obtain $\operatorname{supp}\mu \cap G_{\mu,\alpha}(\mathcal{F})^c \subset J_{erg}(\mathcal{F}) \cap J_\mu(\mathcal{F})$ for each $\alpha \in \mathcal{H} \cap C_0(M)$. If $J_{erg}(\mathcal{F}) \cap J_\mu(\mathcal{F}) = \emptyset$, then $\operatorname{supp}\mu \cap G_{\mu,\alpha}(\mathcal{F})^c = \emptyset$ for each $\alpha \in \mathcal{H} \cap C_0(M)$. Thus we have

$$\operatorname{supp}\mu \subset \bigcap_{\alpha \in \mathcal{H} \cap C_0(M)} G_{\mu,\alpha}(\mathcal{F}). \tag{4.7}$$

Note that $G_{\mu,\alpha}(f^t) = G_{\mu,\alpha}(\mathcal{F})$ for each $\alpha \in \mathcal{H} \cap C_0(M)$. Similarly, we can prove that the conclusion holds if $J_{erg}(f^t) \cap J_\mu(f^t) = \emptyset$. However, $J_{erg}(f^t) = \emptyset$ iff f^t is ergodic. If M is compact, $J_\mu(f^t) = \emptyset$ iff f^t is uniquely ergodic. Thus if μ is a ergodic measure, then (4.7) must be held. Also we can prove that (4.7) holds if, for some p with $0 \leq p \leq \infty$, $J_\mu^p(\mathcal{F}) = \emptyset$ or $J_\mu^p(f^t) = \emptyset$ for some t.

Let (M, g) be a Riemannian manifold. Consider the symplectic manifold $(T^*(M), \varphi^*)$ with $\varphi^* = -d\theta$. As a Hamiltonian function take

$$H^*(X^*) = \frac{1}{2}g^*(X^*, X^*).$$

Here g^* is the scalar product on $T^*(M)_p$ induced from the scalar product g on $T(M)_p$. The flow of the system $(T^*(M), \varphi^*, H^*)$ is called the *co-geodesic flow*. From the Riemannian metric g on M we get a bundle isomorphism:

$$
\begin{array}{ccc}
T(M) & \xrightarrow{A_g} & T^*(M) \\
\downarrow \pi & & \downarrow \pi \\
M & \xrightarrow{id} & M
\end{array}
\qquad
\begin{array}{ccc}
(x_i, \dot{x}_i) & \longmapsto & (x_i, \sum_k g_{ik}\dot{x}_i) \\
\downarrow & & \downarrow \\
(x_i) & \longmapsto & (x_i) \; .
\end{array}
$$

Here $A_g|_{T(M)_p}$ is given by $X \mapsto g(X, \ ) = g_X$. Define φ by

$$\varphi(X, Y) = \varphi^*((dA_g)X, (dA_g)Y).$$

The pullback H of H^* yields the kinetic energy

$$H : T(M) \longrightarrow \mathbb{R}; \quad X \mapsto \frac{1}{2}g(X, X) = \frac{1}{2}g^*(A_g X, A_g X).$$

Definition 4.1 *The geodesic (Hamiltonian) system (associated with M) is the Hamiltonian system $(T(M), \varphi, H)$. Its flow $f^t : T(M) \longrightarrow T(M)$ is called a geodesic flow.*

Proposition 4.1 *Let M be a complete Riemannian manifold. Then*

1) The flow lines $f^t X$ of the geodesic system $(T(M), \varphi, H)$ are defined for all $t \in \mathbb{R}$. The projection $\pi(f^t X), t \in \mathbb{R}$, of such a flow line is the geodesic $\gamma(t)$ determined by $f^t X = \dot{\gamma}(t)$.

2) The non-constant periodic flow lines $f^t X$ with $\|X\| \neq 0$ are in $1:1$ correspondence with the non-constant closed geodesics on M.

A proof can be found in [139].

We consider a smooth fiber bundle $\pi : M \longrightarrow B$ and describe the averaging method. We shall assume that the fibers of the bundle M are m-dimensional tori. In the neighborhood of every point of the base B, the fibering is assumed to be a direct product. We restrict ourselves to such a neighborhood and shall describe a point of the fiber space M by a pair (I, φ), where I is a point of the base and φ is a point of an m-dimensional torus $\mathbb{T}^m$ which is given by a collection of m angular coordinates $(\varphi_1, ..., \varphi_m) \bmod 2\pi$. We fix a coordinate system (I, φ). The *unperturbed equation* of the averaging method is the equation

$$\dot{\varphi} = \omega(I), \quad \dot{I} = 0,$$

where ω is a vertical vector field given by a frequency vector $(\omega_1(I), ..., \omega_m(I))$ depending on the point I of the base. The *perturbed equation* of the averaging method is the equation

$$\dot{\varphi} = \omega(I) + \varepsilon f(I, \varphi, \varepsilon), \quad \dot{I} = \varepsilon g(I, \varphi, \varepsilon),$$

where f and g are 2π-periodic in φ and $\varepsilon < 1$ is a small parameter. The angular coordinates φ_i are called *fast variables* and the coordinates I_j on the base are called *slow variables*.

Now we discuss a Hamiltonian system and assume that in the unperturbed Hamiltonian system, action-angle variables are introduced, i.e., canonical conjugate variables $(I_1, ..., I_m; \varphi_1, ..., \varphi_m \bmod 2\pi)$ such that the unperturbed Hamiltonian H_0 depends only on the action variables I, where the coordinates (I, φ) are said to be canonically conjugate if these are the Darboux coordinates. Hamilton's canonical equations take the form

$$\dot{\varphi} = \omega(I), \quad \dot{I} = 0,$$

where the frequency vector $\omega(I)$ is equal to $\frac{\partial H_0}{\partial I}$. The perturbed system is given by the Hamiltonian $H = H_0 + \varepsilon H_1(I, \varphi, \varepsilon)$, where the function H_1 has period 2π with respect to the angular variables φ. Consequently, the equation of the perturbed motion have the form

$$\dot{\varphi} = \omega(I) + \varepsilon \frac{\partial H_1}{\partial I}, \quad \dot{I} = -\varepsilon \frac{\partial H_1}{\partial \varphi}.$$

The following is the main theorem of the KAM-theory (the Kolmogorov-Arnold-Moser theory):

Theorem 4.3 ([20],[178],[230]) *Consider an integrable Hamiltonian system with m degrees of freedom. Let $\mathcal{O} = U \times \mathbb{T}^m$ be the open set in the phase space M, where U is the neighborhood in B. Assume that $H = H_0(I) + \varepsilon H_1(I, \varphi, \varepsilon)$ is analytic in the domain $\mathcal{O}' = U' \times S$, where U' is a complex neighborhood of U in $\mathbb{C}^m$ and S is a complex neighborhood of $\mathbb{T}^m$ in $\mathbb{C}^m$. Let $\det\left(\frac{\partial^2 H}{\partial I_i \partial I_j}\right) \neq 0$ in U. Then for all sufficiently small $\varepsilon > 0$, one can* *find a subset $K \subset \mathcal{O}$ such that mes $K \to$ mes $\mathcal{O}$ as $\varepsilon \to 0$ and there exists a measurable partition of K into invariant m-dimensional tori. The Hamiltonian dynamical system with the Hamiltonian H reduces on every such torus to the quasi-periodic motion with pure point spectrum and m basic frequencies.*

This theorem shows that a small perturbation of an integrable system is nonergodic and has an invariant subset of positive measure. Ergodic components belonging to this subset have pure point spectrum. In particular it disproved completely the hypothesis which appeared often in physical works, that a generic multi-dimensional nonlinear Hamiltonian system is ergodic.

Let (M, φ) be an almost Hamiltonian manifold. Then M is even-dimensional and φ^m is a volume form ($m = \frac{1}{2} \dim M$). In particular, M is orientable. Thus we can define a function v_{f,φ^p} by

$$f^*(\varphi^p) \wedge \varphi^{m-p} = v_{f,\varphi^p} \varphi^m \ (1 \leq p \leq m).$$

Definition 4.2 ([126]) *A compact forward invariant set $\Lambda \subset M$ is said to be a p-type contracting (resp., expanding) hyperbolic set of a cascade $\{f^n\}_{n \in \mathbb{Z}_+}$ (or $f \in C^1(M,M)$) if, for each point $z \in \Lambda$*

$$|v_{f^n,\varphi^p}(z)| \leq ae^{-cn} \ (resp., \ \geq \frac{1}{a}e^{cn}),$$

where a, c are positive constants that are independent of z and n, and is called p-type hyperbolic if each forward invariant component of Λ is either p-type contracting or expanding hyperbolic.

Also we can define functions by

$$h^s_{\varphi^p}(f; x) = \limsup_{n \to \infty} \frac{1}{n} \log |v_{f^n,\varphi^p}(x)|,$$

$$h^u_{\varphi^p}(f; x) = \liminf_{n \to \infty} \frac{1}{n} \log |v_{f^n,\varphi^p}(x)|.$$

If Λ is p-type contracting (or expanding) hyperbolic, then $h^s_{\varphi^p}(f; x) \leq -c < 0$ (or $h^u_{\varphi^p}(f; x) \geq c > 0$) on Λ, that is, $h^s_{\varphi^p}(f; x)$ is bounded above by a negative constant (resp., $h^u_{\varphi^p}(f; x)$ is bounded low by a positive constant) on Λ. If M is compact, these functions don't depend on the choice of the forms. So we also write

$$h^s_p(f; x) = h^s_{\varphi^p}(f; x), \quad h^u_p(f; x) = h^u_{\varphi^p}(f; x).$$

Further, define

$$h^{ss}_p(f) = \limsup_{n \to \infty} \frac{1}{n} \sup_{x \in M} \log |v_{f^n,\varphi^p}|,$$

$$h^s_p(f, \mu) = \int_M h^s_p(f; x) d\mu(x)$$

for some $\mu \in \Sigma(M)$. Then $h^s_p(f, \mu) \leq h^{ss}_p(f)$. These notions can be extend to the case of flows similarly.

A symplectic diffeomorphism $f : M \longrightarrow M$ is said to be a *Hamiltonian mapping* if f can be interpolated by a Hamiltonian vector field X_H associated with some Hamiltonian H, that is, the Hamiltonian flow $\{f^t\}$ of some Hamiltonian H exists such that $f = f^1$. In the sixties V. I. Arnold conjectured that the number of fixed points of a Hamiltonian mapping f on a compact Hamiltonian manifold M can be estimated in terms of the topology of M

$$\#\mathrm{Fix}(f) \geq \sum_{r \geq 0} b_r,$$

provided that the fixed points are all nondegenerate [19], [20], where b_r is the r-th Betti number. In the case that the integral of φ vanishes over $\pi_2(M)$ this conjecture has been proved by A. Floer [80], [81] and H. Hofer [113]. As for the history of this conjecture which originated in old questions of celestial mechanics related to the Poincaré-Birkhoff fixed point theorem we refer to [283]. Here we suggest the following question:

Conjecture 4.1 *If $f : M \longrightarrow M$ is a Hamiltonian mapping, then $J(f)$ is 1-type hyperbolic.*

4.2 Hamiltonian systems near an equilibrium solution

Let H be a smooth real valued function on a manifold M and fix a critical point $p \in M$ of H. If we choose a local coordinate system $(u_1, ..., u_n)$ $(n = \dim M)$ in a neighborhood U of p, then p is a critical point of H if and only if

$$\frac{\partial H}{\partial u_1}(p) = \cdots = \frac{\partial H}{\partial u_n}(p) = 0.$$

A critical point p is called *non-degenerate* if and only if the matrix

$$\mathcal{J} = \mathcal{J}_H = \left(\frac{\partial^2 H}{\partial u_i \partial u_j}(p) \right)$$

is non-singular. It can be checked directly that non-degeneracy does not depend on the coordinate system. This will follow also from the following intrinsic definition.

Let ∇ be an affine connection on M. By the definition, we easily obtain

$$\nabla H = dH, \quad \nabla^2 H(X, Y) = YXH - (\nabla_Y X)H,$$

for $X, Y \in \Gamma(T(M))$. If $X, Y \in T(M)_p$, then X and Y have extensions $\tilde{X}$ and $\tilde{Y}$ to vector fields. We know that

$$\nabla^2 H(X, Y) = \tilde{Y}_p(\tilde{X}H) - ((\nabla_{\tilde{Y}}\tilde{X})H)(p)$$

is well-defined. Then

$$\nabla^2 H(X, Y) = \tilde{Y}_p(\tilde{X}H),$$

since $((\nabla_{\tilde{Y}}\tilde{X})H)(p) = 0$ because p is a critical point of H. Also we have

$$\nabla^2 H(Y, X) - \nabla^2 H(X, Y) = [\tilde{X}, \tilde{Y}]_p H = 0,$$

i.e., $\nabla^2 H$ is a symmetric bilinear function on $T(M)_p$, called the *Hessian* of H at p. If $(u_1, ..., u_n)$ is a local coordinate system, and

$$X = \sum_{i=1}^n a_i \frac{\partial}{\partial u_i}\big|_p, \quad Y = \sum_{i=1}^n b_i \frac{\partial}{\partial u_i}\big|_p,$$

we can take $\tilde{X} = \sum_{i=1}^n a_i \frac{\partial}{\partial u_i}$, where a_i now denotes a constant function. Then

$$\nabla^2 H(X, Y) = Y(\tilde{X}H) = \sum_{i,j=1}^n a_i b_j \frac{\partial^2 H}{\partial u_i \partial u_j}(p),$$

so that the matrix $\mathcal{J}$ represents the bilinear function $\nabla^2 H$ with respect to the basis $\frac{\partial}{\partial u_1}\big|_p, ..., \frac{\partial}{\partial u_n}\big|_p$. Thus the point p is a non-degenerate critical point of H if and only if $\nabla^2 H$ on $T(M)_p$ has nullity equal to 0. The index of the bilinear function $\nabla^2 H$ on $T(M)_p$ will be referred to the *index* of H at p, denoted by $\iota_H(p)$.

Now we calculate X_H for a 2-form φ and a local coordinate system $(u_1, ..., u_n)$ on M. Write

$$\varphi = \sum_{i,j=1}^n \varphi_{ij} du_i \wedge du_j, \quad X_H = \sum_{i=1}^n \xi_i \frac{\partial}{\partial u_i}.$$

Then we have

$$
\begin{aligned}
X_H \lrcorner \varphi &= \sum_{i,j=1}^n \varphi_{ij}\langle X_H, du_i\rangle du_j - \sum_{i,j=1}^n \varphi_{ij}\langle X_H, du_j\rangle du_i \\
&= \sum_{i,j=1}^n \varphi_{ij}\xi_i du_j - \sum_{i,j=1}^n \varphi_{ij}\xi_j du_i \\
&= 2\sum_{j=1}^n \left(\sum_{i=1}^n \varphi_{ij}\xi_i\right) du_j \\
&= dH = \sum_{j=1}^n \frac{\partial H}{\partial u_j} du_j.
\end{aligned}
$$

Hence

$$\frac{\partial H}{\partial u_j} = 2\sum_{i=1}^n \varphi_{ij}\xi_i, \quad j = 1, ..., n,$$

or

$$^t\left(\frac{dH}{du}\right) = 2\Phi\xi, \tag{4.8}$$

where

$$\xi = {}^t(\xi_1, ..., \xi_n), \quad \Phi = (\varphi_{ij}).$$

Now assume that (M, φ) is a Hamitonian manifold. Then p is a zero of X_H, i.e., $\xi(p) = 0$, if and only if p is a critical point of H. If so, by differentiating (4.8) we obtain

$$\mathcal{J} = \left(\frac{d}{du}{}^t\left(\frac{dH}{du}\right)\right)(p) = 2\Phi(p)\mathbf{A},$$

where $\mathbf{A} = \frac{d\xi}{du}(p)$ is just the Hessian of the vector field X_H at p. Since

$$\mathbf{A} = \frac{1}{2}\Phi(p)^{-1}\mathcal{J},$$

then p is a non-degenerate critical point of H if and only if p is a non-degenerate zero point of X_H. Thus if p is non-degenerate, we have

$$\iota_{X_H}(p) = \operatorname{sign} \det \mathbf{A} = \operatorname{sign} \det \Phi(p) \cdot \operatorname{sign} \det \mathcal{J}.$$

If u is the local coordinate system in Darboux theorem, we have

$$\Phi(p) = \frac{1}{2}\begin{pmatrix} 0 & I_m \\ -I_m & 0 \end{pmatrix},$$

where $m = \frac{1}{2}n$, and hence

$$\operatorname{sign} \det \Phi(p) = (-1)^{m^2+m} = 1.$$

To calculate $\operatorname{sign} \det \mathcal{J}$, we will use the lemma of Morse:

Lemma 4.1 (Morse lemma, cf.[173]) *Let H be a smooth real valued function on a manifold M. Let p be a non-degenerate critical point for H. Then there is a local coordinate system $(u_1, ..., u_n)$ in a neighborhood U of p with $u_i(p) = 0$ for all i and such that the identity*

$$H = H(p) - u_1^2 - \cdots - u_\iota^2 + u_{\iota+1}^2 + \cdots + u_n^2$$

holds throughout U, where $\iota = \iota_H(p)$ is the index of H at p.

By using the local coordinate system $(u_1, ..., u_n)$ in Morse lemma, we have

$$\frac{\partial^2 H}{\partial u_i \partial u_j}(p) = \begin{cases} -2 &: \quad \text{if } i = j \leq \iota \\ 2 &: \quad \text{if } i = j > \iota \\ 0 &: \quad \text{otherwise} \end{cases}$$

which implies

$$\operatorname{sign} \det \mathcal{J} = (-1)^\iota.$$

Therefore

$$\iota_{X_H}(p) = (-1)^\iota.$$

Thus if $\{f^t\}$ is the Hamiltonian flow of H, we also have

$$\iota_{f^t}(p) = \iota_{X_H}(p) = (-1)^\iota.$$

By Hopf index theorem, we obtain

Theorem 4.4 *Let (M, φ) be a compact Hamitonian manifold of dimension $2m$. Let H be a smooth real valued function on M with isolated non-degenerate critical points. Then*

$$\sum_{p \in C_H} (-1)^{\iota_H(p)} = \chi(M).$$

In fact, the Hamitonian condition in this theorem is not necessary. By considering the gradient vector field grad(H), in a similar fashion we can prove the following result:

Theorem 4.5 *Let H be a smooth real valued function on a compact Riemannian manifold M with isolated non-degenerate critical points. Then*

$$\sum_{p \in C_H} (-1)^{\iota_H(p)} = \chi(M).$$

Definition 4.3 *A critical point p of H will be called hyperbolic for φ if p is a hyperbolic zero point of X_H.*

Thus a hyperbolic critical point p of H is isolated, non-degenerate, and $\mathbf{A} = \frac{1}{2}\Phi(p)^{-1}\mathcal{J}$ has no eigenvalues of pure imaginaries. Here we describe dynamical properties near a hyperbolic critical point p of H for φ on a compact Hamitonian manifold M. Let $\mathcal{F} = \{f^t\}$ be the Hamiltonian flow of H. According to the discussion in §A.3, we have

$$(df^t)_p = \exp(\mathbf{A}t).$$

Note that $\exp(\mathbf{A}t)$ has an eigenvalue $\exp(\lambda t)$ if λ is an eigenvalue of $\mathbf{A}$. Thus p is a hyperbolic fixed point of f^t for all $t \in \mathbb{R}$. By the stable manifold theorem, then there exist immersions b^s, b^u of class C^∞:

$$b^s : \mathbb{R}^l \longrightarrow M, \quad b^s(0) = p, \quad b^s(\mathbb{R}^l) = W^s(p),$$

$$b^u : \mathbb{R}^{2m-l} \longrightarrow M, \quad b^u(0) = p, \quad b^u(\mathbb{R}^{2m-l}) = W^u(p),$$

where $l = \dim E_p^s$. Thus by Corollary 1.4, we obtain

$$p \in J(f^t) \ (l < 2m),$$

for all $t \in \mathbb{R}$. If $l = 2m$ and under some conditions, we can prove $W^s(p) = \text{Att}(p) \subset F(f^t)$ (see § 6.6).

Now we shall assume that the eigenvalues of the matrix $\mathbf{A}$ are purely imaginary and distinct, and shall discuss the system (4.3) on $M = \mathbb{R}^{2m}$. Without loss of generality we may assume $p = 0$. Assuming H to be a real analytic function we represent H as a power series without constant and linear terms:

$$H(x,y) = \sum_{n=2}^{\infty} P_n(x,y),$$

where $P_n(x,y)$ is a homogeneous polynomial of degree n in $x = (x_1, ..., x_m)$ and $y = (y_1, ..., y_m)$. Note that the matrix $\mathbf{A}$ has the property that if λ is an eigenvalue, so are $\overline{\lambda}, -\lambda, -\overline{\lambda}$ eigenvalues. It is well known that one can find a linear canonical coordinate transformation such that P_2 takes the form

$$P_2 = \sum_{k=1}^{m} \frac{\alpha_k}{2} \left(x_k^2 + y_k^2 \right). \tag{4.9}$$

The following Birkhoff's theorem [45] (also see [178]) describes solutions near the equilibrium.

Theorem 4.6 *If the $\alpha_1, ..., \alpha_m$ are rationally independent, then there exists a formal canonical transformation*

$$x = \phi(\xi, \eta) = \xi + \cdots, \quad y = \psi(\xi, \eta) = \eta + \cdots$$

which are given by formal power series such that $H(x, y)$ is took into a formal Hamiltonian $\Gamma = \Gamma(\rho_1, ..., \rho_m)$ which is a power series in $\rho_k = \xi_k^2 + \eta_k^2$.

Thus if all series in question would converge, the system

$$\dot{\xi}_k = \frac{\partial \Gamma}{\partial \eta_k}, \quad \dot{\eta}_k = -\frac{\partial \Gamma}{\partial \xi_k}$$

would be explicitly integrated by the formula

$$\xi_k + i\eta_k = c_k \exp(-2it\frac{\partial \Gamma}{\partial \rho_k}) \tag{4.10}$$

and the solution of the original system would have the form

$$x = \phi(\xi, \eta), \quad y = \psi(\xi, \eta).$$

For two degrees of freedom, the following convergent theorem was discovered by Rüssmann [222] (or see [178]):

Theorem 4.7 *If the Hamiltonian system possesses an analytic integral*

$$G = \sum_{k=1}^{2} \beta_k(x_k^2 + y_k^2) + \cdots$$

satisfying

$$\alpha_1\beta_2 - \alpha_2\beta_1 \neq 0,$$

then Birkhoff's transformation is convergent.

Let Ω be an open set in $\mathbb{R}^n$ which contains the origin. Let H_ε be a smooth real valued function on M depending on a parameter $\varepsilon \in \Omega$ such that $H_\varepsilon \to H_0 = H$ as $\varepsilon \to 0$ in the C^1-topology. Let $\mathcal{F}_\varepsilon = \{f_\varepsilon^t\}_{t \in \mathbb{R}}$ be the Hamiltonian flow of H_ε.

Problem 4.1 *When we have $d_H(J(\mathcal{F}_\varepsilon), J(\mathcal{F})) \to 0$ and $\dim_H J(\mathcal{F}_\varepsilon) \to \dim_H J(\mathcal{F})$ as $\varepsilon \to 0$? We conjecture that it is true if $\{f^t\}_{t \in \mathbb{R}}$ is structurally stable.*

4.3 Linearization of equations

Let M be a differential manifold of dimension m and X a differential vector field on M. Let $p \in M$ be a zero of X and f^t the (local) 1-parameter group generated by X in a neighborhood of p. Then p is a fixed point of f^t. In terms of local coordinates $x = {}^t(x_1, ..., x_m)$ centered at p, and in terms of the components of X, we can write

$$X(x) = \mathbf{A}x + O(\|x\|^r), \quad (r \geq 2)$$

in a coordinate neighborhood U of p. Then the *linearization* of the differential equation

$$\dot{x} = X(x)$$

at p is the linear differential equation

$$\dot{y} = \mathbf{A}y$$

where $y = {}^t(y_1, ..., y_m)$ are local coordinates at p. Let $y = h(x)$ be the change of coordinates. Then the linearization is equivalent to

$$h_*X(x) = \mathbf{A}h(x)$$

which is equivalent to

$$h(f^t(x)) = \exp(\mathbf{A}t)h(x),$$

i.e., f^t is conjugate to the linear flow $\exp(\mathbf{A}t)x$.

Theorem 4.8 (Hartman-Grobman, cf. [21]) *Let p be a hyperbolic fixed point of f^t. Then there is a neighborhood U of p on which f^t is topologically conjugate to the linear flow $\exp(\mathbf{A}t)x$.*

Now we look at the normal form calculations reducing the vector field X to its linear part, i.e., to transform $\dot{x} = X(x)$ into $\dot{y} = \mathbf{A}y$. Since the linearization is a local problem, we start on $\mathbb{R}^m$.

First we make some preparation. Recall that the canonical Hermitian structure of $\mathbb{C}^m$ is given by

$$(z, w) = \sum_{j=1}^{m} z_j \bar{w}_j$$

for $z = (z_1, ..., z_m)$, $w = (w_1, ..., w_m) \in \mathbb{C}^m$.

Definition 4.4 *A point $z = (z_1, ..., z_m) \in \mathbb{C}^m$ is said to be linearly resonant if there exists an integral relation of the form*

$$z_i = (z, \alpha) = \sum_{k=1}^{m} \alpha_k z_k, \quad \alpha \in (\mathbb{Z}_+)^m, \quad |\alpha| \geq 2.$$

Such a relation is called a linearly resonance. The number $|\alpha|$ is called the order of the linearly resonance.

Fix $\alpha \in (\mathbb{Z}_+)^m$ with $|\alpha| \geq 2$. A hyperplane in $\mathbb{C}^m$ given by an equation

$$z_i - (z, \alpha) = 0, \quad z = (z_1, ..., z_m) \in \mathbb{C}^m$$

is called a *resonant plane*. Varying the integral vector α and the index i, we obtain countably many resonant planes, which are the total of linearly resonant points.

Let $\mathbf{A} : \mathbb{R}^m \longrightarrow \mathbb{R}^m$ be a linear mapping and let $\mathcal{L}_r = \mathcal{L}_r(\mathbb{R}^m) \subset \mathbb{R}[x_1, ..., x_m]$ be the real vector space of all homogeneous polynomials of degree r on $\mathbb{R}^m$. Denote the k-fold product of $\mathcal{L}_r$ by $\mathcal{L}_r^k = \mathcal{L}_r^k(\mathbb{R}^m) \subset \mathbb{R}[x_1, ..., x_m]^k$. Given a vector field $v \in \mathcal{L}_r^m$ on $\mathbb{R}^m$ we consider the *homological equation*

$$L_{\mathbf{A}} h + v = 0$$

associated with $\mathbf{A}$, where

$$L_{\mathbf{A}} h(x) = \mathbf{A} h(x) - h_* \mathbf{A} x$$

is the Lie derivative of the vector field $h(x)$ with respect to $\mathbf{A} x$.

The Lie derivative $L_{\mathbf{A}} : \mathcal{L}_r^m \longrightarrow \mathcal{L}_r^m$ is a linear map and its eigenvalues can be expressed in terms of those of $\mathbf{A}$. We now calculate the eigenvalues and eigenvectors of $L_{\mathbf{A}}$. When $\mathbf{A}$ has distinct eigenvalues $\lambda_1, ..., \lambda_m$, denoting by e_i an eigenvector of $\mathbf{A}$ with eigenvalue λ_i, then $e_1, ..., e_m$ form a basis of $\mathbb{R}^m$. We denote by $(x_1, ..., x_m)$ coordinates with respect to the basis and write

$$x^\alpha = x_1^{\alpha_1} \cdots x_m^{\alpha_m}.$$

Relative to the basis $e_1, ..., e_m$, $\mathbf{A}$ is diagonal. Then

$$L_{\mathbf{A}} x^\alpha e_i = \{\lambda_i - (\lambda, \alpha)\} x^\alpha e_i. \tag{4.11}$$

Thus the monomial $x^\alpha e_i$ is an eigenvector of $L_{\mathbf{A}}$ with eigenvalue $\lambda_i - (\lambda, \alpha)$. Even when $\mathbf{A}$ cannot be diagonalised, by using Jordan blocks of $\mathbf{A}$, it can be shown that the eigenvalues of $L_{\mathbf{A}}$ are still given by the above expression. Therefore, $L_{\mathbf{A}}$ is invertible iff

$$\lambda_i - (\lambda, \alpha) \neq 0, \quad i = 1, ..., m$$

for every allowed α. If $\lambda_i = (\lambda, \alpha)$ with $|\alpha| \geq 2$, the vector-valued monomial $x^\alpha e_i$ is said to be *resonant*. Thus one obtains the following result:

Proposition 4.2 *If the m-tuple $\lambda = (\lambda_1, ..., \lambda_m)$ of the eigenvalues of $\mathbf{A}$ has no linearly resonances of order $r(\geq 2)$ and if $v \in \mathcal{L}_r^m$, then the homological equation $L_{\mathbf{A}} h + v = 0$ is solvable in $\mathcal{L}_r^m$, that is, $h = -L_{\mathbf{A}}^{-1} v \in \mathcal{L}_r^m$.*

Let X be a vector field on $\mathbb{R}^m$ with $X(0) = 0$. We now consider the differential equation

$$\dot{x} = X(x).$$

We denote X by a formal vector-valued power series

$$X(x) = \mathbf{A} x + v_r(x) + \cdots,$$

where $v_r(x) \in \mathcal{L}_r^m$ with $r \geq 2$, and where the dots denote terms of degree greater than r. We solve the homological equation

$$L_{\mathbf{A}} h_r + v_r = 0$$

on the basis of the proposition above. Substituting

$$x = h(y) = y + h_r(y),$$

the original differential equation can be transformed as follows:

$$
\begin{aligned}
y &= x - h_r(x) + \cdots, \\
\dot{y} &= (id - h_{r*} + \cdots)\dot{x} = (id - h_{r*} + \cdots)(\mathbf{A}y + \mathbf{A}h_r(y) + v_r(y) + \cdots) \\
&= \mathbf{A}y + \{L_\mathbf{A}h_r(y) + v_r(y)\} + \cdots \\
&= \mathbf{A}y + w_{r+1} + \cdots.
\end{aligned}
$$

Consequently, we have annihilated the terms of degree r on the right-hand side of the original equation.

Successively eliminating the terms of degree $2, 3, ...$, we construct a sequence of substitutions. The product of these substitutions stabilizes in the class of formal series, i.e., terms of any fixed degree do not change from a certain step. The limit substitution converts our formal equation into $\dot{y} = \mathbf{A}y$. This is the fundamental result of Poincaré's dissertation.

Theorem 4.9 (Poincaré) *If the eigenvalues of the matrix $\mathbf{A}$ is not linearly resonant, then the equation*

$$
\dot{x} = \mathbf{A}x + \sum_{r \geq 2} v_r(x)
$$

can be reduced to a linear equation

$$
\dot{y} = \mathbf{A}y
$$

by a formal change of variable

$$
x = h(y) = y + \sum_{r \geq 2} h_r(y), \quad h_r(y) \in \mathcal{L}_r^m.
$$

In the case of resonance, all nonresonant terms in the equation can be annihilated by a formal change of variables in the same way as in nonresonance case. This is the following Poincaré-Dulac theorem (see Arnold [18]):

Theorem 4.10 *The equation*

$$
\dot{x} = \mathbf{A}x + \sum_{r \geq 2} v_r(x)
$$

can be reduced to the canonical form

$$
\dot{y} = \mathbf{A}y + w(y)
$$

by a formal change of variables $x = y + \cdots$, where all monomials in the series w are resonant.

Definition 4.5 *A point $z \in \mathbb{C}^m$ is said to belong to the Poincaré domain P_I^m if the convex hull of z does not contain zero, and belong to the Siegel domain S_I^m if zero lies inside the convex hull of z.*

Lemma 4.2 (Arnold[18]) *Every point of P_I^m satisfies not more than a finite number of linear resonance relations, and has a neighborhood not intersecting the other resonant planes. The resonant planes are everywhere dense in the Siegel domain S_I^m.*

Lemma 4.2 and Theorem 4.10 imply the following result:

Theorem 4.11 *If the m-tuple of eigenvalues of the matrix* $\mathbf{A}$ *belongs to* P_I^m*, then the equation*

$$\dot{x} = \mathbf{A}x + \sum_{r \geq 2} v_r(x)$$

can be reduced to the polynomial normal form

$$\dot{y} = \mathbf{A}y + w(y)$$

by a formal change of variables $x = y + \cdots$*, where all monomials in the polynomial* w *are resonant.*

Let X be a holomorphic vector field on $\mathbb{C}^m$. Now we consider the differential equation

$$\dot{z} = X(z). \tag{4.12}$$

Let 0 be a zero of X. Then X can be denoted by a convergent vector-valued power series

$$X(z) = \mathbf{A}z + \sum_{r=2}^{\infty} v_r(z),$$

where $v_r(z) \in \mathcal{L}_r^m(\mathbb{C}^m)$. It is a consequence of well-known results of A. Lyapunov[160] that a necessary condition for future stability of $z = 0$ is that the eigenvalues $\lambda_1, ..., \lambda_m$ of $\mathbf{A}$ satisfy

$$\mathrm{Re}(\lambda_k) \leq 0, \quad k = 1, ..., m.$$

On the other hand, the condition $\mathrm{Re}(\lambda_k) < 0$ for $k = 1, ..., m$ is sufficient for future stability. The following surprising statement is due to Carathéodory and Cartan(1932) which can be derived from the results of [46]:

Theorem 4.12 *Necessary and sufficient conditions for the stability of the solution* $z = 0$ *of (4.12) for all real* t *is that*

1) $\mathbf{A}$ *is diagonalizable with purely imaginary eigenvalues and*
2) there exists a holomorphic mapping

$$z = h(w) = w + \cdots$$

taking (4.12) into the linear system

$$\dot{w} = \mathbf{A}w.$$

Theorem 4.13 (Siegel) *If* 0 *is a zero of a holomorphic vector field* X *on* $\mathbb{C}^m$*, then* X *is biholomorphically equivalent to* $\mathbf{A}z$ *in the neighborhood of* 0 *if one of the following conditions holds:*

1) the m-tuple of eigenvalues of the matrix $\mathbf{A}$ *belongs to* P_I^m *and is not linearly resonant (Poincaré);*

2) there exist $c, \nu \in \mathbb{R}^+$ *such that the eigenvalues of the matrix* $\mathbf{A}$ *satisfy*

$$|\lambda_i - (\lambda, \alpha)| \geq \frac{c}{|\alpha|^\nu},$$

for all i *and for all* $\alpha \in (\mathbb{Z}_+)^m$ *with* $|\alpha| \geq 2$.

In other words, the formal Poincaré series are convergent which also is true even in the case of resonance:

Theorem 4.14 (Poincaré-Dulac theorem) *If 0 is a zero of a holomorphic vector field X on $\mathbb{C}^m$ and if the m-tuple of eigenvalues of the matrix $\mathbf{A}$ belongs to P_I^m, then X is biholomorphically equivalent to a polynomial field in which all vector-valued monomials with coefficients of degree greater than 1 are linearly resonant.*

All nonresonant vectors in the Poincaré domain P_I^m satisfy the Siegel's condition for some $c > 0$. The proof of Theorem 4.13 can be obtained from the proof of the analogous theorem for mappings in §4.4.

4.4   Linearization of mappings

Definition 4.6 *A point $z = (z_1, ..., z_m) \in \mathbb{C}^m$ is said to be algebraically resonant if there exists an algebraically relation of the following form*

$$z_i = z^\alpha, \quad \alpha \in (\mathbb{Z}_+)^m, \quad |\alpha| \geq 2.$$

Such a relation is called an algebraically resonance. The number $|\alpha|$ is called the order of the algebraically resonance.

Definition 4.7 *A point $z = (z_1, ..., z_m) \in \mathbb{C}^m$ is said to belong to the Poincaré domain P_II^m if either $|z_i| < 1$ for all i or $|z_i| > 1$ for all i. The complement of P_II^m is the Siegel domain S_II^m.*

Obviously, S_II^1 is just the unit circle $|z| = 1$. In $\mathbb{C}^m$, the algebraically resonance equation $z_i = z^\alpha$ determines a complex hypersurface, called *resonant surface*. The resonant surfaces lie discretely in P_II^m, but in S_II^m, both the algebraically resonant and non-algebraically resonant points are everywhere dense.

Let M be a differential manifold of dimension m and $f : M \longrightarrow M$ a differential mapping with an isolated fixed point p. In terms of local coordinates $x = {}^t(x_1, ..., x_m)$ centered at p, we can write

$$f(x) = Ax + O(\|x\|^r), \quad (r \geq 2)$$

in a coordinate neighborhood U of p, where $A = (df)_p$. Then the linearization of f at p is the change $y = \phi(x)$ of coordinates such that

$$\phi(f(x)) = A\phi(x),$$

i.e., f is conjugate to the linear mapping A on U.

Theorem 4.15 (Hartman-Grobman, cf. [21]) *Let p be a hyperbolic fixed point of f. Then there is a neighborhood U of p on which f is topologically conjugate to the linear mapping A.*

In terms of local coordinates $x = {}^t(x_1, ..., x_m)$ centered at p, we denote the mapping f by a formal power series

$$f(x) = Ax + \sum_{r=2}^{\infty} f_r(x), \quad f_r \in \mathcal{L}_r^m.$$

First we solve the homological equation

$$P_A \phi_r + f_r = 0$$

associated with A, where

$$P_A \phi_r(x) = A\phi_r(x) - \phi_r(Ax).$$

The operator $P_A : \mathcal{L}_r^m \longrightarrow \mathcal{L}_r^m$ is a linear map whose eigenvalues can be expressed in terms of those of A. If e_i is an eigenvector of A with eigenvalue λ_i such that $e_1, ..., e_m$ is a basis, the monomial $x^\alpha e_i$ is an eigenvector of P_A, and the eigenvalues of P_A are given by

$$P_A x^\alpha e_i = \{\lambda_i - \lambda^\alpha\} x^\alpha e_i \quad i = 1, ..., m. \tag{4.13}$$

If $\lambda_i = \lambda^\alpha$ with $|\alpha| \geq 2$, the vector-valued monomial $x^\alpha e_i$ is said to be *algebraically resonant*. Therefore we have

Proposition 4.3 *If the m-tuple $\lambda = (\lambda_1, ..., \lambda_m)$ of the eigenvalues of A has no algebraically resonances of order $r(\geq 2)$ and if $f_r \in \mathcal{L}_r^m$, then the homological equation $P_A \phi_r + f_r = 0$ is solvable in $\mathcal{L}_r^m$.*

Next we look at the normal form calculations reducing the mapping f to the linear mapping A. Solve the homological equation

$$P_A \phi_r + f_r = 0$$

on the basis of the proposition above. Substituting

$$x = \phi(y) = y + \phi_r(y),$$

the mapping f can be transformed as follows:

$$\begin{aligned} \phi^{-1}(x) &= x - \phi_r(x) + \cdots, \\ \phi \circ f \circ \phi^{-1}(x) &= Ax - \{P_A \phi_r + f_r\} + \cdots \\ &= Ax + g_{r+1} + \cdots. \end{aligned}$$

Consequently, we have annihilated the terms of degree r on the right-hand side of the original mapping. Successively eliminating the terms of degree $2, 3, ...$, a sequence of substitutions is constructed.

Theorem 4.16 (Poincaré) *If the m-tuple of eigenvalues of the matrix $A = (df)_p$ is not algebraically resonant, then the mapping $x \mapsto f(x)$ can be reduced to its linear part $x \mapsto Ax$*

$$\phi \circ f = A\phi,$$

by a formal change of variable

$$x = \phi(y) = y + \sum_{r \geq 2} \phi_r(y), \quad \phi_r(y) \in \mathcal{L}_r^m.$$

In the case of resonance, Poincaré's method can be used to deduce the following Poincaré-Dulac theorem:

Theorem 4.17 *Any formal mapping $x \mapsto Ax + \cdots$ can be reduced to the normal form $y \mapsto Ay + w(y)$ by a formal change of variables $x = y + \cdots$, where all monomials in the series w are algebraically resonant.*

To discuss convergence problems, we turn on complex manifolds. Let M be a complex manifold of dimension m. Take $f \in \text{Hol}(M, M)$ such that $f(p) = p$ for a point $p \in M$. In terms of local holomorphic coordinates $z = {}^t(z_1, ..., z_m)$ centered at p, the mapping f can be given by a convergent power series

$$f(z) = Az + \sum_{r=2}^{\infty} f_r(z), \quad f_r \in \mathcal{L}_r^m(\mathbb{C}^m),$$

where $A = f'(p)$. The mapping f is said to be linearized at the point p if there exist a local holomorphic coordinate system ϕ of p with $\phi(p) = 0$ and $\phi'(p) = I$, where I is the identity, such that the Schröder equation:

$$\phi \circ f = A\phi, \tag{4.14}$$

holds near the point p.

Schröder [224], [225] introduced this equation in dimension 1 which is related to the well-known Abel's equation introduced by Abel in [5]. Poincaré [198], Sternberg [242], and others studied the functional equation in the case of several variables. The following results can be found in [18] and [279]:

Theorem 4.18 (Siegel) *Let f be a holomorphic self-mapping on a complex manifold M of dimension m with $f(p) = p$ for a point $p \in M$. Then f can be linearized at the point p if one of the following conditions is satisfied:*

1) the m-tuple of eigenvalues of $f'(p)$ belongs to P_{II}^m and is not algebraically resonant (Poincaré);

2) there exist $c, \nu \in \mathbb{R}^+$ such that the m-tuple λ of eigenvalues λ_i of $f'(p)$ satisfy some Diophantine approximation conditions:

$$|\lambda_i - \lambda^\alpha| \geq \frac{c}{|\alpha|^\nu}, \tag{4.15}$$

for all $i = 1, ..., m$ and for all $\alpha \in (\mathbb{Z}_+)^m$ with $|\alpha| \geq 2$.

In other words, the formal Poincaré series are convergent which also is true even in the case of resonance:

Theorem 4.19 (Poincaré-Dulac theorem) *If p is a fixed point of a holomorphic self-mapping f on M and if the m-tuple of eigenvalues of the matrix $A = f'(p)$ belongs to P_I^m, then by a biholomorphic substitution f can be reduced to a polynomial normal form in which all vector-valued monomials with coefficients of degree greater than 1 are algebraically resonant.*

For the proofs, see Arnold[18]. Now we discuss some global results. First we note that the linearization of mappings is closely related to an older topic, namely the Fatou-Bieberbach method (cf. [79], [42]) of constructing biholomorphic images of $\mathbb{C}^m$ in $\mathbb{C}^m$, starting with an automorphism that has an attractive fixed point. Here is the basic theorem:

Theorem 4.20 *If $f \in \mathrm{Aut}(\mathbb{C}^m)$ has an attractive fixed point $p \in \mathbb{C}^m$, then there exists a biholomorphic mapping ψ from the basin $\mathrm{Att}(p)$ of attraction of p onto $\mathbb{C}^m$.*

As remarked in [216], a large part of the paper [70] by Dixon and Esterle depends on this theorem. The mapping ψ is obtained as a solution of the functional equation

$$G^{-1} \circ \psi \circ f = \psi, \tag{4.16}$$

where G is a "normal form" for f. On p.142 they refer to Reich's papers [206], [207] for the solution of (4.16). Reich [207] claims to prove that

$$\psi = \lim_{n \to \infty} G^{-n} \circ f^n$$

solves (4.16). However, a counterexample was given by Rosay and Rudin in [216].

We first introduce a weak form of the basic theorem, which is sufficient to solve the functional equation

$$\psi \circ f = f'(p) \circ \psi$$

in the cases which arise in the constructions of Fatou [79], Bieberbach [42], Sadullaev [223], and Kodaira [147]. The proof given below is due to Rosay and Rudin. For the result, also see [70] and Stehlé [241].

Theorem 4.21 *Suppose $f \in \mathrm{Hol}(M, M)$, $p \in M$, $f(p) = p$, and the eigenvalues λ_i of $A = f'(p)$ satisfy $|\lambda_1| \geq |\lambda_2| \geq \dots \geq |\lambda_m|$. Assume that*

$$|\lambda_1|^2 < |\lambda_m|. \tag{4.17}$$

Then there exists a surjective holomorphic mapping $\psi : \mathrm{Att}(p) \longrightarrow \mathbb{C}^m$ such that $\psi(p) = 0, \psi'(p) = I$ and such that

$$\psi \circ f = A\psi. \tag{4.18}$$

Proof. Pick constants $\alpha, \beta_1, \beta_2, \beta$ so that $\alpha < |\lambda_m|, |\lambda_1| < \beta_1 < \beta_2 < \beta$, and $\beta^2 < \alpha$. The spectral radius formula gives an k so that $\|A^{-n}\| < \alpha^{-n}$ and $\|A^n\| < \beta_1^n$ for all $n \geq k$. Taking a local coordinate system ϕ at p with $\phi(p) = 0$. According to the proof of Theorem 1.6, we can obtain (1.9). Now pick a compact set $K \subset \mathrm{Att}(p)$. For some s, $f^s(K) \subset \phi^{-1}(\mathbb{C}^m(r))$. Hence (1.9) shows that

$$\|\phi \circ f^n(\xi)\| < \beta^{n-s} \|\phi \circ f^s(\xi)\| \text{ for all } \xi \in K, \ n \geq s + n_0.$$

Since $(A^{-1}\phi \circ f \circ \phi^{-1})'(0) = I$, there is a constant b so that

$$\|z - A^{-1}\phi \circ f \circ \phi^{-1}(z)\| \leq b\|z\|^2 \quad (\|z\| \leq r).$$

Thus if $\xi \in K$ and if we set $z_n = \phi \circ f^n(\xi)$, we get the estimate

$$
\begin{aligned}
\|A^{-n}\phi \circ f^n(\xi) - A^{-n-1}\phi \circ f^{n+1}(\xi)\| &\leq \|A^{-n}\| \cdot \|z_n - A^{-1}\phi \circ f \circ \phi^{-1}(z_n)\| \\
&\leq \alpha^{-n}b\|z_n\|^2 \leq b\beta^{-2s}(\beta^2/\alpha)^n
\end{aligned}
$$

for all $n \geq s + n_0$. Since $\beta^2/\alpha < 1$, it follows that

$$
\psi = \lim_{n \to \infty} A^{-n}\phi \circ f^n \tag{4.19}
$$

is uniform on compact subsets of $\mathrm{Att}(p)$. One immediate consequence of (4.19) is the functional equation (4.18). It is clear that $\psi : \mathrm{Att}(p) \longrightarrow \mathbb{C}^m$ is holomorphic. Obviously, we have $\psi(p) = 0, \psi'(p) = I$. Since $A^{-n}\phi \circ f^n$ is one-to-one near p, so does ψ. Hence we have obtained a linearization of f. Note that $f(\mathrm{Att}(p)) = \mathrm{Att}(p)$. We see that ψ and $A^{-1}\psi$ have the same range. Since the linear operator A^{-1} is an expansion, it follows that $\psi(\mathrm{Att}(p))$ is all of $\mathbb{C}^m$. □

If $f \in \mathrm{Aut}(M)$, the mapping $\psi : \mathrm{Att}(p) \longrightarrow \mathbb{C}^m$ is biholomorphic. If $\mathcal{J}f$ is constant, then $\mathcal{J}f = \det(A)$ because $A = f'(p)$. For this case, if we apply the chain rule to $\psi \circ f = A\psi$, we obtain, for $\xi \in \mathrm{Att}(p)$,

$$
(\mathcal{J}\psi)(f(\xi))(\mathcal{J}f)(\xi) = \det(A)(\mathcal{J}\psi)(\xi).
$$

Hence

$$
(\mathcal{J}\psi)(\xi) = (\mathcal{J}\psi)(f(\xi)) = \dots = (\mathcal{J}\psi)(f^n(\xi)) = \dots .
$$

Since $f^n(\xi) \to p$ as $n \to \infty$ we conclude that

$$
(\mathcal{J}\psi)(\xi) = (\mathcal{J}\psi)(p) = 1
$$

for all $\xi \in \mathrm{Att}(p)$. Hence $\mathcal{J}\psi \equiv 1$ on $\mathrm{Att}(p)$.

Example 4.1 ([216]) *Define* $f \in \mathrm{Aut}(\mathbb{C}^2)$ *by*

$$
f(z,w) = (\alpha z, \beta w + z^2), \quad \text{where } 0 < \beta < \alpha < 1.
$$

This f fixes the origin, and

$$
A = f'(0,0) = \begin{pmatrix} \alpha & 0 \\ 0 & \beta \end{pmatrix}.
$$

By induction

$$
f^n(z,w) = (\alpha^n z, \beta^n w + \beta^{n-1}(1 + c + \dots + c^{n-1})z^2),
$$

where $c = \alpha^2/\beta$. *Thus*

$$
(A^{-n}f^n)(z,w) = (z, w + \beta^{-1}(1 + c + \dots + c^{n-1})z^2).
$$

The coefficient of z^2 in the second component of $A^{-n}f^n$ tends to infinity, except when $c < 1$, i.e., when $\alpha^2 < \beta$.

This example shows that *the sequence* (4.19) *may fail to converge (even locally, and even on the level of formal power series) if assumption* (4.17) *is violated.*

To introduce the proof of the basic theorem given by Rosay and Rudin, which is much shorter and simpler than the work in [206] and [207], we begin with some facts about holomorphic mappings $G = (g_1, ..., g_m)$ from $\mathbb{C}^m$ into $\mathbb{C}^m$ of the form

$$g_1 = \lambda_1 z_1, \quad g_i = \lambda_i z_i + h_i(z_1, ..., z_{i-1}), \quad i = 2, 3, ..., m$$

where $\lambda_1, ..., \lambda_m$ are scalars and each h_i is a holomorphic function of $(z_1, ..., z_{i-1})$ which vanishes at the origin. We call each such a mapping *lower triangular*. The matrix that represents the linear operator $G'(0)$ is then lower triangular. Thus $G'(0)$ is invertible iff no λ_i is 0. It follows that $G \in \text{Aut}(\mathbb{C}^m)$ (a composition of an invertible linear map and $m - 1$ shears) iff no λ_i is 0. If $g_1, ..., g_m$ are polynomials, the mapping $G = (g_1, ..., g_m)$ will be called a *polynomial mapping* and the integer $\max_i \deg g_i$ is usually called the "*degree*" of G. If no confusion with one defined in §1.6, we also write

$$\deg G = \max_i \deg g_i. \tag{4.20}$$

Lemma 4.3 ([216]) *Let G be a lower triangular polynomial automorphism of $\mathbb{C}^m$.*

1) The degree of the iterates G^n of G are then bounded, and there is a constant $\beta < \infty$ so that

$$G^n(\mathbf{\Delta}^m) \subset \beta^n D^m \quad (n = 1, 2, 3, ...).$$

Here $\mathbf{\Delta}^m$ is the unit polydisc in $\mathbb{C}^m$.

2) If also $|\lambda_i| < 1$ for $1 \leq i \leq m$, then $G^n \to 0$, uniformly on compact subsets of $\mathbb{C}^m$, and

$$\bigcup_{n=1}^{\infty} G^{-n}(U) = \mathbb{C}^m$$

for every neighborhood U of 0.

Lemma 4.4 ([216]) *Suppose that U is a neighborhood of 0 in $\mathbb{C}^m$, that $\hat{f} : U \longrightarrow \mathbb{C}^m$ is holomorphic, $\hat{f}(0) = 0$, and that all eigenvalues λ_i of $A = \hat{f}'(0)$ satisfy $0 < |\lambda_i| < 1$. Then there exist*

1) a lower triangular polynomial automorphism G of $\mathbb{C}^m$, with $G(0) = 0, G'(0) = A$, and

2) polynomial mappings $T_n : \mathbb{C}^m \longrightarrow \mathbb{C}^m$, with $T_n(0) = 0, T_n'(0) = I$, so that

$$G^{-1} \circ T_n \circ \hat{f}(z) - T_n(z) = O(\|z\|^n) \quad (n = 2, 3, 4, ...).$$

Theorem 4.22 *Suppose $f \in \text{Hol}(M, M)$, $p \in M$, $f(p) = p$, and the eigenvalues λ_i of $A = f'(p)$ satisfy $1 > |\lambda_1| \geq |\lambda_2| \geq ... \geq |\lambda_m| > 0$ ($m = \dim M$). Then there exist a surjective holomorphic mapping $\psi : \text{Att}(p) \longrightarrow \mathbb{C}^m$, a lower triangular polynomial automorphism G of $\mathbb{C}^m$ such that $\psi(p) = 0, \psi'(p) = I, G(0) = 0, G'(0) = A$ and such that*

$$G^{-1} \circ \psi \circ f = \psi. \tag{4.21}$$

Furthermore, if $f \in \text{Aut}(M)$, then ψ is biholomorphic. Moreover, ψ can be chosen so that $\mathcal{J}\psi \equiv 1$ if $\mathcal{J}f$ is constant.

Proof. Fix $\beta, \|A\| < \beta < 1$. According to the proof of Theorem 1.6, there are $r > 0, n_0 > 0$, such that

$$\|\phi \circ f^n \circ \phi^{-1}(z)\| < \beta^n \|z\| \text{ for all } z \in \mathbb{C}^m(r), \ n \geq n_0.$$

where ϕ is some local coordinate system at p with $\phi(p) = 0$. It follows that $\phi^{-1}(\mathbb{C}^m(r)) \subset \text{Att}(p)$, that $\text{Att}(p)$ is a region, and that $f(\text{Att}(p)) = \text{Att}(p)$.

Next, we associate G to $\hat{f} = \phi \circ f \circ \phi^{-1}$ as in Lemma 4.4, and apply Lemma 4.3 $1)$ to G^{-1} in place of G to conclude (with the aid of the Schwarz lemma) that there is a constant $\gamma < \infty$ so that

$$\|G^{-n}(w) - G^{-n}(w')\| \leq \gamma^n \|w - w'\| \quad (n = 1, 2, 3, ...) \tag{4.22}$$

for all $w, w' \in \mathbb{C}^m$ with $\|w\| \leq 1/2, \|w'\| \leq 1/2$.

Fix a positive integer k so that $\beta^k < 1/\gamma$. Lemma 4.4 gives us a polynomial map $T = T_k$, with $T(0) = 0, T'(0) = I$, and it gives us constants $\delta > 0, c_1 < \infty$, so that $\|w\| \leq \delta$ implies

$$\|G^{-1} \circ T \circ \hat{f}(w) - T(w)\| \leq c_1 \|w\|^k. \tag{4.23}$$

Now let $E \subset \text{Att}(p)$ be compact. Then $\phi \circ f^s(E) \subset \mathbb{C}^m(r)$ for some integer s. Hence $\phi \circ f^{s+n}(E) \subset \hat{f}^n(\mathbb{C}^m(r)) \subset \mathbb{C}^m(\beta^n r)$, for all $n \geq n_0$. Thus there exists $n_1 \geq s + n_0$ such that $\beta^n r \leq \delta$ when $n \geq n_1$. Then for $x \in E, n \geq n_1$,

$$\|G^{-1} \circ T \circ \phi \circ f^{n+1}(x) - T \circ \phi \circ f^n(x)\| \leq c_1 \|\phi \circ f^n(x)\|^k \leq c_1 \beta^{nk} \|\phi(x)\|^k. \tag{4.24}$$

For large n, $\|G^{-1} \circ T \circ \phi \circ f^{n+1}(x)\|$ and $\|T \circ \phi \circ f^n(x)\|$ are $< 1/2$, for all $x \in E$. Hence (4.22) can be applied to (4.24), and we conclude that for $n \geq n_2$ and for all $x \in E$,

$$\|G^{-n-1} \circ T \circ \phi \circ f^{n+1}(x) - G^{-n} \circ T \circ \phi \circ f^n(x)\| \leq c_1 r^k (\gamma \beta^k)^n.$$

Since $\gamma \beta^k < 1$, we have proved: *The limit*

$$\psi(x) = \lim_{n \to \infty} G^{-n} \circ T \circ \phi \circ f^n(x) \tag{4.25}$$

exists uniformly on compact subsets of $\text{Att}(p)$, *and defines a holomorphic mapping* $\psi : \text{Att}(p) \longrightarrow \mathbb{C}^m$ *which satisfies* $\psi(p) = 0, \psi'(p) = I$, *as well as the functional equation* (4.21) Note that $f(\text{Att}(p)) = \text{Att}(p)$. We see that ψ and $G^{-1} \circ \psi$ have the same range. Thus

$$\psi(\text{Att}(p)) = G^{-1}(\psi(\text{Att}(p))) = ... = G^{-n}(\psi(\text{Att}(p))) = ...$$

and since $\psi(\text{Att}(p))$ contains a neighborhood of 0, it follows that $\psi(\text{Att}(p))$ is all of $\mathbb{C}^m$ from Lemma 4.3 $2)$.

Assume $f \in \text{Aut}(M)$ and that $x, y \in \text{Att}(p)$ with $\psi(x) = \psi(y)$. By (4.21), $\psi(f(x)) = \psi(f(y))$. Continuing, we see that $\psi(f^n(x)) = \psi(f^n(y))$ for all positive n. But when n is sufficiently large, both $f^n(x)$ and $f^n(y)$ are in a neighborhood of p in which ψ is one-to-one. Thus $f^n(x) = f^n(y)$, and this implies $x = y$. So ψ is one-to-one in $\text{Att}(p)$.

Final, assume that $\mathcal{J}f$ is constant. Since G is a polynomial automorphism of $\mathbb{C}^m$, the polynomial $\mathcal{J}G$ has no zero in $\mathbb{C}^m$, hence is also constant. In fact, $\mathcal{J}G = \mathcal{J}f$ because $G'(0) = f'(0)$. If we apply the chain rule to $\psi \circ f = G \circ \psi$, we obtain, for $\xi \in \text{Att}(p)$,

$$(\mathcal{J}\psi)(f(\xi))(\mathcal{J}f)(\xi) = \mathcal{J}G(\psi(\xi))(\mathcal{J}\psi)(\xi).$$

Hence

$$(\mathcal{J}\psi)(\xi) = (\mathcal{J}\psi)(f(\xi)) = \ldots = (\mathcal{J}\psi)(f^n(\xi)) = \ldots .$$

Since $f^n(\xi) \to p$ as $n \to \infty$ we conclude that

$$(\mathcal{J}\psi)(\xi) = (\mathcal{J}\psi)(p) = 1$$

for all $\xi \in \mathrm{Att}(p)$. Hence $\mathcal{J}\psi \equiv 1$ on $\mathrm{Att}(p)$. $\square$

For Theorem 4.22, also see Karlin and McGregor [132]. Here we adopt Rosay and Rudin's method in proving some of the above results. By using Siegel linearization theorem, He Wu [269] proved the following theorem.

Theorem 4.23 *Suppose that $f \in \mathrm{Aut}(\mathbb{C}^m)$, the point 0 is a hyperbolic fixed point of f, and that there exist $c, \nu \in \mathbb{R}^+$ such that the m-tuple λ of eigenvalues λ_i of $f'(p)$ satisfy some Diophantine approximation conditions:*

$$|\lambda_i - \lambda^\alpha| \geq \frac{c}{|\alpha|^\nu},$$

for all $i = 1, \ldots, m$ and for all $\alpha \in (\mathbb{Z}_+)^m$ with $|\alpha| \geq 2$. Then the stable manifold

$$W^s(0) = \left\{ z \in \mathbb{C}^m \mid \lim_{n \to \infty} f^n(z) = 0 \right\}$$

is an injectively immersed complex submanifold and is biholomorphically equivalent to $\mathbb{C}^k$, where k is the cardinal number of the eigenvalues λ_i with $|\lambda_i| < 1$.

Some similar results were obtained by Bedford and Smillie [34] and Wu [270] when f is the so-called "generalized Hénon" mapping of $\mathbb{C}^2$.

Theorem 4.24 *Suppose $f \in \mathrm{Hol}(M, M)$, $p \in M$, $f(p) = p$, and $A = f'(p)$ is unitary. If $p \in F(f)$, then f is linearizable at the point p.*

Proof. Observe first that $\{A^n\}$ and $\{A^{-n}\}$ are bounded. Let ϕ be some local coordinate system at p with $\phi(p) = 0$. Define

$$\psi_n = \frac{1}{n} \sum_{j=0}^{n-1} A^{-j} \phi \circ f^j.$$

W.l.o.g. assume that $\{f^n\}$ is uniformly convergent on $\phi^{-1}(\mathbb{C}^m(r))$. Thus local bounded sequence $\{\psi_n\}$ contains a convergent subsequence. Since

$$\psi_n \circ f = A\psi_n + O(\frac{1}{n}),$$

any limit ψ of the $\{\psi_n\}$ satisfies

$$\psi \circ f = A\psi.$$

From $A = f'(p)$ we have $\psi_n'(p) = I$, and $\psi'(p) = I$. $\square$

An open subset $D \subset \mathbb{C}^m$ is known as a *Fatou-Bieberbach domain* if D is biholomorphically equivalent to $\mathbb{C}^m$, but $D \neq \mathbb{C}^m$. Thus if f is a holomorphic diffeomorphism of $\mathbb{C}^m$ with an attractive fixed point p, then the basin $\mathrm{Att}(p)$ of attraction of p is a Fatou-Bieberbach domain iff $\mathrm{Att}(p)$ is not all of $\mathbb{C}^m$ (also see Bochner and Martin [46]).

As remarked in [216], the result of M. Green [98] shows that a Fatou-Bieberbach domain in $\mathbb{C}^2$ must intersect any set of three complex lines. Dektyarev [68] points out that a Fatou-Bieberbach domain in $\mathbb{C}^2$ must intersect every pair of disjoint lines, and he asks if it must intersect every complex line. In [216], a counterexample is constructed as a basin of attraction of a holomorphic diffeomorphism of $\mathbb{C}^2$ but the diffeomorphism is not a polynomial automorphism. The following theorem of Bedford and Smillie [36] shows that Dektyarev's question has a positive answer for polynomial cases.

Theorem 4.25 *Let* $\mathrm{Att}(p)$ *be a basin of attraction for a polynomial automorphism in* $\mathbb{C}^2$. *Then* $\mathrm{Att}(p)$ *intersects every algebraic curve* V. *Furthermore* $\overline{\mathrm{Att}(p)} \cap V$ *is compact.*

4.5 L_p-Normality

Let M and N be manifolds with distances d_g and d_h respectively, and let μ be a Borel measure on M such that $\mu(U) > 0$ for any non-empty open set U of M. Fixed $p \in \mathbb{R}^+, O \in N$, write $\hbar(x) = d_h(x, O)$, and define

$$\mathcal{L}_p(M, N) = \{f \in \mathcal{L}^0(M, N) \mid \int_K \hbar(f)^p d\mu < \infty \text{ for compact subsets } K \subset M\},$$

$$\mathcal{L}_\infty(M, N) = \{f \in \mathcal{L}^0(M, N) \mid \hbar(f) \text{ is bounded on compact subsets } K \subset M \text{ a.e. } (\mu)\},$$

and for $f \in \mathcal{L}_p(M, N)$ define

$$\|f\|_{p,K} = \begin{cases} \int_K \hbar(f)^p d\mu & : \quad 0 < p < 1 \\ \{\int_K \hbar(f)^p d\mu\}^{\frac{1}{p}} & : \quad 1 \leq p < \infty \\ \inf\{C \mid \hbar(f) \leq C \text{ on } K \text{ a.e. } (\mu)\} & : \quad p = \infty. \end{cases}$$

We will identify mappings equivalent under the equivalence relation: $f \sim g$ if f and g differ only on a set of zero measure. We can write

$$L_0(M, N) = L^0(M, N) = \mathcal{L}^0(M, N)/\sim,$$

$$L_p(M, N) = \mathcal{L}_p(M, N)/\sim \quad (0 < p \leq \infty).$$

Obviously, $C(M, N) \subset L_p(M, N)$. For $f_1, f_2 \in L_p(M, N)$, define

$$\|f_1, f_2\|_{p,K} = \begin{cases} \int_K d_h(f_1, f_2)^p d\mu & : \quad 0 < p < 1 \\ \{\int_K d_h(f_1, f_2)^p d\mu\}^{\frac{1}{p}} & : \quad 1 \leq p < \infty \\ \inf\{C \mid d_h(f_1, f_2) \leq C \text{ on } K \text{ a.e. } (\mu)\} & : \quad p = \infty. \end{cases}$$

On M, there exist compact subsets E_k such that $E_k \subset E_{k+1}$, and $\cup_{k=1}^{\infty} E_k = M$. For $f_1, f_2 \in L_p(M, N)$ we set

$$\rho_p(f_1, f_2) = \sum_{k=1}^{\infty} \frac{\|f_1, f_2\|_{p,E_k}}{2^k(1 + \|f_1, f_2\|_{p,E_k})}. \tag{4.26}$$

It is trivial to verify that $\rho_p(f_1, f_2)$ is finite and satisfies the triangle inequality.

A sequence $\{f_n\} \subset L_p(M, N)$ is called *compactly L_p-convergent* on M to some mapping $f : M \longrightarrow N$ if for any compact subset K in M,

$$\lim_{n\to\infty} \|f_n, f\|_{p,K} = 0,$$

i.e., $\{f_n\}$ is L_p-convergent to f on K. Obviously, $f \in L_p(M, N)$. If $p < q$, Hölder's inequality yields

$$\int_K d_h(f_n, f)^p d\mu \le \mu(K)^{1-\frac{p}{q}} \{\int_K d_h(f_n, f)^q d\mu\}^{\frac{p}{q}}$$

so that L_q-convergence implies L_p-convergence.

For $f_n, f \in L_0(M, N)$, the compact set $K \subset M$, and any $\delta > 0$, define

$$K_{n,\delta} = \{x \in K \mid d_h(f_n(x), f(x)) \ge \delta\},$$

$$\|f_n, f\|_{0,K} = \|f_n, f\|_{0,K,\delta} = \mu(K_{n,\delta}).$$

If for any $\delta > 0$,

$$\lim_{n\to\infty} \|f_n, f\|_{0,K,\delta} = 0,$$

then it is said that f_n *converges metrically* to f on K. A sequence $\{f_n\} \subset L_0(M, N)$ is called *compactly L_0-convergent* on M to some mapping $f : M \longrightarrow N$ if for any compact subset K in M, f_n converges metrically to f on K. Noting that

$$\int_K d_h(f_n, f)^p d\mu \ge \int_{K_{n,\delta}} d_h(f_n, f)^p d\mu \ge \delta^p \mu(K_{n,\delta}),$$

we have the relation:

Theorem 4.26 *If $f_n \in L_p(M, N)$ is L_p-convergent to f on K, then f_n converges metrically to f on K.*

Assume that f_n converges metrically to f on K. Then for each $k \in \mathbb{Z}^+$, there exists $n_k(n_1 < n_2 < ...)$ such that

$$\mu(K_{n_k, 2^{-k}}) < \frac{1}{2^k}.$$

Setting

$$E = \bigcap_{s=1}^{\infty} \bigcup_{k=s}^{\infty} K_{n_k, 2^{-k}},$$

then for each $s \geq 1$,

$$\mu(E) \leq \sum_{k=s}^{\infty} \frac{1}{2^k} = \frac{1}{2^{s-1}},$$

which implies $\mu(E) = 0$ as $s \to \infty$. Obviously, if

$$x \in K - E = \bigcup_{s=1}^{\infty} \bigcap_{k=s}^{\infty} \{x \in K \mid d_h(f_{n_k}(x), f(x)) < \frac{1}{2^k}\},$$

we have $f_{n_k}(x) \to f(x)$ as $k \to \infty$. Hence we obtain

Theorem 4.27 *If f_n converges metrically to f on K, then there exists a subsequence $\{f_{n_k}\}$ of $\{f_n\}$ which converges to f on K a.e. (μ).*

We say that a sequence $\{f_n\} \subset L_p(M, N)$ is *compactly L_p-divergent* on M if for any compact subset K in M for which

$$\mu(K) = \int_K d\mu > 0,$$

we have

$$\lim_{n \to \infty} \|f_n\|_{p,K} = \infty,$$

if $0 < p \leq \infty$, or for any $R > 0$

$$\lim_{n \to \infty} \mu(K^{n,R}) = 0,$$

if $p = 0$, where

$$K^{n,R} = \{x \in K \mid \hbar(f_n(x)) \leq R\}.$$

We also write

$$\|f_n\|_{0,K} = \|f_n\|_{0,K,R} = \mu(K^{n,R}).$$

Noting that for $p > 0$

$$\int_K \hbar(f_n)^p d\mu \geq \int_{K-K^{n,R}} \hbar(f_n)^p d\mu \geq R^p \{\mu(K) - \mu(K^{n,R})\},$$

and noting that if $p < q$, Hölder's inequality yields

$$\int_K \hbar(f_n)^p d\mu \leq \mu(K)^{1-\frac{p}{q}} \{\int_K \hbar(f_n)^q d\mu\}^{\frac{p}{q}},$$

then L_p-divergence implies L_q-divergence for $0 \leq p < q \leq \infty$.

Lemma 4.5 *Convergence of sequences in $L_p(M, N)$ with respect to the distance function ρ_p means precisely the same thing as compact L_p-convergence on M.*

Proof. Suppose first that $f_n \to f$ in the sense of the ρ_p-distance. For large n we have then $\rho_p(f_n, f) < \varepsilon$ and consequently, by the definition of ρ_p,

$$\delta_{k,p}(f_n, f) = \frac{\|f_n, f\|_{p,E_k}}{1 + \|f_n, f\|_{p,E_k}} < 2^k \varepsilon.$$

But this implies that $f_n \to f$ on E_k in the sense of L_p-convergence with respect to the d_h-metric. Since every compact E is contained in an E_k, it follows that the L_p-convergence can be on E.

Conversely, suppose that $f_n \to f$ in the sense of L_p-convergence on every compact set. Then $\delta_{k,p}(f_n, f) \to 0$ for every k, and because the series $\sum \delta_{k,p}(f_n, f) 2^{-k}$ has a convergent majorant with terms independent of n it follows readily that $\rho_p(f_n, f) \to 0$. $\square$

Definition 4.8 *A subset $\mathcal{F}$ of $C(M, N)$ is called L_p-normal, or an L_p-normal family, in M iff every sequence of $\mathcal{F}$ contains a subsequence which is either compactly L_p-convergent on M or compactly L_p-divergent on M.*

Obviously, if N is complete and if the subset $\mathcal{F}$ is normal on M, then $\mathcal{F}$ is L_p-normal for $0 \le p \le \infty$.

A sequence $\{f_n\}$ of mappings from a metric space (M, d_g) into a metric space (N, d_h) is called L_p-convergent locally on M to some mapping f if each point z_0 of M has a disc $M(z_0; r)$ such that

$$\lim_{n \to \infty} \|f_n, f\|_{p, M(z_0; r)} = 0,$$

if $0 < p \le \infty$, or for any $\delta > 0$,

$$\lim_{n \to \infty} \|f_n, f\|_{0, M(z_0; r), \delta} = 0,$$

if $p = 0$. In these circumstances, $\{f_n\}$ is L_p-convergent on each compact subset of M. Conversely, since M is a locally compact connected metric space, then f_n is L_p-convergent locally on M if it is L_p-convergent on each compact subset of M.

We say that a sequence $\{f_n\} \subset C(M, N)$ is L_p-divergent locally on M if each point z_0 of M has a disc $M(z_0; r) \subset M$ such that

$$\lim_{n \to \infty} \|f_n\|_{p, M(z_0; r)} = \infty,$$

if $0 < p \le \infty$, or for any $R > 0$

$$\lim_{n \to \infty} \|f_n\|_{0, M(z_0; r), R} = 0,$$

if $p = 0$. The following fact is obvious.

Lemma 4.6 *A subset $\mathcal{F}$ of $C(M, N)$ is L_p-normal in M iff every sequence of $\mathcal{F}$ contains a subsequence which is either L_p-convergent locally on M or L_p-divergent locally on M.*

A point $z_0 \in M$ is called L_p-normal for $\mathcal{F}$, or $\mathcal{F}$ is L_p-normal at z_0, if the family $\mathcal{F}$ is L_p-normal in some disc $M(z_0; r) \subset M$. Just like the proof of Theorem 1.9, we can prove

Theorem 4.28 *A family $\mathcal{F}$ in $C(M, N)$ is L_p-normal in M iff the family $\mathcal{F}$ is L_p-normal at each point of M.*

Thus we have the following general principle

Theorem 4.29 *Let $\mathcal{F}$ be any family in $C(M, N)$. Then there is a maximal open subset $F_p(\mathcal{F})$ of M on which $\mathcal{F}$ is L_p-normal. In particular, if f maps a metric space (M, d_g) into itself, then there is a maximal open subset $F_p(f) = F_p(f, d_g, \mu)$ of M on which the family of iterates $\{f^n\}$ is L_p-normal.*

As usual, define

$$J_p(\mathcal{F}) = M - F_p(\mathcal{F}), \quad J_p(f) = J_p(f, d_g, \mu) = M - F_p(f).$$

If M is complete, we have

$$J_p(\mathcal{F}) \subset J(\mathcal{F}), \quad J_p(f) \subset J(f).$$

If $p < q$ and if M is compact, we have

$$J_p(\mathcal{F}) \subset J_q(\mathcal{F}), \quad J_p(f) \subset J_q(f).$$

The following result is basic:

Theorem 4.30 *Take $f \in \mathrm{Diff}^\infty(M, M)$. Suppose that M is compact, orientable and that f, f^{-1} are orientation preserving. Let μ be the measure induced by a volume form Ω of M. Then*

$$f^{-1}(F_p(f)) = F_p(f), \quad f^{-1}(J_p(f)) = J_p(f) \quad (0 < p \leq \infty). \tag{4.27}$$

Therefore we also have

$$f(F_p(f)) = F_p(f), \quad f(J_p(f)) = J_p(f). \tag{4.28}$$

Theorem 4.31 *If $f \in \mathrm{Hom}(M, M)$, then*

$$f^{-1}(F_0(f)) = F_0(f), \quad f^{-1}(J_0(f)) = J_0(f).$$

The proof can be given after the fashion of Theorem 2.18 and Theorem 2.20.

Theorem 4.32 *For each positive integer $k \geq 2$,*

$$F_p(f) \subseteq F_p(f^k), \quad J_p(f^k) \subseteq J_p(f). \tag{4.29}$$

Further, if M is compact, and if f satisfies some Lipschitz condition

$$d_g(f(x), f(y)) \leq \lambda d_g(x, y)$$

on the manifold M, we have

$$F_p(f) = F_p(f^k), \quad J_p(f^k) = J_p(f). \tag{4.30}$$

Proof. It suffices to prove the assertion for the set F_p. Since the family $\{f^{kn}\}$ is contained in the family $\{f^n\}$, we thus obtain (4.29). Assume that M is compact. Given any domain $D \subset M$ we set

$$\mathcal{F} = \{f^n|_D \mid n \geq 0\}, \quad \mathcal{F}_j = \{f^j \circ f^{kn}|_D \mid n \geq 0\}.$$

Then obviously,

$$\mathcal{F} = \mathcal{F}_0 \cup ... \cup \mathcal{F}_{k-1},$$

and since f^j satisfy Lipschitz conditions on M, $\mathcal{F}$ is L_p-normal iff $\mathcal{F}_0$ is L_p-normal. $\quad\square$

For results related to this section, we refer the reader to Hu and Yang [122] and [124].

4.6 L_p-Equicontinuity

Let M and N be manifolds with distances d_g and d_h respectively, and let μ be a Borel measure on M such that $\mu(U) > 0$ for any non-empty open set U of M.

Definition 4.9 *A family $\mathcal{F}$ of mappings of a metric space (M, d_g) into a metric space (N, d_h) is called L_p-equicontinuous or an L_p-equicontinuous family at $z_0 \in M$ if and only if for every positive ε there exists a positive δ such that for all z in M, and for all f in $\mathcal{F}$,*

$$d_g(z, z_0) < \delta \Longrightarrow \left\{ \begin{array}{ll} \|f_z, f\|_{p, M(z_0;\delta)} < \varepsilon & : \quad p > 0 \\ \|f_z, f\|_{p, M(z_0;\delta),\alpha} < \varepsilon & : \quad p = 0, \alpha > 0 \end{array} \right.$$

where f_z is a constant mapping defined by $f_z(x) = f(z)$ for all $x \in M$. The family $\mathcal{F}$ is said L_p-equicontinuous on M iff $\mathcal{F}$ is L_p-equicontinuous at each point of M.

We have the following general principle.

Theorem 4.33 *Let $\mathcal{F}$ be any family of mappings, which map (M, d_g) into (N, d_h). Then there is a maximal open subset $F_{equ,p}(\mathcal{F})$ of M on which $\mathcal{F}$ is L_p-equicontinuous. In particular, if f maps a metric space (M, d_g) into itself, then there is a maximal open subset $F_{equ,p}(f) = F_{equ,p}(f, d_g, \mu)$ of M on which the family of iterates $\{f^n\}$ is L_p-equicontinuous.*

Define

$$J_{equ,p}(\mathcal{F}) = M - F_{equ,p}(\mathcal{F}), \quad J_{equ,p}(f) = J_{equ,p}(f, d_g, \mu) = M - F_{equ,p}(f).$$

If $z_0 \in F_{equ}(f)$, then for every positive ε there exists a positive δ such that for all z in M, and for all $n \in \mathbb{Z}_+$,

$$d_g(z, z_0) < \delta \Longrightarrow d_g(f^n(z), f^n(z_0)) < \varepsilon.$$

Thus if $0 < p \leq 1$, by using an elementary inequality, we have

$$\|f_z^n, f^n\|_{p, M(z_0;\delta)} \leq \|f_z^n, f_{z_0}^n\|_{p, M(z_0;\delta)} + \|f_{z_0}^n, f^n\|_{p, M(z_0;\delta)} \leq 2\mu(M)\varepsilon^p,$$

and if $p > 1$, by using Minkowski's inequality, we see

$$\|f_z^n, f^n\|_{p, M(z_0;\delta)} \leq \|f_z^n, f_{z_0}^n\|_{p, M(z_0;\delta)} + \|f_{z_0}^n, f^n\|_{p, M(z_0;\delta)} \leq 2\mu(M)^{\frac{1}{p}}\varepsilon,$$

for all $z \in M(z_0; \delta)$. Therefore $z_0 \in F_{equ,p}(f)$, i.e.,

$$F_{equ}(f) \subset F_{equ,p}(f) \; (p > 0),$$

and hence

$$J_{equ,p}(f) \subset J_{equ}(f) \; (p > 0).$$

Now we define a quantity which is closely related to L^p-equicontinuity. Let $f : M \longrightarrow M$ be a continuous mapping of a compact space M. Fix $p \in \mathbb{R}^+$. For any $n \geq 1, \varepsilon > 0$, we call a finite set $K \subset M$ an (n, ε)-*covering set of order* p if for each $x \in K$, there exists a positive number $\rho = \rho(x, n, \varepsilon)$ such that

$$\max_{0 \leq k \leq n-1} \|f_x^k, f^k\|_{p, M(x;\rho)} \leq \varepsilon,$$

and such that

$$\bigcup_{x \in K} M(x; \rho) = M.$$

Set

$$n_p(f, \mu, d_g; n, \varepsilon) = \min\{\#K \mid K \text{ is a } (n, \varepsilon)\text{-covering set of order } p\}.$$

Notice that for each $n \geq 1, \varepsilon \mapsto n_p(f, \mu, d_g; n, \varepsilon)$ is monotone decreasing. Define

$$h_p(f, \mu, d_g) = \lim_{\varepsilon \to 0} \limsup_{n \to \infty} \frac{1}{n} \log n_p(f, \mu, d_g; n, \varepsilon) \geq 0. \tag{4.31}$$

If μ is a probability measure and if $p \geq 1$, then each (n, ε)-covering set also is a (n, ε)-covering set of order p, and hence

$$n_p(f, \mu, d_g; n, \varepsilon) \leq n(f, d_g; n, \varepsilon).$$

Therefore we have

$$h_p(f, \mu, d_g) \leq h_{equ}(f), \quad p \geq 1. \tag{4.32}$$

If $0 < p < 1$, then each $(n, \varepsilon^{\frac{1}{p}})$-covering set is a (n, ε)-covering set of order p, and hence

$$n_p(f, \mu, d_g; n, \varepsilon) \leq n(f, d_g; n, \varepsilon^{\frac{1}{p}}).$$

Therefore we also have

$$h_p(f, \mu, d_g) \leq h_{equ}(f), \quad 0 < p < 1. \tag{4.33}$$

For any $n \geq 1, \varepsilon > 0, \delta > 0$, we call a finite set $K \subset M$ an (n, ε)-*covering set of order* 0 *for* δ if for each $x \in K$, there exists a positive number $\rho = \rho(x, n, \varepsilon, \delta)$ such that

$$\max_{0 \leq k \leq n-1} \|f_x^k, f^k\|_{p, M(x;\rho), \delta} \leq \varepsilon,$$

and such that

$$\bigcup_{x \in K} M(x; \rho) = M.$$

Set

$$n_0(f, \mu, d_g; n, \varepsilon, \delta) = \min\{\#K \mid K \text{ is a } (n, \varepsilon)\text{-covering set of order } 0 \text{ for } \delta\}.$$

Notice that for fixed δ, $n \geq 1$, the mapping $\varepsilon \mapsto n_0(f, \mu, d_g; n, \varepsilon, \delta)$ is monotone decreasing, and for fixed ε, $n \geq 1$, the mapping $\delta \mapsto n_0(f, \mu, d_g; n, \varepsilon, \delta)$ also is monotone decreasing. Define

$$h_0(f, \mu, d_g) = \lim_{\delta \to 0} \lim_{\varepsilon \to 0} \limsup_{n \to \infty} \frac{1}{n} \log n_0(f, \mu, d_g; n, \varepsilon, \delta) \geq 0. \tag{4.34}$$

Theorem 4.34 *Let M be a compact smooth manifold and let $f \in C(M, M)$ such that $h_p(f, \mu, d_g) > 0$. Then $J_{equ,p}(f) \neq \emptyset$.*

Proof. Here we only prove the theorem for the case $p > 0$. The case $p = 0$ can be proved similarly. Assume $J_{equ,p}(f) = \emptyset$. Then for every positive number ε and each $x \in M$, there exist a positive number $\delta = \delta(x)$ such that for $n \geq 0$

$$\|f_x^n, f^n\|_{p, M(x; \delta)} < \varepsilon.$$

Note that M is compact and note that $\{M(x; \delta(x))\}_{x \in M}$ is an open covering of M. Then there exists a finite set $K \subset M$ such that

$$\bigcup_{x \in K} M(x; \delta(x)) = M.$$

Then for $n \geq 1$, the set K is an (n, ε)-covering set of order p. Hence we have

$$n_p(f, \mu, d_g; n, \varepsilon) \leq \#K \quad (n \geq 1)$$

which yields $h_p(f, \mu, d_g) = 0$. This is a contradiction. □

Problem 4.2 *Are there some relations between $J_p(f)$ and $J_{equ,p}(f)$? When $h_p(f, \mu, d_g) > 0$, $J_p(f) \neq \emptyset$?*

For results related to this section, we refer the reader to Hu and Yang [122] and [124].

4.7 C^r-hyperbolicity

Definition 4.10 *Let D be a manifold with a Riemannian metric g. Let M be a manifold and take a family $\mathcal{F} \subset C(D, M)$. Let $x, y \in M$ be arbitrary points. A $\mathcal{F}$-chain α from x to y is the collection of mappings $f_i \in \mathcal{F}$ and $p_i, q_i \in D$ for $i = 0, ..., l$ such that*

$$f_0(p_0) = x, \quad f_i(q_i) = f_{i+1}(p_{i+1}) \quad (0 \leq i \leq l - 1), \quad f_l(q_l) = y.$$

Assume that the space M is $\mathcal{F}$-chain connected, that is, for arbitrary points $x, y \in M$, there exists a $\mathcal{F}$-chain from x to y. Then a pseudo-distance $d_{\mathcal{F}, g}$ between x and y is given by

$$d_{\mathcal{F}, g}(x, y) = \inf_{\alpha} \{ \sum_{i=0}^{l} d_g(p_i, q_i) \}, \tag{4.35}$$

where the infimum is taken for all $\mathcal{F}$-chains α from x to y, and where d_g is the distance function induced by g.

Then it is easy to see that

$$d(x, x) = 0, d(x, y) = d(y, x), d(x, z) \leq d(x, y) + d(y, z) \tag{4.36}$$

for $x, y, z \in M$, $d = d_{\mathcal{F}, g}$. In general, a mapping

$$d : M \times M \longrightarrow \mathbb{R}_+$$

satisfying the relation above is called a *pseudo-distance* which may identically vanish.

If $\mathcal{F} = C^r(D, M)$, then a $\mathcal{F}$-chain will be called a C^r-*chain* on D and a $\mathcal{F}$-chain connected space M will be called C^r-chain connected on D. For this case, we write

$$d^r_{M, D, g} = d_{\mathcal{F}, g}.$$

The most fundamental property of the pseudo-distance is described by the following so called *distance decreasing principle* .

Theorem 4.35 *If M and N are $\mathcal{F}$-chain connected and $\mathcal{G}$-chain connected, respectively, where $\mathcal{F} \subset C(D, M)$ and $\mathcal{G} \subset C(D, N)$, such that for $f \in C(M, N)$*

$$f(\mathcal{F}) = \{f \circ h \mid h \in \mathcal{F}\} \subset \mathcal{G},$$

then we have

$$d_{\mathcal{G}, g}(f(x), f(y)) \leq d_{\mathcal{F}, g}(x, y)$$

for all $x, y \in M$. In particular, if $\mathcal{G} = f(\mathcal{F})$ for $f \in \mathrm{Hom}(M, N)$,

$$d_{\mathcal{G}, g}(f(x), f(y)) = d_{\mathcal{F}, g}(x, y).$$

Furthermore, for every $f \in \mathcal{F}$, we have

$$d_{\mathcal{F}, g}(f(x), f(y)) \leq d_g(x, y).$$

The proof follows directly from the definition of the pseudo-distance $d_{\mathcal{F}, g}$. Take $\mathcal{F} = C^r(D, M)$, $\mathcal{G} = C^r(D, N)$ and $f \in C^r(M, N)$. Note that $f(\mathcal{F}) \subset \mathcal{G}$ and that $\mathcal{G} = f(\mathcal{F})$ if $f \in \mathrm{Diff}^r(M, N)$. We obtain the following result:

Corollary 4.1 *Let M and N be C^r-chain connected manifolds on D and let $f : M \longrightarrow N$ be a C^r mapping. Then we have*

$$d^r_{N, D, g}(f(x), f(y)) \leq d^r_{M, D, g}(x, y)$$

for all $x, y \in M$. In particular, if f is a C^r diffeomorphism, then

$$d^r_{N, D, g}(f(x), f(y)) = d^r_{M, D, g}(x, y).$$

Furthermore, for every C^r mapping $f : D \longrightarrow M$, we have

$$d^r_{M, D, g}(f(x), f(y)) \leq d_g(x, y).$$

It is known for the complex manifolds, see [143]. The following theorem says that $d_{\mathcal{F},g}$ is the largest pseudo-distance on M such that every mapping $f \in \mathcal{F}$ is distance decreasing.

Theorem 4.36 *Let M be a $\mathcal{F}$-chain connected manifold and d' any pseudo-distance on M such that*

$$d'(f(p), f(q)) \leq d_g(p, q), \quad p, q \in D,$$

for every mapping $f \in \mathcal{F}$. Then

$$d_{\mathcal{F},g}(x, y) \geq d'(x, y), \quad x, y \in M.$$

Proof. Let p_i, q_i, f_i $(i = 0, ..., l)$ as in the definition of $d_{\mathcal{F},g}$. Then

$$
\begin{aligned}
d'(x, y) &\leq \sum_{i=0}^{l} d'(w_{i-1}, w_i) = \sum_{i=0}^{l} d'(f_i(p_i), f_i(q_i)) \\
&\leq \sum_{i=0}^{l} d_g(p_i, q_i),
\end{aligned}
$$

where

$$w_{-1} = x, \quad w_i = f_i(q_i) \quad (0 \leq i \leq l).$$

Hence

$$d'(x, y) \leq \inf \sum_{i=0}^{l} d_g(p_i, q_i) = d_{\mathcal{F},g}(x, y).$$

$\square$

Corollary 4.2 *Let M be a C^r-chain connected manifold on D and d' any pseudo-distance on M such that*

$$d'(f(p), f(q)) \leq d_g(p, q), \quad p, q \in D,$$

for every C^r mapping $f : D \longrightarrow M$. Then

$$d^r_{M,D,g}(x, y) \geq d'(x, y), \quad x, y \in M.$$

For more properties of distance decreasing mappings, we refer the following:

Theorem 4.37 ([143]) *Let M be a connected, locally compact, separable space with a pseudodistance and N a connected, locally compact, complete metric space. The set $\mathcal{F}$ of distance decreasing mappings $f : M \longrightarrow N$ is locally compact with respect to the compact-open topology. In fact, if p is a point of M and K is a compact subset of N, then the subset $\mathcal{F}(p, K) = \{f \in \mathcal{F} \mid f(p) \in K\}$ of $\mathcal{F}$ is compact.*

Here we introduce an important space and its main metrics. For $x = (x_1, ..., x_n) \in \mathbb{R}^n$, recall that

$$\|x\| = \left(\sum_{i=1}^{n} x_i^2 \right)^{\frac{1}{2}}.$$

Then the a-ball $\mathbb{R}^n(a) = \{x \in \mathbb{R}^n \mid \|x\| < a\}$ is well-defined. With the Riemannian metric

$$g = \frac{4a^2}{(a^2 - \|x\|^2)^2} \sum_{i=1}^{n} dx_i \otimes dx_i, \tag{4.37}$$

$\mathbb{R}^n(a)$ is a space of constant curvature with sectional curvature -1. With respect to the Riemannian metric given by

$$g = \frac{4}{a^2 - \|x\|^2} \sum_{i=1}^{n} dx_i \otimes dx_i + \frac{4}{(a^2 - \|x\|^2)^2} \left(\sum_{i=1}^{n} x_i dx_i \right) \otimes \left(\sum_{i=1}^{n} x_i dx_i \right), \tag{4.38}$$

$\mathbb{R}^n(a)$ is a space of constant curvature with sectional curvature -1.

Assume that M is complete and consider the family $\mathcal{G}$ of geodesics $\gamma : \mathbb{R} \longrightarrow M$. We know that arbitrary points $x, y \in M$ can be connected by a geodesic. By arc parametrization on $\mathbb{R}(a)$ given by

$$s(t) = \int_0^t \frac{2a}{a^2 - x^2} dx = \log \frac{a+t}{a-t},$$

we obtain a family

$$\mathcal{F} = \{\, \gamma \circ s \mid \gamma \in \mathcal{G} \,\}$$

in $C^\infty(\mathbb{R}(a), M)$. Hence M is C^∞-chain connected on $\mathbb{R}(a)$. Consequently, M is C^∞-chain connected on $\mathbb{R}^n(a)$, so $d_{M,\mathbb{R}^n(a),g}^\infty$ is finite.

Remark. The *Carathéodory pseudodistance* $\rho_{M,D,g}^r$ of M is defined by

$$\rho_{M,D,g}^r(x,y) = \sup_{f \in C^r(M,D)} d_g(f(x), f(y))$$

for $x, y \in M$. The pseudodistance $\rho_{M,D,g}^r$ shares many properties with $d_{M,D,g}^r$. For example, the distance decreasing principle holds, that is, if $f : M \longrightarrow N$ is a C^r mapping, then

$$\rho_{N,D,g}^r(f(x), f(y)) \le \rho_{M,D,g}^r(x,y)$$

for all $x, y \in M$. In particular, if f is a C^r diffeomorphism, then

$$\rho_{N,D,g}^r(f(x), f(y)) = \rho_{M,D,g}^r(x,y).$$

Furthermore, for every C^r mapping $f : M \longrightarrow D$, we have

$$d_g(f(x), f(y)) \le \rho_{M,D,g}^r(x,y).$$

Also $\rho_{M,D,g}^r$ is the smallest pseudo-distance on M for which every C^r mapping $f : M \longrightarrow D$ is distance decreasing, that is, if ρ' is any pseudo-distance on M such that

$$d_g(f(p), f(q)) \le \rho'(p,q), \quad p, \ q \in D,$$

for every C^r mapping $f : M \longrightarrow D$, then

$$\rho_{M,D,g}^r(x,y) \le \rho'(x,y), \quad x, y \in M.$$

If each $f \in C^r(D, D)$ is distance decreasing for d_g, we have

$$d^r_{M,D,g}(x,y) \geq \rho^r_{M,D,g}(x,y)$$

for $x, y \in M$. Generally, for a family $\mathcal{H} \subset C(M, D)$, we can define

$$\rho_{\mathcal{H},g}(x,y) = \sup_{f \in \mathcal{H}} d_g(f(x), f(y))$$

for $x, y \in M$.

With respect to the pseudo-distance $d_{\mathcal{F},g}$, we obtain a set

$$J_{Kob}(\mathcal{F}, g)_x = \{y \in M \mid y \neq x, d_{\mathcal{F},g}(y, x) = 0\}$$

for each $x \in M$. Set

$$J_{Kob}(\mathcal{F}) = J_{Kob}(\mathcal{F}, g) = \bigcup_{x \in M} J_{Kob}(\mathcal{F}, g)_x,$$

$$F_{Kob}(\mathcal{F}) = F_{Kob}(\mathcal{F}, g) = M - J_{Kob}(\mathcal{F}, g).$$

Then for any $x, y \in F_{Kob}(\mathcal{F}, g)$ with $x \neq y$, we have $d_{\mathcal{F},g}(x,y) > 0$. By Theorem 4.35, we see that $J_{Kob}(\mathcal{F}, g)$ is invariant under C^r diffeomorphisms of M if $\mathcal{F} = C^r(D, M)$. If g' is another Riemannian metric on D with $c' d_g \leq d_{g'} \leq c d_g$, where c', c are constants, then

$$c' d_{\mathcal{F},g} \leq d_{\mathcal{F},g'} \leq c d_{\mathcal{F},g},$$

and hence $J_{Kob}(\mathcal{F}, g) = J_{Kob}(\mathcal{F}, g')$. The case $J_{Kob}(\mathcal{F}, g) = \emptyset$ is important which reduces to the following definition.

Definition 4.11 *Let D be a Riemannian manifold and let M be a $\mathcal{F}$-chain connected manifold. We say that M is $\mathcal{F}$-hyperbolic if there exists a Riemannian metric g on D such that the pseudo-distance $d_{\mathcal{F},g}$ is a distance, that is $x \neq y$ in M implies $d_{\mathcal{F},g}(x,y) > 0$. Such metric g is called a $\mathcal{F}$-hyperbolic frame of M. We say that M is complete $\mathcal{F}$-hyperbolic if there is a $\mathcal{F}$-hyperbolic frame g of M such that M is complete with respect to the distance $d_{\mathcal{F},g}$. Such metric g is called a (complete) $\mathcal{F}$-hyperbolic frame of M.*

If $\mathcal{F} = C^r(D, M)$, we say that M is (complete) C^r-hyperbolic on D if M is (complete) $\mathcal{F}$-hyperbolic, and a (complete) $\mathcal{F}$-hyperbolic frame of M is said to be a (complete) C^r-hyperbolic frame of M on D. We also say that M is (complete) C^r-hyperbolic if M is (complete) C^r-hyperbolic on some manifold D.

An interesting question is to exhibit some examples of $\mathcal{F}$-hyperbolic manifolds for some space D and some family $\mathcal{F} \subset C(D, M)$. Are there some examples of $\mathcal{F}$-hyperbolic manifolds with different $\mathcal{F}$-hyperbolic frames? Usually one has special interesting for the case of the space D of constant curvature with negative sectional curvature with respect to a Riemannian metric g which will serve as a hyperbolic frame. We will return to this question for the case of complex manifolds. The following proposition is immediate from Theorem 4.36.

Proposition 4.4 *Let M be a $\mathcal{F}$-chain connected manifold. If M admits a (complete) distance d' such that*

$$d'(f(p), f(q)) \leq d_g(p, q), \quad p, \ q \in D,$$

for every mapping $f \in \mathcal{F}$, then g is (complete) $\mathcal{F}$-hyperbolic frame of M.

If M is $\mathcal{F}$-hyperbolic, i.e., $d_{\mathcal{F},g}$ is a distance on M, or $J_{Kob}(\mathcal{F}, g) = \emptyset$, then we have $J_{equ}(\mathcal{F}) = \emptyset$. Conversely, if $J_{equ}(\mathcal{F}) = \emptyset$, can we find a $\mathcal{F}$-hyperbolic frame of M? We will discuss the question lately. Theorem 1.16 and Theorem 4.35 imply the following fact:

Proposition 4.5 *Let g be a $\mathcal{F}$-hyperbolic frame of M. If f is a continuous self-mapping on M such that $f(\mathcal{F}) \subset \mathcal{F}$, then*

$$J_{equ}(f, d_{\mathcal{F},g}) = \emptyset.$$

Corollary 4.3 *If f is a continuous self-mapping on a $\mathcal{F}$-hyperbolic compact manifold M such that $f(\mathcal{F}) \subset \mathcal{F}$, then $h_{top}(f) = 0$.*

Conjecture 4.2 *If M is compactly oriented, and if $f \in C^\infty(M, M)$ with $h_{top}(f) > 0$, then connected components of the Fatou set $F(f)$ are C^∞-hyperbolic on $\mathbb{R}^n(1)$.*

If the conjecture is true, the conclusion in Theorem 1.20 can be strengthened as $J(f) = J(g)$ under the conditions of Theorem 1.20 and Conjecture 4.2.

Problem 4.3 *If M is a compact manifold such that $J(f) = \emptyset$ for all $f \in C^r(M, M)$, can we find some C^r-hyperbolic frame of M on some manifold D?*

If $f \in C(M, M)$ is surjective, then $d_{\mathcal{F},g}$ can be defined for $\mathcal{F} = \{f^n\}_{n \in \mathbb{Z}_+}$ such that

$$d_{\mathcal{F},g}(f(x), f(y)) \leq d_{\mathcal{F},g}(x, y)$$

for all $x, y \in M$ since $f(\mathcal{F}) \subset \mathcal{F}$. For this case, we write

$$J_{Kob}(f) = J_{Kob}(\mathcal{F}).$$

If $J_{Kob}(f) = \emptyset$, then $d_{\mathcal{F},g}$ is a distance on M, and hence $J_{equ}(f, d_{\mathcal{F},g}) = \emptyset$. Further if M is compact, then $J(f) = \emptyset$. Here we suggest the following question:

Problem 4.4 *Let f be a continuous surjective self-mapping on a compact manifold M with $J(f) = \emptyset$. Can we have $J_{Kob}(f) = \emptyset$?*

Recall that a regular measure satisfies the property that the measure of a set is the inf of the measures of the open sets containing it. Take $f \in C(M, N)$ and let μ, ν be regular measures on M and N respectively. Recall that f is *measure decreasing* if

$$\nu(f(A)) \leq \mu(A)$$

for all measurable A. Instead of measurable A it would suffice to take open sets U.

Definition 4.12 *Let D and M be two manifolds of the same dimension m. Let A be a Borel measurable subset of M. A $\mathcal{F}$-chain α for A is the collection of mappings $f_i \in \mathcal{F} \subset C(D, M)$ and open sets U_i in D for $i = 0, ..., $ such that*

$$A \subset \bigcup_i f_i(U_i).$$

The space M is said to be covered by $\mathcal{F}$-chains if there exists a $\mathcal{F}$-chain for M. Then a measure $\mu_{\mathcal{F},\nu}$ is given by

$$\mu_{\mathcal{F},\nu}(A) = \inf_\alpha \sum_{i=1}^\infty \nu(U_i), \tag{4.39}$$

where the infimum is taken for all $\mathcal{F}$-chains α for A, where ν is a regular measure on D.

If $\mathcal{F} = C^r(D, M)$, a $\mathcal{F}$-chain for A will be called a C^r-chain on D for A, and the space M is said to be covered by C^r-chains on D if it is covered by $\mathcal{F}$-chains. For this case, we write $\mu^r_{M,D,\nu} = \mu_{\mathcal{F},\nu}$. By the definition, we can prove easily the measure decreasing principle:

Theorem 4.38 *Let M and N be manifolds covered by C^r-chains on D and let $f : M \longrightarrow N$ be a C^r mapping. Then we have*

$$\mu^r_{N,D,\nu}(f(A)) \leq \mu^r_{M,D,\nu}(A)$$

for all measurable A. In particular, if f is a C^r diffeomorphism, then

$$\mu^r_{N,D,\nu}(f(A)) = \mu^r_{M,D,\nu}(A).$$

Furthermore, for every C^r mapping $f : D \longrightarrow M$, we have

$$\mu^r_{M,D,\nu}(f(A)) \leq \nu(A).$$

For the complex manifold cases, see [143]. If M and N are covered by $\mathcal{F}$-chains and $\mathcal{G}$-chains, respectively, where $\mathcal{F} \subset C(D, M)$ and $\mathcal{G} \subset C(D, N)$, such that for $f \in C(M, N)$ $f(\mathcal{F}) \subset \mathcal{G}$, then we have

$$\mu_{\mathcal{G},\nu}(f(A)) \leq \mu_{\mathcal{F},\nu}(A)$$

for all measurable A. In particular, if $\mathcal{G} = f(\mathcal{F})$ for $f \in \text{Hom}(M, N)$,

$$\mu_{\mathcal{G},\nu}(f(A)) = \mu_{\mathcal{F},\nu}(A).$$

Furthermore, for every $f \in \mathcal{F}$, we have

$$\mu_{\mathcal{F},\nu}(f(A)) \leq \nu(A).$$

The following theorem is trivial.

Theorem 4.39 *Assume that M is covered by $\mathcal{F}$-chains and let μ' be a measure on M such that every mapping $f \in \mathcal{F}$ satisfies*

$$\mu'(f(A)) \leq \nu(A),$$

for every Borel measurable subset A of D. Then

$$\mu_{\mathcal{F},\nu} \geq \mu'.$$

Corollary 4.4 *Assume that M is covered by C^r-chains on D and let μ' be a measure on M such that every C^r mapping $f : D \longrightarrow M$ satisfies*

$$\mu'(f(A)) \leq \nu(A),$$

for every Borel measurable subset A of D. Then

$$\mu^r_{M,D,\nu} \geq \mu'.$$

Definition 4.13 *We define M to be measure $\mathcal{F}$-hyperbolic if there is a regular measure ν on D such that $\mu_{\mathcal{F},\nu}(U) > 0$ for all non-empty open subsets U of M. The space M is said to be measure C^r-hyperbolic on D if it is measure $C^r(D, M)$-hyperbolic.*

Theorem 4.40 *If M is $\mathcal{F}$-hyperbolic, then it is also measure $\mathcal{F}$-hyperbolic.*

Proof. The distances $d_{\mathcal{F},g}$ and d_g induce the m-dimensional Hausdorff measures μ' and ν of M and D, respectively, where $m = \dim M$. Since $d_{\mathcal{F},g}$ do not increase under mappings in $\mathcal{F}$, the induced Hausdorff measure do not increase under mappings in $\mathcal{F}$, that is, every mapping $f \in \mathcal{F}$ satisfies

$$\mu'(f(A)) \leq \nu(A),$$

for every Borel measurable subset A of D. From Theorem 4.39, we obtain

$$\mu_{\mathcal{F},\nu}(U) \geq \mu'(U).$$

On the other hand, $\mu'(U)$ is positive for every nonempty open set U (see Hurewicz and Wallman [128], Chapter VII).   □

An interesting question is to study the converse of the theorem.

Chapter 5

Hyperbolicity in complex dynamics

The convention in force throughout this chapter is that all complex manifolds are *locally compact connected spaces* and all objects defined on them (differential forms, Hermitian metrics, etc) are C^∞ unless stated to the contrary. It is well known that such complex manifolds under consideration are metrizable. A customary and useful device is to metrize these by imposing on them a Hermitian metric h, from which one derives a distance function $d(\ ,\ ) \equiv d_h(\ ,\ )$ which converts the manifold into a metric space.

5.1 Complex dynamics

Let M be a complex manifold of dimension m with a Hermitian metric h. We consider a dynamical system $\mathcal{F} = \{f^t\}_{t \in \kappa}$ on M, where κ is an additive semigroup, and $f^t \in \mathrm{Hol}(M, M)$ for all $t \in \kappa$. As usual, the system is called a complex dynamics.

Example 5.1 *If $\kappa = \mathbb{Z}_+$, the complex dynamics $\mathcal{F}$ is a cascade formed by iterate of a mapping $f \in \mathrm{Hol}(M, M)$, i.e.,*

$$f^0 = id, \quad f^n = f^{n-1} \circ f = f \circ f^{n-1}(n > 0).$$

If $\kappa = \mathbb{Z}$, then $\mathcal{F}$ is given by iterate of the mappings f and f^{-1}, where $f \in \mathrm{Aut}(M)$.

Example 5.2 *If $\kappa = \mathbb{R}_+$, the complex dynamics $\mathcal{F}$ is a semiflow or a 1-parameter semigroup of holomorphic mappings on M such that $f : \mathbb{R}_+ \longrightarrow \mathrm{Hol}(M, M)$ is a continuous semigroup homomorphism. If the image of f is contained in $\mathrm{Aut}(M)$, then f extends to a continuous group homomorphism of $\mathbb{R}$ into $\mathrm{Aut}(M)$, which results a flow or a 1-parameter group of holomorphic mappings on M. We also can study the case $\kappa = \mathbb{C}$.*

First of all we discuss a cascade $\mathcal{F} = \{f^n\}_{n \in \mathbb{Z}_+}$ generated by a mapping $f \in \mathrm{Hol}(M, M)$. Obviously, $\mathrm{Fix}(f)$ is a complex analytic set of M. Let $z = (z_1, ..., z_m)$ be holomorphic coordinates centered at $p \in M$. Note that in terms of the bases $\{\frac{\partial}{\partial z_j}, \frac{\partial}{\partial \bar{z}_j}\}$ for $T_{\mathbb{C}}(M)_p$, $(df)_p$ is given by

$$(df)_{p,\mathbb{C}} = \begin{pmatrix} f'(p) & 0 \\ 0 & \overline{f'(p)} \end{pmatrix}.$$

Then $p \in \text{Fix}(f)$ is non-degenerate if and only if $\det(f'(p) - I) \neq 0$. If M is compact, and if f has only non-degenerate fixed points, the *holomorphic Lefschetz Fixed-Point Formula*

$$\text{Lef}(f, \mathcal{O}) = \sum_{p \in \text{Fix}(f)} \frac{1}{\det(I - f'(p))}$$

holds, where $\text{Lef}(f, \mathcal{O})$ is the *holomorphic Lefschetz number* of f defined by

$$\text{Lef}(f, \mathcal{O}) = \sum_n (-1)^n \text{trance}(f^*|_{H^{0,n}_{\bar{\partial}}(M)}).$$

We regard M as the underlying differentiable manifold, and f as a C^∞ mapping on the differentiable manifold. Then we can define hyperbolic sets for the C^∞ mapping. If a compact forward invariant set $\Lambda \subset M$ is a *hyperbolic set* of a cascade $\{f^n\}_{n \in \mathbf{Z}_+}$ (or f), then for each point $z \in \Lambda$ the holomorphic tangent space $\mathbf{T}(M)_z$ decomposes into a direct sum

$$\mathbf{T}(M)_z = \mathbf{E}^s_z \oplus \mathbf{E}^u_z$$

of two subspaces, namely, a *stable space* $\mathbf{E}^s_z$ and an *unstable space* $\mathbf{E}^u_z$, where

$$\mathbf{E}^s_z = \Re^{-1}(E^s_z), \quad \mathbf{E}^u_z = \Re^{-1}(E^u_z),$$

where $\Re : \mathbf{T}(M)_z \longrightarrow T(M)_z$ is the real representation. Note that for $Z \in \mathbf{T}(M)_z$,

$$\Re(f'(z)Z) = (df)_{z, \mathbf{R}} \Re(Z).$$

Then the following properties are trivial: for $\xi \in \mathbf{E}^s_z, \eta \in \mathbf{E}^u_z, n \geq 0$

$$f'(z)\mathbf{E}^s_z \subset \mathbf{E}^s_{f(z)}, \quad f'(z)\mathbf{E}^u_z = \mathbf{E}^u_{f(z)},$$

$$\|(f^n)'(z)\xi\| \leq ae^{-cn}\|\xi\|, \quad \|(f^n)'(z)\eta\| \geq \frac{1}{a}e^{cn}\|\eta\|,$$

where a, c are positive constants that are independent of z, ξ, η, n. Conversely, if the properties above are true, then Λ is a hyperbolic set.

Note that dimensions of the subspaces $\mathbf{E}^s_z, \mathbf{E}^u_z$ are locally constant (as functions of $z \in \Lambda$), while the subspaces themselves depend continuously on z. The unions

$$\mathbf{E}^s = \bigcup_{z \in \Lambda} \mathbf{E}^s_z, \quad \mathbf{E}^u = \bigcup_{z \in \Lambda} \mathbf{E}^u_z$$

are vector subbundles of the restriction $\mathbf{T}(M)|_\Lambda$ of the holomorphic tangent bundle of M to Λ, and

$$\mathbf{T}(M)|_\Lambda = \mathbf{E}^s \oplus \mathbf{E}^u.$$

These subbundles are invariant with respect to f' and are called (holomorphic) *stable* and *unstable bundles*, respectively (for Λ, f and $\{f^n\}$). If the dimension $\dim \mathbf{E}^u_z$ is constant for a hyperbolic set Λ, then it is called the (complex) *Morse index* of Λ, denoted by $\mathbf{u}_\Lambda$. Obviously, $u_\Lambda = 2\mathbf{u}_\Lambda$.

Definition 5.1 ([126]) *A compact forward invariant set $\Lambda \subset M$ is said to be a (p,q)-type hyperbolic set of a cascade $\{f^n\}_{n \in \mathbb{Z}_+}$ (or f) if, for each point $z \in \Lambda$ the holomorphic tangent space $\mathbf{T}(M)_z$ decomposes into a direct sum $\mathbf{T}(M)_z = \mathbf{E}^s_z \oplus \mathbf{E}^u_z$ of two subspaces such that*

$$f'(z)\mathbf{E}^s_z \subset \mathbf{E}^s_{f(z)}, \quad f'(z)\mathbf{E}^u_z = \mathbf{E}^u_{f(z)},$$

$$(f^n)^*\omega^p \leq ae^{-cn}\omega^p \text{ on } \mathbf{E}^s_z, \quad (f^n)^*\omega^q \geq \frac{1}{a}e^{cn}\omega^q \text{ on } \mathbf{E}^u_z,$$

where a, c are positive constants that are independent of z and n, ω is the associated $(1,1)$-form of the metric h, and where $p = p(z), q = q(z)$ are non-negative integers with $0 \leq p(z) \leq \mathbf{s}(z)$ and $0 \leq q(z) \leq \mathbf{u}(z)$ for all $z \in \Lambda$, here

$$\mathbf{s}(z) = \dim \mathbf{E}^s_z, \quad \mathbf{u}(z) = \dim \mathbf{E}^u_z.$$

If a set $\Lambda \subset M$ is hyperbolic, then it is $(1,0)$, $(0,1)$ or $(1,1)$-type hyperbolic according to $\mathbf{E}^u_z = \{0\}$, $\mathbf{E}^s_z = \{0\}$ or $0 < \dim \mathbf{E}^u_z < m$ for all $z \in \Lambda$. The converse also is true. A compact forward invariant set $\Lambda \subset M$ is volume contracting (resp., expanding) hyperbolic iff it is $(m, 0)$-type (resp., $(0, m)$-type) hyperbolic.

Let Λ be (p, q)-type hyperbolic. If there exist positive integers k and l such that $0 \leq kp \leq \mathbf{s}$ and $0 \leq lq \leq \mathbf{u}$ hold on Λ, noting that

$$(f^n)^*\omega^{kp} \leq a'e^{-c'n}\omega^{kp} \text{ on } \mathbf{E}^s_z, \quad (f^n)^*\omega^{lq} \geq \frac{1}{a'}e^{c'n}\omega^{lq} \text{ on } \mathbf{E}^u_z,$$

where $a' = \max\{a^k, a^l\}$, $c' = \min\{kc, lc\}$, then Λ is (kp, lq)-type hyperbolic. Thus a hyperbolic set must be $(\mathbf{s}, \mathbf{u})$-type hyperbolic.

Here we define a function v_{f,ω^p} by

$$f^*(\omega^p) \wedge \omega^{m-p} = v_{f,\omega^p}\omega^m.$$

Take $x \in M$. A local coordinates z_k exist such that at point x

$$\omega = \frac{i}{2\pi}\sum_{k=1}^m dz_k \wedge d\bar{z}_k, \quad f^*(\omega) = \frac{i}{2\pi}\sum_{k,l} a_{kl} dz_k \wedge d\bar{z}_l,$$

then clearly $v_{f,\omega^p}(x)$ is the p-th elementary symmetric function of the eigenvalues of the Hermitian matrix (a_{kl}). The only general pointwise relation among v_{f,ω^p}'s is just that proved by Newton's inequality

$$v_{f,\omega^q}(x)^{1/q} \leq c_{pq}v_{f,\omega^p}(x)^{1/p} \quad (q \geq p).$$

Thus by the Newton's inequality, we see that if $q > p$, a p-type contracting hyperbolic set is q-type contracting hyperbolic, and that a q-type expanding hyperbolic set is p-type expanding hyperbolic

For $z \in M$, $Z \in \mathbf{T}(M)_z$, we also define the *(upper) Lyapunov exponent* (possibly infinite) of (z, Z) by

$$\chi(z, Z) = \limsup_{n \to \infty} \frac{1}{n}\log \|(f^n)'(z)Z\|.$$

By the definition, we clearly have

$$\chi(z, Z) = \chi(z, \Re(Z)) = \limsup_{n\to\infty} \frac{1}{n} \log \|(df^n)_{z,\mathbb{R}}\Re(Z)\|.$$

Thus for each point $z \in M$, the holomorphic tangent space $\mathbf{T}(M)_z$ decomposes into a direct sum

$$\mathbf{T}(M)_z = \mathbf{E}_z^s \oplus \mathbf{E}_z^u \oplus \mathbf{E}_z^c$$

according to

$$\chi(z, Z) \begin{cases} < 0 & : & Z \in \mathbf{E}_z^s \\ = 0 & : & Z \in \mathbf{E}_z^c \\ > 0 & : & Z \in \mathbf{E}_z^u. \end{cases}$$

Obviously, we have

$$\Re(\mathbf{E}_z^s) = E_z^s, \quad \Re(\mathbf{E}_z^u) = E_z^u, \quad \Re(\mathbf{E}_z^c) = E_z^c.$$

Note that $\chi(z, Z) = \chi(f(z), f'(z)Z)$ for $Z \in \mathbf{T}(M)_z$. We obtain

$$f'(z)\mathbf{E}_z^s \subset \mathbf{E}_{f(z)}^s, \quad f'(z)\mathbf{E}_z^u \subset \mathbf{E}_{f(z)}^u, \quad f'(z)\mathbf{E}_z^c \subset \mathbf{E}_{f(z)}^c.$$

If $f'(z) : \mathbf{T}(M)_z \longrightarrow \mathbf{T}(M)_{f(z)}$ is nondegenerate, we also have

$$\chi(f(z), W) = \chi(z, f'(z)^{-1}W),$$

for $W \in \mathbf{T}(M)_{f(z)}$. Thus in this case, we obtain

$$f'(z)\mathbf{E}_z^s = \mathbf{E}_{f(z)}^s, \quad f'(z)\mathbf{E}_z^u = \mathbf{E}_{f(z)}^u, \quad f'(z)\mathbf{E}_z^c = \mathbf{E}_{f(z)}^c.$$

Define the *stable* and *unstable functions* $\mathbf{s} : M \longrightarrow \mathbb{Z}_+$ and $\mathbf{u} : M \longrightarrow \mathbb{Z}_+$, respectively, by setting

$$\mathbf{s}(z) = \dim \mathbf{E}_z^s, \quad \mathbf{u}(z) = \dim \mathbf{E}_z^u.$$

Thus the set $\mathcal{K}_{Lya}(f)$ is given by

$$\mathcal{K}_{Lya}(f) = \{z \in M \mid \mathbf{s}(z) + \mathbf{u}(z) = \dim M\}.$$

Obviously, if Λ is a hyperbolic set, we have

$$\mathbf{s}|_\Lambda + \mathbf{u}|_\Lambda = \dim M,$$

and hence $\Lambda \subset \mathcal{K}_{Lya}(f)$. If $\mathbf{u}|_\Lambda$ is constant, the constant is the (complex) Morse index. We suspect $\mathrm{supp}\,\mathbf{u} \subset J_{equ}(f)$.

Next we discuss a semiflow $\mathcal{F} = \{f^t\}_{t\in\mathbb{R}_+}$. The following result is the Proposition 2.5.17 of Abate [2]:

Lemma 5.1 *Let* $\sigma : \mathbb{R}_+ \longrightarrow GL(m; \mathbb{C})$ *be a continuous semigroup homomorphism. Then* $\sigma(t) = \exp(\mathbf{A}t)$ *for some* $m \times m$ *complex matrix* $\mathbf{A}$.

Let $f : \mathbb{R}_+ \longrightarrow \mathrm{Hol}(M, M)$ be a continuous semigroup homomorphism on a complex manifold M. If p is an equilibrium point of the DS, then $\{(f^t)'(p)\}$ is a linear semigroup on $\mathbf{T}(M)_p$. The lemma above implies that there exists a linear operator $\mathbf{A} = \mathbf{A}_f$ acting on $\mathbf{T}(M)_p$ such that

$$(f^t)'(p) = \exp(\mathbf{A}t). \tag{5.1}$$

The operator $\mathbf{A}$ is the *spectral generator* of f at p, and the eigenvalues of $\mathbf{A}$ are the *spectral values* of f at p.

Lemma 5.2 *For a complex dynamics $\{f^t\}_{t\in\mathbb{R}_+}$ on a complex manifold M, then f^t is injective for all $t \geq 0$.*

Proof. First of all note that, since $(f^t)' \to id$ as $t \to 0$, for t small enough, every f^t is locally injective. Assume, by contradiction, that $f^{t_0}(z_1) = f^{t_0}(z_2)$ for some $t_0 > 0$ and $z_1, z_2 \in M$ with $z_1 \neq z_2$. If $t > t_0$, we have

$$f^t(z_1) = f^{t-t_0}(f^{t_0}(z_1)) = f^{t-t_0}(f^{t_0}(z_2)) = f^t(z_2).$$

Thus the motions $t \mapsto f^t(z_1)$ and $t \mapsto f^t(z_2)$ start at distinct points, meet at $t = t_0$, and coincide thereafter. Let t_0 be the least $t > 0$ such that $f^t(z_1) = f^t(z_2)$. Then no f^t can be injective in a neighborhood of $z_0 = f^{t_0}(z_1)$, and this is a contradiction. $\square$

Finally assume that $\kappa = \mathbb{R}$. Then each phase flow $\{f^t\}$ induces a holomorphic vector field X as follows: For every point $p \in M$, $Z(p) \in \mathbf{T}(M)_p$ is the vector tangent to the curve $z(t) = f^t(p)$ at $t = 0$, i.e.,

$$Z(p) = \frac{dz(t)}{dt}|_{t=0} = \dot{z}(0).$$

The vector field Z is the phase velocity field of the flow, or the generating field of the 1-parameter group $\{f^t\}$. The latter completely determines the flow: for a fixed p and variable t, $z(t) = f^t(p)$ satisfies

$$\dot{z}(t) = \frac{dz(t)}{dt} = Z(z(t)). \tag{5.2}$$

so that $f^t(p)$ is a solution of the differential equation

$$\dot{z} = Z(z) \tag{5.3}$$

with initial condition p.

Let $U \subset M$ be a coordinate neighborhood with holomorphic coordinates $(z_1, ..., z_m)$ and write

$$Z = \sum_{k=1}^{m} \zeta_k \frac{\partial}{\partial z_k}. \tag{5.4}$$

If our initial point p is represented by coordinates $(z_1^0, ..., z_m^0)$, then the differential equation (A.9) with initial condition p is the system of first-order ordinary differential equations

$$\frac{dz_k}{dt} = \zeta_k(z_1, ..., z_m) \tag{5.5}$$

with initial conditions

$$z_k(0) = z_k^0, \quad k = 1, ..., m. \tag{5.6}$$

If setting $z_k = x_k + iy_k$, $z_k^0 = x_k^0 + iy_k^0$, and write

$$X = \sqrt{2}\Re(Z) = \sum_{k=1}^{m}\left(\xi_k\frac{\partial}{\partial x_k} + \eta_k\frac{\partial}{\partial y_k}\right),$$

then the system is of the form

$$\frac{dx_k}{dt} = \xi_k(x_1, ..., x_m, y_1, ..., y_m),$$

$$\frac{dy_k}{dt} = \eta_k(x_1, ..., x_m, y_1, ..., y_m),$$

with initial conditions

$$x_k(0) = x_k^0, \quad y_k(0) = y_k^0, \quad k = 1, ..., m.$$

Conversely, let Z be a holomorphic vector field on a complex manifold M. Recall that a curve $z(t)$ in M is called an integral curve of Z if, for every parameter value t, the vector $Z(z(t))$ is tangent to the curve at $z(t)$. We know from the standard theory of ordinary differential equations that for any point $p \in M$, there is an unique integral curve $z(t)$ of Z, defined for $|t| < \varepsilon$ for some $\varepsilon > 0$, such that $z(0) = p$. More generally, one has the following properties:

Proposition 5.1 *Let Z be a holomorphic vector field on a complex manifold M. For any point $p \in M$, there exist a neighborhood U of p, a positive number ε and unique holomorphic mappings $f^t : U \longrightarrow M$ such that*
   1) for each $t \in (-\varepsilon, \varepsilon)$, f^t is a biholomorphic mapping of U onto the open set $f^t(U)$ of M;
   2) if $t, s, t + s \in (-\varepsilon, \varepsilon)$, and if $z, f^s(z) \in U$, then $f^{t+s}(z) = f^t(f^s(z))$.

The result can be derived by following arguments used in the proof of Proposition A.1. Also see Narasimhan [180] or Hörmander [115].

Similarly, a compact invariant set $\Lambda \subset M$ is a *hyperbolic set* of a flow $\{f^t\}_{t\in\mathbb{R}}$ on M if and only if, for each point $z \in \Lambda$ the holomorphic tangent space $\mathbf{T}(M)_z$ decomposes into a direct sum

$$\mathbf{T}(M)_z = \mathbf{E}_z^s \oplus \mathbf{E}_z^u \oplus \mathbf{E}_z^c$$

of subspaces with the following properties: for $\xi \in \mathbf{E}_z^s, \eta \in \mathbf{E}_z^u, t \geq 0$

$$(f^t)'(z)\mathbf{E}_z^s = \mathbf{E}_{f^t(z)}^s, \quad (f^t)'(z)\mathbf{E}_z^u = \mathbf{E}_{f^t(z)}^u, \quad (f^t)'(z)\mathbf{E}_z^c = \mathbf{E}_{f^t(z)}^c,$$

$$\|(f^t)'(z)\xi\| \leq ae^{-ct}\|\xi\|, \quad \|(f^t)'(z)\eta\| \geq \frac{1}{a}e^{ct}\|\eta\|, \tag{5.7}$$

where a, c are positive constants that are independent of z, ξ, η, t. Here $\mathbf{E}^c$ is spanned by the vector field generating f^t. Here we also have

$$\Re(\mathbf{E}_z^s) = E_z^s, \quad \Re(\mathbf{E}_z^u) = E_z^u, \quad \Re(\mathbf{E}_z^c) = E_z^c.$$

If $\kappa = \mathbb{C}$, the condition (5.7) will be replaced by

$$\|(f^t)'(z)\xi\| \le ae^{-c\operatorname{Re}(t)}\|\xi\|, \quad \|(f^t)'(z)\eta\| \ge \frac{1}{a}e^{c\operatorname{Re}(t)}\|\eta\|. \tag{5.8}$$

We also can define (p, q)-type hyperbolic set for a flow.

Remark. For complex dynamics, the family $\{C_x\}$ in Definition 1.8 will be taken as the family of analytic sets.

5.2   Schwarz lemma

By a *Riemann surface* we mean a connected complex manifold of complex dimension one. Two such surfaces M and M' are *conformally isomorphic* if there is a biholomorphic mapping from M onto M', i.e., a homeomorphism from M onto M' which is holomorphic, with holomorphic inverse. According to Poincaré and Koebe, there are only three kinds of simply connected Riemann surfaces, up to isomorphism.

Theorem 5.1 (Uniformization Theorem) *Any simply connected Riemann surface is conformally isomorphic either to the plane $\mathbb{C}$, or to the open unit disk Δ, or to the Riemann sphere $\mathbb{P}^1 \cong \mathbb{C} \cup \{\infty\}$.*

The proof may be found in Springer, Farkas & Kra, Ahlfors, or in Beardon.

Theorem 5.2 (Schwarz-Pick lemma, cf. [143]) *Assume $f \in \operatorname{Hol}(\Delta, \Delta)$. Then*

$$\frac{|f'|}{1 - |f|^2} \le \frac{1}{1 - |z|^2} \quad \text{for } z \in \Delta,$$

and the equality at a single point z implies that $f \in \operatorname{Aut}(\Delta)$.

Let D be an open subset on $\mathbb{C}$. A Hermitian metric h on D is given by

$$h = a(z)dz \otimes d\bar{z},$$

where $a(z)$ is a positive C^∞-function on D. The induced Riemannian metric is of the following form:

$$ds_D^2 = 2\operatorname{Re}(h) = 2a(z)dzd\bar{z}.$$

The associated $(1,1)$-form of h is given by

$$\omega = ia(z)dz \wedge d\bar{z}.$$

The *Gaussian curvature* of h is defined by

$$K(z) = -\frac{1}{a(z)}\frac{\partial^2 \log a(z)}{\partial z \partial \bar{z}}.$$

Specially we consider the Käehler metric h on Δ given by

$$h = \frac{2}{(1 - |z|^2)^2}dz \otimes d\bar{z},$$

then the inequality in Theorem 5.2 may be written as follows:

$$f^* ds_\Delta^2 \leq ds_\Delta^2,$$

or

$$d_h(f(z), f(z')) \leq d_h(z, z')$$

for the associated distance function d_h. The metric h (or ds_Δ^2) is called the *Poincaré metric* or the *Poincaré-Bergman metric* of Δ. We note that the Gaussian curvature of the metric h is equal to -1 everywhere.

Generally, the metric

$$h_r = \frac{2r^2}{(r^2 - |z|^2)^2} dz \otimes d\bar{z},$$

on $\mathbb{C}(r)$ has Gaussian curvature -1. We will abbreviate as

$$ds_r^2 = ds_{\mathbb{C}(r)}^2 = \frac{4r^2}{(r^2 - |z|^2)^2} dz d\bar{z}.$$

By a simple calculation we have

$$d_{h_r}(z, w) = \log \frac{r + |\alpha|}{r - |\alpha|} \quad (z, w \in \mathbb{C}(r)),$$

where

$$\alpha = \frac{r^2(w - z)}{r^2 - \bar{z}w}.$$

The following theorem of Ahlfors[9] generalizes the Schwarz-Pick lemma.

Theorem 5.3 *Let N be a Riemann surface with a Käehler metric ds_N^2 whose Gaussian curvature is bounded above by a negative constant $-B$. Then every $f \in \mathrm{Hol}(\mathbb{C}(r), N)$ satisfies*

$$f^* ds_N^2 \leq \frac{1}{B} ds_r^2.$$

A proof can be found in [143].

If M is a completely arbitrary Riemann surface, then the universal covering $\tilde{M}$ is a well defined simply connected Riemann surface with a canonical projection mapping

$$\pi : \tilde{M} \longrightarrow M.$$

According to the Uniformization Theorem, this universal covering $\tilde{M}$ must be conformally isomorphic to one of the three model surfaces. Thus one has the following

Lemma 5.3 ([172]) *Every Riemann surface M is conformally isomorphic to quotient of the form $\tilde{M}/\Gamma$, where $\tilde{M}$ is a simply connected Riemann surface which is conformally isomorphic to either Δ, $\mathbb{C}$, or $\mathbb{P}^1$. Here Γ is a discrete subgroup of $\mathrm{Aut}(\tilde{M})$ such that every non-identity element of Γ acts without fixed points on $\tilde{M}$.*

This discrete subgroup $\Gamma \subset \text{Aut}(\tilde{M})$ can be identified with the fundamental group $\pi_1(M)$ (cf. [172]). The elements of Γ are called *deck transformations*. They can be characterized as mappings $\gamma : \tilde{M} \longrightarrow \tilde{M}$ which satisfy $\pi \circ \gamma = \pi$.

If $\tilde{M} \cong \Delta$, M is said to be *Poincaré hyperbolic* in this book. Every Poincaré hyperbolic surface M possesses a unique *Poincaré metric*, which is complete, with Gaussian curvature identically equal to -1. To construct this metric, we note that the Poincaré metric on Δ is invariant under the action Γ. Hence there is one and only one metric on M such that the $\pi : \Delta \longrightarrow M$ is a local isometry.

Let M be a complex manifold of dimension m. By a *volume form* on M, we mean a form of type (m, m), which locally in terms of complex coordinates z_k can be written as

$$\Psi(z) = \rho(z) \prod_{k=1}^{m} \frac{i}{2\pi} dz_k \wedge d\bar{z}_k,$$

where ρ is a positive C^∞ function. Thus a volume form is a metric on the *canonical bundle*

$$K_M = \bigwedge_m \mathbf{T}^*(M).$$

We define the *Ricci form* of Ψ to be the Chern form of this metric, so $\text{Ric}(\Psi)$ is the real $(1, 1)$-form given by

$$\text{Ric}(\Psi) = dd^c \log \rho = -\frac{i}{2\pi} \sum_{k,l} K_{kl}(z) dz_k \wedge d\bar{z}_l,$$

where

$$K_{kl}(z) = -\frac{\partial^2 \log \rho(z)}{\partial z_k \partial \bar{z}_l}$$

is the *Ricci tensor* of M with respect to Ψ. In the case when the *Griffiths function*

$$G(\Psi) = \frac{1}{m!} \text{Ric}(\Psi)^m / \Psi$$

associated with the original volume form Ψ is constant and $\text{Ric}(\Psi)$ is positive, i.e., the matrix $(-K_{kl}(z))$ is Hermitian and positive definite for all values of z, the manifold is called *Einsteinian*. This condition is independent of the choice of holomorphic coordinates $z_1, ..., z_m$.

Example 5.3 *For a Käehler metric $h = a(z)dz \otimes d\bar{z}$ on an open subset of $\mathbb{C}$, the associated volume element is*

$$\Theta = ia(z)dz \wedge d\bar{z}.$$

Then we have

$$G(\Theta) = \text{Ric}(\Theta)/\Theta = \frac{1}{2\pi a} \frac{\partial^2 \log a}{\partial z \partial \bar{z}} = -\frac{1}{2\pi} K,$$

where K is the Gaussian curvature of h. In particular, for the volume element

$$\Theta_r = i \frac{2r^2}{(r^2 - |z|^2)^2} dz \wedge d\bar{z},$$

of the metric h_r on $\mathbb{C}(r)$, we have

$$\mathrm{Ric}(\Theta_r) = \frac{1}{2\pi}\Theta_r.$$

Example 5.4 *Let*

$$\omega = \frac{i}{2\pi}\sum_{k,l} h_{kl}(z)dz_k \wedge d\bar{z}_l,$$

be a positive $(1,1)$-form. Also ω has an associated Hermitian metric h_ω defined by the matrix $(\frac{1}{2\pi}h_{kl})$, and the associated volume element

$$\Theta = \frac{1}{m!}\omega^m = \det(h_{kl}) \prod_{k=1}^{m} \frac{i}{2\pi}dz_k \wedge d\bar{z}_k.$$

Then the Ricci tensor is given by

$$K_{kl} = -\frac{\partial^2 \log \det(h_{kl})}{\partial z_k \partial \bar{z}_l}.$$

The closed 2-form $-\mathrm{Ric}(\Theta)$ is known to represent the first Chern class $c_1(M)$ of M. The Hermitian metric h_ω is called an Einstein-Kähler metric if there exists a constant c such that

$$\mathrm{Ric}(\Theta) = c\,\omega.$$

Then a necessary condition for the existence of an Einstein-Kähler metric on M is that

$$c_1(M) = -c[\omega],$$

where $[\omega]$ is the cohomology of ω, which is equivalent to

$$c_1(M) > 0, \quad c_1(M) = 0 \quad or \quad c_1(M) < 0.$$

Example 5.5 *Let M be a complex manifold of dimension m and $\mathcal{H}$ the Hilbert space of holomorphic m-forms θ such that*

$$i_m \int_M \theta \wedge \bar{\theta} < \infty,$$

where

$$i_m = (\frac{i}{2\pi})^m (-1)^{\frac{m(m-1)}{2}} m!.$$

The inner product in $\mathcal{H}$ is defined by

$$(\theta, \eta) = \frac{i_m}{m!} \int_M \theta \wedge \bar{\eta}$$

for $\theta, \eta \in \mathcal{H}$. Let $\theta_0, \theta_1, \ldots$ be an orthonormal basis for $\mathcal{H}$. Assume that for every point $x \in M$, there is a $\theta \in \mathcal{H}$ with $\theta(x) \neq 0$. Now the (m,m)-form

$$\Psi = \frac{i_m}{m!}\sum_k \theta_k \wedge \bar{\theta}_k \tag{5.9}$$

is an intrinsically defined volume form on M with $\mathrm{Ric}(\Psi) \geq 0$. If $\mathrm{Ric}(\Psi) > 0$, the associated Hermitian metric $h_{\mathrm{Ric}(\Psi)}$ is called the Bergman metric of M. Further, if M is homogeneous, the Griffiths function $G(\Psi)$ is constant. Hence M is Einsteinian.

For a bounded domain M of $\mathbb{C}^m$, $\mathrm{Ric}(\Psi)$ is always positive. If M is a complete canonical algebraic manifold, then $\mathrm{Ric}(\Psi)$ is positive. Hence there is a positive constant $c > 0$ such that

$$cG(\Psi) \geq 1,$$

(see Griffiths[99]).

Example 5.6 *On $\Delta^m(r)$ with $r = (r_1, ..., r_m)$, consider the following positive $(1,1)$-form*

$$\omega = \frac{i}{2\pi} \sum_k \frac{2r_k^2}{(r_k^2 - |z_k|^2)^2} dz_k \wedge d\bar{z}_k, \tag{5.10}$$

with the associated volume element

$$\Theta(r) = \frac{1}{m!}\omega^m = \prod_{k=1}^m \frac{2r_k^2}{(r_k^2 - |z_k|^2)^2} \frac{i}{2\pi} dz_k \wedge d\bar{z}_k. \tag{5.11}$$

We can obtain

$$\mathrm{Ric}(\Theta(r)) = \sum_{k=1}^m dd^c \log \frac{2r_k^2}{(r_k^2 - |z_k|^2)^2} = \omega. \tag{5.12}$$

Thus the Hermitian metric h_ω is an Einstein-Kähler metric.

Example 5.7 *On $\mathbb{C}^m(r)$ there is the standard positive $(1,1)$-form*

$$\omega = \frac{i}{2\pi} a\{\sum_k \frac{1}{r^2 - \|z\|^2} dz_k \wedge d\bar{z}_k + \frac{4\|z\|^2}{(r^2 - \|z\|^2)^2} \partial\|z\| \wedge \bar{\partial}\|z\|\} \quad (a > 0) \tag{5.13}$$

with

$$\Theta_{r,a} = \frac{1}{m!}\omega^m = \frac{a^m r^2}{(r^2 - \|z\|^2)^{m+1}} \prod_{k=1}^m \frac{i}{2\pi} dz_k \wedge d\bar{z}_k, \tag{5.14}$$

$$\mathrm{Ric}(\Theta_{r,a}) = -(m+1)dd^c \log(r^2 - \|z\|^2) = \frac{m+1}{a}\omega. \tag{5.15}$$

Hence h_ω is an Einstein-Kähler metric.

The components h_{kl} of the metric h_ω in Example 5.7 is given by

$$\frac{2\pi}{a}(r^2 - \|z\|^2)^2 h_{kl} = (r^2 - \|z\|^2)\delta_{kl} + \bar{z}_k z_l.$$

We differentiate this identity with respect to $\frac{\partial}{\partial z_i}$ and $\frac{\partial^2}{\partial z_i \partial \bar{z}_j}$ and set $z_1 = \cdots = z_m = 0$. Then we obtain

$$h_{kl} = \frac{a}{2\pi r^2}\delta_{kl}, \quad \frac{\partial h_{kl}}{\partial z_i} = 0,$$

$$\frac{\partial^2 h_{kl}}{\partial z_i \partial \bar{z}_j} = \frac{a}{2\pi r^4}\{\delta_{ij}\delta_{kl} + \delta_{kj}\delta_{li}\},$$

$$K_{klij} = -\frac{\partial^2 h_{kl}}{\partial z_i \partial \bar{z}_j} + \sum_{p,q} h^{pq} \frac{\partial h_{kp}}{\partial z_i} \frac{\partial h_{ql}}{\partial \bar{z}_j} = -\frac{a}{2\pi r^4}\{\delta_{ij}\delta_{kl} + \delta_{kj}\delta_{li}\},$$

at the origin. We see that h_ω is of constant holomorphic sectional curvature $-\frac{4\pi}{a}$ at the origin. Since we know that $\mathbb{C}^m(r)$ admits a transitive group of holomorphic isometric transformations, we may conclude that h_ω is of constant holomorphic sectional curvature $-\frac{4\pi}{a}$ everywhere.

Take $a = 4\pi$ in Example 5.7 and denote the Einstein-Kähler metric h_ω on $\mathbb{C}^m(r)$ by $h_{r,m}$, which is given by

$$h_{r,m} = \sum_{k=1}^m \frac{2}{r^2 - \|z\|^2} dz_k \otimes d\bar{z}_k + \frac{2}{(r^2 - \|z\|^2)^2} \left(\sum_{k=1}^m \bar{z}_k dz_k \right) \otimes \left(\sum_{k=1}^m z_k d\bar{z}_k \right). \qquad (5.16)$$

Then $h_{r,m}$ is of constant holomorphic sectional curvature -1 everywhere. Obviously, $h_{r,1} = h_r$ is just the Poincaré metric on $\mathbb{C}(r)$. Abbreviate

$$ds_{r,m}^2 = 2\mathrm{Re}(h_{r,m}) = ds_{\mathbb{C}^m(r)}^2.$$

We will use the following generalization of Schwarz lemma:

Lemma 5.4 ([143]) *If N is a Hermitian manifold whose holomorphic sectional curvature is bounded above by a negative constant $-B$, then every holomorphic mapping $f : \mathbb{C}(r) \longrightarrow N$ satisfies*

$$f^* ds_N^2 \le \frac{1}{B} ds_r^2.$$

Theorem 5.4 *If N is a Hermitian manifold whose holomorphic sectional curvature is bounded above by a negative constant $-B$, then every holomorphic mapping $f : \mathbb{C}^m(r) \longrightarrow N$ satisfies*

$$f^* ds_N^2 \le \frac{1}{B} ds_{r,m}^2.$$

Proof. Take $\xi = (\xi_1, ..., \xi_m) \in \mathbb{C}^m$ with $\|\xi\| = 1$ and let $\iota : \mathbb{C}(r) \longrightarrow \mathbb{C}^m(r)$ be the embedding defined by $\iota(z) = z\xi$. Then we have $\iota^* h_{r,m} = h_r$, i.e., ι is isometric. Let X be a tangent vector of $\mathbb{C}^m(r)$ at the origin. For a suitable ξ, we can find a tangent vector Y of $\mathbb{C}(r)$ at the origin such that $\iota_* Y = X$. Then for any $f \in \mathrm{Hol}(\mathbb{C}^m(r), N)$, by Lemma 5.4 we see

$$\|f_* X\|^2 = \|(f \circ \iota)_* Y\|^2 \le \frac{1}{B} \|Y\|^2 = \frac{1}{B} \|X\|^2.$$

Since $\mathbb{C}(r)$ is homogeneous, the inequality

$$\|f_* X\|^2 \le \frac{1}{B} \|X\|^2,$$

holds for all tangent vectors X of $\mathbb{C}(r)$.   □

Corollary 5.1 *Assume $f \in \mathrm{Hol}(\mathbb{C}^m(r), \mathbb{C}^m(r))$. Then*

$$f^* ds_{r,m}^2 \le ds_{r,m}^2.$$

Consequently, we obtain $J_{equ}(f) = \emptyset$ for $f \in \mathrm{Hol}(\mathbb{C}^m(r), \mathbb{C}^m(r))$. Similarly, by using Lemma 5.4 one can prove

Theorem 5.5 ([143]) *Let Δ_r^m be the direct product of m copies of disk $\mathbb{C}(r)$ of radius r with the Hermitian metric*

$$h_{r;m} = \sum_{k=1}^{m} \frac{2r^2}{(r^2 - |z_k|^2)^2} dz_k \otimes d\bar{z}_k. \tag{5.17}$$

If N is a Hermitian manifold whose holomorphic sectional curvature is bounded above by a negative constant $-B$, then every holomorphic mapping $f : \Delta_r^m \longrightarrow N$ satisfies

$$f^* ds_N^2 \le \frac{1}{B} ds_{r;m}^2,$$

where $ds_{r;m}^2 = 2\mathrm{Re}(h_{r;m}) = ds_{\Delta_r^m}^2$.

Since ds_r^2 has constant curvature -1, the holomorphic sectional curvature of $ds_{r;m}^2$ varies between -1 and $-\frac{1}{m}$. In particular, if $f \in \mathrm{Hol}(\Delta_r^m, \Delta_r^m)$, then

$$f^* ds_{r;m}^2 \le m ds_{r;m}^2.$$

Yau [276] proved the following generalization of Schwarz lemma.

Theorem 5.6 *Let M be a complete Kähler manifold with Ricci curvature bounded from below by a constant K_1. Let N be another Hermitian manifold with holomorphic bisectional curvature bounded from above by a negative constant K_2. Then if there is a non-constant holomorphic mapping f from M into N, we have $K_1 \le 0$ and*

$$f^* ds_N^2 \le \frac{K_1}{K_2} ds_M^2.$$

Lemma 5.5 *Let M be a complex manifold of dimension m and let Ψ be a volume form on M such that $\mathrm{Ric}(\Psi)$ is positive, and such that there exists a constant $c > 0$ satisfying $cG(\Psi) \ge 1$. Then for all holomorphic mappings $f : \mathbb{C}^m(r) \longrightarrow M$, we have*

$$.^*\Psi \le c(\frac{m+1}{a})^m \Theta_{r,a},$$

where $\Theta_{r,a}$ is given in Example 5.7.

Proof. Write

$$f^*\Psi = u\Theta_{r,a},$$

where u is a non-negative function on $\mathbb{C}^m(r)$. We first show that u has a maximum in $\mathbb{C}^m(r)$. To see this we restrict f to the smaller ball $\mathbb{C}^m(t)$ where $0 < t < r$. Write

$$f^*\Psi = u_t \Theta_{t,a}.$$

From (5.15) it follows that for fixed $z \in \mathbb{C}^m(t)$,

$$\lim_{t \to r} u_t(z) = u(z),$$

and that for fixed t,

$$\lim_{z \to \mathbb{C}^m \langle t \rangle} u_t(z) = 0,$$

since $\Theta_{t,a}$ goes to infinity everywhere at $\mathbb{C}^m \langle t \rangle$. Thus u_t has an interior maximum point for $t < r$. Hence u has a maximum in $\mathbb{C}^m(r)$.

Let z_0 be a maximum point for u. If $u(z_0) = 0$, we are done, so we may assume $u(z_0) \neq 0$. Then

$$dd^c \log u = \text{Ric}(f^*\Psi) - \text{Ric}(\Theta_{r,a}) = f^*\text{Ric}(\Psi) - \text{Ric}(\Theta_{r,a}).$$

Since at a maximum point we have

$$dd^c \log u(z_0) \leq 0,$$

we obtain the inequality

$$f^*\text{Ric}(\Psi)(z_0) \leq \text{Ric}(\Theta_{r,a})(z_0).$$

We take the m-th power of each side and divide by $m!$. By definition of the Griffiths function, we know that

$$\Psi \leq \frac{c}{m!}\text{Ric}(\Psi)^m.$$

This shows that

$$f^*\Psi(z_0) \leq c(\frac{m+1}{a})^m \Theta_{r,a}(z_0).$$

Hence $u(z_0) \leq c(\frac{m+1}{a})^m$, and the lemma follows.   □

For more detail of the lemma, see Chern[61], Kobayashi[144], and Lang[153]. Note that $G(\Theta_{r,a}) = \left(\frac{m+1}{a}\right)^m$. Thus we have

$$f^*\Theta_{r,a} \leq \Theta_{r,a}$$

for $f \in \text{Hol}(\mathbb{C}^m(r), \mathbb{C}^m(r))$. Hence f is neither volume expanding nor p-type expanding hyperbolic $(1 \leq p \leq m)$ on any subset of $\mathbb{C}^m(r)$.

By a *pseudo volume form* Ψ we shall mean a continuous (m, m)-form which is C^∞ outside a proper complex subspace, and which locally in terms of complex coordinates z_k can be expressed as

$$\Psi(z) = |g(z)|^{2q}\rho(z) \prod_{k=1}^m \frac{i}{2\pi}dz_k \wedge d\bar{z}_k,$$

where q is some fixed positive rational number, g is holomorphic not identically zero, and ρ is a positive C^∞ function. We can define $\text{Ric}(\Psi)$ for a pseudo volume form just as we did for a volume form, by the formula

$$\text{Ric}(\Psi) = dd^c \log \rho,$$

since g is assumed to be holomorphic, $dd^c \log |g(z)|^{2q} = 0$ wherever $g \neq 0$.

Lemma 5.6 *Let M be a complex manifold of dimension m and let Ψ be a pseudo volume form on M such that $\mathrm{Ric}(\Psi)$ is positive, and such that there exists a constant $c > 0$ satisfying $cG(\Psi) \geq 1$. Then for all holomorphic mappings $f : \mathbb{C}^m(r) \longrightarrow M$, we have*

$$f^*\Psi \leq c(\frac{m+1}{a})^m \Theta_{r,a}.$$

The proof is identical with the previous one.

5.3 Kobayashi hyperbolicity

We continue to study the pseudo-distance $d^r_{M,D,g}$ of Definition 4.10 in Section 4.7. Here M and D are complex manifolds. We replace the Riemannian metric g on D by a Hermitian metric h on D, and replace the C^r properties by holomorphic properties, say, *holomorphic chains, holomorphic hyperbolic* and so on. Thus the pseudo-distance will be denoted by $d_{M,D,h}$. For this case, the measure in Section 4.7 is denoted by $\mu_{M,D,\nu}$.

With respect to the metric $h = h_{1,n}$ defined by (5.16), we will abbreviate as

$$d_{M,n} = d_{M,\Delta^n,h}, \quad d_M = d_{M,1},$$

and call d_M the Kobayashi pseudodistance. With respect to the Einstein-Kähler metric $h = h_{1;n}$ on the polydisk Δ^n, abbreviate as

$$d_{M;n} = d_{M,\Delta^n,h}.$$

We know the following fact: If M is connected, then for any $x, y \in M$, there exists a holomorphic chain on Δ joining x to y (see [153] and [190]). Note that each holomorphic mapping $f : \Delta \longrightarrow M$ can lift a holomorphic mapping $f : \Delta^n \longrightarrow M$ in the natural way. Hence connected complex manifolds are holomorphic chain connected on Δ^n. Thus $d_{M,n}(x,y)$ is finite. Further, the Kobayashi pseudodistance d_M is continuous. If d_M is an actual distance, it induces the standard topology on M (see [28]).

Theorem 5.7 ([123]) *For a complex manifold M, we always have*

$$d_{M,n} = d_M, \quad d_{M;n} = d_M, \quad n = 1, 2, \ldots .$$

Proof. Define the embedding $\iota : \Delta \longrightarrow \Delta^n$ by

$$\iota(z) = (z, 0, ..., 0) \in \Delta^n, \quad z \in \Delta.$$

Note that

$$\iota^* h_{1,n} = h_1,$$

where h_1 is just the Poincaré metric on Δ. We have

$$d_{h_{1,n}}(\iota(z), \iota(w)) = d_{h_1}(z, w), \quad z, w \in \Delta.$$

Denote the projection $\hat{\iota} : \Delta^n \longrightarrow \Delta$ given by

$$\hat{\iota}(z_1, ..., z_n) = z_1.$$

For a holomorphic chain $\alpha = \{p_i, q_i; f_i\}_{i=0}^{l}$ of order 1 from x to y, we have a holomorphic chain $\alpha' = \{\iota(p_i), \iota(q_i); f_i \circ \hat{\iota}\}$ of order n from x to y. Then

$$d_M = \inf_{\alpha} \sum_{i=0}^{l} d_{h_1}(p_i, q_i) = \inf_{\alpha'} \sum_{i=0}^{l} d_{h_{1,n}}(\iota(p_i), \iota(q_i)) \geq d_{M,n}, \quad n = 1, 2, \dots . \tag{5.18}$$

Let $U(n, 1)$ be the group of $(n+1) \times (n+1)$ matrices

$$H = \begin{pmatrix} A & B \\ C & d \end{pmatrix}$$

where $d \in \mathbb{C}$ and A is a $n \times n$ matrix, such that

$$\overline{{}^tA}A - \overline{{}^tC}C = I_n, \quad \overline{{}^tA}B = d\,\overline{{}^tC}, \quad \overline{{}^tB}B = |d|^2 - 1,$$

where I_n denotes the identity matrix of order n, ${}^tA$ is the transpose of A. If we set

$$H(z) = \frac{1}{C\,{}^tz + d}(A\,{}^tz + B),$$

where ${}^tz$ is the transpose of the vector $z = (z_1, \dots, z_n)$, then

$$1 - \|H(z)\|^2 = \frac{1 - \|z\|^2}{|C\,{}^tz + d|^2},$$

which shows that $H(z) \in \Delta^n$ if $z \in \Delta^n$. We can show easily that the group $U(n, 1)$ acts transitively on Δ^n, and $H^* h_{1,n} = h_{1,n}$ for $H \in U(n, 1)$.

Now for a given holomorphic chain $\beta = \{p_i, q_i; f_i\}_{i=0}^{l}$ of order n from x to y and each i, we can choose a $H_i \in U(n, 1)$ such that $H_i(p_i) = 0$. Note that the unitary group $U(n) \subset \mathrm{Hol}(\Delta^n, \Delta^n)$. There exist $\ell_i \in U(n)$, $a_i \in \Delta$ such that

$$\ell_i \circ H_i(q_i) = \iota(a_i).$$

Thus we obtain holomorphic mappings

$$F_i = f_i \circ H_i^{-1} \circ \ell_i^{-1} \circ \iota : \Delta \longrightarrow M, \quad i = 0, 1, \dots, l$$

with

$$F_i(0) = f_i(p_i), \quad F_i(a_i) = f_i(q_i).$$

Hence $\beta' = \{0, a_i; F_i\}_{i=0}^{l}$ is a holomorphic chain of order 1 from x to y. Finally, we obtain

$$\begin{aligned} d_{M,n}(x, y) &= \inf_{\beta} \sum_{i=0}^{l} d_{h_{1,n}}(p_i, q_i) = \inf_{\beta} \sum_{i=0}^{l} d_{h_{1,n}}(\ell_i \circ H_i(p_i), \ell_i \circ H_i(q_i)) \\ &= \inf_{\beta} \sum_{i=0}^{l} d_{h_{1,n}}(\iota(0), \iota(a_i)) = \inf_{\beta'} \sum_{i=0}^{l} d_{h_1}(0, a_i) \\ &\geq d_M(x, y). \end{aligned}$$

Similarly, we can prove $d_{M;n} = d_M$, $n = 1, 2, \dots$. $\square$

Example 5.8 *If $M = \Delta^n$, then d_{Δ^n} coincides with the distance function $d_{h_{1,n}}$ on Δ^n since a holomorphic mapping of Δ^n into itself is distance decreasing by Corollary 5.1. Theorem 4.36 implies $d_{h_{1,n}} \leq d_{\Delta^n,n}$ and the definition of $d_{\Delta^n,n}$ implies $d_{\Delta^n,n} \leq d_{h_{1,n}}$. For $M = \Delta^n$, we only obtain*

$$\frac{1}{\sqrt{n}} d_{h_{1;n}} \leq d_{\Delta^n} \leq d_{h_{1;n}}.$$

Example 5.9 *$d_{\mathbb{C}}(x,y) = 0$ for all $x, y \in \mathbb{C}$. Indeed, given $x \neq y$ there exists a disc $\mathbb{C}(r)$ of arbitrarily large radius r such that $y - x \in \mathbb{C}(r)$. Set $f_0(z) = z + x$. Then $f_0(0) = x, f_0(y - x) = y$. Define $f : \Delta \longrightarrow \mathbb{C}$ by $f(z) = f_0(rz)$. Then $f(0) = x, f(\frac{y-x}{r}) = y$, and*

$$d_{\mathbb{C}}(x,y) = d_{\mathbb{C}}(f(0), f(\frac{y-x}{r})) \leq d_{\Delta}(0, \frac{y-x}{r}) = d_{h_1}(0, \frac{y-x}{r}) \to 0 \text{ as } r \to \infty$$

so that $d_{\mathbb{C}}(x,y) = 0$.

Let X, M be complex manifolds. Then one has the following formula (see [190])

$$d_{X \times M}((x_1, z_1), (x_2, z_2)) = \max\{d_X(x_1, x_2), d_M(z_1, z_2)\}$$

for all $x_1, x_2 \in X; z_1, z_2 \in M$. Thus we obtain $d_{\mathbb{C}^m} = 0$. As a consequence of Theorem 4.35, we can get: If $f : \mathbb{C}^m \longrightarrow M$ is holomorphic, then

$$d_M(x,y) = 0, \quad \text{for all} \quad x, y \in f(\mathbb{C}^m). \tag{5.19}$$

Definition 5.2 *Let M be a complex manifold. Then d_M is called the Kobayashi pseudo-distance on M. If d_M is a (complete) distance, then M is called a (complete) Kobayashi hyperbolic manifold.*

Directly from the definition, we note that Kobayashi hyperbolic is a biholomorphic invariant. Here we exhibit some examples of Kobayashi hyperbolic manifolds. If X, M are Kobayashi hyperbolic, so is $X \times M$. If X is a complex subspace of M; or if $f : X \longrightarrow M$ is holomorphic and injective, then if M is Kobayashi hyperbolic, so is X. Discs and polydiscs are Kobayashi hyperbolic. A bounded domain in $\mathbb{C}^m$ is Kobayashi hyperbolic, since it is an open subset of a polydisc.

Theorem 5.8 *Let M be a complex manifold and $\tilde{M}$ a covering manifold. Then $\tilde{M}$ is (complete) Kobayashi hyperbolic iff M is (complete) Kobayashi hyperbolic.*

The proof which is longer is given in Kobayashi [143] (or see Abate [2]). Now (5.19) shows that any holomorphic mapping from $\mathbb{C}$ into a Kobayashi hyperbolic manifold is constant. The converse also is true for compact manifolds. This is the Brody's theorem [54]:

Theorem 5.9 *Let M be a compact complex manifold. Then M is Kobayashi hyperbolic iff every $f \in \text{Hol}(\mathbb{C}, M)$ is constant.*

A complete proof also can be found in S. Lang [153] and [155]. Here we introduce the sketch of proof. Let M be a compact complex manifold with a given Hermitian metric. Define

$$c(M) = \sup_{f \in \text{Hol}(\Delta, M)} \|f'(0)\|.$$

Then M is Kobayashi hyperbolic iff $c(M)$ is finite (see [54]). If M is not Kobayashi hyperbolic, then there exists a sequence $\{f_n\}$ of $\text{Hol}(\Delta, M)$ such that $\|f'_n(0)\| \to \infty$. W. l. o. g., by making a dilation, we may consider a sequence $\{f_n\}$ of $\text{Hol}(\mathbb{C}(r_n), M)$ such that $\|f'_n(0)\| = 1$ and the radii r_n increase to infinity. According to Brody's reparametrization lemma [54], there exist holomorphic mappings $g_n \in \text{Hol}(\mathbb{C}(r_n), M)$ such that

$$\sup_{z \in \mathbb{C}(r_n)} \|g'_n(z)\| = \|g'_n(0)\| = 1.$$

It is a simple matter to show that given a compact subset K of $\mathbb{C}$ every sequence of $\{g_n\}$ contains a subsequence which converges uniformly on K. Thus we get a subsequence of $\{g_n\}$ converging uniformly on $\mathbb{C}(1)$ to a mapping f. A further subsequence converges uniformly on $\mathbb{C}(2)$, and so on. We can then extend f analytically to all of $\mathbb{C}$. Furthermore f is not constant since

$$\|f'(0)\| = \lim_{n \to \infty} \|g'_n(0)\| = 1.$$

The following theorem is immediate from Proposition 4.4 and Lemma 5.4.

Theorem 5.10 ([143]) *A (complete) Hermitian manifold M whose holomorphic sectional curvature is bounded above by a negative constant is (complete) Kobayashi hyperbolic.*

The following fact is given in Kobayashi [143] (or see Abate [2]):

Theorem 5.11 *Let M be a Kobayashi hyperbolic manifold and p a point of M. Let $f \in$ $\text{Hol}(M, M)$ with $f(p) = p$. Then*
   1) The eigenvalues of $f'(p)$ have absolute value ≤ 1;
   2) If $f'(p) = id$, then $f = id$;
   3) If $|\mathcal{J}f(p)| = 1$, then f is a biholomorphic mapping.

Proof. Here we follow Kobayashi [143] to give a sketchy proof. Let r be a positive number such that the closed r-ball $M[p; r]$ about p is compact. Let $\mathcal{F}_p$ denote the set of all continuous mappings of $M[p; r]$ into itself which leave p fixed and are distance decreasing with respect to d_M. Then $\mathcal{F}_p$ is compact with respect to the compact-open topology (cf. [143], Theorem 3.1). Take $f \in \text{Hol}(M, M)$ with $f(p) = p$. Let λ be an eigenvalue of $f'(p)$. The mapping f^n ($n \in \mathbb{Z}^+$) restricted to $M[p; r]$ belongs to $\mathcal{F}_p$, and its differential $(f^n)'(p)$ has an eigenvalue λ^n. If $|\lambda| > 1$, then $|\lambda^n| \to \infty$ as $n \to \infty$, in contradiction to the fact that $\mathcal{F}_p$ is compact, and so (1) follows.

Suppose $f'(p) = id$. If there exists an integer $m \geq 2$ such that $(d^m f)_p \neq 0$, then $(d^m f^n)_p = n(d^m f)_p$ for all positive integers n, and hence $(d^m f^n)_p \to \infty$ as $n \to \infty$, in contradiction to the fact that $\mathcal{F}_p$ is compact. Therefore $(d^m f)_p = 0$ for $m \geq 2$. In conclusion, by using Taylor expansion $f = id$ near p, and by the identity principle, (2) is proved.

Let $\{\lambda_1, ..., \lambda_l\} \subset \overline{\Delta}$ be the eigenvalues of $f'(p)$, and let

$$\mathbf{T}(M)_p = J_{\lambda_1} \oplus \cdots \oplus J_{\lambda_l}$$

be the Jordan decomposition of $\mathbf{T}(M)_p$ with respect to $f'(p)$. Assume $|\mathcal{J}f(p)| = 1$. From (1), it follows that $|\lambda_i| = 1$ for all $1 \le i \le l$. We claim that each J_{λ_i} is diagonal. If not, in the Jordan canonical form of $f'(p)$ there should be a block of the form:

$$\begin{vmatrix} \lambda & 1 & & 0 \\ & \lambda & \ddots & \\ & & \ddots & 1 \\ 0 & & & \lambda \end{vmatrix}, \quad |\lambda| = 1.$$

The corresponding diagonal block of $(f'(p))^n$ is then of the form

$$\begin{vmatrix} \lambda^n & n\lambda^{n-1} & & * \\ & \lambda^n & \ddots & \\ & & \ddots & n\lambda^{n-1} \\ 0 & & & \lambda^n \end{vmatrix}$$

It follows that the entries $n\lambda^{n-1} \to \infty$ as $n \to \infty$, contradicting the compactness of $\mathcal{F}_p$. Since $f'(p)$ in Jordan canonical form is diagonal. Then we can find a subsequence f^{n_ν} converging to a mapping $h \in \mathrm{Hol}(M, M)$ such that $h(p) = p$ and $h'(p) = id$. But we saw that this implies $h = id$, and so f is a biholomorphic mapping. $\qquad\square$

Theorem 5.11 is a generalization of the Cartan-Carathéodory theorem on Poincaré hyperbolic Riemann surface. The following result is contained in the proof of Theorem 5.11.

Proposition 5.2 *Let M be a Kobayashi hyperbolic manifold and p a point of M. Let $f \in \mathrm{Hol}(M, M)$ with $f(p) = p$. Then the holomorphic tangent space $\mathbf{T}(M)_p$ decomposes into the direct sum*

$$\mathbf{T}(M)_p = \mathbf{E}_p^s \oplus \mathbf{E}_p^c$$

such that the spectrum of $f'(p)|_{\mathbf{E}_p^s}$ is contained in Δ, the spectrum of $f'(p)|_{\mathbf{E}_p^c}$ is contained in $\partial\Delta$, and $f'(p)|_{\mathbf{E}_p^c}$ is diagonalizable.

An immediate consequence of the Cartan-Carathéodory theorem is Cartan's uniqueness theorem:

Corollary 5.2 *Let M be a Kobayashi hyperbolic manifold and p a point of M. Then if $f, g \in \mathrm{Aut}(M)$ are such that $f(p) = g(p) = p$ and $f'(p) = g'(p)$, it follows that $f \equiv g$ on M.*

Now suppose that d_h is a distance on X defining the topology, and that d_X is a pseudodistance which is continuous for the topology. Let d_M be the given distance on M. The set of mappings $f : X \longrightarrow M$ which are distance decreasing from d_X to d_M is equicontinuous with respect to the given topology on X and the given distance on M. Indeed, given ε and $x_0 \in X$ there exists δ such that if $d_h(x, x_0) < \delta$ then $d_X(x, x_0) < \varepsilon$, because d_X is continuous; and if f is distance decreasing as above, then $d_M(f(x), f(x_0)) < \varepsilon$, which proves the equicontinuity. Thus if M is Kobayashi hyperbolic, $\mathrm{Hol}(\Delta, M)$ is equicontinuous.

Theorem 5.12 *If M is a completely arbitrary Riemann surface, then M is Kobayashi hyperbolic iff M is Poincaré hyperbolic.*

Proof. Assume first that M is Poincaré hyperbolic. The universal covering $\tilde{M}$ must be conformally isomorphic to Δ. Since Δ is Kobayashi hyperbolic, Theorem 5.8 implies that M is Kobayashi hyperbolic. Conversely, assume that M is Kobayashi hyperbolic. Theorem 5.8 implies that the universal covering $\tilde{M}$ is Kobayashi hyperbolic. According to the Uniformization Theorem, $\tilde{M}$ must be conformally isomorphic to one of the three model surfaces. Note that $\mathbb{C}$ is not Kobayashi hyperbolic since $d_{\mathbb{C}} = 0$. Since $\mathrm{Hol}(\mathbb{C}, \mathbb{P}^1)$ contain nonconstant holomorphic mappings, by Theorem 5.9, $\mathbb{P}^1$ also is not Kobayashi hyperbolic. Hence $\tilde{M} \cong \Delta$, so that M is Poincaré hyperbolic. $\qquad\square$

Since $\mathrm{Aut}(M)$ is a closed subgroup of the isometry group $\mathrm{Iso}^\infty(M)$ with respect to the invariant distance d_M, it follows from Lemma A.7 that $\mathrm{Aut}(M)$ is locally compact with respect to the compact-open topology. By a theorem of Bochner and Montgomery [47], a locally compact group of differentiable transformations of a manifold is a Lie transformation group. Hence $\mathrm{Aut}(M)$ is a Lie transformation group. For more detail, see the following

Theorem 5.13 ([143]) *Let M be a Kobayashi hyperbolic manifold. Then $\mathrm{Aut}(M)$ is a Lie transformation group, and its isotropy subgroup $\mathrm{Aut}_p(M)$ at $p \in M$ is compact. If M is moreover compact, then $\mathrm{Aut}(M)$ is finite.*

Definition 5.3 *Let M be a complex manifold and $\mathbf{T}(M)$ be the holomorphic tangent bundle of M. A mapping $\kappa : \mathbf{T}(M) \longrightarrow \mathbb{R}_+$ is called a differential metric if the following conditions are satisfied:*
1) $\kappa(O_x) = 0$, where O_x is the zero vector of $\mathbf{T}(M)_x$;
2) $\kappa(a\xi_x) = |a|\kappa(\xi_x)$, for any $a \in \mathbb{C}$ and for any $\xi_x \in \mathbf{T}(M)_x$.
Moreover, if κ is continuous and $\kappa(\xi_x) \neq 0$ for any $\xi_x \in \mathbf{T}(M)_x - \{O_x\}$, then κ is especially called a Finsler metric.

Let M be a complex manifold with a Finsler metric κ. Any real vector $v_x \in T(M)_x, x \in M$, can be uniquely expressed as $v_x = \xi_x + \bar{\xi}_x$ with $\xi_x \in \mathbf{T}(M)_x$. Set

$$\kappa(v_x) = 2\kappa(\xi_x).$$

Then $\kappa(av_x) = |a|\kappa(v_x)$, for any $a \in \mathbb{R}$ and for any $v_x \in T(M)_x$. Let $\gamma : [a, b] \longrightarrow M$ be a piecewise C^∞-curve. One defines the length of γ by

$$\mathrm{Length}(\gamma) = \int_a^b \kappa(\dot{\gamma}(t))dt. \qquad (5.20)$$

For arbitrary two points $x, y \in M$, define the distance

$$d_\kappa(x, y) = \inf\{\mathrm{Length}(\gamma)\}, \qquad (5.21)$$

where the infimum is taken for all piecewise C^∞-curves joining x and y. Each Hermitian metric h of M induces a Finsler metric κ as follows:

$$\kappa(\xi_x) = \frac{1}{\sqrt{2}}\|\xi_x\| = \frac{1}{\sqrt{2}}\sqrt{h(\xi_x, \xi_x)}, \quad \xi_x \in \mathbf{T}(M)_x$$

with $d_\kappa = d_h$.

Now one defines a mapping $\kappa_M : \mathbf{T}(M) \longrightarrow \mathbb{R}_+$ as follows: for any $\xi_x \in \mathbf{T}(M)_x$, and set

$$\kappa_M(\xi_x) = \inf\{\frac{1}{r} \mid \exists f \in \mathrm{Hol}(\mathbb{C}(r), M), f(0) = x, \dot{f}(0) = \xi_x\}. \tag{5.22}$$

Here for the sake of simplicity, we have used

$$\dot{f}(0) = f'(0)\left(\frac{\partial}{\partial z}\right).$$

The mapping κ_M is a differential metric called *Kobayashi differential metric* on M. For arbitrary two points $x, y \in M$, one has (see [190])

$$d_{\kappa_M}(x, y) = d_M(x, y). \tag{5.23}$$

In case $M = \Delta$, κ_Δ is the Poincaré metric, more precisely,

$$\kappa_M(\xi_z) = \frac{|\xi_z|}{1 - |z|^2}.$$

If $M = \mathbb{C}^m$ or $\mathbb{P}^m$, the number r can be arbitrarily large, so $\kappa_M = 0$ in either case.

Theorem 5.14 *Let D and M be complex manifolds and $f : D \longrightarrow M$ a holomorphic mapping. Then we have $f^*\kappa_M \leq \kappa_D$, i.e., for any $\xi_x \in \mathbf{T}(D)_x$,*

$$\kappa_M(f'(x)\xi_x) \leq \kappa_D(\xi_x).$$

In particular, if f is biholomorphic, $f^*\kappa_M = \kappa_D$. If $f : D \longrightarrow M$ is a holomorphic unramified covering of a complex manifold M, then we also have $f^*\kappa_M = \kappa_D$.

Theorem 5.14 can be referred to the differential form of distance decreasing principle (see [190]). Kobayashi hyperbolicity can be characterized by differential metrics as follows (cf. [190]):

Theorem 5.15 *Let M be a complex manifold and let $\kappa : \mathbf{T}(M) \longrightarrow \mathbb{R}_+$ be a Finsler metric. Then M is Kobayashi hyperbolic iff for any $x \in M$, there are a neighborhood U of x and a constant $c > 0$ such that*

$$\kappa_M(\xi_y) \geq c\kappa(\xi_y) \text{ for all } \xi_y \in \mathbf{T}(M)_y \text{ with } y \in U. \tag{5.24}$$

Definition 5.4 *Let X be a complex manifold and M a locally closed complex submanifold of X. Then M is said to be hyperbolically embedded into X if M satisfies the following three conditions:*

1) M is relatively compact in X;

2) M is Kobayashi hyperbolic;

3) for any point $p \in \partial M = \bar{M} - M$ and any open neighborhood U of p in X, there exists an open neighborhood V of p in X such that $V \subset U$ and

$$\inf\{d_M(x, y) \mid x \in M \cap V, y \in M - U\} > 0.$$

Lemma 5.7 (cf. [190]) *Let M be relatively compact in X and κ a Finsler metric on X. Then M is hyperbolically imbedded into X iff there is a positive constant c such that*

$$\kappa_M(\xi) \geq c\kappa(\xi) \text{ for all } \xi \in \mathbf{T}(M).$$

Let Ψ be a pseudo volume form on M. Then Ψ defines a positive functional on $C_0(M)$ by

$$\phi \mapsto \int_M \phi\Psi.$$

Hence by measure theory, there is an unique measure μ_Ψ such that for all $\phi \in C_0(M)$ we have

$$\int_M \phi\Psi = \int_M \phi d\mu_\Psi.$$

Assume that M is covered by holomorphic chains on $\mathbb{C}^n(r)$. The *Kobayashi measure* μ_M is defined by

$$\mu_M = \mu_{M,\mathbb{C}^n(r),\mu_{\Theta_{r,a}}},$$

where $\Theta_{r,a}$ is the associated volume element of ω in Example 5.7. Obviously, we have

$$\mu_{\mathbb{C}^n(r)} = \mu_{\Theta_{r,a}}.$$

The following properties are immediate from the definitions.

Theorem 5.16 *Let $f : M \longrightarrow N$ be a holomorphic mapping between complex manifolds of dimension m. Then f is Kobayashi measure decreasing. If μ is a measure on M such that every holomorphic mapping $f : \mathbb{C}^n(r) \longrightarrow M$ is measure decreasing from $\mu_{\mathbb{C}^n(r)}$ to μ, then $\mu \leq \mu_M$.*

We define M to be *Kobayashi measure hyperbolic* if $\mu_M(U) > 0$ for all non-empty open subsets U of M. Then Lemma 5.6 implies the following results:

Theorem 5.17 *Let M be a complex manifold of dimension m and let Ψ be a pseudo volume form on M such that $\mathrm{Ric}(\Psi)$ is positive, and such that there exists a constant $c > 0$ satisfying $cG(\Psi) \geq 1$. Then M is Kobayashi measure hyperbolic.*

5.4   Tautness and tightness

In § 4.7, we asked that are there $\mathcal{F}$-hyperbolic frames if $J_{equ}(\mathcal{F}) = \emptyset$? Here we discuss this question on complex manifolds, and begin with the following notions of Wu[272].

Definition 5.5 ([272]) *A complex manifold M is called taut iff for every complex manifold X, $J(\mathrm{Hol}(X, M)) = \emptyset$, i.e., $\mathrm{Hol}(X, M)$ is normal on X. If d is a metric on M inducing its topology, then (M, d) is tight iff for every complex manifold X, $J_{equ}(\mathrm{Hol}(X, M)) = \emptyset$, i.e., $\mathrm{Hol}(X, M)$ is equicontinuous on X.*

Example 5.10 *Let $M = D$ be a bounded domain in $\mathbb{C}^n$, the Montel theorem shows that $\mathrm{Hol}(X, D)$ is equicontinuous for every complex manifold X in the usual metric. Hence D is a tight manifold.*

Tautness is an intrinsic property of the complex structure of M, tightness is dependent on the given metric d. It can happen that two metrics d and d' both induce the same topology of M, but (M,d) is tight while (M,d') is not. If M is compact, tautness and tightness coincide. Clearly both concepts are entirely local questions:

Lemma 5.8 ([272]) *A complex manifold M is taut if $J(\mathrm{Hol}(\Delta^m, M)) = \emptyset$ for all m. Let d be a metric on M inducing its topology. Then (M,d) is tight if $J_{equ}(\mathrm{Hol}(\Delta^m, M)) = \emptyset$ for all m.*

Theorem 5.18 ([27]) *Let M be a complex manifold and d be a metric on M inducing its topology. If $J_{equ}(\mathrm{Hol}(\Delta, M)) = \emptyset$, then (M,d) is tight.*

Proof. Suppose that (M,d) is not tight. Then $\mathrm{Hol}(\Delta^m, M)$ is not equicontinuous for some m. This means that there exist $\varepsilon_0 > 0, z_0 \in \Delta^m$, and sequence $\{z_k\}$ in Δ^m, $\{f_k\}$ in $\mathrm{Hol}(\Delta^m, M)$ such that $z_k \to z_0$ as $k \to \infty$ and

$$d(f_k(z_k), f_k(z_0)) \geq \varepsilon_0 \quad \text{for all} \quad k.$$

We may assume that $z_0 = 0$. For $z \in \Delta$, define $g_k(z) = f_k(zz_k/\|z_k\|)$. Then $g_k \in \mathrm{Hol}(\Delta, M)$, $\|z_k\| \to 0$ as $k \to \infty$, and

$$d(g_k(\|z_k\|), g_k(0)) = d(f_k(z_k), f_k(z_0)) \geq \varepsilon_0 \quad \text{for all} \quad k.$$

Thus $\mathrm{Hol}(\Delta, M)$ is not equicontinuous. $\qquad\square$

One-point (or *Alexandroff*) *compactification* (M^*, τ^*) of a noncompact connected Hausdorff locally compact topological space (M, τ) is the set $M \cup \{\infty\}$, where ∞ is a point not in M, endowed with the topology

$$\tau^* = \tau \cup \{(M - K) \cup \{\infty\} | K \subset M \text{ compact }\}.$$

We shall also use ∞ to denote the mappings into M^* which are identically ∞. It is easy to check (see [135], p.150) that (M^*, τ^*) is a connected Hausdorff compact topological space, with M as dense subspace. Furthermore, if M is second countable, then so is M^*, which, therefore, is metrizable (see [135], p.125). In particular, if X is another locally compact metrizable second countable space then $C(X, M^*)$ is still metrizable and a subset of $C(X, M^*)$ is compact iff it is sequentially compact. A moment's thought shows that a sequence $\{f_n\} \subset C(X, M)$ is compactly divergent iff it converges, in $C(X, M^*)$, to the constant mapping ∞; hence Lemma 1.5 implies

Lemma 5.9 *Let X and M be connected locally compact metric spaces, and let $\mathcal{F}$ be a closed subset of $C(X, M)$. Then $\mathcal{F}$ is normal iff $\mathcal{F} \cup \{\infty\}$ is compact in $C(X, M^*)$.*

Theorem 5.19 ([27]) *Let M be a complex manifold. If $J(\mathrm{Hol}(\Delta, M)) = \emptyset$, then M is taut.*

Proof. Since M^* is metrizable, we can take a metric d inducing the topology on M^*. By Lemma 1.11 and Lemma 5.9, $\mathrm{Hol}(\Delta, M) \cup \{\infty\}$ is closed and equicontinuous in $C(\Delta, M^*)$. According to Theorem 5.18, (M, d) is tight.

Suppose that M is not taut. Then $\mathcal{F} = \mathrm{Hol}(\Delta^m, M)$ is not a normal family for some m. By Lemma 5.9, $\mathcal{F} \cup \{\infty\}$ is not a compact subset of $C(\Delta^m, M^*)$. Since (M, d) is tight, $\mathcal{F} \cup \{\infty\}$ is equicontinuous; thus(Lemma 1.11) $\mathcal{F} \cup \{\infty\}$ is not closed in $C(\Delta^m, M^*)$. Let $\{f_n\}$ be a sequence in $\mathcal{F}$ with

$$f_n \to f \in C(\Delta^m, M^*) - (\mathcal{F} \cup \{\infty\}) \quad \text{as} \quad n \to \infty.$$

Note that $f \notin \mathrm{Hol}(\Delta^m, M) \cup \{\infty\}$. If $f(0) = \infty$, take $a \in \Delta^m$ with $f(a) \neq \infty$; if $f(0) \neq \infty$, take $a \in \Delta^m$ with $f(a) = \infty$. For $z \in \Delta$, define

$$g_n(z) = f_n(za/\|a\|) \quad \text{and} \quad g(z) = f(za/\|a\|).$$

Then we have

$$g_n \in \mathrm{Hol}(\Delta, M), g \in C(\Delta, M^*), g \notin \mathrm{Hol}(\Delta, M) \cup \{\infty\}, \text{and} \quad g_n \to g \quad \text{as} \quad k \to \infty.$$

Thus $\mathrm{Hol}(\Delta, M) \cup \{\infty\}$ is not closed in $C(\Delta, M^*)$, a contradiction. $\qquad \square$

Corollary 5.3 *A complex manifold M is taut iff $\mathrm{Hol}(\Delta, M) \cup \{\infty\}$ is compact in $C(\Delta, M^*)$.*

Corollary 5.4 $\mathrm{Hol}(\Delta, M) \cup \{\infty\}$ *is compact in $C(\Delta, M^*)$ iff $\mathrm{Hol}(X, M) \cup \{\infty\}$ is compact in $C(X, M^*)$ for every complex manifold X.*

For a domain D in M, one has the following results:

Lemma 5.10 ([2]) *Let D be a domain in the complex manifold M. Then $\mathrm{Hol}(\Delta, D)$ is relatively compact in $\mathrm{Hol}(\Delta, M)$ iff $\mathrm{Hol}(X, D)$ is relatively compact in $\mathrm{Hol}(X, M)$ for every complex manifold X.*

Proof. One direction is obvious. Conversely, assume that $\mathrm{Hol}(\Delta, D)$ is relatively compact in $\mathrm{Hol}(\Delta, M)$; in particular, since $\mathrm{Hol}(\Delta, D)$ contains the constant maps, D is relatively compact in M. Fix a distance d on M inducing the manifold topology. Since the closure of $\mathrm{Hol}(X, D)$ in $\mathrm{Hol}(X, M)$ is contained in $C(X, \overline{D})$ and $\overline{D}$ is compact, by the Arzelà-Ascoli theorem, for every complex manifold X, $\mathrm{Hol}(X, D)$ is relatively compact in $\mathrm{Hol}(X, M)$ iff it is equicontinuous with respect to d, iff D is tight. However, the tightness of D follows from the equicontinuity of $\mathrm{Hol}(\Delta, D)$ by the assumption. $\qquad \square$

Lemma 5.11 ([2]) *Let M be a taut manifold, and D is relatively compact domain. Then $\mathrm{Hol}(\Delta, D)$ is relatively compact in $\mathrm{Hol}(\Delta, M)$.*

Proof. Since no sequence in $\mathrm{Hol}(\Delta, D)$ can be compactly divergent in M, the lemma follows. $\qquad \square$

Lemma 5.12 ([2]) *Let D be a domain in the complex manifold M such that $\mathrm{Hol}(\Delta, D)$ is relatively compact in $\mathrm{Hol}(\Delta, M)$. Then D is taut iff for every $f \in \overline{\mathrm{Hol}(\Delta, D)} \subset \mathrm{Hol}(\Delta, M)$ we have either*

$$f(\Delta) \subset D \text{ or } f(\Delta) \subset \partial D.$$

Proof. Assume D taut, and take a sequence $\{f_n\} \subset \mathrm{Hol}(\Delta, D)$ which is converging toward $f \in \mathrm{Hol}(\Delta, M)$. Since D is taut, either $f \in \mathrm{Hol}(\Delta, D)$ or, up to a subsequence, $\{f_n\}$ is compactly divergent, and thus $f(\Delta) \subset \partial D$.

Conversely, take a sequence $\{f_n\} \subset \mathrm{Hol}(\Delta, D)$. Since $\mathrm{Hol}(\Delta, D)$ is relatively compact in $\mathrm{Hol}(\Delta, M)$, there is a subsequence $\{f_{n_j}\}$ converging toward a mapping $f \in \mathrm{Hol}(\Delta, M)$. Hence, by assumption, either $f(\Delta) \subset D$, and so $f_{n_j} \to f$ in $\mathrm{Hol}(\Delta, M)$, or $f(\Delta) \subset \partial D$, and thus $\{f_{n_j}\}$ is compactly divergent. □

Repeating the first paragraph of the proof of Theorem 5.19, one has

Theorem 5.20 ([27]) *Let M be a taut manifold. Then there exists a metric d inducing the topology on M such that (M, d) is tight.*

Definition 5.6 *A complex manifold M is called L_p-taut (resp., L_p-chaotic) iff for every complex manifold X,*

$$J_p(\mathrm{Hol}(X, M)) = \emptyset \,(resp., J_p(\mathrm{Hol}(X, M)) = X).$$

Theorem 5.21 *Let M be a complex manifold, and d a distance on M compatible with its topology. Assume $J_p(\mathrm{Hol}(\Delta, M)) = \Delta$ with respect to d for $0 \le p \le \infty$; then*

$$J_p(\mathrm{Hol}(X, M)) = X$$

with respect to d for every complex manifold X.

Proof. Assume, by contradiction, that there is a complex manifold X such that

$$J_p(\mathrm{Hol}(X, M)) \ne X.$$

Hence there exists a point $z_0 \in X$ such that $\mathrm{Hol}(X, M)$ is L_p-normal at z_0. Choosing a suitable local coordinate system we can then assume X to be the Euclidean unit ball Δ^m of some $\mathbb{C}^m$, and take $z_0 = 0$. Since we can embed $\mathrm{Hol}(\Delta, M)$ into $\mathrm{Hol}(\Delta^m, M)$ in the usual fashion: if $f \in \mathrm{Hol}(\Delta, M)$, define $\hat{f} \in \mathrm{Hol}(\Delta^m, M)$ by

$$\hat{f}(z_1, z_2, ..., z_m) = f(z_1).$$

Thus the L_p-normality of $\mathrm{Hol}(\Delta^m, M)$ at $0 \in \Delta^m$ implies the L_p-normality of $\mathrm{Hol}(\Delta, M)$ at $0 \in \Delta$. □

Definition 5.7 *A complex manifold (M, d_h) is said to be L_p-tight if*

$$J_p(f, d_h) = \emptyset$$

for every $f \in \mathrm{Hol}(M, M)$. A complex manifold (M, d_h) is said to be L_p-light if

$$J_p(f, d_h) = M$$

for every $f \in \mathrm{Hol}(M, M) - \mathrm{Aut}(M)$.

According to [272], L_∞-tautness is called *tautness* and L_0-tautness is called tightness. Obviously, if M is L_p-taut, then M is L_p-tight. What about converse?

5.5   Relations between hyperbolicity and tautness

Lemma 5.13 ([138]) *Let x and y be distinct points in the complex manifold M and let Δ^n be a coordinate neighborhood centered at x, such that $y \notin \Delta^n$. If there exists a pair (r, δ) with $0 < r, \delta < 1$ such that for every $f \in \mathrm{Hol}(\Delta, M)$, $f(0) \in \mathbb{C}^n(r)$ implies $f(\mathbb{C}(\delta)) \subset \Delta^n$, then $d_M(x, y) > 0$.*

Remark. Here Δ^n is said to be a coordinate neighborhood centered at x iff there exists a pair (U, φ) such that U is a neighborhood of x, and $\varphi : U \longrightarrow \Delta^n$ is a biholomorphic mapping with $\varphi(x) = 0$. In this lemma, we identify U with Δ^n under the coordinates φ.

Proof. Choose a constant $c > 0$ such that

$$d_\Delta(0, z) \geq c d_{\mathbb{C}(\delta)}(0, z) \text{ for all } z \in \mathbb{C}(\delta/2).$$

Let α be a holomorphic chain from x to y defined in Definition 5.2. Without loss of generality, we can assume that $z_0, z_1, ..., z_k \in \mathbb{C}(\delta/2)$; $f_0(0), f_1(0), ..., f_k(0) \in \mathbb{C}^n(r)$ for some $k \leq l$ and that $f_{k+1}(0) \in \partial \mathbb{C}^n(r)$. Now

$$\sum_{i=0}^{l} d_g(0, z_i) \geq \sum_{i=0}^{k} d_\Delta(0, z_i) \geq c \sum_{i=0}^{k} d_{\mathbb{C}(\delta)}(0, z_i)$$

$$\geq c \sum_{i=0}^{k} d_{\Delta^n}(f_i(0), f_i(z_i)) \geq c d_{\Delta^n}(0, f_k(z_k)) = c',$$

where c' is constant > 0. Thus $d_M(x, y) \geq c' > 0$.   □

Theorem 5.22 ([138]) *If (M, d) is tight, M is Kobayashi hyperbolic. If M is Kobayashi hyperbolic, then (M, d_M) is tight.*

Proof. Assume (M, d) is tight. Let x and y be distinct points in the complex manifold M and let Δ^n be a coordinate neighborhood centered at x, such that $y \notin \Delta^n$. There exists $\varepsilon > 0$ such that $M(x; 2\varepsilon) \subset \Delta^n$. Since $\mathrm{Hol}(\Delta, M)$ is equicontinuous, there exists $\delta > 0$ such that if $f \in \mathrm{Hol}(\Delta, M)$ with $f(0) \in M(x; \varepsilon)$, then $f(\mathbb{C}(\delta)) \subset M(x; 2\varepsilon) \subset \Delta^n$. Choose $r > 0$ such that $\mathbb{C}^n(r) \subset M(x; \varepsilon)$. Then (r, δ) satisfies the property in Lemma 5.13, so that $d_M(x, y) > 0$. Since x and y were arbitrary distinct points, M is Kobayashi hyperbolic. The second statement is trivial.   □

Theorem 5.23 ([138]) *Every taut manifold is Kobayashi hyperbolic and every Kobayashi complete hyperbolic manifold is taut.*

Proof. Assume M is not Kobayashi hyperbolic. Then there exist distinct points x and y with $d_M(x, y) = 0$. By Lemma 5.13, $(1/2, 1/k)$ does not satisfy the property in Lemma 5.13 for any k. Thus there exists $f_k \in \mathrm{Hol}(\Delta, M)$ with $f_k(0) \in \mathbb{C}^n(1/2)$ and $f_k(\mathbb{C}(1/k)) \not\subset \Delta^n$. The sequence $\{f_k\}$ has no subsequence which is either uniformly convergent on compact sets or compactly divergent. Thus M is not taut.

Let X be another manifold. Since (M, d_M) is tight, $\mathrm{Hol}(X, M)$ is equicontinuous. Since M is complete hyperbolic, every bounded set in M is relatively compact. This implies that $\mathrm{Hol}(X, M)$ is normal by Lemma 1.12. Thus M is taut.   □

Theorem 5.24 ([1]) *A complex manifold M is Kobayashi hyperbolic iff $\mathrm{Hol}(\Delta, M)$ is relatively compact in $C(\Delta, M^*)$.*

Actually, it was conjectured that taut were equivalent to Kobayashi complete hyperbolic, until Rosay's counter-example ([215]). Thus if M is a Kobayashi complete hyperbolic complex manifold, then M is taut so that $J(f) = \emptyset$ for any holomorphic self-mapping f on M. If M is not complete, we propose the following question:

Conjecture 5.1 *If M is a Kobayashi hyperbolic complex manifold, then $J(f) = \emptyset$ for any holomorphic self-mapping f on M.*

A *holomorphic retraction* of a complex manifold M is a holomorphic mapping $\rho : M \longrightarrow M$ such that $\rho^2 = \rho$. The image of a holomorphic retraction is said to be a *holomorphic retract* of M. If $f \in \mathrm{Hol}(M, M)$ is such that $\{f^n\}$ converges to $\rho \in \mathrm{Hol}(M, M)$, it follows that $\rho^2 = \rho$ since $\{f^{2n}\}$ tends to ρ too.

Theorem 5.25 ([2]) *Let M be a taut manifold, and $f \in \mathrm{Hol}(M, M)$. Assume that the sequence $\{f^n\}$ of iterates of f is not compactly divergent. Then $\{f^n\}$ is relatively compact in $\mathrm{Hol}(M, M)$ and there exist a submanifold S of M and a holomorphic retraction $\rho : M \longrightarrow S$ such that every limit point $h \in \mathrm{Hol}(M, M)$ of $\{f^n\}$ is of the form*

$$h = \gamma \circ \rho, \tag{5.25}$$

where γ is an automorphism of S. Moreover, even ρ is a limit point of the sequence $\{f^n\}$.

Under the conditions of Theorem 5.25, $\{f^n\}$ is relatively compact in $\mathrm{Hol}(M, M)$, that is, it do not contain compactly divergent subsequences (see [2]). Denote by $\Gamma(f)$ the set of limit points of $\{f^n\}$ in $\mathrm{Hol}(M, M)$. Then $\Gamma(f)$ is a compact topological semigroup, and by Theorem 5.25, it is isomorphic to a compact topological semigroup of $\mathrm{Aut}(S)$. Noting that

$$f \circ \rho = \rho \circ f,$$

it is clear that $f(S) \subset S$. Let $\{f^{n_k}\}$ be a subsequence of the iterates converging to ρ. Then $\{f^{n_k+1}\}$ tends to $(f|_S) \circ \rho$. Thus Theorem 5.25 implies that $f|_S$ is automorphism of S. For a limit point $h = \gamma \circ \rho \in \Gamma(f)$, let $\{f^{m_k}\}$ be a subsequence of the iterates converging to h. W. l. o. g., we may assume that $n_k - m_k \to +\infty$ and $f^{n_k-m_k} \to h_1 = \gamma_1 \circ \rho$ as $k \to +\infty$. Then $h \circ h_1 = \rho = h_1 \circ h$, that is, $\gamma_1 = \gamma^{-1}$. Hence $\Gamma(f)$ is isomorphic to a compact Abelian subgroup of $\mathrm{Aut}(S)$, which is the closed subgroup generated by $f|_S \in \mathrm{Aut}(S)$.

Let p be a fixed point of f and let $A = f'(p)$. We have the splitting

$$\mathbf{T}(M)_p = \mathbf{E}_p^s \oplus \mathbf{E}_p^c.$$

Noting that

$$A^{n_k} = (f^{n_k})'(p) \to \rho'(p),$$

then $\rho'(p)|_{\mathbf{E}_p^s} = 0$ and $\rho'(p)|_{\mathbf{E}_p^c} = id$ since $f'(p)$ acts diagonally on $\mathbf{E}_p^c$ and $\mathrm{sp}(f'(p)|_{\mathbf{E}_p^c}) \subset \partial \Delta$. Hence $\mathrm{sp}(\rho'(p)) \subset \{0, 1\}$, and

$$\mathbf{E}_p^c = \mathbf{T}(S)_p.$$

Theorem 5.26 ([2]) *Let M be a taut manifold, and $f \in \mathrm{Hol}(M, M)$. Then the sequence $\{f^n\}$ of iterates of f converges in $\mathrm{Hol}(M, M)$ iff f has a fixed point $p \in M$ such that $\mathrm{sp}(f'(p)) \subset \Delta \cup \{1\}$.*

Proof. Assume first that the sequence $\{f^n\}$ converges, necessarily to a holomorphic retraction $\rho : M \longrightarrow S$ by Theorem 5.25. Note that

$$f \circ \rho = \lim_{n \to \infty} f \circ f^n = \lim_{n \to \infty} f^{n+1} = \rho.$$

Then $f|_S = id$. Take $z_0 \in M$ and set $p = \rho(z_0)$. Then $f(p) = f(\rho(z_0)) = \rho(z_0) = p$. Since

$$A^n = (f^n)'(p) \to \rho'(p) \quad (A = f'(p)),$$

it follows that if $\lambda \in \mathrm{sp}(A)$, then

$$\lim_{n \to \infty} \lambda^n \in \mathrm{sp}(\rho'(p)) \subset \{0, 1\}.$$

Thus $\lambda \in \Delta \cup \{1\}$, and the first part of the assertion is proved.

Conversely, assume f has a fixed point $p \in M$ such that $\mathrm{sp}(f'(p)) \subset \Delta \cup \{1\}$; in particular, $\{f^n\}$ is not compactly divergent, and hence relatively compact in $\mathrm{Hol}(M, M)$. Since $f'(p)|_{\mathbf{E}_p^c} = id$ by hypothesis and the arguments after Theorem 5.25, it follows that $A^n \to \rho'(p)$ as $n \to \infty$, where ρ is a limit point of the sequence $\{f^n\}$. In particular, p is fixed by every limit point $h \in \mathrm{Hol}(M, M)$ of $\{f^n\}$, and $h'(p) = \rho'(p)$. Thus, by Theorem 5.11, 2), $h|_S = id$ and, by Corollary 5.2, $h = \rho$, that is, ρ is the unique limit point of the sequence $\{f^n\}$ and, being $\{f^n\}$ relatively compact in $\mathrm{Hol}(M, M)$, $f^n \to \rho$. □

Corollary 5.5 ([2]) *Let M be a taut manifold, and $f \in \mathrm{Hol}(M, M)$. Then the sequence $\{f^n\}$ of iterates of f converges to a point $p \in M$ iff p is a attractive fixed point of f.*

Proposition 5.3 *Let $\{f^t\}_{t \in \mathbb{R}_+}$ be a semiflow on a compact Kobayashi hyperbolic manifold M. Then $f^t = id$ for all $t \geq 0$.*

Proof. By Lemma 5.2, every f^t is injective, and hence open. Since M is compact, every f^t is an automorphism of M. But $\mathrm{Aut}(M)$ is finite, and the assertion follows from Lemma A.1. □

Lemma 5.14 *Let $\{f^t\}_{t \in \mathbb{R}_+}$ be a semiflow on a compact Kobayashi hyperbolic manifold M. If $f^{t_0} \in \mathrm{Aut}(M)$ for some $t_0 > 0$, then $\{f^t\}_{t \in \mathbb{R}_+}$ can be extended to a flow.*

Proof. Since $(f^{at_0/b})^b = f^{at_0} \in \mathrm{Aut}(M)$ for all $a, b \in \mathbb{N}$, we have $f^{kt_0} \in \mathrm{Aut}(M)$ for all $k \in \mathbb{Q}_+$. By continuity, $f^{kt_0} \in \mathrm{Aut}(M)$ for all $k \in \mathbb{R}_+$ since $\mathrm{Aut}(M)$ is closed in $\mathrm{Hol}(M, M)$ by Theorem 5.13. Finally, it follows from Lemma A.1. □

Theorem 5.27 *Let $f : \mathbb{R}_+ \longrightarrow \mathrm{Hol}(M, M)$ be a continuous semigroup homomorphism on a taut manifold M. Then f^t converges as $t \to +\infty$ to a mapping $\rho \in \mathrm{Hol}(M, M)$ iff f has an equilibrium point $p \in M$ such that its spectral values at p are contained in $\{z \in \mathbb{C} \mid \mathrm{Re}(z) < 0\} \cup \{0\}$.*

For the proof, see Theorem 2.5.21 of Abate [2]. By (5.1), f has an equilibrium point $p \in M$ such that its spectral values at p are contained in $\{z \in \mathbb{C} \mid \text{Re}(z) < 0\} \cup \{0\}$ iff f has an equilibrium point $p \in M$ such that

$$\text{sp}((f^t)'(p)) \subset \Delta \cup \{1\},$$

for $t > 0$. Hence Theorem 5.27 is the correspondent of Theorem 5.26 for semiflows.

As a consequence of Theorem 4.35 and Theorem 1.16, we have the following result:

Theorem 5.28 *Let* $f : M \longrightarrow M$ *be a holomorphic mapping on a Kobayashi hyperbolic manifold* M. *Then* $J_{equ}(f) = \emptyset$.

Thus Theorem 5.28, Theorem 3.14 and Corollary 3.2 directly imply the following

Corollary 5.6 *Each hyperbolic set of a holomorphic mapping on a Kobayashi hyperbolic manifold* M *is of the Morse index* $= 0$, *and is asymptotically stable.*

We transfer Conjecture 4.2 as follows:

Conjecture 5.2 *If* f *is a holomorphic mapping on a complex manifold* M *with* $h_{top}(f) > 0$, *then each component of the Fatou set is Kobayashi hyperbolic.*

Finally, we state a normality criterion of Wu. According to Wu [272], a *pseudo-hermitian metric* h^* on a Riemann surface M is a C^∞ covariant tensor of rank two such that, (1) h^* is a Hermitian metric on $M - S$, where S is a subset of M consisting only of isolated points, and (2) h^* is zero on S. In the sequel, it will be understood that by the curvature of h^* on M, we mean the curvature of h^* on $M - S$.

Let $\mathcal{F}$ be a family of holomorphic mappings from a complex manifold M into a Hermitian manifold N with a Hermitian metric h. Then $\mathcal{F}$ is called a *strongly negatively curved family* (*of order* $-B < 0$) iff for any $f \in \mathcal{F}$ and for any holomorphically embedded disc D in M, the curvature of the pseudo-hermitian metric $(f|_D)^* h$ is bounded above by $-B < 0$. H. Wu [272] proved the following criterion:

Theorem 5.29 *If* $\mathcal{F}$ *is a strongly negatively curved family, then* $\mathcal{F}$ *is equicontinuous. Further, if* N *is complete, then* $\mathcal{F}$ *is normal.*

This result entails as a corollary the main theorem of Grauert-Reckxiegel [96], Satz 1:

Corollary 5.7 *If* N *is a strongly negatively curved Hermitian manifold, then it is tight. If* N *is furthermore complete, then it is taut.*

Here a Hermitian manifold is called *strongly negatively curved* (*of order* $-B < 0$) iff its holomorphic curvature of all complex lines is bounded above by $-B < 0$. By Theorem 5.22 and Corollary 5.7, strongly negatively curved Hermitian manifolds are Kobayashi hyperbolic.

5.6   Julia sets of meromorphic mappings

We first define the Fatou and Julia sets of meromorphic mappings. Here we begin from a more general case. Let $S \neq \emptyset$ be an analytic subset of a complex manifold M such that $A = M - S \neq \emptyset$. Let $f_A : A \longrightarrow N$ be a holomorphic mapping into a complex manifold N. Then there exists a subset $\mathcal{I}_f \subset S$ such that the holomorphic mapping $f_A : A \longrightarrow N$ extends to a "maximal" holomorphic mapping $f_{M-\mathcal{I}_f} : M - \mathcal{I}_f \longrightarrow N$, that is, f_A can not be extended holomorphically at any point of $\mathcal{I}_f$. For convenience, we regard the "maximal" holomorphic mapping $f_{M-\mathcal{I}_f}$ as a mapping $f : M \longrightarrow N$ with the indeterminacy $\mathcal{I}_f$. Let $\mathcal{F}$ be a family of such mappings from M into N and set

$$\mathcal{I}_\mathcal{F} = \overline{\bigcup_{f \in \mathcal{F}} \mathcal{I}_f}.$$

Let $F(\mathcal{F})$ (resp. $F_{uc}(\mathcal{F})$, or $F_{equ}(\mathcal{F})$) be the maximal open subset of $M - \mathcal{I}_\mathcal{F}$ on which $\mathcal{F}$ is normal (resp. uc-normal, or equicontinuous). Let $J(\mathcal{F})$ (resp. $J_{equ}(\mathcal{F})$) be the complement of the set $F(\mathcal{F})$ (resp. $F_{equ}(\mathcal{F})$). By the definition, we see the indeterminacy of the family

$$\mathcal{I}_\mathcal{F} \subset J(\mathcal{F}) \cap J_{equ}(\mathcal{F}).$$

Now we consider a meromorphic mapping $f : M \longrightarrow N$. Assume that M is embedded into N. Define

$$\mathcal{I}_f = I_f \cup f^{-1}(f(M - I_f) - M).$$

We can obtain the following iterations of f. Define $S_1 = \mathcal{I}_f$, $f^1 = f$ and set

$$f^2 = f \circ f : M - S_2 \longrightarrow M, \quad S_2 = S_1 \cup f^{-1}(S_1),$$

$$f^{n+1} = f \circ f^n : M - S_{n+1} \longrightarrow M, \quad S_{n+1} = S_n \cup f^{-n}(S_1) \ (n = 0, 1, 2, ...).$$

Obviously, we have $\mathcal{I}_{f^n} \subset S_n$, and obtain a family

$$\mathcal{F} = \{f^n\}_{n=1}^\infty \subset \mathrm{Hol}(M - S, M), \quad S = \overline{\bigcup_{j=1}^\infty S_j},$$

with $\mathcal{I}_\mathcal{F} \subset S$. If $M - \mathcal{I}_\mathcal{F} \neq \emptyset$, define Fatou set and Julia set respectively by

$$F(f) = F(\mathcal{F}), \quad J(f) = M - F(f).$$

Similarly define

$$F_{equ}(f) = F_{equ}(\mathcal{F}), \quad J_{equ}(f) = M - F_{equ}(f).$$

Example 5.11 *If $f : M \longrightarrow M$ is a meromorphic mapping, then $\mathcal{I}_f = I_f$. The Fatou and Julia sets are well-defined.*

Example 5.12 *If f is a meromorphic function on $\mathbb{C}$ with some poles, but f is not a rational function, then $\mathcal{I}_f = f^{-1}(\infty)$ is just the set of pole points of f. The definitions introduced here coincide with the usual definitions of Fatou and Julia sets of meromorphic functions on $\mathbb{C}$.*

Similarly, one can study the dynamical properties of these sets. For example, see Berg-weiler [39] and Büger [56] in the case of one variable, and Fornaess and Sibony [85] for several variables. It is an interesting problem to study the properties of value distribution of the indeterminacy $\mathcal{I}_{\mathcal{F}}$ of the family.

Obviously, $J(f) \neq \emptyset$ if $\mathcal{I}_f \neq \emptyset$. However, how to measure $J(f)$ if $\mathcal{I}_f = \emptyset$ when M is not compact? Assume that the iteration f^n constructed by f above all are meromorphic on M and that M has a logarithmic convex exhaustion function $\tau : M \longrightarrow \mathbb{R}[0, \infty)$ with Levi form $v = dd^c\tau$. For $p \in \mathbb{Z}[1, m]$, where $m = \dim M$, we introduce a quantity

$$d_p[f] = \limsup_{n \to \infty} \limsup_{r \to \infty} \left(\frac{T_{p,f^n}(r, \omega)}{T_{p,f}(r, \omega)} \right)^{\frac{1}{n}},$$

where ω is the associated $(1, 1)$-form of an Hermitian metric on N.

Conjecture 5.3 *If $d_p[f] > 1$ for some $p \in \mathbb{Z}[1, m]$, then $J(f) \neq \emptyset$.*

We will discuss further the problem for a special case in § 7.2.

Conjecture 5.4 *Assume that $f \in \mathrm{Hol}(M, M)$ with $d_p[f] > 1$ for some $p \in \mathbb{Z}[1, m]$. Then for each positive integer n, f has a n-cycle at least, except for at most finite many n.*

If $M = \mathbb{C}$ and if f is a polynomial with $\deg(f) \geq 2$, then for each positive integer n, f has a n-cycle at least, except for at most one integer n (see Baker [22]). If f is a transcendental entire function, then $d_1[f] = +\infty$, and for each positive integer n, f has infinitely many n-cycles, except for at most one integer n (see Baker [23]). Thus we also suggest the following problem:

Conjecture 5.5 *Assume that $f \in \mathrm{Hol}(M, M)$ with $d_p[f] = +\infty$ for some $p \in \mathbb{Z}[1, m]$. Then for each positive integer n, f has infinitely many n-cycles, except for at most finite many n.*

Next let M and N be connected complex manifolds of dimension m and let $f : M \longrightarrow N$ be a holomorphic mapping. For $x \in M$, the differential $f'(x) : \mathbf{T}(M)_x \longrightarrow \mathbf{T}(N)_{f(x)}$ is a linear mapping. Then the set of critical points

$$C_f = \{x \in M \mid \mathcal{J}f(x) = 0\}$$

is analytic. By Sard's theorem, $f(C_f)$ is then of measure zero. Here $C_f \neq M$ iff $\mathrm{rank}(f) = m$, i.e., f is non-degenerate, or equivalently, $f(M)$ contains an open subset of N. Now the mapping

$$f : M - f^{-1}(f(C_f)) \longrightarrow f(M) - f(C_f) \subset N$$

is a locally biholomorphic mapping.

Now suppose that $f \in \mathrm{Hol}(M, N)$ is proper and surjective. Then

$$f : M - f^{-1}(f(C_f)) \longrightarrow N - f(C_f)$$

is a proper, surjective, locally biholomorphic mapping, hence a covering space of finite sheet number $\deg(f)$. By a theorem of Remmert[208] (see also [15]) the *singularity set*

$$S_f = \{x \in M \mid \dim_x f^{-1}(f(x)) > 0\}$$

is analytic with $S_f \subseteq C_f$ and

$$\dim f(S_f) \leq m - 2.$$

Also $\dim f(C_f) \leq m - 1$ and

$$\dim_x f(C_f) = m - 1 \text{ if } x \in f(C_f) - f(S_f).$$

Lemma 5.15 ([243]) $\#f^{-1}(z) \leq \deg(f)$ *for all* $z \in N - f(S_f)$.

Take $x \in M - S_f$. Let U be an open relative compact neighborhood of x in $M - S_f$ such that $\overline{U} \cap f^{-1}(f(x)) = \{x\}$, then the *mapping degree* $\mu_f(x)$ of f at x is given by

$$1 \leq \mu_f(x) = \lim_{z \to x} \sup \#(\overline{U} \cap f^{-1}(f(z))) < \infty, \tag{5.26}$$

where $\mu_f(x)$ does not depend on U (Stoll [244]). Here $x \in M - C_f$ iff $\mu_f(x) = 1$.

Lemma 5.16 ([243]) *For all* $a \in N - f(S_f)$, *we have*

$$\sum_{x \in f^{-1}(a)} \mu_f(x) = \deg(f). \tag{5.27}$$

The following can be deduced from Lemma B.7 and fiber integrations:

Lemma 5.17 (cf.[188]) *Assume that M and N are compact. Let $\{f_n\}_{n=1}^{\infty}$ be a sequence in $\mathrm{Hol}(M, N)$ converging to $f \in \mathrm{Hol}(M, N)$. If $\mathrm{rank}(f_n) = k$ for all n, then $\mathrm{rank}(f) = k$.*

Theorem 1.30 implies the following result:

Theorem 5.30 *If M is a compact complex manifold and if $f \in \mathrm{Hol}(M, M)$ with $\deg(f) \geq 2$, then the Julia set $J(f)$ is always nonempty.*

Conjecture 5.6 *Assume that M is a compact complex manifold and take $f \in \mathrm{Hol}(M, M)$ with $\deg(f) \geq 2$. Then for each positive integer n, f has a n-cycle at least, except for at most finite many n.*

If $M = \mathbb{P}^1$ and if f is rational with $\deg(f) \geq 2$, then for each positive integer n, f has a n-cycle at least, except for at most two integer n (see Baker [22]). Now we discuss Problem 1.1 for complex cases.

Theorem 5.31 *Let M be a complex manifold and take $f \in \mathrm{Hol}(M, M)$. If p is an attractive fixed point of f such that $\mathrm{Att}(p)$ is contained in a coordinate neighborhood of p, then $\mathrm{Att}(p)$ is contained in $F(f) \cap F_{equ}(f)$.*

Note that if p is an attractive fixed point of f, then each $O^+(z)$ is bounded for $z \in Att(p)$. By Montel theorem, we see

$$Att(p) \subset F(f) \cap F_{equ}(f).$$

The theorem follows. In particular, if $M = \mathbb{C}^m$, then $F(f)$ contains all attractive fixed points and its basins of attraction. Similarly, we can prove that the Fatou set of a holomorphic mapping contains all plus asymptotical stable fixed points, but we are not sure whether an attractive fixed point is plus asymptotically stable.

We will consider the set $\mathrm{Hol}^*(M)$ of holomorphic mappings, consisting of those f in $\mathrm{Hol}(M, M)$ which have maximal rank m on some nonempty open sets. By the chain rule,

$$\mathcal{J}(f \circ g)(x) = \mathcal{J}f(g(x)) \cdot \mathcal{J}g(x),$$

then $f, g \in \mathrm{Hol}^*(M)$ imply $f \circ g \in \mathrm{Hol}^*(M)$. Let $\mathrm{Hol}_*(M) \subset \mathrm{Hol}^*(M)$ be the set of surjective holomorphic mappings. Assume that M is compact. Then each $f \in \mathrm{Hol}(M, M)$ is proper. Note that $\mathrm{Hol}(M, M)$ is also closed in $C(M, M)$.

Theorem 5.32 *If a sequence $\{f_n\} \subset \mathrm{Hol}_*(M)$ converge uniformly on M to a mapping f, then $f \in \mathrm{Hol}_*(M)$ and for all sufficiently large n,*

$$\deg(f_n) = \deg(f).$$

Proof. We only sketch the proof. First, the uniform convergence of f_n to f guarantees that f is holomorphic throughout M and so $f \in \mathrm{Hol}^*(M)$ by Lemma 5.17. For any $p \in M$, there exist $x_{n_j} \in f_{n_j}^{-1}(p)$ which converge to a point $x_0 \in M$ so that

$$p = \lim_{j \to \infty} f_{n_j}(x_{n_j}) = f(x_0),$$

i.e., $f \in \mathrm{Hol}_*(M)$.

Take $a \in M - f(C_f)$. Then $C_f \cap f^{-1}(a) = \emptyset$ so that $\mu_f(x) = 1$ if $x \in f^{-1}(a)$. Thus $f^{-1}(a)$ contains $d = \deg(f)$ distinct elements $x_1, ..., x_d$. Take $r > 0$ such that the discs $M(x_j; r)$ are mutually disjoint and such that $M(x_j; r)$ lie on a local coordinate neighborhood of x_j. As f_n and f are uniformly close on $\partial M(x_j; r)$, Rouché's principle shows that f_n and f have the same number of a-valued points in each $M(x_j; r)$. Finally, f is bounded away from a on the compact $M - \cup_j M(x_j; r)$, hence so are the f_n (for large n). Thus for all sufficiently large n, f_n and f have the same number of a-valued points and so they have the same degree. $\square$

Corollary 5.8 *Let M be a compact complex manifold and suppose that $\mathrm{Hol}_*(M)$ is finite. Then for each $f \in \mathrm{Hol}_*(M)$, we have $J(f) = \emptyset$ so that $\deg(f) = 1$ and*

$$f : M - f^{-1}(f(S_f)) \longrightarrow M - f(S_f)$$

is a proper, surjective, biholomorphic mapping

Proof. For $f \in \mathrm{Hol}_*(M)$, then $\{f^n\} \subset \mathrm{Hol}_*(M)$ is finite so that $F(f) = M$. Now the corollary follows from Theorem 5.30 and Lemma 5.16. $\square$

Related to this result, we have Horst's theorem:

Theorem 5.33 ([116],[117]) *Let N be a Kobayashi hyperbolic Kähler manifold and M a complex manifold. Then there are only finitely many surjective holomorphic mappings from M onto N.*

If M, N are complex projective, this result was conjectured by Lang[154]. Assuming additionally that K_N carries a metric with non-positive curvature form, Noguchi[189] proved the above finiteness theorem. Instead of the hyperbolicity assumption on N, Kobayashi and Ochiai[146] assumed that N is of general type, and showed the above finiteness theorem. Thus $\mathrm{Hol}_*(M)$ is finite if M is of general type or Kobayashi hyperbolic Kähler manifold.

Problem 5.1 *If $\mathrm{Hol}_*(M)$ is finite, is $J(f) = \emptyset$ for all $f \in \mathrm{Hol}(M, M)$?*

Conjecture 5.7 *If f is a holomorphic self-mapping on a compact complex manifold M with $h_{top}(f) > 0$, then each component of the Fatou set is eventually periodic, i.e., for each component U of the Fatou set, there exist some positive integers m and n such that $f^n(f^m(U)) = f^m(U)$.*

If M is the Riemann sphere, this is the Sullivan's theorem.

Theorem 5.34 ([172]) *Every holomorphic mapping f on a torus $\mathbb{C}/\Gamma$ is an affine mapping*

$$f(z) \equiv az + b \pmod{\Gamma}.$$

The Julia set $J(f)$ is either the empty set or the entire torus according as $|a| \leq 1$ or $|a| > 1$.

Proof. Assume that the lattice $\Gamma \subset \mathbb{C}$ is spanned by the two numbers 1 and $\tau \notin \mathbb{R}$. Then f lifts to a holomorphic mapping $\tilde{f} : \mathbb{C} \longrightarrow \mathbb{C}$ on the universal covering surface $\mathbb{C}$ with

$$\tilde{f}(z+1) \equiv \tilde{f}(z) \ (\bmod \ \Gamma), \quad \tilde{f}(z+\tau) \equiv \tilde{f}(z) \ (\bmod \ \Gamma),$$

for all $z \in \mathbb{C}$. Then the difference functions

$$\tilde{f}(z+1) - \tilde{f}(z), \tilde{f}(z+\tau) - \tilde{f}(z) : \mathbb{C} \longrightarrow \Gamma$$

must be constants since $\mathbb{C}$ is connection and the target space Γ is discrete. Thus there exist $a, a' \in \Gamma$ such that

$$\tilde{f}(z+1) = \tilde{f}(z) + a, \quad \tilde{f}(z+\tau) = \tilde{f}(z) + a'.$$

Define

$$g(z) = \tilde{f}(z) - az$$

so that

$$g(z+1) = g(z), \quad g(z+\tau) = g(z) + a' - a\tau.$$

Thus g gives rise to a mapping from the torus $\mathbb{C}/\Gamma$ to the infinite cylinder

$$\mathbb{C}/(a' - a\tau)\mathbb{Z} \cong \mathbb{C} - \{0\},$$

or from the torus $\mathbb{C}/\Gamma$ to $\mathbb{C}$ if $a' - a\tau = 0$. The mapping must be constant since $\mathbb{C}/\Gamma$ is compact, by using the maximum modulus principle. Hence $g(z) \equiv b$ (constant), so that

$$\tilde{f}(z) = az + b.$$

In particular, we see $a' = a\tau \in \Gamma$.

Write

$$a = m_1 + n_1\tau \ (m_1, n_1 \in \mathbb{Z}), \quad a\tau = m_2 + n_2\tau \ (m_2, n_2 \in \mathbb{Z}).$$

If $a \notin \mathbb{Z}$, then $n_1 \neq 0, m_2 \neq 0$, and τ satisfies a quadratic equation with integer coefficients

$$n_1\tau^2 + (m_1 - n_2)\tau - m_2 = 0.$$

Such a torus is said to admit *complex multiplications*. Also a satisfies a quadratic equation with integer coefficients

$$a^2 - (m_1 + n_2)a + m_1n_2 - m_2n_1 = 0.$$

Note that this equation has no real roots if $a \notin \mathbb{Z}$, and hence $a \notin \mathbb{R}$. Therefore $|a + n|^2$ always is an integer for each $n \in \mathbb{Z}$ so that $|a| \geq 1$.

If $|a| = 1$, setting $a = e^{i\theta}$ ($0 \leq \theta < 2\pi$), then we find that

$$\theta \in \{0, \frac{\pi}{6}, \frac{\pi}{2}, \frac{5\pi}{6}, \pi, \frac{7\pi}{6}, \frac{3\pi}{2}, \frac{11\pi}{6}\}.$$

It follows that

$$f^n(z) \equiv a^n z + \sum_{k=0}^{n-1} ba^k \ (\text{mod } \Gamma), \quad n = 1, 2, \dots$$

is normal on $\mathbb{C}/\Gamma$, and consequently, $J(f)$ is empty.

If $|a| > 1$, the equation $f^n(z) = z$ on $\mathbb{C}/\Gamma$ is equivalent to

$$(a^n - 1)z \equiv -\sum_{k=0}^{n-1} ba^k \ (\text{mod } \Gamma),$$

which has exactly $|a^n - 1|^2$ solutions. Obviously, the set of the solutions for all $n > 0$ is dense in $\mathbb{C}$, and hence the projective image of the set in $\mathbb{C}/\Gamma$, which is the set of all periodic points of f, is dense. Note that all periodic points of f are repelling since $|(f^n)'| = |a|^n > 1$. It follows that the Julia set $J(f)$ is the entire torus.   □

In this theorem, the unstable index function $\mathbf{u} \equiv 1$ if $|a| > 1$, and hence

$$\text{supp}\,\mathbf{u} = \mathbb{C}/\Gamma = J(f).$$

Let $f : \mathbb{C}^m/\Gamma \to \mathbb{C}^m/\Gamma$ be a holomorphic mapping. Then f is a homomorphism modulo a translation in $\mathbb{C}^m/\Gamma$ which is induced by a mapping

$$\tilde{f} = (f_1, \dots, f_m) : \mathbb{C}^m \longrightarrow \mathbb{C}^m$$

of the form

$$f_k = \sum_{j=1}^{m} a_k^j z_j + b_k, \quad b_k \in \mathbb{C}, \quad k = 1, \dots, m,$$

such that $\pi \circ \tilde{f} = f \circ \pi$, where $\pi : \mathbb{C}^m \to \mathbb{C}^m/\Gamma$ is the projections. Let $\lambda_1, \dots, \lambda_m$ be the eigenvalues of $\mathbf{A} = (a_k^j)$. Thus if the spectral radius $r(\mathbf{A}) > 1$, then

$$\text{supp}\,\mathbf{u} = \mathbb{C}^m/\Gamma.$$

Now f is Anosov if and only if $\mathbf{A}$ is hyperbolic, i.e., $|\lambda_i| \neq 1$ for all $i \in \mathbb{Z}[1, m]$. Moreover any Anosov diffeomorphism f on the torus is topologically conjugate to some hyperbolic automorphism $\mathbf{A}$. Thus, by Theorem 3.16 we obtain the following result:

Theorem 5.35 *Suppose that f is a holomorphic mapping on a torus $\mathbb{C}^m / \Gamma$ with the eigenvalues λ_i of f' satisfying $|\lambda_1| \geq |\lambda_2| \geq \dots \geq |\lambda_m|$. Then $J(f) = \emptyset$ if $|\lambda_1| < 1$, or $J(f) = \mathbb{C}^m / \Gamma$ if $|\lambda_m| > 1$.*

Let f be a holomorphic Anosov diffeomorphism of a complex torus $\mathbb{C}^m / \Gamma$ such that the eigenvalues λ_i of f' satisfy $|\lambda_1| \geq |\lambda_2| \geq \dots \geq |\lambda_m|$. If the (complex) Morse index of M is zero, then $|\lambda_1| < 1$. Now the eigenvalues $\frac{1}{\lambda_i}$ of $(f^{-1})'$ are of $\frac{1}{|\lambda_i|} > 1$. Theorem 5.35 implies $J(f) = \emptyset$ and $J(f^{-1}) = \mathbb{C}^m / \Gamma$, and hence $J(f) \cap J(f^{-1}) = \emptyset$. If the Morse index of M is m, similarly we see $J(f^{-1}) = \emptyset$ and $J(f) = \mathbb{C}^m / \Gamma$, and hence $J(f) \cap J(f^{-1}) = \emptyset$. If the Morse index satisfies $0 < \mathbf{u}_{\mathbb{C}^m / \Gamma} < m$, Corollary 3.2 and Theorem 1.24 imply

$$J(f) = J(f^{-1}) = \mathbb{C}^m / \Gamma.$$

Note that the following result was obtained by Ghys [95]:

Theorem 5.36 *Let f be a holomorphic Anosov diffeomorphism of a compact complex surface M. Then M is a complex torus $\mathbb{C}^2 / \Gamma$ and f is holomorphically conjugate to a linear automorphism of $\mathbb{C}^2 / \Gamma$.*

5.7 A generalization of Marty's criterion

If M and N are complex manifolds with Finsler metrics κ and κ', respectively, and if $f \in \text{Hol}(M, N)$, we can define the norm

$$\|f'(z)\| = \sup\{\kappa'(f'(z)Z) \mid Z \in \mathbf{T}(M)_z, \kappa(Z) = 1\}.$$

For Hermitian manifolds, we will use the Finsler metrics induced by Hermitian metrics.

Proposition 5.4 *Assume that M and N are complex manifolds with Finsler metrics κ and κ', respectively, and take $\mathcal{F} \subset \text{Hol}(M, N)$. If $\{\|f'(z)\| \mid f \in \mathcal{F}\}$ is bounded uniformly on any compact subset of M, then the family $\mathcal{F}$ is equicontinuous on M.*

Proof. Take $z_0 \in M$. Note that M is locally compact. We can choose relatively compact neighborhood U of z_0 such that $\{\|f'(z)\| \mid f \in \mathcal{F}\}$ is bounded uniformly on U by a constant C. By the geometric theory (see [221]), we can choose a positive number ε with $M_{d_\kappa}(z_0; \varepsilon) \subset U$ such that arbitrary two points of $M_{d_\kappa}(z_0; \varepsilon)$ can be joined by unique geodesic in $M_{d_\kappa}(z_0; \varepsilon)$. For any $z \in M_{d_\kappa}(z_0; \varepsilon)$, let $\gamma : [0, 1] \longrightarrow M$ with $\gamma(0) = z_0, \gamma(1) = z$ be the geodesic joining z_0 and z in $M_{d_\kappa}(z_0; \varepsilon)$. Then for any $f \in \mathcal{F}$, λ uniquely defined by $\lambda = f \circ \gamma$ is piecewise differentiable curves joining $f(z_0)$ and $f(z)$. Set

$$\gamma'(t) = \frac{1}{\sqrt{2}} \Re^{-1}(\dot{\gamma}(t))$$

and note that

$$\lambda'(t) = f'(\gamma(t))\gamma'(t).$$

We have

$$
\begin{aligned}
d_{\kappa'}(f(z_0), f(z)) &\leq \text{Length}(\lambda) = \int_0^1 \kappa'(\dot\lambda(t))dt \\
&= 2\int_0^1 \kappa'(\lambda'(t))dt = 2\int_0^1 \kappa'(f'(\gamma(t))\gamma'(t))dt \\
&\leq \int_0^1 \|f'(\gamma(t))\| \cdot \kappa(\dot\gamma(t))dt \\
&\leq C\text{Length}(\gamma) = Cd_\kappa(z_0, z),
\end{aligned}
$$

which implies that $\mathcal{F}$ is equicontinuous at z_0, and hence $\mathcal{F}$ is equicontinuous on M since z_0 is arbitrary. $\qquad\square$

Conversely, we have the following result:

Theorem 5.37 *Assume that M and N are complex manifolds with Finsler metrics κ and κ', respectively, and take $\mathcal{F} \subset \text{Hol}(M, N)$. If the family $\mathcal{F}$ is uc-normal on M, then $\{\|f'(z)\| \mid f \in \mathcal{F}\}$ is bounded uniformly on any compact subset of M.*

Proof. Assume, to the contrary, that there exists a compact subset $K \subset M$ such that $\{\|f'(z)\| \mid f \in \mathcal{F}\}$ is not bounded uniformly on K. Then there are a sequence $\{f_n\} \subset \mathcal{F}$ and points $z_n \in K$ such that

$$\lim_{n\to\infty} \|f_n'(z_n)\| = +\infty.$$

Since K is compact and by the uc-normal family hypothesis, without loss of generality we may suppose that the sequence $\{f_n\}$ converges uniformly to a holomorphic mapping $f \in \text{Hol}(M, N)$ on any compact subset of M. Thus for each point $z \in M$ we can choose relatively compact coordinate neighborhood U_z of z such that $\{f_n\}$ converges uniformly to f on U_z. By shrinking U_z if it is necessary, and by using Weierstrass theorem, we know that $\|f_n'(z)\|$ are bounded on U_z. Since K is compact, we may cover K with finitely many of the sets U_z. It follows that $\|f_n'(z)\|$ are bounded on K, contradicting our assumption. $\qquad\square$

If M is a Kobayashi hyperbolic domain in $\mathbb{C}^m$ ($m > 1$) with the Kobayashi differential metric κ_M (see § 5.3), and if N is a complete Hermitian manifold, this result appears in [11].

Corollary 5.9 *Assume that M and N are complex manifolds with Finsler metrics κ and κ', respectively, such that N is compact and take $\mathcal{F} \subset \text{Hol}(M, N)$. Then the family $\mathcal{F}$ is normal on M iff $\{\|f'(z)\| \mid f \in \mathcal{F}\}$ is bounded uniformly on any compact subset of M.*

Corollary 5.10 *Assume that M is a compact complex manifold with a Finsler metric κ and take $f \in \text{Hol}(M, M)$. If $h^{ss}(f) > 0$, then $J(f) \neq \emptyset$.*

Corollary 5.11 *Assume that M is a compact complex manifold with a Finsler metric κ and take $f \in \text{Hol}(M, M)$. Then $J(f) \neq \emptyset$ if there exists a ergodic f-invariant probability measure μ on M with a positive Lyapunov exponent.*

Since $\dim Mh^{ss}(f) \geq h^{ss}_{vol}(f)$ (see § 3.6), we obtain

Corollary 5.12 *Let M be a compact Hermitian manifold and take $f \in \text{Hol}(M, M)$. If $h^{ss}_{vol}(f) > 0$, then $J(f) \neq \emptyset$.*

Note that if M is compact, then $\{f^n\}$ is uc-normal on $F(f)$. Thus $\{\|(f^n)'(z)\|\}$ is bounded uniformly on any compact subset of $F(f)$. Hence we have

Corollary 5.13 *Assume that M is a compact complex manifold with a Finsler metric κ and take $f \in \text{Hol}(M, M)$. Then*

$$\limsup_{n \to \infty} \sup_{z \in J(f)} \|(f^n)'(z)\| = +\infty.$$

Now we consider a special case, that is, $N = \mathbb{P}^1$ is the Riemannian sphere and M is a domain in $\mathbb{C}$, and hence $\mathcal{F}$ is a family of meromorphic functions on M. Let Ω be the Fubini-Study form on $\mathbb{P}^1$ and note that (see § B.6)

$$\Omega(z) = -dd^c \log |z, \infty|^2 = \frac{1}{(1 + |z|^2)^2} \frac{i}{2\pi} dz \wedge d\bar{z}$$

on $\mathbb{C}$. Then for $f \in \mathcal{F}$, one obtains

$$(f^*\Omega)(z) = \frac{|f'(z)|^2}{(1 + |f(z)|^2)^2} \frac{i}{2\pi} dz \wedge d\bar{z}.$$

Let g be the Euclidean metric on $\mathbb{C}$ and let h be the induced Riemannian metric by the Hermitian metric of the associated $(1,1)$-form $\pi\Omega$ on $\mathbb{P}^1$. Then on $\mathbb{C}$ we have

$$f^*h = \frac{|f'(z)|^2}{(1 + |f(z)|^2)^2} g,$$

which easily implies

$$\|f'(z)\| = \frac{|f'(z)|}{1 + |f(z)|^2}.$$

Thus we obtain the classic Marty's criterion:

Corollary 5.14 ([167]) *Let $\mathcal{F}$ be a family of meromorphic functions on a domain M in $\mathbb{C}$. Then $\mathcal{F}$ is normal on M iff the family $\{\frac{|f'|}{1+|f|^2} \mid f \in \mathcal{F}\}$ is uniformly bounded on any compact subset of M.*

For this special case, we compare with Theorem 1.25. First we fix a compact set K and assume that the inequality

$$\frac{|f'(z)|}{1 + |f(z)|^2} \leq C_K, \quad z \in K, f \in \mathcal{F},$$

holds for a constant C_K. Without loss of generality we may assume that K is the closure of a connected convex open set U. By using the method of the proof in Theorem 1.25, we can prove the inequality

$$|f(z), f(w)| \leq C_K |z - w|, \quad z, w \in K, f \in \mathcal{F},$$

and hence

$$\text{Lip}_{\mathcal{F}}(K) \leq C_K,$$

where

$$\text{Lip}(f, K) = \sup \left\{ \frac{|f(z), f(w)|}{|z - w|} \mid z, w \in K, \ z \neq w \right\}.$$

Conversely, assume that $\text{Lip}_{\mathcal{F}}(K) = C_K < +\infty$. Then

$$\frac{|f(z), f(w)|}{|z - w|} = \left| \frac{f(z) - f(w)}{z - w} \right| \frac{1}{(1 + |f(z)|^2)^{\frac{1}{2}}(1 + |f(w)|^2)^{\frac{1}{2}}}$$
$$\leq C_K, \quad z, w \in K, f \in \mathcal{F}.$$

Letting $w \to z$, we have

$$\frac{|f'(z)|}{1 + |f(z)|^2} \leq C_K, \quad z \in K, f \in \mathcal{F}.$$

Hence $\text{Lip}_{\mathcal{F}}(K)$ is finite for any compact subset K of M iff the family $\{ \frac{|f'|}{1 + |f|^2} \mid f \in \mathcal{F}\}$ is uniformly bounded on any compact subset of M. Thus Theorem 1.25 may be referred as the integrated form of the Marty's criterion. Now we generalize the Zalcman's criterion [282] as follows:

Theorem 5.38 *Assume that M and N are complex manifolds with Finsler metrics κ and κ', respectively, and take $\mathcal{F} \subset \text{Hol}(M, N)$. If the family $\mathcal{F}$ is not equicontinuous on M, then there exist a local coordinate system $(U; \varphi)$, a compact set $K \subset U$ and sequences $\{z_n\} \subset \varphi(K)$, $\{f_n\} \subset \mathcal{F}$, $\{\rho_n\}$ with $\rho_n > 0$ and $\rho_n \to 0$, and $\{\xi_n\} \subset \mathbb{C}^m$ ($m = \dim M$) Euclidean unit vectors, such that $\|g_n'(\zeta)\|$ are uniformly bounded on compact sets on $\mathbb{C}$ with $\|g_n'(0)\| = 1$, where*

$$g_n(\zeta) = f_n \circ \varphi^{-1}(z_n + \rho_n \xi_n \zeta), \quad \zeta \in \mathbb{C}.$$

Conversely, the family $\mathcal{F}$ is not uc-normal on M.

Proof. Assume that $\mathcal{F}$ is not equicontinuous on M, by Proposition 5.4, $\{\|f'(z)\| \mid f \in \mathcal{F}\}$ is not bounded uniformly on a compact subset K of M. Thus there exist sequences $\{f_n\} \subset \mathcal{F}$, $\{p_n\} \subset K$, and $Z_n \in \mathbf{T}(M)_{p_n} - \{0\}$ for $n \geq 1$, such that

$$\kappa'(f_n'(p_n)Z_n) \geq n\kappa(Z_n), \quad n \geq 1.$$

Since K is compact, w. l. o. g., we may assume $p_n \to p_0$ for some $p_0 \in K$, and assume that

$$K \subset V \subset K_0 = \overline{V} \subset U$$

for a coordinate neighborhood U of p_0 and a relative compact neighborhood V of p_0. Let $\varphi : U \longrightarrow D \subset \mathbb{C}^m$ is the coordinate system with $\varphi(p_0) = 0$. Since K is compact, so does $\varphi(K)$, and hence there exist positive constants c_1 and c_2 such that

$$c_1 \|\xi\| \leq \kappa((\varphi^{-1})'(z)\xi) \leq c_2 \|\xi\|, \quad z \in \varphi(K), \quad \xi \in \mathbb{C}^m$$

where $\|\xi\|$ is taked for the Euclidean norm. Set

$$w_n = \varphi(p_n), \quad \eta_n = \varphi'(p_n)Z_n \in \mathbb{C}^m.$$

Then

$$
\begin{aligned}
\kappa'((f_n \circ \varphi^{-1})'(w_n)\eta_n) &= \kappa'(f_n'(p_n)Z_n) \\
&\geq n\kappa(Z_n) = n\kappa((\varphi^{-1})'(w_n)\eta_n) \\
&\geq c_1 n\|\eta_n\|, \quad n \geq 1.
\end{aligned}
$$

Define

$$
\rho(z) = \begin{cases} d(z, D - \varphi(K_0)) & \text{if } z \in \varphi(K_0) \\ 0 & \text{if } z \notin \varphi(K_0), \end{cases}
$$

where d is the Euclidean distance on $\mathbb{C}^m$. Then $\rho \in C(\overline{D}, \mathbb{R}_+)$ with $c = \min_{z \in \varphi(K)} \rho(z) > 0$.
Let

$$a_n = \max_{z \in \varphi(K), \|\xi\|=1} \{\rho(z)\kappa'((f_n \circ \varphi^{-1})'(z)\xi)\},$$

and take z_n and ξ_n which maximize the expression above. W. l. o. g., we assume $z_n \to z_0$
for some $z_0 \in \varphi(K)$, and define

$$\rho_n = \frac{\rho(z_n)}{a_n}.$$

Note that

$$a_n \geq c\frac{\kappa'((f_n \circ \varphi^{-1})'(w_n)\eta_n)}{\|\eta_n\|} \geq cc_1 n \to \infty$$

as $n \to \infty$. Then

$$\frac{\rho_n}{\rho(z_n)} = \frac{1}{a_n} \to 0.$$

Therefore the function

$$g_n(\zeta) = f_n \circ \varphi^{-1}(z_n + \rho_n \xi_n \zeta), \quad \zeta \in \mathbb{C}$$

are defined for $|\zeta| < R_n = \frac{\rho(z_n)}{\rho_n}$. Moreover

$$
\begin{aligned}
\kappa'(g_n'(0)1) &= \kappa'((f_n \circ \varphi^{-1})'(z_n)\rho_n \xi_n) \\
&= \rho_n \kappa'((f_n \circ \varphi^{-1})'(z_n)\xi_n) = \frac{\rho_n a_n}{\rho(z_n)} = 1.
\end{aligned}
$$

Fix $R > 0$ and n large such that $R < R_n$. If $|\zeta| < R$, then

$$
\begin{aligned}
\|g_n'(\zeta)\| &= \kappa'(g_n'(\zeta)1) = \kappa'((f_n \circ \varphi^{-1})'(z_n + \rho_n \xi_n \zeta)\rho_n \xi_n) \\
&\leq \frac{\rho_n a_n}{\rho(z_n + \rho_n \xi_n \zeta)} = \frac{\rho(z_n)}{\rho(z_n + \rho_n \xi_n \zeta)} \to 1.
\end{aligned}
$$

So the g_n are holomorphic on larger and larger discs in $\mathbb{C}$ such that $\|g_n'(\zeta)\|$ are uniformly
bounded on compact subsets.

Conversely, suppose, to the contrary, that $\mathcal{F}$ is uc-normal on M. Then by Theorem 5.37, there exists a number C_K such that

$$\|f'(z)\| \le C_K, \quad z \in K, \quad f \in \mathcal{F}.$$

Therefore

$$
\begin{aligned}
\|g_n'(0)\| &= \rho_n \kappa'((f_n \circ \varphi^{-1})'(z_n)\xi_n) \\
&= \rho_n \kappa'(f_n'(\varphi^{-1}(z_n))(\varphi^{-1})'(z_n)\xi_n) \\
&\le c_1 C_K \rho_n \to 0.
\end{aligned}
$$

This is a contradiction.   □

Thus by Theorem 5.38 and Corollary 5.9, one has the following generalization of Zalcman's theorem obtained by Aladro and Krantz [12].

Corollary 5.15 *Assume that (M, d_g) and (N, d_h) are Hermitian manifolds such that N is compact and take $\mathcal{F} \subset \mathrm{Hol}(M, N)$. The family $\mathcal{F}$ is not normal on M if and only if there exist a local coordinate system $(U; \varphi)$, a compact set $K \subset U$ and sequences $\{z_n\} \subset \varphi(K)$, $\{f_n\} \subset \mathcal{F}$, $\{\rho_n\}$ with $\rho_n > 0$ and $\rho_n \to 0$, and $\{\xi_n\} \subset \mathbb{C}^m$ ($m = \dim M$) Euclidean unit vectors, such that*

$$g_n(\zeta) = f_n \circ \varphi^{-1}(z_n + \rho_n \xi_n \zeta), \quad \zeta \in \mathbb{C}$$

converges uniformly on compact subsets of $\mathbb{C}$ to a nonconstant holomorphic mapping $g \in \mathrm{Hol}(\mathbb{C}, N)$.

Chapter 6

Iteration theory on $\mathbb{P}^m$

In this chapter, we will introduce the Fatou-Julia theory on $\mathbb{P}^m$ obtained by Fornaess and Sibony. Here we mainly introduce the theory on holomorphic mappings. For the case of meromorphic mappings, see their paper [85]. We also prove the Ueda's theorem related to Conjecture 5.2, and simply introduce the Newton's method.

6.1 Meromorphic self-mappings on $\mathbb{P}^m$

Let $\mathrm{Pol}(\mathbb{C}^m, \mathbb{C}^n)$ be the space of polynomial mappings from $\mathbb{C}^m$ into $\mathbb{C}^n$. Let $\mathcal{L}_k(\mathbb{C}^m)$ be the subvector space of $\mathrm{Pol}(\mathbb{C}^m, \mathbb{C})$ consisting of all homogeneous polynomials of degree k. Then $\mathcal{L}_k(\mathbb{C}^m)$ has dimension

$$\dim \mathcal{L}_k(\mathbb{C}^m) = \binom{m-1+k}{k} = \frac{(m-1+k)!}{(m-1)!k!}.$$

If $d = (d_1, ..., d_n)$, we will write

$$\mathcal{L}_d(\mathbb{C}^m) = \mathcal{L}_{d_1}(\mathbb{C}^m) \times \cdots \times \mathcal{L}_{d_n}(\mathbb{C}^m).$$

In particular, if $d_1 = \cdots = d_n = k$, write

$$\mathcal{L}_k^n(\mathbb{C}^m) = \mathcal{L}_d(\mathbb{C}^m).$$

Take $f = (f_1, ..., f_n) \in \mathrm{Pol}(\mathbb{C}^m, \mathbb{C}^n)$. Denote the degree of f_j by $\deg(f_j)$ and define the "degree" of the polynomial mapping f by

$$\deg(f) = \max_{1 \le j \le n} \{\deg(f_j)\}.$$

Set $d_j = \deg(f_j)$ and let $d = (d_1, ..., d_n)$. Then we obtain a mapping $\hat{f} = (\hat{f}_1, \hat{f}_2, ..., \hat{f}_n) \in \mathcal{L}_d(\mathbb{C}^m)$, where $\hat{f}_j$ consist of the homogeneous terms of degree d_j of f_j, that is, $\hat{f}_j \in \mathcal{L}_{d_j}(\mathbb{C}^m)$ with $\deg(f_j - \hat{f}_j) < d_j$. The following result is well-known:

Lemma 6.1 (Bézout theorem) *For $f_j \in \mathrm{Pol}(\mathbb{C}^m, \mathbb{C})$, $j = 1, 2, ..., m$, with $d_j = \deg(f_j)$, if $\{f_1 = f_2 = \cdots = f_m = 0\}$ is a discrete variety in $\mathbb{C}^n$, then the number of common*

179

zeros of $f_1, f_2, ..., f_m$, counted with their multiplicities, is at most $d_1 d_2 \cdots d_m$. If $\{\hat{f}_1 = \hat{f}_2 = \cdots = \hat{f}_m = 0\} = \{0\}$, then $f = (f_1, f_2, ..., f_m)$ has just $d_1 \cdot d_2 \cdots d_m$ zero points counting multiplicities.

Take $f \in \mathcal{L}_d^m(\mathbb{C}^n)$ with $\{f = 0\} = \{0\}$. Write

$$M_f = \max_{\|z\|=1} \|f(z)\|, \quad m_f = \min_{\|z\|=1} \|f(z)\| > 0.$$

Then we have

$$m_f \|z\|^d \le \|f(z)\| \le M_f \|z\|^d.$$

For any $a \in \mathbb{C}^n - \{0\}$, take large $r > 0$ such that $r^d m_f > \|a\|$. By Rouché principle,

$$n(\mathbb{C}^n(r), f - a) = n(\mathbb{C}^n(r), f) = d^m.$$

Hence f is a surjective and d^m to one mapping.

If $d \ge 2$, letting $R_1 = M_f^{-\frac{1}{d-1}}, R_2 = m_f^{-\frac{1}{d-1}}$, we have

$$R_2 \left(\frac{\|z\|}{R_2} \right)^{d^n} \le \|f^n(z)\| \le R_1 \left(\frac{\|z\|}{R_1} \right)^{d^n},$$

so that

$$\mathbb{C}^n(R_1) \subset \mathrm{Att}(0) \subset F(f), \quad \mathbb{C}^n - \mathbb{C}^n[R_2] \subset F(f).$$

Also we obtain

$$\mathrm{Att}(0) \subset \mathbb{C}^n(R_2), \quad J(f) \subset \mathbb{C}^n[R_2] - \mathbb{C}^n(R_1),$$

and hence $J(f) \ne \emptyset$. Obviously, $\mathrm{Att}(0)$ is a *star-shaped circular region* (or *balanced*), i.e., $cz \in \mathrm{Att}(0)$ whenever $z \in \mathrm{Att}(0)$ and $c \in \mathbb{C}, |c| \le 1$. Note that

$$\mathrm{Att}(0) = \bigcup_{n=0}^{\infty} f^{-n}(\mathbb{C}^n(R_1)).$$

Hence $\mathrm{Att}(0)$ is pseudoconvex.

Lemma 6.2 *Let ϕ be a function on $\mathbb{C}^m - \{0\}$ such that $\phi(z) - \log\|z\|$ is bounded. Then*

$$\lim_{n \to \infty} \frac{1}{d^n} \phi \circ f^n(z) = G(z)$$

exists, and is independent of the choice of ϕ. Further G satisfies

$$G \circ f = d \cdot G. \tag{6.1}$$

Proof. Define a bounded function on $\mathbb{C}^m - \{0\}$ by

$$\gamma(z) = \phi \circ f(z) - d \cdot \phi(z).$$

Then we have

$$\frac{1}{d^j} \phi \circ f^j(z) = \frac{1}{d^{j-1}} \phi \circ f^{j-1}(z) + \frac{1}{d^j} \gamma \circ f^{j-1}(z).$$

Summing those for $j = 1, ..., n$ to get

$$\frac{1}{d^n}\phi \circ f^n(z) = \phi(z) + \frac{1}{d}\gamma(z) + \cdots + \frac{1}{d^n}\gamma \circ f^{n-1}(z)$$

which implies the limit

$$\lim_{n\to\infty} \frac{1}{d^n}\phi \circ f^n(z) = G(z)$$

exists, since γ is bounded and $d \geq 2$. The limit above implies (6.1). If ϕ_1 is another choice of ϕ, then $\phi_1 - \phi$ is bounded so that

$$\lim_{n\to\infty} \frac{1}{d^n}(\phi_1 \circ f^n(z) - \phi \circ f^n(z)) = 0.$$

Hence the function $G(z)$ is independent of the choice of ϕ. $\qquad\square$

Particularly, letting $\phi = \log\|z\|$, we obtain

$$\lim_{n\to\infty} \frac{1}{d^n} \log\|f^n(z)\| = G(z) \tag{6.2}$$

which implies

$$\mathrm{Att}(0) = \{z \in \mathbb{C}^m \mid G(z) < 0\}.$$

Note that the function $\frac{1}{d^n}\log\|f^n(z)\|$ is plurisubharmonic, continuous on $\mathbb{C}^m - \{0\}$, and that $\frac{1}{d^n}\log\|f^n(z)\| - \log\|z\|$ is homogeneous of degree 0. Then G is plurisubharmonic, continuous on $\mathbb{C}^m - \{0\}$, and $G(z) - \log\|z\|$ is homogeneous of degree 0, and we have

$$\frac{1}{d-1}\log m_f \leq G(z) - \log\|z\| \leq \frac{1}{d-1}\log M_f.$$

The function G is unique, and is called the *Green function* of f.

We now describe the holomorphic mappings and the meromorphic mappings from $\mathbb{P}^m$ to $\mathbb{P}^m$.

Theorem 6.1 ([82]) *Let f be a non constant holomorphic mapping from $\mathbb{P}^m$ to $\mathbb{P}^m$. Then f is given in homogeneous coordinates by $[f_0 : f_1 : \cdots : f_m]$ with $\tilde{f}^{-1} = \{0\}$, where $\tilde{f} = (f_0, f_1, ..., f_m) \in \mathcal{L}_d^{m+1}(\mathbb{C}^{m+1})$. The mapping $\tilde{f} : \mathbb{C}^{m+1} \longrightarrow \mathbb{C}^{m+1}$ is called the reduced lifted mapping of f on $\mathbb{C}^{m+1}$.*

Proof. Let $[z_0 : z_1 : \cdots : z_m]$ be homogeneous coordinates in $\mathbb{P}^m$. Assume that $f(\mathbb{P}^m)$ is not contained in any $\{z_j = 0\}$, otherwise rotate coordinates. By the Weierstrass-Hurwitz Theorem [cf. [101], or [102]] it follows that each of the meromorphic functions $\frac{z_j}{z_0} \circ f$ is a quotient of two homogeneous polynomials $\frac{P_j}{Q_j}$ of the same degree. Thus we obtain a holomorphic mapping

$$\tilde{f} = (f_0, f_1, ..., f_m) : \mathbb{C}^{m+1} \longrightarrow \mathbb{C}^{m+1},$$

where $f_j \in \mathcal{L}_d(\mathbb{C}^{m+1})(j = 0, ..., m)$ are obtained by dividing out common factors from the polynomials $\frac{P_j}{Q_j}\prod Q_k$. Obviously, we have

$$\dim \tilde{f}^{-1}(0) \leq m - 1.$$

We only need to show that the f_j have no common zero except the origin. Suppose to the contrary that $p \in \mathbb{C}^{n+1} - \{0\}$ is a common zero. Choose a local lifting $\tilde{F} = (F_0, F_1, ..., F_m)$ of f in a neighborhood of p. We may assume that one of the F_j is equal to 1, say, $F_0 \equiv 1$. Then it follows that $f_j = f_0 F_j$ and that $f_0(p) = 0$. But this implies that the common zero set of the f_j is a complex hypersurface, which implies that they have a common factor, contradicting the fact that all the common factors have been eliminated.   □

Let $\mathcal{H}_d = \mathcal{H}_d(\mathbb{P}^m)$ denote the space of the holomorphic self-mappings on $\mathbb{P}^m$ given by homogeneous polynomials of degree d. Then

$$\text{Hol}(\mathbb{P}^m, \mathbb{P}^m) = \bigcup_{d=0}^{\infty} \mathcal{H}_d,$$

where $\mathcal{H}_0$ is the space of constant holomorphic mappings on $\mathbb{P}^m$. Obviously, $\mathcal{H}_1 = \text{Aut}(\mathbb{P}^m)$. By Bézout Theorem and Theorem 6.1, $f \in \mathcal{H}_d$ is a d^m to one mapping. Thus we have

$$\deg(f) = d^m \quad \text{for } f \in \mathcal{H}_d.$$

Now Theorem 5.32 yields:

Proposition 6.1 *The mapping* $\deg : \text{Hol}(\mathbb{P}^m, \mathbb{P}^m) \longrightarrow \mathbb{Z}$ *is continuous. In particular, if a sequence* $\{f_n\} \subset \text{Hol}(\mathbb{P}^m, \mathbb{P}^m)$ *converge uniformly on the* $\mathbb{P}^m$ *to a mapping* f*, then* $f \in \text{Hol}(\mathbb{P}^m, \mathbb{P}^m)$ *and for all sufficiently large* n*,* $\deg(f_n) = \deg(f)$*.*

Now Theorem 5.30 gives

Proposition 6.2 ([82]) *The Julia set of a holomorphic mapping* f *from* $\mathbb{P}^m$ *to* $\mathbb{P}^m$ *given by any homogeneous polynomials of degree* $d \geq 2$ *is always nonempty.*

The following fact is obvious:

Theorem 6.2 *For any* $f \in \mathcal{H}_d$ *with* $d \geq 2$*, the Julia set* $J(f)$ *is completely invariant, and for* $p \geq 2$*,* $J(f^p) = J(f)$*.*

It is proved in [84] that the mapping $f : \mathbb{P}^2 \longrightarrow \mathbb{P}^2$ defined by

$$f([z : w : t]) = [(z - 2w)^2 : (z - 2t)^2 : z^2]$$

has $J(f) = \mathbb{P}^2$.

Theorem 6.3 *Let* f *be a non constant meromorphic mapping from* $\mathbb{P}^m$ *to* $\mathbb{P}^m$*. Then* f *is given in homogeneous coordinates by* $[f_0 : f_1 : ... : f_m]$ *with* $\dim \tilde{f}^{-1}(0) \leq m - 1$*, where* $\tilde{f} = (f_0, f_1, ..., f_m) \in \mathcal{L}_d^{m+1}(\mathbb{C}^{m+1})$*. The mapping* $\tilde{f} : \mathbb{C}^{m+1} \longrightarrow \mathbb{C}^{m+1}$ *is called the reduced lifted mapping of* f *on* $\mathbb{C}^{m+1}$*.*

The proof follows from the proof of Theorem 6.1. Conversely, if $\tilde{f} = (f_0, f_1, ..., f_m) \in \mathcal{L}_d^{m+1}(\mathbb{C}^{m+1})$, by dividing out common factors from the polynomials f_j, we obtain a meromorphic mapping $f = \mathbb{P} \circ \tilde{f} \in \text{Mer}(\mathbb{P}^m, \mathbb{P}^m)$. Hence we have

$$\text{Mer}(\mathbb{P}^m, \mathbb{P}^m) = \bigcup_{d=0}^{\infty} \mathcal{M}_d$$

where

$$\mathcal{M}_d = \mathcal{M}_d(\mathbb{P}^m) = \mathbb{P}(\mathcal{L}_d^{m+1}(\mathbb{C}^{m+1}))$$

is easily identified with $\mathbb{P}^N$ where

$$N = N_d = (m+1)\binom{m+d}{d} - 1.$$

If $\tilde{f} : \mathbb{C}^{m+1} \longrightarrow \mathbb{C}^{m+1}$ is a reduced lifted mapping of $f \in \text{Mer}(\mathbb{P}^m, \mathbb{P}^m)$ on $\mathbb{C}^{m+1}$, the *indeterminacy*

$$I_f = \mathbb{P}(\tilde{f}^{-1}(0)).$$

Throughout the following discussion of this chapter, we will assume $d \geq 2$. We will also consider the space $\mathcal{M}_d^*$ of meromorphic mappings, consisting of those f in $\mathcal{M}_d$ which have maximal rank on some nonempty open sets. Obviously,

$$\mathcal{H}_d \subset \mathcal{M}_d^* \subset \mathcal{M}_d \cong \mathbb{P}^N.$$

As a consequence of Proposition 6.1, $\mathcal{H}_d$ is an open subset of $\text{Hol}(\mathbb{P}^m, \mathbb{P}^m)$ for it is the inverse image of the open subset $\{d^m\}$ of $\mathbb{Z}$ under the continuous mapping deg. More precisely one has the following result.

Theorem 6.4 ([82]) *The sets $\mathcal{H}_d$ and $\mathcal{M}_d^*$ are Zariski open sets of $\mathcal{M}_d \cong \mathbb{P}^N$. In particular $\mathcal{H}_d$ and $\mathcal{M}_d^*$ are connected. If $f \in \mathcal{H}_d$, then the critical set of f is an algebraic variety of degree $(m+1)(d-1)$.*

proof. Let $\pi : \mathbb{P}^N \times \mathbb{P}^m \longrightarrow \mathbb{P}^N$ be the projection and define an analytic set

$$Z = \{(f, z) \in \mathbb{P}^N \times \mathbb{P}^m \mid z \in I_f\}.$$

Since π is proper, by Tarski Theorem, we infer that $\pi(Z)$ is an analytic set. Obviously, $f \in \pi(Z)$ iff $I_f \neq \emptyset$. Hence $\mathcal{H}_d$ is a Zariski open set of $\mathbb{P}^N$ since $\pi(Z) = \mathbb{P}^N - \mathcal{H}_d$. The fact that $\mathcal{M}_d^*$ is Zariski open follows from the equation

$$\mathbb{P}^N - \mathcal{M}_d^* = \bigcap_{z \in \mathbb{P}^m} \{f \mid J\tilde{f}(z) = 0\}.$$

Let $f \in \mathcal{H}_d$ with a reduced lifted mapping $\tilde{f}$. Let $C_{\tilde{f}}$ be the critical set of $\tilde{f}$. Then the critical set C_f of f is the projection $\mathbb{P}(C_{\tilde{f}})$. Clearly,

$$\deg(C_{\tilde{f}}) \leq (m+1)(d-1)$$

so that $\deg(C_f) \leq (m+1)(d-1)$. On the other hand for the mapping

$$h = [z_0^d : z_1^d : \cdots : z_m^d] : \mathbb{P}^m \longrightarrow \mathbb{P}^m$$

the critical set C_h has degree $(m+1)(d-1)$ and therefore since $\mathcal{H}_d$ is connected we infer that for any $f \in \mathcal{H}_d$ the critical set C_f has degree exactly $(m+1)(d-1)$. $\square$

Note that the Fubini-Study form on $\mathbb{P}(\mathcal{L}_d^{m+1})$ determines the Fubini-Study Käehler metric which induces a distance d_{FS} on $\mathcal{M}_d$. Fix $f \in \mathcal{M}_d$ and take a random perturbation $\mathcal{F} = \{f_j\}_{j=1}^\infty \subset \mathcal{M}_d$ of f. We obtain the random perturbation

$$Ds(\mathcal{F}) = \{f_1 \circ f_2 \circ \cdots \circ f_n \mid n = 1, 2, ...\}$$

of the DS $\{f^n\}$. An interesting question is to compare $J(Ds(\mathcal{F}))$ and $J(f)$ when $\mathcal{F} \to f$, that is, $d_{FS}(f_j, f) \to 0$ for all $j \geq 1$. We translate Conjecture 3.11 into the following case:

Conjecture 6.1 *Let $\mathcal{F}$ be a random perturbation of f on $\mathcal{M}_d$. Then*

$$d_H(J(Ds(\mathcal{F})), J(f)) \to 0 \quad as \ \mathcal{F} \to f,$$

if and only if each component of $F(f)$ is a basin of attraction.

6.2   Fatou sets in $\mathbb{P}^m$

Lemma 6.3 *For a plurisubharmonic ϕ on $\mathbb{C}^m$, define*

$$D_\phi = \{z \in \mathbb{C}^m \mid \phi \text{ pluriharmonic in a neighborhood of } z\}. \tag{6.3}$$

Then D_ϕ is pseudoconvex.

Proof. The pseudoconvexity of D_ϕ follows from the following observation: For a (local) coordinates $z = (z_1, ..., z_m)$ on $\mathbb{C}^m$, if the Hartogs domain

$$U = \{x \in \mathbf{\Delta}^m \mid r < |z_1(x)| < 1, \max_{2 \leq j \leq m} |z_j(x)| < r'\} \subset D_\phi,$$

where $0 \leq r, r' \leq 1$, and where

$$\mathbf{\Delta}^m = \{x \in \mathbb{C}^m \mid \left|z_j(x)\right| < 1, j = 1, ..., m\},$$

then $\mathbf{\Delta}^m \subset D_\phi$.

By definition, $U \subset D_\phi$ implies that $\phi|_U$ is pluriharmonic, and hence is the real part of a holomorphic function on U. By the Hartogs theorem, the holomorphic function on U can be extended to a holomorphic function on $\mathbf{\Delta}^m$. Thus there exists a pluriharmonic $\hat{\phi}$ on $\mathbf{\Delta}^m$ such that $\hat{\phi}|_U = \phi$. We obtain a plurisubharmonic function $u = \phi - \hat{\phi}$ on $\mathbf{\Delta}^m$ with $u|_U \equiv 0$. As a function of one variable z_1, u is subharmonic on $|z_1| < 1$, vanishes on $r < |z_1| < 1$. By the maximal principle, $u \leq 0$ on $\mathbf{\Delta}^m$. Note that u takes the value 0 in $\mathbf{\Delta}^m$. The maximal principle implies $u \equiv 0$, i.e., $\phi \equiv \hat{\phi}$ is pluriharmonic on $\mathbf{\Delta}^m$. Therefore $\mathbf{\Delta}^m \subset D_\phi$. $\square$

Remark. The result is due to Cegrell[60], however, it is true on any complex manifold.

Lemma 6.4 *Let $f \in \mathcal{H}_d$ with a reduced lifted mapping $\tilde{f}$ of $d \geq 2$. Let G be the Green function of $\tilde{f}$. Then $p \in \mathbb{P}(D_G)$ iff there are a neighborhood U of p and $s \in \mathrm{Hol}(U, \mathbb{C}^{m+1})$ such that $\mathbb{P} \circ s = id$, and $s(U) \subset \partial \mathrm{Att}(0)$.*

Proof. Let U be a local coordinate open ball centred at p. We can identify $\mathbb{P}^{-1}(U)$ with $U \times (\mathbb{C} - \{0\})$ and denote the point of $\mathbb{P}^{-1}(U)$ by (w, z). Then on $\mathbb{P}^{-1}(U)$ the Green function G has the form

$$G(w, z) = \log |z| + \eta(w),$$

where $\eta(w)$ is plurisubharmonic on U. Then $\eta(w)$ is pluriharmonic on U iff $G(w, z)$ is pluriharmonic on $\mathbb{P}^{-1}(U)$. The mapping $s \in \mathrm{Hol}(U, \mathbb{C}^{m+1})$ in the lemma can be written as the form $s(w) = (w, \sigma(w))$ such that

$$0 = G(s(w)) = \log |\sigma(w)| + \eta(w),$$

i.e., $\eta(w) = -\log |\sigma(w)|$ is pluriharmonic.

Conversely, assume that $\eta(w)$ is plurisubharmonic. Choose a pluriharmonic function η^* on U such that $\eta + i\eta^*$ is holomorphic and define

$$\sigma(w) = \exp(-\eta(w) - i\eta^*(w)), \quad s(w) = (w, \sigma(w)).$$

Then $G(s(w)) = \log |\sigma(w)| + \eta(w) = 0$, i.e., $s(U) \subset \partial \mathrm{Att}(0)$. $\square$

Theorem 6.5 *Let $f \in \mathcal{H}_d$ with a reduced lifted mapping $\tilde{f}$ of $d \geq 2$. Let G be the Green function of $\tilde{f}$. Then $F(f) = \mathbb{P}(D_G)$.*

Proof. If $p \in F(f)$, there exists a subsequence $\{f^{n_j}\}$ which converges uniformly on a neighborhood U of p. Set

$$g = \lim_{j \to \infty} f^{n_j}|_U.$$

Note that $g(p) \notin H$ for some hyperplane H. We can choose a homogeneous coordinate system $[z_0 : z_1 : \cdots : z_m]$ in $\mathbb{P}^m$ such that $H = \{z_0 = 0\}$. Then there exists a neighborhood

$$H_\varepsilon = \{|z_0| < \varepsilon \|z\|\}$$

of H such that $g(p) \notin H_\varepsilon$. Thus if U is small enough, and if j is large enough, then $f^{n_j}(U) \cap H_\varepsilon = \emptyset$. Define a function ϕ on $\mathbb{C}^{m+1} - \{0\}$ by

$$\phi(z) = \begin{cases} \log \|z\| & : \quad z \in \mathbb{P}^{-1}(H_\varepsilon) \\ \log \dfrac{|z_0|}{\varepsilon} & : \quad z \in \mathbb{P}^{-1}(\mathbb{P}^m - H_\varepsilon) \end{cases}$$

which obviously satisfies

$$0 \leq \phi(z) - \log \|z\| \leq \log \frac{1}{\varepsilon}.$$

By Lemma 6.2, we have

$$\lim_{j \to \infty} \frac{1}{d^{n_j}} \phi \circ \tilde{f}^{n_j}(z) = G(z).$$

Note that $z \in \mathbb{P}^{-1}(U)$ implies $\tilde{f}^{n_j}(z) \in \mathbb{P}^{-1}(\mathbb{P}^m - H_\varepsilon)$. Then the limit G is pluriharmonic on $\mathbb{P}^{-1}(U)$ since $\phi \circ \tilde{f}^{n_j}$ is pluriharmonic on $\mathbb{P}^{-1}(U)$. Therefore $p \in \mathbb{P}(D_G)$, i.e., $F(f) \subset \mathbb{P}(D_G)$.

If $p \in \mathbb{P}(D_G)$, by Lemma 6.4, there is a neighborhood U of p, and $s \in \text{Hol}(U, \mathbb{C}^{m+1})$ such that $\mathbb{P} \circ s = id$, and $s(U) \subset \partial \text{Att}(0)$. Then the sequence $\{\tilde{f}^n \circ s\}$ is uniformly bounded on U since $\tilde{f}^n \circ s(U) \subset \partial \text{Att}(0)$, and consequently by Montel theorem, there exists a subsequence $\{\tilde{f}^{n_j} \circ s\}$ which converges uniformly to a holomorphic mapping $s' \in \text{Hol}(U, \mathbb{C}^{m+1})$ on U. Note that

$$s'(U) \subset \partial \text{Att}(0) \subset \mathbb{C}^{m+1} - \{0\}.$$

Hence the sequence $\{f^{n_j} = \mathbb{P} \circ \tilde{f}^{n_j}\}$ converges uniformly to $\mathbb{P} \circ s'$. Thus $\{f^n\}$ is normal on U, that is, $p \in F(f)$.   □

Theorem 6.6 (Ueda [257]) *For any $f \in \mathcal{H}_d$ with $d \geq 2$, the Fatou set $F(f)$ is pseudo-convex, and its connected components are Kobayashi hyperbolic.*

Proof. Since the open subset $F(f) = \mathbb{P}(D_G) \neq \mathbb{P}^m$, the pseudoconvexity of $F(f)$ follows from that of $D_G = \mathbb{P}^{-1}(F(f))$.

Let W be a connected component of $F(f)$, i.e., a Fatou component. Fix a point $p \in W$ and let $s \in \text{Hol}(U, \mathbb{C}^{m+1})$ be the mapping defined in Lemma 6.4. Continue s analytically along any curve in W which results a holomorphic mapping $\tilde{s} : \tilde{W} \longrightarrow \mathbb{C}^{m+1}$ and a covering mapping $\alpha : \tilde{W} \longrightarrow W$ with $\mathbb{P} \circ \tilde{s} = \alpha$ such that $\tilde{s}$ is injective, and $\tilde{s}(\tilde{W}) \subset \partial \text{Att}(0)$ is bounded. Then $\tilde{W}$ is Kobayashi hyperbolic, and consequently W is Kobayashi hyperbolic.   □

Corollary 6.1 ([126]) *If $f, g \in \mathcal{H}_d$ with $d \geq 2$, $f \circ g = g \circ f$, then $J(f) = J(g)$.*

Proof. Note that f and g satisfy some Lipschitz condition on $\mathbb{P}^m$ (see § 1.5). By Theorem 1.20, we see that $f^n(F_{equ}(g)) \subset F_{equ}(g)$ and $g^n(F_{equ}(f)) \subset F_{equ}(f)$ for all $n \in \mathbb{Z}_+$. Since connected components of $F(g) = F_{equ}(g)$ and $F(f) = F_{equ}(f)$ are Kobayashi hyperbolic, then Proposition 4.5 and the facts above imply that $\{f^n\}$ and $\{g^n\}$ are equicontinuous on $F_{equ}(g)$ and $F_{equ}(f)$, respectively. Therefore we have $F_{equ}(g) \subset F_{equ}(f)$ and $F_{equ}(f) \subset F_{equ}(g)$, respectively, that is, $F_{equ}(g) = F_{equ}(f)$. Thus we obtain $J(g) = J_{equ}(g) = J_{equ}(f) = J(f)$.   □

For the rational function case, Corollary 6.1 is nothing but Theorem 4.2.9 of [31].

Theorem 6.7 *For any $f \in \mathcal{H}_d$ with $d \geq 2$, its Fatou set is a domain of holomorphy and its Julia set $J(f)$ is connected.*

Proof. Since the Fatou set $F(f)$ is pseudoconvex, by the solution of the Levi Problem in $\mathbb{P}^m$, it follows that the Fatou set is a domain of holomorphy.

Assume, by contradiction, that $J(f)$ is not connected. Since $J(f)$ is closed, and hence compact in $\mathbb{P}^m$, we can write $J(f) = J_1 \cup J_2$ for disjoint nonempty compact subsets J_1 and J_2. Then $D = \mathbb{P}^m - J_2$ contains the compact subset J_1. By Hartogs theorem, all holomorphic functions on $D - J_1 = F(f)$ can be extended across J_1. This is a contradiction since $F(f)$ is a domain of holomorphy.   □

For the case $m = 2$, see Fornaess and Sibony [85]. Let G be the Green function in $\mathbb{C}^{m+1}$ associated to $f \in \mathcal{H}_d$ with $d \geq 2$. Then a current ω is defined on $\mathbb{P}^m$ by the relation

$$\mathbb{P}^*\omega = dd^c G.$$

Theorem 6.5 implies that $\text{supp}\omega = J(f)$. The closed positive currents ω^k of bidegree (k, k) are defined by the relation

$$\mathbb{P}^*\omega^k = (dd^c G)^k, \quad 1 \leq k \leq m.$$

Fornaess and Sibony [87] proved that $\text{supp}\omega^k$ is backward invariant for every $1 \leq k \leq m$, nonempty for $k = m$, connected if $2k \leq m$, and of

$$\text{supp}\omega^k \subset J_{equ}(f; k - 1), \quad 1 \leq k \leq m.$$

They also showed that f cannot be prehyperbolic on $\mathbb{P}^m$ nor on $\text{supp}\omega^k$ for $k < m$.

Problem 6.1 (Hubbard and Papadopol[127]) *Is the set of all repelling periodic points of $f \in \mathcal{H}_d(\mathbb{P}^m)$ dense in $\text{supp}\omega^m$?*

We end this section by the following open question:

Conjecture 6.2 *Mappings with volume hyperbolic Julia sets (resp., hyperbolic $\text{supp}\omega^m$) are dense in $\mathcal{H}_d(\mathbb{P}^m)$ for $d \geq 2$.*

For the case $m = 1$, this is the Fatou's conjecture which is unsolved even for polynomials of degree 2 on $\mathbb{C}$. Also see Smale[235]. Also it is natural to ask: For some integer p with $0 \leq p < k$, are mappings with $(p, m - k + 1)$-type hyperbolic $\text{supp}\omega^k$ dense in $\mathcal{H}_d(\mathbb{P}^m)$ for $d \geq 2$?

6.3 Periodic points

We show that the fixed point set of $f \in \mathcal{H}_d$ is discrete. More precisely we have:

Theorem 6.8 ([82]) *Let $f : \mathbb{P}^m \longrightarrow \mathbb{P}^m$ be a holomorphic mapping of degree $d \geq 2$, and g be a meromorphic map of degree $d' < d$. There can be no compact algebraic curve Z such that $f = g$ on $Z \cap \{\mathbb{P}^m - I_g\}$ and $Z \cap \{\mathbb{P}^m - I_g\} \neq \emptyset$. If g is holomorphic, the number of points where $f = g$ equals $(d^{m+1} - d'^{m+1})/(d - d')$ counted with multiplicity.*

Proof. Suppose that $E = \{ x \mid f(x) = g(x), x \in \mathbb{P}^m \}$ contains an open set of a compact complex subvariety Z of dimension one. We will arrive at a contradiction. First we write $f = [f_0 : f_1 : ... : f_m]$ and $g = [g_0 : g_1 : ... : g_m]$, where $f_j \in \mathcal{L}_d(\mathbb{C}^{m+1})$ and $g_j \in \mathcal{L}_{d'}(\mathbb{C}^{m+1})$ with $0 \leq d' < d$. Hence we can lift f and g to mappings

$$\tilde{f} = (f_0, f_1, ..., f_m), \tilde{g} = (g_0, g_1, ..., g_m) : \mathbb{C}^{m+1} \longrightarrow \mathbb{C}^{m+1}.$$

Also the variety Z lifts to conic two dimensional surface $\tilde{Z}$ in $\mathbb{C}^{m+1}$. Introduce one more complex variable t and consider the $m + 1$ equations

$$f_j(z_0, z_1, ..., z_m) - t^{d-d'} g_j(z_0, z_1, ..., z_m) = 0, \quad j = 0, 1, ..., m. \tag{6.4}$$

These are homogeneous equations of degree d in $\mathbb{C}^{m+2}$. Hence the common zero set is a conic complex variety Y. Consider at first the intersection with the hyperplane $\{t = 0\}$. Then the equations reduce to $f_0 = f_1 = \ldots = f_m = 0$. Since f is a well defined holomorphic mapping, this zero set consists only of the origin, that is, $Y \cap \{t = 0\} = \{0\}$. Note that

$$\mathbb{P}^{m+1} = \mathbb{C}^{m+1} \cup \mathbb{P}^m, \quad \{t = 0\} = \mathbb{P}^m.$$

The natural projection $\mathbb{P}(Y)$ to $\mathbb{P}^{m+1}$ is therefore a compact complex space which does not intersect the hyperplane $\{t = 0\}$ at infinity. Hence the image ($\simeq Y \cap \{t = 1\}$) is a compact subvariety in $\mathbb{C}^{m+1}$ and hence, by Lemma B.6, it must be finite. This means that Y consists of a finite number of complex lines in $\mathbb{C}^{m+2}$ through the origin. Suppose next that x is in $Z \cap \{\mathbb{P}^m - I_g\}$, so $f(x) = g(x)$. Then there exists a complex value $t \neq 0$ and $(z_0, z_1, \ldots, z_m) \neq 0$ such that $x = [z_0 : \ldots : z_m]$ and $f_j(z_0, z_1, \ldots, z_m) = t^{d-d'} g_j(z_0, z_1, \ldots, z_m)$. Hence the point $(t, z_0, z_1, \ldots, z_m)$ belongs to Y. But this implies that Y is two dimensional, a contradiction. Hence we have shown that there is no such Z.

In case g is holomorphic this implies that E is finite. Next we need to count the number of points. First we count the number of solutions using Bezout's theorem on the equations (6.4). There are d^{m+1} of these. However d'^{m+1} of these occur at the point $[1 : 0 : \ldots : 0]$, so this gives $d^{m+1} - d'^{m+1}$ solutions, but rotation of t by a $d - d'$ root of unity produces an equivalent solution, so the total number of solutions to $f = g$ is $(d^{m+1} - d'^{m+1})/(d - d')$. This complete the proof of the Theorem. $\qquad\square$

Applying the above theorem in the case $g = id$, we obtain the number of periodic points as follows.

Corollary 6.2 *Let* $f : \mathbb{P}^m \longrightarrow \mathbb{P}^m, f \in \mathcal{H}_d, d \geq 2$. *The number of periodic points of order* n *counted with multiplicity is* $(d^{n(m+1)} - 1)/(d^n - 1)$.

Lemma 6.5 ([82]) *Let* 0 *be a fixed point for a germ of a local holomorphic mapping* $f : \mathbb{C}^2 \longrightarrow \mathbb{C}^2$ *at* 0. *Assume that* 0 *is an isolated point of* $\mathrm{Fix}(f^n)$ *for all* $n \geq 1$. *Then there exists an integer* N *such that for all* $n \geq 1$, *the inequality,*

$$\|f^n(z) - z\| \geq c_n \|z\|^N \quad (c_n > 0),$$

holds in some neighborhood (depending on n) of 0.

Theorem 6.9 *Let* $f : \mathbb{P}^2 \longrightarrow \mathbb{P}^2$ *be a holomorphic mapping of degree* $d \geq 2$. *Then there exists infinitely many distinct periodic orbits.*

Proof. Suppose that there are only finitely many periodic orbits. By Corollary 6.2, for any n, f^n has $d^{2n} + d^n + 1$ fixed points counted with multiplicity. Then for some point p the multiplicity of $f^n - id$ at p can be chosen arbitrarily large. Taking local coordinates z with $z(p) = 0$. From the lemma we have

$$\|f^n(z) - z\| \geq c_n \|z\|^N.$$

Let P_N denote the Taylor polynomial of $f^n - id$ of order N. Then for r sufficiently small

$$\|f^n - id - P_N\| < \|f^n - id\|$$

on $\mathbb{P}^2(p;r)$. Hence by Rouché's principle, the multiplicity of $f^n - id$ at p is at most N^2, a contradiction. □

The following theorem shows that periodic orbits of holomorphic self-mappings of $\mathbb{P}^m$ are non-attractive in the complement of the critical orbits under the hypothesis of Kobayashi hyperbolicity.

Theorem 6.10 ([82]) *Let $f : \mathbb{P}^m \longrightarrow \mathbb{P}^m$ be a holomorphic mapping with critical set C_f. Let*

$$C = C_f = \overline{O^+(C_f)} = \bigcup_{j=0}^{\infty} f^j(C_f).$$

Assume that $\mathbb{P}^m - C$ is Kobayashi hyperbolic and hyperbolically embedded. If p is a periodic point for f, $f^l(p) = p$, with eigenvalues $\lambda_1, \lambda_2, ..., \lambda_m$ and $p \notin C$, then

$$|\lambda_i| \geq 1, \quad 1 \leq i \leq m.$$

Also $|\lambda_1...\lambda_m| > 1$ or f is an automorphism of the component of $\mathbb{P}^m - C$ containing p.

Proof. Let $M = \mathbb{P}^m - C$ and let $M_1 = M - f^{-1}(C)$. As $f : M_1 \longrightarrow M$ is a covering mapping we see that the Kobayashi differential metric satisfy

$$\kappa_M(f'(x)\xi) = \kappa_{M_1}(\xi) \geq \kappa_M(\xi),$$

for a point $x \in M$ and for a tangent vector $\xi \in \mathbf{T}(M)_x$. So if $x \in M$, and if $f^l(x) = x$, then all eigenvalues of $(f^l)'(x)$ have modulus at least one.

Let U be a component of M, $p \in U$, and let $U_l \subset U$ be the connected component of $f^{-l}(U)$ containing p. Let N be the universal covering of U_l and $\pi : N \longrightarrow U_l$ the projection. Observe that N is hyperbolic and that for the Kobayashi metric biholomorphic mappings are isometries. Also observe that $(N, f^l \circ \pi)$ is the universal covering of U. Pick any non-vanishing holomorphic m-form α at p. Fix a Hermitian metric on $\mathbf{T}(N)$. Let $\| \ \|$ be a volume form on the space of $(0, m)$-forms, such that holomorphic automorphisms preserve the volume. Fix a point $q \in N$ with $\pi(q) = p$. Define

$$E_U(p, q, \alpha) = \inf\{\|\gamma\|_q^2 \mid g(q) = p, g_*(\gamma) = \alpha\},$$

where g runs through all holomorphic mappings with nonvanishing Jacobian from N to U with $g(q) = p$.

Now we prove that the extremal mapping exists and is surjective. Let g_n be a minimizing sequence. Consider g_n as mappings from N to M which is hyperbolically embedded. Then g_n is equicontinuous with respect to a metric on $\mathbb{P}^m$. Hence by Arzela-Ascoli theorem, there exists a subsequence $g_{n_j} \to g$ and $g(q) = p$ and $\mathcal{J}g \neq 0$ and hence g attains values in U. Let $\tilde{g}$ be such that $f^l \circ \pi \circ \tilde{g} = g$ and $\tilde{g}(q) = q$. If $|\mathcal{J}\tilde{g}(q)| < 1$, then by the chain rule, this will contradict that g is extremal. Since N is Kobayashi hyperbolic, we must have that $|\mathcal{J}\tilde{g}(q)| \leq 1$ (Theorem 5.11). Hence it follows that $\tilde{g}$ is an automorphism (Theorem 5.11), and hence g is surjective.

Similarly define

$$E_{U_l}(p, q, \alpha) = \inf\{\|\gamma\|_q^2 \mid g(q) = p, \pi(q) = p, g_*(\gamma) = \alpha\},$$

where g runs through all holomorphic mappings with non-vanishing Jacobian from N to U_l with $g(q) = p$, $\pi(q) = p$. Also the extremal mapping exists and is surjective.

Since f^l is a covering mapping from U_l to U and since $f^l(p) = p$, then

$$E_{U_l}(p, q, \alpha) = E_U(f^l(p), q, (f^l)_*(p)(\alpha)). \qquad (6.5)$$

If $U_l = U$, then (6.5) implies that $|\mathcal{J} f^l(p)| = 1$. Hence f^l is an automorphism of U (Theorem 5.11). If U_l is a proper subset of U, then the fact above implies

$$E_{U_l}(p, q, \alpha) > E_U(p, q, \alpha).$$

Hence $|\mathcal{J} f^l(p)| > 1$. $\square$

Conjecture 6.3 *If $J(f) \cap \mathcal{C}_f = \emptyset$, then $J(f)$ is volume hyperbolic.*

Definition 6.1 *Let f be a continuous self-mapping on a manifold M. A Fatou component is a connected component of the Fatou set $F(f)$. A Fatou component D is a Siegel domain if there exists a subsequence $\{f^{n_i}\}$ of $\{f^n\}$ converging uniformly on compact sets of D to identity.*

Ueda [258] prove that if $f \in \mathcal{H}_d(\mathbb{P}^m)$ with $d \geq 2$, and if Ω is a Siegel domain, then the boundary of Ω is contained in $\mathcal{C}$. Under the assumptions of Theorem 6.10, Fornaess and Sibony [82] prove that if there is a Fatou component U such that $f^n(U)$ does not converge uniformly on compact sets to $\mathcal{C}$, then U is preperiodic to a Siegel domain.

Conjecture 6.4 *If $f \in \mathcal{H}_d(\mathbb{P}^m)$ with $d \geq 2$, then the set of attractive cycles is finite.*

6.4   Classification of recurrent domains on $\mathbb{P}^2$

We will use the following lemmas in this section:

Lemma 6.6 *If M is a Poincaré hyperbolic Riemann surface, then for every holomorphic mapping $f : M \longrightarrow M$ the Julia set $J(f)$ is empty. Furthermore either:*

  1) every forward orbit converges towards a unique attractive fixed point $f(p) = p$;

  2) every forward orbit diverges to infinity with respect to the Poincaré metric on M;

  3) f is an automorphism of finite order; or

  4) M is isomorphic either to a disc Δ, a punctured disc $\Delta_* = \Delta - \{0\}$, or an annulus $\{1 < |z| < r\}$, and f corresponds to an irrational rotation: $z \mapsto e^{2\pi i t} z$ with $t \notin \mathbb{Q}$.

The proof can be found in [172].

Lemma 6.7 ([82]) *Let $C \subset \mathbb{P}^2$ be an algebraic curve of degree r. Let $f \in \mathcal{H}_d(\mathbb{P}^2)$ and assume that $f(C) \subset C$. If f is an l to 1 mapping on C, then $l \geq d$.*

Take $M = \mathbb{P}^m$, $f \in \mathrm{Hol}(\mathbb{P}^m, \mathbb{P}^m)$ in the definition. Let D be a recurrent Fatou component. Recall that D is *recurrent* if there exists $p_0 \in D$ such that $\{f^{n_i}(p_0)\}$ is relatively compact in D for some subsequence n_i. Without loss of generality, we assume $f^{n_i}(p_0) \to p, n_{i+1} - n_i \to \infty$. Taking a subsequence $\{i = i_j\}$ and recalling that we are in the Fatou set, we can assume that the sequence $\{f^{n_{i+1}-n_i}\}_i$ converges uniformly on compact sets in D to a $h \in \mathrm{Hol}(D, \overline{D})$. Let $p_i = f^{n_i}(p_0)$. Then

$$f^{n_{i+1}-n_i}(p_i) = f^{n_{i+1}}(p_0) = p_{i+1}.$$

Hence

$$d(f^{n_{i+1}-n_i}(p), p_{i+1}) = O(d(p_i, p)) \to 0.$$

It follows that $h(p) = p$.

Consider the set $\mathcal{D}_f$ of all mappings $h \in \mathrm{Hol}(D, \overline{D})$ with $h(p) = p$ for some $p \in D$ and $h = \lim f^{k_j}$ for some subsequence k_j. Then $\mathcal{D}_f \neq \emptyset$. Since h commutes with f, it follows that f maps $\mathrm{Fix}(h)$ to itself.

Further, assume that $f(D) = D$. We show that f is a surjective self mapping of $h(D)$. If $x \in h(D)$, then $x = h(y)$ for some $y \in D$ and $f(x) = f(h(y)) = h(f(y)) \in h(D)$ so that

$$f(h(D)) \subset h(D).$$

Choose $y' \in D$ such that $f(y') = y$. Define $x' = h(y')$. Then $x' \in h(D)$ and

$$f(x') = f(h(y')) = h(f(y')) = h(y) = x.$$

Define

$$M^0 = h(D) \cap D.$$

Note $f(D) = D$ and $f(M^0) \subset M^0$. Since f is an open mapping, f maps the boundary of D to itself and hence $f(M^0) = M^0$.

If f is a rational map in $\mathbb{P}^1$, there are only finitely many recurrent domains, and the recurrent Fatou components are basins of attraction, Siegel discs and Herman rings. A recent theorem by E. Gavosto [94] shows that holomorphic mappings on $\mathbb{P}^2$ can have infinitely many basins of attraction, hence recurrent domains.

Theorem 6.11 ([83]) *Let* $f \in \mathcal{H}_d(\mathbb{P}^2)$ *with* $d \geq 2$. *Let* D *be a recurrent Fatou component such that* $f(D) = D$. *Then one of the following statements holds:*

1) There is an attractive fixed point $p \in D$, *the eigenvalues* λ_1, λ_2 *of* $f'(p)$ *satisfy* $|\lambda_1| < 1, |\lambda_2| < 1$.

2) There exists a Riemann surface M *which is a closed complex submanifold of* D *and* $f|M \longrightarrow M$ *is an automorphism, moreover* $d(f^n(K), M) \to 0$ *for any compact set* K *in* D. *The Riemann surface* M *is biholomorphic to a disc, a punctured disc or an annulus and* $f|M$ *is conjugate to a rotation. The limit* h *of any convergent subsequence* $\{f^{n_i}\}$ *has the same image. Any two limits* h_1, h_2 *differ only by a rotation in* M.

3) The domain D *is a Siegel domain. Any limit of a convergent subsequence of* $\{f^n\}$ *is an automorphism of* D.

Proof. If, for some $h \in \mathcal{D}_f$, the rank of h is 0, then $h(D) = p$ and necessarily $f(p) = p$. Also both eigenvalues of $f'(p)$ must have modulus strictly less than one since some iterates of f converge to the constant mapping. Hence this leads to case 1).

Assume that for some $h \in \mathcal{D}_f$, the rank of h is two. Then for some sequence $\{k_j\}$, $f^{k_{j+1}-k_j} \to id$, and hence D is a Siegel domain. The restriction $f|_D$ is clearly an automorphism of D. We are then in case 3). Now we show that if $\{f^{k_j}\}$ converge to a mapping h, then $h \in \text{Aut}(D)$. From Theorem 6.6, D is Kobayashi hyperbolic, so $\text{Aut}(D)$ has the structure of a Lie group (Theorem 5.13). Let $\tilde{G}$ be the closed subgroup generated by f in $\text{Aut}(D)$. Then $\tilde{G}$ is a Lie group. Let G^0 be the connected component containing id in $\tilde{G}$, it is also a Lie group. Since $\{f^{k_{j+1}-k_j}\}$ converge to id, then G^0 is not reduced to id. But G^0 is clearly commutative, hence we have an isomorphism

$$\Phi : \mathbb{T}^k \times \mathbb{R}^l \longrightarrow G^0.$$

For some $(a, b) \in \mathbb{T}^k \times \mathbb{R}^l$, we have $\Phi(a, b) = f$. If $b \neq 0$, we cannot have subsequences of $\{f^n\}$ converging to id. So $b = 0$ and hence G^0 is isomorphic to $\mathbb{T}^k$, consequently G^0 is compact. It follows that each convergent subsequence of $\{f^n\}$ tend to an element of $\text{Aut}(D)$.

Assume that for all $h \in \mathcal{D}_f$, the maximal rank of h is one. Fix an $h \in \mathcal{D}_f$ with $h(p) = p$. Then $h(D) \subset \overline{D}$. For $x \in D$, there is an irreducible piece $M_x \subset \overline{D}$ of a Riemann surface with singularities and a neighborhood $U(x)$ so that $h(U(x)) = M_x$. We define an abstract Riemann surface R as the union $\cup M_{x_i}$ for a covering $U(x_i)$ of D, with the identifications at $y \in M_{x_i} \cap M_{x_j}$ if the two pieces agree as germs. Then R is Hausdorff by the identity theorem. The mapping $h : D \longrightarrow h(D)$ factors naturally as a mapping $h = \pi \circ \overline{h}$ where $\overline{h} : D \longrightarrow R$ and $\pi : R \longrightarrow h(D)$. Now we show that h is not constant on the irreducible component M_p of M^0 which contains p. Assume not. Since $h(p) = p$, then $h|_{M_p} \equiv p \in h(D)$. But $f^{2k_j} \to p$ so we are in case 1).

Since $f^{k_{j+1}-k_j}(f^{k_j}) = f^{k_{j+1}}$, we can assume, using a diagonal process, that for a subsequence m_i, $\{f^{m_i}\}$ converges to a new mapping h and $h = id$ on M_p. Since $f^l \circ h = h \circ f^l$ it follows that $h = id$ on each $f^l(M_p)$, $l \geq 1$. We use this new h from now on. We know that

$$\bigcup_{l \geq 0} f^l(M_p) \subset \text{Fix}(h) \subset D.$$

Since $id - h'$ has at least rank one, $\text{Fix}(h)$ is a countable union of disjoint irreducible components each of which is a point or a smooth complex curve. It follows that M_p is a component of $\text{Fix}(h)$ and since f is a proper self mapping of D, $\bigcup_{l \geq 0} f^l(M_p)$ is a closed countable union of irreducible curves in $\text{Fix}(h)$.

Suppose $h(D)$ is a torus. Then $f : h(D) \longrightarrow h(D)$ is an l to 1 mapping so that $l \geq 2$ by Lemma 6.7. Hence repelling points for $f|_{h(D)}$ are dense in $h(D)$, which contradicts normality in D.

If $h(D)$ is $\mathbb{P}^1, \mathbb{C}$ or $\mathbb{C}_* = \mathbb{C} - \{0\}$, we next show that $f|_{h(D)}$ is an automorphism. Suppose not, then $f : h(D) \longrightarrow h(D)$ is an l to 1 surjective mapping with $l \geq 2$. From the Fatou-Julia theory in one variable, repelling periodic points for $f|_{h(D)}$ are dense in the Julia set of $f|_{h(D)}$. Choose q a repelling periodic point for $f|_{h(D)}$, say, $f^s(q) = q$ and $h(z_0) = q$ with $z_0 \in D$. Recall that

$$h = \lim_{i \to \infty} f^{m_i}.$$

We can assume $h'(z_0) \neq 0$ in some direction. For each $l \geq 2$, choose $r_l > 0$ such that

$$f^{ls}(h(D)(q; r_l)) \subset h(D)(q; \delta)$$

for a sufficient small $0 < \delta < 1$. Choose $m_{i(l)}$ such that $f^{m_{i(l)}}(z_0) \in h(D)(q; r_l)$. Then $f^{ls+m_{i(l)}}(z_0) \in h(D)(q; \delta)$. The sequence $\{f^{ls+m_{i(l)}}\}_l$ is equicontinuous, hence we can assume that in a ball B_1 containing z_0, $f^{ls+m_{i(l)}}(B_1) \subset h(D)(q; \delta)$. We can always increase $m_{i(l)}$ so that $f^{ls+m_{i(l)}}$ is close to $f^{ls} \circ h$. Then the derivative of $f^{ls+m_{i(l)}}$ at z_0 is not bounded in all directions, a contradiction. Hence we have shown that if $h(D)$ is $\mathbb{P}^1, \mathbb{C}$ or $\mathbb{C}_*$, $f|_{h(D)}$ is an automorphism. This proves our claim.

As a consequence, $h(D)$ cannot be $\mathbb{P}^1$ since by Lemma 6.7 $f|_{h(D)}$ cannot be an automorphism. If $h(D)$ is $\mathbb{C}$ or $\mathbb{C}_*$, since $f^{m_{i(l)}}|_{M_p} \to id$, and $\bigcup_{l>0} f^l(M_p)$ is closed, then necessarily f (or f^2) is conjugate to an irrational rotation. We will rule out this case in the following.

If $h(D)$ is Poincaré hyperbolic, we use the classification of holomorphic mappings $g : h(D) \longrightarrow h(D)$ in Lemma 6.6. Since $f^{m_{i(l)}}|_{M_p} \to id$, we know that not all forward orbits converge to an attractive fixed point, nor do all forward orbits diverge to infinity. From Lemma 6.7, we know also that f is not of finite order, hence Lemma 6.6 implies that $h(D)$ is isomorphic to the unit disc Δ, Δ_* or an annulus and f is conjugate to an irrational rotation.

We prove next that $h(D)$ is independent of h. Assume $f^{k_j} \to h_0$ uniformly on compact sets of D. We have $f^{k_j}|_{M^0} \to h_0|_{M^0}$, but since f is conjugate to a rotation on M^0, $h_0(M^0) \subset M^0$, and $h_0(D)$ is an extension of M^0, one can prove similarly that f is conjugate to a rotation on $h_0(D)$. Similarly M^0 is an extension of $h_0(D) \cap D$, so

$$M^0 = h_0(D) \cap D.$$

Let M be the maximal extension of $h(D)$ in $\overline{D}$ such that f is conjugate to a rotation on M. We then get that f^n converges u.c.c. on D to M, i.e., $d(f^n, M) \to 0$. We would like to show next that $h(D) = h_0(D)$. Pick a point p in M^0. Then we can find a holomorphic coordinate system in a neighborhood of p such that in that neighborhood

$$h(D) = \{w = 0; a < |z| < b\},$$

and

$$f(z, w) = (e^{i\theta}z + wg_1(z, w), wa_1(z) + w^2 k_1(z, w)).$$

Then

$$f^n(z, w) = (e^{ni\theta}z + O(w), wa_n(z) + O(w^2)),$$

where

$$a_n(z) = \prod_{j=0}^{n-1} a_1(e^{ij\theta}z).$$

Since we are in a Fatou component, the function $a_n(z)$ are necessarily uniformly bounded on any smaller set $\{a < a' < |z| < b' < b\}$. We must even have that $a_n \to 0$ uniformly, since all limits have rank 1. It follows that a neighborhood of M^0 is attracted to M^0.

An easy estimate gives that if we start with small enough w, and consider tangent vectors $v = (1, \alpha)$ based at (z, w), $|\alpha|$ small enough, then

$$(f^n)'v = c_n(1, \alpha_n), \quad \alpha_n \to 0,$$

$||c_n| - 1|$ as small as we want. For each n, let $\mathcal{F}_n$ be the "vertical" foliation consisting of leaves L with

$$f^n(L) \subset \{z = \text{ const }\}.$$

We show next that $\mathcal{F}_n \to \mathcal{F}$, a foliation with leaves of the form $z = g(w)$. The above observation shows that in order to compute the horizontal distance between leaves of $\mathcal{F}_n$ and corresponding leaves of $\mathcal{F}_{n+1}$, it is enough to compute the distance after applying f^n. If one considers the leaves of $\mathcal{F}_n$ as almost vertical discs of radius ρ, then after applying f^n, the discs have radius at most $c\rho r^n$ for some $r < 1$. Since the discs of $\mathcal{F}_1 = f^n(\mathcal{F}_{n+1})$ and $\{z_1 = \text{ const }\} = f^n(\mathcal{F}_n)$ start at the same point, they can be at most at $c\rho r^n$ away from each other. So the horizontal distance between $\mathcal{F}_n$ and $\mathcal{F}_{n+1}$ is at most $c\rho r^n$, so $\mathcal{F}_n$ converges to a foliation $\mathcal{F}$ with leaves of the form $z = g(w)$. Moreover f maps leaves of $\mathcal{F}$ to leaves of $\mathcal{F}$. It follows that these leaves are in level sets of h. In particular, in a neighborhood of $h(D) \cap D$, the level sets of h are independent of h.

By connectivity reasons of $h(D)$ and $h_0(D)$, one must contain the other, say, $h(D) \subset h_0(D)$. Let λ_θ denote rotation by θ in $h_0(D)$. Then near one component of $h(D) \cap D$, there must exist a θ so that $h \equiv \lambda_\theta \circ h_0$. But this must hold everywhere. So

$$h(D) = h_0(D).$$

In particular, the level sets of h and h_0 are the same (even globally), and f maps level sets to level sets (globally). It follows that if $p_0 \in D$, then either $\{f^n(p_0)\}$ converges to the boundary or is a relatively compact set in D.

Note that using the local coordinates above, it follows from the maximum principle that $h(D) \cap D$ can not have more than one component. Indeed, let A be a subannulus of $h(D)$ whose boundary with respect to $h(D)$ is in D. Assume A intersects ∂D. Then A has a Stein neighborhood isomorphic to $A \times \Delta$, and we can apply the maximum principle there. Since $\{f^n\}$ converges towards $h(D)$ near the boundary of A, we still have convergence in a neighborhood of A, so $A \subset D$.

Next we prove that actually $h(D)$ is a closed complex submanifold of D. Namely, let us assume not. We consider a circle in $h(D)$ so that one side is in D and the other side is in the boundary. We then choose a local coordinate system as above. consider the coefficient $a_1(z)$. For each radius r let $A(r)$ denote the average of $\log|a_1(z)|$ over the circle of radius r. Similarly let $A_n(r)$ denote the average of $\log|a_n(z)|$. Then $A_n(r)$ and $A(r)$ have the same sign always and they are continuous and monotonic. Also note that the rotation by θ on the circle is ergodic. Hence it follows that

$$\frac{1}{n}\log|a_n(z)| \to A(r)$$

in L^2 on the circle $|z| = r$. Note that the functions $\frac{1}{n}\log|a_n(z)|$ are equicontinuous so they converge uniformly to $A(r)$ except near circles where a_1 has a zero.

In particular, it follows that if $A(r) < 0$, then the circle with radius r is in the Fatou component. Since $A(|z|)$ is subharmonic, it follows that $A(r) > 0$ on the side which belongs to the boundary. But then it follows from ergodicity that for large n, $|a_n(z)| > 1$ uniformly, on circles $|z| = r$. But this implies that these points repel points from D. Hence there can be no points in D converging to them. So

$$h(D) = M^0 \subset D.$$

From Theorem 6.6, D is Kobayashi hyperbolic so $h(D)$ which is contained in D cannot be $\mathbb{C}$ or $\mathbb{C}_*$. This completes the proof. $\qquad\square$

6.5 Exceptional varieties and critical sets

We know that a rational mapping f of degree at least two on $\mathbb{P}^1$ has at most two exceptional points, and that a closed, backward invariant subset E of $\mathbb{P}^1$ is either E has at most two elements and $E \subset \mathrm{Exc}(f) \subset F(f)$; or E is infinite and $J(f) \subset E$. Thus a subset E of $\mathbb{P}^1$ is contained the subset $\mathrm{Exc}(f)$ iff E is a compact analytic subset of $\mathbb{P}^1$ with $f^{-1}(E) = E$.

Definition 6.2 *Let $f \in \mathrm{Hol}(M, M)$ and E a compact analytic subset of M. Then E is exceptional if $f^{-1}(E) = E$.*

Obviously, if $x \in \mathrm{Exc}(f)$, then $[x]$ is exceptional, and any finite exceptional set is contained in $\mathrm{Exc}(f)$. In particular, if $M = \mathbb{C}^m$, $\mathrm{Exc}(f)$ contains all exceptional sets. Generally, is $E \subset \mathrm{Exc}(f)$ if E is exceptional?

Let $f \in \mathrm{Hol}(M, M)$ be surjective. If E is an exceptional set of f such that $M - E$ is taut, then we also have $f^{-1}(M - E) = M - E$, and hence $f(M - E) = M - E$. Then $M - E \subset F(f)$ or $J(f) \subset E$. Hence the Julia set of f is the smallest closed, backward invariant set with a taut complement.

Let $f \in \mathcal{H}_d(\mathbb{P}^m)$ have an exceptional hypersurface E. Here $d \geq 2$. Note that replacing f with an iterate we may assume that each irreducible branch of E is mapped to itself. Hence any collection of irreducible branches of E is also exceptional. We denote by $\tilde{E}$ the pull back of E to $\mathbb{C}^{m+1}$ under the natural projection $\mathbb{P} : \mathbb{C}^{m+1} \longrightarrow \mathbb{P}^m$. Then $\tilde{E}$ is a homogeneous complex hypersurface, so we can write it as

$$\tilde{E} = \{h(z_0, z_1, \cdots, z_m) = 0\}$$

for a homogeneous polynomial h. Since E is completely invariant, the polynomial $h \circ \tilde{f}$ only vanishes on $\tilde{E}$. Note that the degree of $h \circ \tilde{f}$ is $(\deg h)d$. Then there exists a non-zero constant c such that the Böttcher functional equation

$$h \circ \tilde{f} = ch^d$$

holds, which implies $\tilde{E}$ is a part of the critical set $C_{\tilde{f}}$ of $\tilde{f}$, so $E \subset C_f$. Let $E_1, \cdots, E_\ell$ denote irreducible branches of E. Then

$$\tilde{E}_i = \{h_i(z_0, z_1, \cdots, z_m) = 0\}$$

for an irreducible homogeneous polynomial h_i. Note that $\mathcal{J}\tilde{f}$ has $(\prod h_i)^{d-1}$ as a factor. From Theorem 6.4 it follows that

$$\sum \deg h_i \leq m + 1.$$

In particular there are at most $m + 1$ irreducible branches of the exceptional hypersurface E, and if the number is $m + 1$, they are all linear.

Proposition 6.3 ([82]) *The set of holomorphic mappings without exceptional hypersurfaces is a non-empty Zariski open set in* $\mathcal{H}_d$.

Proof. Define

$$D_\ell = \{(f, \alpha) \in \mathcal{H}_d \times \mathbb{P}(\mathcal{L}_\ell) \mid h \circ \tilde{f} = ch^d \text{ for some constant } c, \alpha = \mathbb{P}(h)\}.$$

If f has an exceptional hypersurface then $(f, \mathbb{P}(h))$ is in D_ℓ for some $\ell \leq m+1$ and some $h \in \mathcal{L}_\ell$. The projection of D_ℓ on $\mathcal{H}_d$ is again an analytic variety. Since there exists one mapping in each $\mathcal{H}_d$ which is not exceptional, the proposition follows.   □

Theorem 6.12 ([82]) *For fixed* $d \geq 2$, *the set of holomorphic self-mappings on* $\mathbb{P}^m$ *without exceptional finite set is a non-empty Zariski open set in* $\mathcal{H}_d$.

Proof. Consider the analytic set

$$A = \{(f, a) \in \mathcal{H}_d \times \mathbb{P}^m \mid f^{-1}(a) \text{ is one point }\}.$$

If E is an exceptional finite set of $f \in \mathcal{H}_d$, then f induces to a bijection of E. Take $a \in \mathbb{P}^m$. If $a \in E$, then $f^{-1}(a)$ is one point, i.e., all solutions of $f(z) = a$ coincide so that f is in the projection of A. Since there exists mappings without exceptional points, the theorem follows.   □

Theorem 6.13 ([82]) *There exist constants* $c(d)$ *so that for any* $f \in \mathcal{H}_d(\mathbb{P}^2)$, *any finite exceptional set has at most* $c(d)$ *points.*

Proof. We can assume that $d \geq 3$ since $c(2) \leq c(4)$. Observe that the degree of the mapping f is d^2, counting multiplicity. Notice that an exceptional point (whether it is fixed or on a periodic orbit) can have only one preimage. Hence all exceptional points necessarily lie in the critical set C_f. Let p be an exceptional point. Assume at first that p is a regular point of C_f and $f(p)$ is a regular point for $f(C_f)$. Then we can choose local coordinates near p and $f(p)$ such that the mapping has the form $(z, w) \mapsto (z, w^l)$ for some integer l. So the mapping is locally l to 1. But by Theorem 6.4,

$$l \leq 3(d-1) + 1 < d^2 \quad (d \geq 3).$$

Hence p cannot be exceptional. It follows that p is a singular point of C_f or $f(p)$ is a singular point of $f(C_f)$. Since $f(C_f)$ has degree at most $3d(d-1)$ and the number of singular points of C_f or $f(C_f)$ is bounded by a constant by Bezout's Theorem, the theorem follows.   □

Theorem 6.14 *Fix an integer* $d \geq 2$. *Then there exists a Zariski dense open set* $\mathcal{H}' \subset \mathcal{H}_d$ *with the following properties: for* $f \in \mathcal{H}'$,
    1) No point of $\mathbb{P}^2$ *lies in* $f^n(C_f)$ *for three different* n, $0 \leq n \leq 4$;
    2) $\mathbb{P}^2 - \{\cup_{n=0}^4 f^n(C_f)\}$ *is Kobayashi complete hyperbolic and hyperbolically embedded in* $\mathbb{P}^2$.

The proof follows from two theorems by M. Green, see [82].

Conjecture 6.5 *Suppose that each critical point of* $f \in \mathcal{H}_d$ $(d \geq 2)$ *has a forward orbit that accumulates at an attracting cycle of* f. *Then* $J(f)$ *is a volume hyperbolic set.*

6.6 The Newton's method

Now we explain the need of studying the iteration of meromorphic mappings (or functions). Given a polynomial equation

$$f(x) = a_n x^n + \cdots + a_0 = 0,$$

in one variable x. Here x is allowed to be complex. The main advantage of the complex number system is that the solutions always exist. The classical Newton method for finding a zero of f is to iterate the equation

$$x_{k+1} = x_k - \frac{f(x_k)}{f'(x_k)}.$$

If the initial value x_0 is sufficiently close to some simple zero of f, this sequence converges to the zero. This method is led to the study of iteration of the rational function

$$N_f(z) = z - \frac{f(z)}{f'(z)},$$

which can be identified with a holomorphic self-mapping on $\mathbb{P}^1$.

Shröder [224] was the first to study Newton's method for complex numbers. Mainly he studied the local behavior of rational functions near attractive fixed points. If f is replaced by an entire function, the method will be led to the study of iteration of the meromorphic function N_f on $\mathbb{C}$.

If instead one considers polynomial equations

$$f_j(z) = 0, \quad j = 1, ..., m$$

in m variables $z = (z_1, ..., z_m)$, the Newton method for finding a zero of mapping

$$f = (f_1, ..., f_m) : \mathbb{C}^n \longrightarrow \mathbb{C}^m$$

is to iterate the equation

$$x_{k+1} = x_k - f(x_k) \, {}^t f'(x_k)^{-1},$$

which is led to the study of iteration of the rational mapping

$$N_f(z) = z - f(z) \, {}^t f'(z)^{-1}.$$

Obviously, provided $f'(\zeta)$ is non-singular, $f(\zeta) = 0$ if and only if $N_f(\zeta) = \zeta$. By Proposition B.11 and Proposition B.10, N_f extends to a meromorphic self-mapping on $\mathbb{P}^m$. Thus to study Newton's method, one has to study the iteration of meromorphic mappings and its convergence near a fixed point.

Here we make a remark on the Bézout theorem. Let $\mathcal{P}_k(\mathbb{C}^m)$ be the subvector space of $\mathrm{Pol}(\mathbb{C}^m, \mathbb{C})$ consisting of all polynomials of degree $\leq k$. Then $\mathcal{P}_k(\mathbb{C}^m)$ decomposes into a direct sum

$$\mathcal{P}_k(\mathbb{C}^m) = \mathcal{L}_0(\mathbb{C}^m) \oplus \mathcal{L}_1(\mathbb{C}^m) \oplus \cdots \oplus \mathcal{L}_k(\mathbb{C}^m),$$

where $\mathcal{L}_0(\mathbb{C}^m) = \mathbb{C}$. Hence

$$\dim \mathcal{P}_k(\mathbb{C}^m) = \sum_{j=0}^{d} \binom{m-1+j}{j} = \binom{m+k}{k}.$$

Let $d = (d_1, ..., d_n)$ and set

$$\mathcal{P}_d(\mathbb{C}^m) = \mathcal{P}_{d_1}(\mathbb{C}^m) \times \cdots \times \mathcal{P}_{d_n}(\mathbb{C}^m).$$

In particular, if $d_1 = \cdots = d_n = k$, write

$$\mathcal{P}_k^n(\mathbb{C}^m) = \mathcal{P}_d(\mathbb{C}^m).$$

Then each $f = (f_1, ..., f_n)$ in $\mathcal{P}_d(\mathbb{C}^m)$ induces a mapping $\tilde{f} = (\tilde{f}_1, ..., \tilde{f}_n) \in \mathcal{L}_d(\mathbb{C}^{m+1})$ by setting

$$\tilde{f}_j(z_0, z_1, ..., z_m) = z_0^{d_j} f_j \left(\frac{z_1}{z_0}, ..., \frac{z_m}{z_0} \right).$$

The process of obtaining $\tilde{f}$ from f is called *homogenization*. In this way, we obtain a natural mapping $\sim: \mathcal{P}_d(\mathbb{C}^m) \longrightarrow \mathcal{L}_d(\mathbb{C}^{m+1})$. The inverse of $\sim$ is obtained by setting $z_0 = 1$. Thus there is a one-to-one correspondence between $\mathcal{P}_d(\mathbb{C}^m)$ and $\mathcal{L}_d(\mathbb{C}^{m+1})$.

Further assume $n = m$. If $f \in \mathcal{P}_d(\mathbb{C}^m)$ and $\zeta \in \mathbb{C}^m$ is a root of f, then $\tilde{\zeta} = (1, \zeta)$ is a solution of $\tilde{f}$. Conversely, if $\tilde{\zeta}$ is a root of $\tilde{f} \in \mathcal{L}_d(\mathbb{C}^{m+1})$, then $\tilde{f}(a\tilde{\zeta}) = 0$ for all $a \in \mathbb{C} - \{0\}$. If $a\tilde{\zeta} = (1, \zeta)$ for some a, then $f(\zeta) = 0$. In other words, we can obtain zeros of f by finding zeros of $\tilde{f}$. Since $\tilde{f}(\zeta) = 0$ iff $a\tilde{f}(\zeta) = 0$ for all $a \in \mathbb{C} - \{0\}$, we will write $\mathbf{f}(x) = 0$ if $\mathbf{f} = \mathbb{P}(\tilde{f})$ and if $x = \mathbb{P}(\zeta)$ with $\tilde{f}(\zeta) = 0$. Define a variety

$$V_d = \{(\mathbf{f}, x) \in \mathbb{P}(\mathcal{L}_d(\mathbb{C}^{m+1})) \times \mathbb{P}(\mathbb{C}^{m+1}) \mid \mathbf{f}(x) = 0\}.$$

Let

$$\pi : \mathbb{P}(\mathcal{L}_d(\mathbb{C}^{m+1})) \times \mathbb{P}(\mathbb{C}^{m+1}) \longrightarrow \mathbb{P}(\mathcal{L}_d(\mathbb{C}^{m+1}))$$

be the projection and let

$$V_d^s = \{(\mathbf{f}, x) \in V_d \mid \pi'(\mathbf{f}, x) \text{ is singular }\}.$$

Then the image $\pi(V_d^s)$ has dimension one. Thus $\mathbb{P}(\mathcal{L}_d(\mathbb{C}^{m+1})) - \pi(V_d^s)$ is path-connected. Hence each $\mathbf{f}$ in $\mathbb{P}(\mathcal{L}_d(\mathbb{C}^{m+1})) - \pi(V_d^s)$ just has $d_1 \cdots d_m$ zeros since the space contains $\mathbb{P}(\tilde{f})$, $\tilde{f}_j(z) = z_j^{d_j} - z_0^{d_j}$, which has $d_1 \cdots d_m$ zeros, and since the number of roots along the arcs in $\mathbb{P}(\mathcal{L}_d(\mathbb{C}^{m+1})) - \pi(V_d^s)$ does not change (see [234]), that is, generically, the number of solutions of $f(z) = 0$ is $d_1 \cdots d_m$. This is another form of the Bézout theorem.

In order to say more about the Newton's method, we introduce some notations and discuss it for general cases. Given two topological spaces M and N, take a family $\mathcal{F} \subset C(M, N)$. Let $F_{lim}(\mathcal{F})$ denote the points in M such that $x \in F_{lim}(\mathcal{F})$ iff $\mathcal{F}(x)$ has unique limit point and each sequence of $\mathcal{F}$ converges to the point at x, and define

$$J_{lim}(\mathcal{F}) = M - F_{lim}(\mathcal{F}).$$

Here we discuss a special case, that is, a cascade $\mathcal{F} = \{f^n\}_{n \in \mathbb{Z}_+}$ generated by $f \in C(M, M)$. Now we also write

$$F_{lim}(f) = F_{lim}(\mathcal{F}), \quad J_{lim}(f) = J_{lim}(\mathcal{F}).$$

Obviously, we have

$$f^{-1}(F_{lim}(f)) \subset F_{lim}(f), \quad f(F_{lim}(f)) \subset F_{lim}(f),$$

that is, $F_{lim}(f)$ is backward invariant, and hence $J_{lim}(f)$ also is backward invariant. By the definition, we see

$$\mathrm{Fix}(f) \subset F_{lim}(f).$$

Also $F_{lim}(f)$ contains the basin of attraction of each attractive fixed point of f.

Now we give a condition which makes the basin of attraction of each attractive fixed point of a differentiable f contained in the Fatou set. Recall the following basic fact in calculus.

Proposition 6.4 *Let $f \in C^1(D, \mathbb{C})$, where D is an open subset of $\mathbb{C}$. Take $a \in D, r \in \mathbb{R}^+$ such that $\overline{\mathbb{C}(a;r)} \subset D$. Then for $z \in \mathbb{C}(a;r)$,*

$$f(z) = \frac{1}{2\pi i} \int_{\partial \mathbb{C}(a;r)} \frac{f(\zeta)}{\zeta - z} d\zeta - \frac{1}{2\pi i} \int_{\mathbb{C}(a;r)} \frac{\partial f(\zeta)}{\partial \bar{\zeta}} \cdot \frac{d\zeta \wedge d\bar{\zeta}}{\zeta - z}.$$

The formula implies the following result:

Proposition 6.5 *Let $f \in C^m(D, \mathbb{C}^n)$, where D is an open subset of $\mathbb{C}^m$. Take $a \in D, r \in (\mathbb{R}^+)^m$ such that $\overline{\Delta^m(a, r)} \subset D$. Then for $z \in \Delta^m(a, r)$,*

$$f(z) = \frac{1}{(2\pi i)^m} \int_\Gamma f(\zeta)(\zeta - z)^{-1} d\zeta + \sum_{k=1}^{m} \sum_{J \in \mathcal{J}_k} \frac{(-1)^k}{(2\pi i)^m} \int_{\Gamma_J} \frac{\partial^k f}{\partial \bar{\zeta}_J}(\zeta)(\zeta - z)^{-1} d\zeta_J,$$

where $\Gamma = \partial \mathbb{C}(a_1; r_1) \times \cdots \times \partial \mathbb{C}(a_m; r_m)$ is the skeleton of $\Delta^m(a, r)$, $d\zeta = d\zeta_1 \wedge \cdots \wedge d\zeta_m$, $\mathcal{J}_k = \{(j_1, ..., j_k) \mid 1 \leq j_1 < \cdots < j_k \leq m\}$, and for $J = (j_1, ..., j_k)$, $\partial \bar{\zeta}_J = \partial \bar{\zeta}_{j_1} \cdots \partial \bar{\zeta}_{j_k}$, $\Gamma_J = \mathbb{C}(a_{j_1}; r_{j_1}) \times \cdots \times \mathbb{C}(a_{j_k}; r_{j_k}) \times \partial \mathbb{C}(a_{j_{k+1}}; r_{j_{k+1}}) \times \cdots \times \partial \mathbb{C}(a_{j_m}; r_{j_m})$, and $d\zeta_J = \left(\prod_{l=1}^{k} d\zeta_{j_l} \wedge d\bar{\zeta}_{j_l} \right) \wedge d\zeta_{j_{k+1}} \wedge \cdots \wedge d\zeta_{j_m}$, where $1 \leq j_{k+1} < \cdots < j_m \leq m$ are such that $\{j_1, ..., j_m\}$ is a permutation of $\{1, ..., m\}$.

Take $f \in C^m(M, M)$, where M is a complex manifold of dimension m. For $x \in F_{lim}(f)$, set

$$\hat{f}(x) = \lim_{n \to +\infty} f^n(x).$$

Then a point $x \in F_{lim}(f)$ is said to be *coordinate stable* if there exist a coordinate neighborhood V of x and an integer n_0 such that for each $n \geq n_0$, $f^n(V)$ is contained in a coordinate neighborhood of $\hat{f}(x)$ and such that the family $\{f^n, \frac{\partial^k f^n}{\partial \bar{\zeta}_J} \mid n \geq n_0, J \in \mathcal{J}_k, 1 \leq k \leq m\}$ is uniformly bounded on V, where ζ are holomorphic coordinates on V. A subset E in $F_{lim}(f)$ is called *coordinate stable* if each point of E is coordinate stable.

Theorem 6.15 *Coordinate stable points are contained in $F(f) \cap F_{equ}(f)$.*

Here a coordinate stable point $p \in F_{equ}(f)$ follows directly from Proposition 6.5, and hence $p \in F(f)$ by Arzela-Ascoli theorem. In particular, if $f \in \text{Hol}(M, M)$ and if $\text{Att}(p)$ is contained in a coordinate neighborhood of p for an attractive fixed point p of f, then $\text{Att}(p)$ is coordinate stable. Thus Theorem 5.31 follows from this result.

Further assume that $(M, \mathcal{B}, \mu)$ is a measure space. For all $B, B' \in \mathcal{B}$, define

$$A_\mu(f; B, B') = \limsup_{n \to \infty} \frac{1}{n} \sum_{k=0}^{n-1} \mu(B \cap f^{-k}(B')).$$

In particular, if B' is backward invariant, then

$$A_\mu(f; B, B') = \mu(B \cap B').$$

If $\mu \in \Sigma_f(M)$, i.e., μ is a f-invariant Borel probability measure on M, and if f is ergodic, then $A_\mu(f; B, B') = \mu(B)\mu(B')$ for all $B, B' \in \mathcal{B}$ by Theorem 2.11. A measure preserving mapping $f : M \longrightarrow M$ is called *mixing* if for all $B, B' \in \mathcal{B}$

$$\mu(B \cap f^{-n}(B')) \to \mu(B)\mu(B') \text{ as } n \to \infty.$$

Thus if f is mixing, then

$$A_\mu(f; B, B') = \lim_{n \to \infty} \frac{1}{n} \sum_{k=0}^{n-1} \mu(B \cap f^{-k}(B')) = \mu(B)\mu(B'),$$

for all $B, B' \in \mathcal{B}$, and hence f is ergodic.

Note that $F_{lim}(f)$ is backward invariant. Thus if $\mu(B) > 0$, the number

$$A_\mu(f, B) = A_\mu(f; B, F_{lim}(f))/\mu(B) = \mu(B \cap F_{lim}(f))/\mu(B)$$

will serves as the probability for a given mapping f that the iteration of f will converge for a random choice of initial point in B.

If S is a subset of attractive fixed points of f, and if $\mu(B) > 0$, we also write

$$A_{\mu,S}(f, B) = A_\mu\left(f; B, \bigcup_{\zeta \in S} \text{Att}(\zeta)\right)/\mu(B).$$

Since $\bigcup_{\zeta \in S} \text{Att}(\zeta)$ is backward invariant, we have

$$
\begin{aligned}
A_{\mu,S}(f, B) &= \mu\left(B \cap \left(\bigcup_{\zeta \in S} \text{Att}(\zeta)\right)\right)/\mu(B) \\
&= \sum_{\zeta \in S} A_\mu(f; B, \text{Att}(\zeta))/\mu(B) = \sum_{\zeta \in S} A_{\mu,\zeta}(f, B).
\end{aligned}
$$

Then $0 \leq A_{\mu,S}(f, B) \leq A_\mu(f, B) \leq 1$.

Now we discuss convergence of the Newton's method. Let S be the subset of zeros of f such that $\zeta \in S$ iff ζ is an attractive fixed point of N_f. Write

$$A_{f,\mu}(B) = A_{\mu,S}(N_f, B).$$

Then $A_{f,\mu}(B)$ serves as the probability for a given mapping f that the iteration of N_f will converge to a zero of f for a random choice of initial point in B.

Problem 6.2 *Under which conditions is $A_{f,\mu}(B)$ positive?*

Here we begin with one complex variable case. If $f = (z - \zeta_1)(z - \zeta_2)$ with distinct numbers $\zeta_1, \zeta_2 \in \mathbb{C}$, Cayley obtained that for each $x_0 \in \mathbb{C}$ not lying on the line

$$l = \left\{ z \in \mathbb{C} \mid z = \frac{1}{2}(\zeta_1 + \zeta_2) + it(\zeta_1 - \zeta_2), \ t \in \mathbb{R} \right\}$$

the sequence $\{x_n\}$ converge to either ζ_1 or ζ_2. If $\zeta_1 = \zeta_2 = \zeta$, the sequence $\{x_n\}$ converge to ζ for each $x_0 \in \mathbb{C}$. Hence for any case, we see $A_{f,\mu}(B) = A_\mu(N_f, B) = 1$ for any $B \subset \mathbb{C}$ with $\mu(B) > 0$, where μ is the Lebesgue measure. However, the case of the cubic polynomials appear different phenomena. For example, we consider the following polynomial exhibited in [151]:

$$f(z) = z^3 + 3z^2 - 5z + 9.$$

A straightforward calculation shows that N_f has $C = \{-1, 1\}$ as an attractive 2-cycle such that the Newton's method fails for each initial value x_0 chosen out of $\text{Att}(C)$. Clearly, $\text{Att}(C)$ is open, non-empty and, in particular, it has positive Lebesgue measure, but, $A_{f,\mu}(B) = 0$ for $B \subset \text{Att}(C)$ with $\mu(B) > 0$.

However, there is a universal set B such that $A_{f,\mu}(B) > 0$ for any "normalized" polynomial f. This is the space of all polynomials $\sum_{k=0}^{d} a_k z^k$ with $a_d = 1$ and $|a_k| \leq 1$, which we denote by $\mathcal{P}_d(\mathbb{C}; 1)$. Given any polynomial $f(z) = \sum_{k=0}^{d} a_k z^k$, $a_d \neq 0$, for appropriate $\alpha \in \mathbb{C}$ the transformation $z \mapsto \alpha z = w$ will transform f into $f_\alpha(w) = \sum_{k=0}^{d} b_k z^k$ with $|b_d| \geq |b_k|$ for all k. Then further division by b_d will reduce the polynomial into $\mathcal{P}_d(\mathbb{C}; 1)$. The roots are all changed by the factor α. Then J. Friedman [92] proved

$$\min_{f \in \mathcal{P}_d(\mathbb{C}; 1)} A_{f,\mu}(\mathbb{C}(2)) \geq d^{-cd^2 \log d}$$

for some positive constant c. The problem is stated by Smale in [233]. Here $B = \mathbb{C}(2)$ is chosen by virtue of the well-known fact that $|\zeta| < 2$ if $f(\zeta) = 0$.

For several variable case, also we can choose "normalized" polynomial mappings from $\mathcal{P}_d(\mathbb{C}^m)$, where $d = (d_1, ..., d_m)$, and denote by $\mathcal{P}_d(\mathbb{C}^m; 1)$ (resp., $\mathcal{P}'_d(\mathbb{C}^m; 1)$) which is defined as follows: $f = (f_1, ..., f_m) \in \mathcal{P}_d(\mathbb{C}^m; 1)$ (resp., $\mathcal{P}'_d(\mathbb{C}^m; 1)$) iff
  (i) $f = (f_1, ..., f_m) \in \mathcal{P}_d(\mathbb{C}^m)$ with $\deg(f_k) = d_k$ $(1 \leq k \leq m)$,
  (ii) one of the coefficients of $\hat{f}_k$ is equal to 1 for each k, and absolute values of all the coefficients of $\hat{f}_k$ are ≤ 1 for each k, where $\hat{f}_k$ is the homogeneous part of f_k with degree d_k, and
  (iii) $\{\hat{f}_1 = \cdots = \hat{f}_m = 0\} = \{0\}$ (resp., $\det(f'(z)) \not\equiv 0$).
By Bézout theorem, each f in $\mathcal{P}_d(\mathbb{C}^m; 1)$ has just $d_1 \cdot d_2 \cdots d_m$ zeros counting multiplicities.

Problem 6.3 *For the Lebesgue measure μ on $\mathbb{C}^m$, does there exist a number $r > 0$ such that $A_{f,\mu}(\mathbb{C}^m(r)) > 0$ for each $f \in \mathcal{P}_d(\mathbb{C}^m; 1)$ (resp., $\mathcal{P}'_d(\mathbb{C}^m; 1)$)?*

If the condition (iii) is replaced by

(iii)* $\{\hat{f}_1 = \cdots = \hat{f}_m = 0\} = \{0\}$ and $\det(f'(z)) \neq 0$ for all $z \in \mathbb{C}^m$,

then a family $\mathcal{P}^*_d(\mathbb{C}^m; 1)$ is obtained. We suspect that the problem should be true for the family $\mathcal{P}^*_d(\mathbb{C}^m; 1)$.

Finally, we introduce some results that relate to the following main problem:

Problem 6.4 (Cayley, 1879) *Under which conditions does the sequence $\{x_n\}$ defined by the Newton's method converge to a root of the function f?*

Let us begin again with one complex variable case. Assume that f is a polynomial on $\mathbb{C}$ with simple zeros. Note that if $f'(z) \neq 0$, $f_z^{-1} : \mathbb{C}(f(z); r) \longrightarrow \mathbb{C}$ is the inverse of f, defined locally by the inverse function theorem, which maps $f(z)$ to z. Let $r(f_z^{-1})$ be the radius of convergence of f_z^{-1} as a power series. According to Smale [233], *approximate zeros* of f is defined to be the set

$$\bigcup_{f(\zeta)=0} f_\zeta^{-1}(\mathbb{C}(r(f_\zeta^{-1})/9)).$$

If ζ is a zero of f and if $x_0 \in f_\zeta^{-1}(\mathbb{C}(r(f_\zeta^{-1})/9))$, then the approximate zero theorem (cf. [233]) shows that there is a number $b < 1$ such that

$$|f(x_n)| \leq b^{2^n-1}|f(x_0)|$$

hold for all $n \in \mathbb{Z}_+$, where b is defined by $|f(x_0)| = br(f_\zeta^{-1})/9$.

The above fact has been extended by Smale [234] to define approximate zeros of mappings in several variables, that is, $x_0 \in \mathbb{C}^m$ is an approximate zero of $f \in \mathrm{Hol}(\mathbb{C}^m, \mathbb{C}^m)$ relative to a zero ζ if

$$\|x_n - \zeta\| \leq \left(\frac{1}{2}\right)^{2^n-1} \|x_0 - \zeta\|$$

hold for all $n \in \mathbb{Z}_+$, and shows that $x_0 \in \mathbb{C}^m$ is an approximate zero of f relative to a zero ζ if either

$$\gamma(f, x_0) = \max_{k \geq 2} \left\| \frac{df}{dz}(x_0)^{-1} \frac{1}{k!} \frac{d^k f}{dz^k}(x_0) \right\|^{\frac{1}{k-1}} < \frac{3 - \sqrt{7}}{2} \cdot \frac{1}{\|x_0 - \zeta\|},$$

where $\frac{d^k f}{dz^k}$ is the k-th Jacobi's matrix of f, or there exists a number α^* such that

$$\gamma(f, x_0) \left\| \frac{df}{dz}(x_0)^{-1} f(x_0) \right\| < \alpha^*.$$

Chapter 7

Complex dynamics in $\mathbb{C}^m$

We will introduce the Fatou-Julia theory on $\mathbb{C}^m$, discuss a problem of Zhang and Ren [285], and prove a fixed point theorem.

7.1 Iteration theory on domains of $\mathbb{C}^m$

Here we collect some facts of complex dynamics on bounded domains in $\mathbb{C}^m$ by first characterizing bounded domains in $\mathbb{C}^m$.

Lemma 7.1 ([2]) *A domain D in $\mathbb{C}^m$ is bounded iff* $\mathrm{Hol}(\Delta, D)$ *is relatively compact in* $\mathrm{Hol}(\Delta, \mathbb{C}^m)$.

Proof. We have already observed that D is relatively compact if $\mathrm{Hol}(\Delta, D)$ is relatively compact in $\mathrm{Hol}(\Delta, \mathbb{C}^m)$ by the proof of Lemma 5.10, and hence is bounded. Conversely, let D is bounded, and take a sequence $\{f_n\} \subset \mathrm{Hol}(\Delta, D)$. Denote by $p_j : \mathbb{C}^m \longrightarrow \mathbb{C}$ be the projection onto the j-th coordinate; then there exists a bounded domain Ω in $\mathbb{C}$ such that $p_j(D) \subset \Omega$ for every $j = 1, ..., m$. Now, applying Montel's theorem to $\{p_j \circ f_n\} \subset \mathrm{Hol}(\Delta, \Omega)$ for $j = 1, ..., m$, we can extract a subsequence $\{f_{n_i}\}$ such that $\{p_j \circ f_{n_i}\}$ is converging in $\mathrm{Hol}(\Delta, \Omega)$ for every $j = 1, ..., m$. Thus $\{f_{n_i}\}$ is converging in $\mathrm{Hol}(\Delta, \mathbb{C}^m)$. □

Consequently, by Arzela-Ascoli theorem or by the theorem of Montel, every bounded domain in $\mathbb{C}^m$ is tight in its usual metric, and hence $J_{equ}(f) = \emptyset$ for any holomorphic self-mapping f on a bounded domain in $\mathbb{C}^m$. It is much difficult to deal with a domain which is taut. We begin with the following fact:

Lemma 7.2 *Let D be a bounded domain in $\mathbb{C}^m$. Then D is pseudo-convex iff for every family $\mathcal{F} \subset \mathrm{Hol}(\Delta, D) \cap C(\overline{\Delta}, \overline{D})$ such that $\bigcup_{f \in \mathcal{F}} f(\partial \Delta)$ is relatively compact in D, we have $\bigcup_{f \in \mathcal{F}} f(\overline{\Delta})$ is relatively compact in D.*

A proof can be found in Krantz[149]. Comparing Lemma 5.12 with Lemma 7.2 one might suspect that every bounded taut domain is pseudoconvex, and hence a domain of holomorphy, which was proved by H. Wu [272] (also see Abate [2]). Conversely, every strongly pseudoconvex domain in $\mathbb{C}^n$ is taut. Further, if D is a bounded convex domain in $\mathbb{C}^m$, then D is Kobayashi complete hyperbolic, and Kobayashi distance can be given by

$$d_D(x,y) = \inf\{d_h(z,w) \mid \exists f \in \mathrm{Hol}(\Delta, D), f(z) = x, f(w) = y\} \qquad (7.1)$$

for all $x, y \in D$, where d_h is the Poincaré distance on Δ. The proof of the results can be found in Abate [2] based on Lemma 7.1. Thus by Theorem 5.23, every bounded convex domain D of $\mathbb{C}^n$ is taut, and hence $J(f) = \emptyset$ for every $f \in \mathrm{Hol}(D, D)$.

Generally, is $\mathrm{Fix}(f)$ smooth? Is it connected? In other words, is $\mathrm{Fix}(f)$ a closed submanifold? To discuss this question, we need some notation. An *affine subspace* of $\mathbb{C}^m$ is the translation of a linear subspace; an *affine subset* of Δ^m is the intersection of Δ^m with an affine subspace of $\mathbb{C}^m$. Every automorphism of Δ^m sends affine subsets into affine subsets.

Theorem 7.1 ([218]) *Let $f \in \mathrm{Hol}(\Delta^m, \Delta^m)$. Then $\mathrm{Fix}(f)$ is either empty or an affine subset.*

Conversely, every affine subset of Δ^m is the fixed point set of some automorphism of Δ^m. Also the holomorphic retracts of Δ^m are exactly the affine subsets of Δ^m. In particular, every fixed point set in Δ^m is a holomorphic retract.

It may happen that an automorphism γ of Δ^m has no fixed points in Δ^m; on the other hand, every $\gamma \in \mathrm{Aut}(\Delta^m)$ has a fixed point in $\overline{\Delta^m}$. This is a consequence of Lemma A.4.

Proposition 7.1 *Assume $\gamma \in \mathrm{Aut}(\Delta^m)$ having no fixed points in Δ^m. Then γ has at least one and at most two fixed points in $\partial \Delta^m$.*

Classically, fixing $\gamma \in \mathrm{Aut}(\Delta^m)$, if γ has some fixed points in Δ^m, it is called *elliptic* ; if γ has no fixed points in Δ^m and one fixed point in $\partial \Delta^m$, then it is called *parabolic* ; if γ has no fixed points in Δ^m and two fixed point in $\partial \Delta^m$, then it is called *hyperbolic* . In 1941, Heins[107] proved that if $\gamma \in \mathrm{Aut}(\Delta)$ is hyperbolic, and if $f \in \mathrm{Hol}(\Delta, \Delta)$ with $f \circ \gamma = \gamma \circ f$, then either $f = id$ or f is a hyperbolic automorphism of Δ with the same fixed points of γ. Some generalizations to several variables are referred to [77] and [78].

Theorem 7.2 *Let $D \subset\subset \mathbb{C}^m$ be a convex domain, and let $f \in \mathrm{Hol}(D, D)$ be such that the fixed point set $\mathrm{Fix}(f) \neq \emptyset$. Then $\mathrm{Fix}(f)$ is a holomorphic retract of D. In particular, $\mathrm{Fix}(f)$ is connected.*

For the proof, see Abate [2]. Let $\mathcal{F}$ be a family of continuous self-mappings of a topological space X. We shall say that $\mathcal{F}$ is a *commuting family* if $f \circ g = g \circ f$ for every $f, g \in \mathcal{F}$. A point $x \in X$ is a *fixed point of the family* $\mathcal{F}$ if $f(x) = x$ for all $f \in \mathcal{F}$.

Theorem 7.3 ([4]) *Let $D \subset\subset \mathbb{C}^m$ be a bounded convex domain, and $\mathcal{F}$ a commuting family of continuous self-mappings of $\overline{D}$ holomorphic in D. Then $\mathcal{F}$ has a fixed point in $\overline{D}$.*

In 1964, Shields [227] proved the theorem for the unit disk $D = \Delta$. In 1972, Eustice [76] extended Shields' theorem to $\Delta^2 = \Delta \times \Delta \subset \mathbb{C}^2$. Later, Suffridge [251] found a proof valid for the Euclidean unit ball Δ^m of $\mathbb{C}^m$, and Heath and Suffridge [104] generalized the theorem to the unit polydisk Δ^m. Abate and Vigué [4] obtained the complete generalization of Shields' theorem to convex domains.

The following fact shows that a semiflow on a domain $D \subset \mathbb{C}^m$ also defines a holomorphic vector field X on D:

Lemma 7.3 ([2]) *Let* $f : \mathbb{R}_+ \longrightarrow \text{Hol}(D, D)$ *be a continuous semigroup homomorphism on a domain* $D \subset \mathbb{C}^m$. *Then there is* $X \in \text{Hol}(D, \mathbb{C}^m)$ *such that*

$$\frac{\partial f}{\partial t} = X \circ f.$$

In particular, f is analytic in t.

Further, according to Corollary 2.2 in [89], every real one parameter subgroup $\{f^t\}_{t \in \mathbb{R}}$ of $\text{Aut}(\mathbb{C}^m)$ extends to a complex one parameter subgroup $\{f^t\}_{t \in \mathbb{C}}$.

The first step in the study of iterates is to decide when $\{f^n\}$ is compactly divergent. If f has a fixed point, clearly the sequence $\{f^n\}$ cannot be compactly divergent. The interesting fact is that the converse is also true (see [2]): If $f \in \text{Hol}(\Delta^m, \Delta^m)$, then the sequence $\{f^n\}$ is not compactly divergent iff f has a fixed point in Δ^m. Generally, the following facts are refereed to Abate [4]:

Theorem 7.4 *Let* $D \subset\subset \mathbb{C}^m$ *be a bounded convex domain, and* $f \in \text{Hol}(D, D)$. *Then the following facts are equivalent:*

1) *f has a fixed point in D;*
2) *the sequence $\{f^n\}$ of iterates of f is not compactly divergent;*
3) *the sequence $\{f^n\}$ of iterates of f is relatively compact in $\text{Hol}(D, D)$;*
4) *for every point $z \in D$ the sequence $\{f^n(z)\}$ is relatively compact in D;*
5) *for one point $z_0 \in D$ the sequence $\{f^n(z_0)\}$ is relatively compact in D.*

Proof. Now we follow Abate to show the equivalence of (1) and (2). For the rest, please refer his book. Obviously, (1) means (2), and so we only need to prove the converse. Let $\rho : D \longrightarrow S$ be a holomorphic retraction. By the notes after Theorem 5.25, we know that the closed subgroup $\Gamma(f)$ of $\text{Aut}(S)$ generated by $\varphi = f|_S$ is compact, and so the orbit

$$\Gamma_{z_0}(f) = \{\gamma(z_0) \mid \gamma \in \Gamma(f)\}, \quad z_0 \in S,$$

is compact and contained in S. Consider a covering of $\Gamma_{z_0}(f)$ as follows

$$\mathcal{K} = \{D[z; r] \mid z \in S, r \in \mathbb{R}^+, \Gamma_{z_0}(f) \subset D[z; r]\},$$

where, with respect to the Kobayashi distance d_D,

$$D[z; r] = \{w \in D \mid d_D(z, w) \le r\}.$$

Since each disc $D[z; r]$ is compact and convex, one obtains a non-empty compact convex subset $K = \cap \mathcal{K}$ of D. For $z \in S, w \in K, r \in \mathbb{R}^+$, one has

$$D[\varphi^{-1}(z); r] \cap S = \varphi^{-1}(D[z; r] \cap S) \supset \varphi^{-1}(\Gamma_{z_0}(f)) = \Gamma_{z_0}(f).$$

Hence $w \in D[\varphi^{-1}(z); r]$ and

$$d_D(z, f(w)) = d_D(f(\varphi^{-1}(z)), f(w)) \leq d_D(\varphi^{-1}(z), w) \leq r,$$

that is, $f(w) \in D[z; r]$, and so $f(w) \in K$. In conclusion, $f(K) \subset K$, by Brouwer's theorem (see Lemma A.4), f must have a fixed point in K. □

Next step to study is the asymptotic behavior of the sequence of the iterates $\{f^n\}$. An early result in this aspect is the Denjoy-Wolff theorem [69], [267]:

Theorem 7.5 *The sequence of the iterates $\{f^n\}$ of a mapping $f \in \mathrm{Hol}(\Delta, \Delta)$ does not converge iff $f \in \mathrm{Aut}(\Delta)$ with exactly one fixed point. Moreover, the limit of $\{f^n\}$, when it exists, is a constant $c \in \overline{\Delta}$.*

Proof. First of all, we prove the following fact: If $h \in \mathrm{Hol}(\Delta, \overline{\Delta})$ is a limit point of the iterates $\{f^n\}$ of a mapping $f \in \mathrm{Hol}(\Delta, \Delta)$, then either
(i) h is a constant $c \in \overline{\Delta}$, or
(ii) h is an automorphism of Δ. In this case, $f \in \mathrm{Aut}(\Delta)$ too.
In fact, let h be the limit of a subsequence $\{f^{n_k}\}$ of $\{f^n\}$, and set $m_k = n_{k+1} - n_k$. Since $\mathrm{Hol}(\Delta, \Delta)$ is normal, up to a subsequence we can assume that $\{f^{m_k}\}$ either converges to an element $g \in \mathrm{Hol}(\Delta, \Delta)$ or is compactly divergent. Suppose h is not constant, and so $h(\Delta)$ is open in Δ. Since

$$\lim_{k \to \infty} f^{m_k}(f^{n_k}(z)) = \lim_{k \to \infty} f^{n_{k+1}}(z)$$

holds for any $z \in \Delta$, then $\{f^{m_k}\}$ cannot be compactly divergent, and g is the identity on $h(\Delta)$ which further implies $g = id$ on Δ. For this case, it is easy to show that f is an automorphism of Δ, and so $h \in \mathrm{Aut}(\Delta)$ by Theorem 5.13.

Now we follow Abate [2] to sketch a proof of the theorem. The sufficiency is obvious. Here one only shows the necessity. To do so, it is sufficient to prove that if f is not an automorphism of Δ with exactly one fixed point, the iterates $\{f^n\}$ converge to a constant $c \in \overline{\Delta}$. If f has a fixed point z_0, by Theorem 5.11, $|f'(z_0)| \leq 1$. If $|f'(z_0)| = 1$, f is an automorphism. This is contrary with our assumption. So assume $|f'(z_0)| < 1$. Since f has a fixed points, $\{f^n\}$ cannot have compactly divergent subsequences. Let h_1 and h_2 be two limit points of $\{f^n\}$. By the claim proved above both h_1 and h_2 are constant since $f \notin \mathrm{Aut}(\Delta)$, but z_0 should be a fixed point for both h_1 and h_2, and so $h_1 = h_2 = z_0$, that is, $f^n \to z_0$.

Next assume f has no fixed points. If $f \in \mathrm{Aut}(\Delta)$ is parabolic, then transferring everything on the upper half-plane $H^+ = \{z \in \mathbb{C} \mid \mathrm{Im}(z) > 0\}$, and noting that

$$\mathrm{Aut}(H^+) = \left\{ f(z) = \frac{az + b}{cz + d} \mid a, b, c, d \in \mathbb{R}; \ ad - bc = 1 \right\},$$

w. l. o. g., we can assume $f(z) = z + b$ with $b \in \mathbb{R} - \{0\}$. Then $f^n(z) = z + nb \to \infty$, the unique fixed point of f. If $f \in \mathrm{Aut}(\Delta)$ hyperbolic, then moving again to H^+, we can assume $f(z) = \lambda z$ for some $\lambda > 0, \lambda \neq 1$. Therefore $f^n(z) = \lambda^n z \to 0$ or ∞ according to $\lambda < 1$ or $\lambda > 1$.

Finally assume that $f \notin \text{Aut}(\Delta)$. If $\{f^n\}$ has a limit point

$$h = \lim_{k \to \infty} f^{n_k}$$

in $\text{Hol}(\Delta, \overline{\Delta})$, then h is a constant $c \in \overline{\Delta}$. If $c \in \Delta$, one has

$$f(c) = \lim_{k \to \infty} f(f^{n_k}(c)) = \lim_{k \to \infty} f^{n_k}(f(c)) = c,$$

which is impossible. Therefore $c \in \partial\Delta$. We will need the following Wolff's lemma:

Lemma 7.4 (cf. [2]) *Let $f \in \text{Hol}(\Delta, \Delta)$ be without fixed points. Then there is a unique $\tau \in \partial\Delta$, called the Denjoy-Wolff point of f, such that for all $z \in \Delta$*

$$\frac{|\tau - f(z)|^2}{1 - |f(z)|^2} \leq \frac{|\tau - z|^2}{1 - |z|^2}.$$

We continue to prove the theorem. Denote the horocycle of center $\tau \in \partial\Delta$ and radius $R > 0$ by

$$E(\tau, R) = \left\{ z \in \Delta \left| \frac{|\tau - z|^2}{1 - |z|^2} < R \right. \right\}.$$

Then Wolff's lemma implies

$$f^{n_k}(E(\tau, R)) \subset E(\tau, R);$$

and hence

$$\{c\} = h(E(\tau, R)) \subset \overline{E(\tau, R)} \cap \partial\Delta = \{\tau\}.$$

Hence $c = \tau$, that is, τ is the unique limit point of $\{f^n\}$. The proof is completed. □

Later, in 1941, Heins [106] extended the result to more general domains in $\mathbb{C}$:

Theorem 7.6 *Let $D \subset \mathbb{C}$ be a finitely connected domain bounded by Jordan curves, and $f \in \text{Hol}(D, D)$. Then the sequence of the iterates $\{f^n\}$ converges iff $f \notin \text{Aut}(\Delta)$. Moreover, the limit of $\{f^n\}$, when it exists, is a constant $c \in \overline{D}$.*

Finally, in 1983, MacCluer [162] and Kubota [152] dealt with the case of the unit ball Δ^m of $\mathbb{C}^n$. By using Theorem 7.1 and combining the method of proof of the Denjoy-Wolff theorem which also works for several variables, they proved the following version:

Theorem 7.7 *Let $f \in \text{Hol}(\Delta^m, \Delta^m)$ be without fixed points. Then the sequence of iterates of f converges to a point of $\partial\Delta^m$.*

The Wolff's lemma for Δ^m is as follows (see [2]):

Lemma 7.5 *Let $f \in \text{Hol}(\Delta^m, \Delta^m)$ be without fixed points. Then there is a unique $\tau \in \partial\Delta^m$ such that for all $z \in \Delta^m$*

$$\frac{|1 - (f(z), \tau)|^2}{1 - \|f(z)\|^2} \leq \frac{|1 - (z, \tau)|^2}{1 - \|z\|^2}.$$

In 1988, Abate [3] described the behavior of the sequence $\{f^n\}$ in a strongly convex domain with C^2 boundary in $\mathbb{C}^n$. Ma [161] noted that the proof of Abate is not complete. The following statement is referred to Abate [2]:

Theorem 7.8 *Let $D \subset\subset \mathbb{C}^n$ be a strongly convex domain and $f \in \text{Hol}(D, D)$ without fixed points. Then the sequence $\{f^n\}$ of iterates of f converges to a point $x_0 \in \partial D$.*

7.2   Holomorphic self-mappings on $\mathbb{C}^m$

Zhang and Ren [285] studied the Fatou-Julia theory on $\mathbb{C}^m$ and asked when a Julia set is nonempty? In this section, we will discuss this problem.

Consider a non-constant holomorphic mapping

$$f = (f_1, ..., f_m) : \mathbb{C}^m \longrightarrow \mathbb{C}^m,$$

and define the quantity

$$d(f) = \limsup_{n\to\infty} \limsup_{r\to\infty} \left(\frac{\log M(r, f^n)}{\log M(r, f)} \right)^{\frac{1}{n}},$$

where

$$M(r, f) = \max_{\|z\| \leq r} \|f(z)\|.$$

According to the relation $(B.73)$, the mapping f can be identified with a holomorphic mapping

$$\mathbf{f} = [1 : f_1 : ... : f_m] : \mathbb{C}^m \longrightarrow \mathbb{P}^m,$$

so the following quantity can be defined:

$$d[f] = \limsup_{n\to\infty} \limsup_{r\to\infty} \left(\frac{T_{\mathbf{f}^n}(r)}{T_{\mathbf{f}}(r)} \right)^{\frac{1}{n}}.$$

Assume that f is a polynomial mapping with $d = \deg(f)$. Define $\tilde{f} \in \mathcal{L}_d^{m+1}(\mathbb{C}^{m+1})$ by

$$\tilde{f}(z_0, z_1, ..., z_m) = \left(z_0^d, z_0^d f_1(\frac{z_1}{z_0}, ..., \frac{z_m}{z_0}), ..., z_0^d f_m(\frac{z_1}{z_0}, ..., \frac{z_m}{z_0}) \right) : \mathbb{C}^{m+1} \longrightarrow \mathbb{C}^{m+1}.$$

Then f extends to a meromorphic mapping

$$\mathbf{f} = \mathbb{P} \circ \tilde{f} : \mathbb{P}^m \longrightarrow \mathbb{P}^m.$$

Thus when $\deg(f) \geq 2$ and when $\mathbf{f}$ is holomorphic, by Theorem 6.2, $J(\mathbf{f})$ is nonempty in $\mathbb{P}^m$. Obviously, we have

$$J(f) = J(\mathbf{f}) \cap \mathbb{C}^m.$$

Assume first that $f \in \mathcal{L}_d^m(\mathbb{C}^m)$ is homogeneous with $d \geq 1$. Then

$$M(r, f) = M(1, f)r^d, \quad M(r, f^n) = M(1, f^n)r^{d^n},$$

and hence

$$d(f) = \deg(f). \tag{7.2}$$

From $(B.85)$, one can derive easily the following prove

$$T_f(r) = d \cdot \log r + O(1). \tag{7.3}$$

Thus we obtain

$$d[f] = d(f) = \deg(f). \tag{7.4}$$

If f is non-homogeneous, it is also easy to prove (7.3) and

$$\log M(r, f) = d \cdot \log r + O(1). \tag{7.5}$$

Therefore we always have

$$d[f] = d(f) = \lim_{n \to \infty} (\deg(f^n))^{\frac{1}{n}}, \tag{7.6}$$

which is a conjugacy invariant called the *dynamical degree* of the polynomial mapping f. The mappings with simple dynamics are those with $d(f) = 1$. The mappings with complicated dynamics are those with $d(f) > 1$.

Example 7.1 *Define $f \in \text{Aut}(\mathbb{C}^2)$ by*

$$f(z, w) = (\alpha z, \beta w + z^d) \quad (2 \le d \in \mathbb{Z}, 0 < \beta < \alpha < 1).$$

This f fixes the origin. By induction,

$$f^n(z, w) = (\alpha^n z, \beta^n w + \beta^{n-1}(1 + c + \dots + c^{n-1})z^d),$$

where $c = \alpha^d/\beta$. If $c < 1$, we obtain $F(f) = \mathbb{C}^2 = \text{Att}(0)$. This example shows that for any integer $d \ge 2$, there exists one $f \in \text{Aut}(\mathbb{C}^2)$ with $\deg(f) = d$ such that $J(f) = \emptyset$. Here f is not holomorphic, and $d(f) = d[f] = 1$.

The example above is a polynomial mapping. How about transcendental cases?

Example 7.2 *Define an entire transcendental mapping $f : \mathbb{C}^2 \longrightarrow \mathbb{C}^2$ by*

$$f(z, w) = (z + e^{w-z}, w + e^{w-z}).$$

Then f is not degenerate since $\mathcal{J}f = 1$. By induction,

$$f^n(z, w) = (z + ne^{w-z}, w + ne^{w-z}).$$

Obviously, f is compactly divergent on $\mathbb{C}^2$, therefore $J(f) = \emptyset$.

For an entire transcendental function $f : \mathbb{C} \longrightarrow \mathbb{C}$, the Fatou-Rosenbloom theorem shows that $f \circ f$ always has some fixed points in $\mathbb{C}$. This fact is not true on $\mathbb{C}^m$ for $m \ge 2$. The example above is just a counterexample, obviously, for any $n \ge 1$, f^n has no fixed point at all.

Let $\pi_j : \mathbb{C}^m \longrightarrow \mathbb{C}$ be the j-th coordinate projection, i.e., $\pi_j(z_1, ..., z_m) = z_j$. Take $f \in \text{Hol}(\mathbb{C}^m, \mathbb{C}^m)$. If there exists some j such that $\pi_j \circ f$ only depend on one variable z_j, and is non-linear, then $J(\pi_j \circ f) \ne \emptyset$ and hence $J(f) \ne \emptyset$.

Conjecture 7.1 *If a holomorphic mapping $f : \mathbb{C}^m \longrightarrow \mathbb{C}^m$ is of $d(f) > 1$, or $d[f] > 1$, then $J(f) \ne \emptyset$.*

In studying polynomial mappings of $\mathbb{C}^m$ it is often useful to make analogies with the theory of polynomial mappings of $\mathbb{C}$. We will restrict our attention to $d = d(f) > 1$. If $f : \mathbb{C} \longrightarrow \mathbb{C}$ is a polynomial of degree $d \geq 2$, we have an attractive fixed point at ∞. The iterates $\{f^n\}$ of f are bounded on the bounded components of $F(f)$, by the maximum principle, so the basin of attraction $\mathrm{Att}(\infty)$ is connected. The Julia set $J(f)$ coincides with $\partial\mathrm{Att}(\infty)$. The *filled-in Julia set* of f, denoted by K^+, is defined to be the union of the Julia set $J(f)$ and the bounded components of $F(f)$. Thus $z \in K^+$ iff the iterates $f^n(z)$ are bounded, and one has

$$J(f) = \partial\mathrm{Att}(\infty) = \partial K^+.$$

Defined

$$f(z) = a_0 + a_1 z + \cdots + a_{d-1}z^{d-1} + a_d z^d, \quad a_d \neq 0.$$

Then

$$\begin{aligned}
G(z) &= \lim_{n \to \infty} \frac{1}{d^n} \log |f^n(z)| \\
&= \log |\phi(z)| + \frac{1}{d-1} \log |a_d|
\end{aligned}$$

is the Green's function for $\mathrm{Att}(\infty)$ with pole at ∞. Recall that a Böttcher's function ϕ is defined to be the unique solution of Böttcher's functional equation

$$\phi \circ f = a_d \phi^d,$$

normalized by

$$\phi(z) = z + b_0 + \frac{b_1}{z} + \cdots,$$

which always exists in some neighborhood of ∞ and may be continued analytically along any arc in $\mathrm{Att}(\infty)$.

Analogous terminology for holomorphic mappings of $\mathbb{C}^m$ has been introduced by J. H. Hubbard and R. Oberste-Vorth, where the authors defined

$$K^+ = \{z \in \mathbb{C}^m \mid \{f^n(z)\}_{n=1}^{\infty} \text{ is bounded }\}, \tag{7.7}$$

and furthermore $J^+ = \partial K^+$. In fact, we see $K^+ = \mathcal{K}(f)$, where $\mathcal{K}(f)$ is defined in § 1.4. Thus, $K^+ \cap F_{dc}(f) = \emptyset$, and hence

$$K^+ \subset J(f) \cup F_{uc}(f).$$

By Montel theorem, we see

$$(K^+)^o \subset F_{equ}(f) \cap F_{uc}(f).$$

It is easy to prove that

$$J^+ \subset J(f).$$

Further if f is a diffeomorphism, they also defined

$$\begin{aligned}
K^- &= \{z \in \mathbb{C}^m \mid \{f^{-n}(z)\}_{n=1}^{\infty} \text{ is bounded }\}, \tag{7.8}\\
J^- &= \partial K^-, \quad K = K^+ \cap K^-, \quad \mathbf{J} = J^+ \cap J^-. \tag{7.9}
\end{aligned}$$

For a general polynomial mapping f of degree d on $\mathbb{C}^m$, define

$$G^+(z) = \lim_{n\to\infty} \frac{1}{d^n} \log^+ \|f^n(z)\|,$$

$$G(z) = \lim_{n\to\infty} \frac{1}{d^n} \log \|f^n(z)\|.$$

Further if f is a diffeomorphism, also define

$$G^-(z) = \lim_{n\to\infty} \frac{1}{d^n} \log^+ \|f^{-n}(z)\|.$$

Note that

$$\frac{1}{d}G^+ \circ f = G^+, \quad \frac{1}{d}G \circ f = G.$$

The positive closed $(1,1)$ currents dd^cG^+ and dd^cG satisfy the equations

$$\frac{1}{d}f^*(dd^cG^+) = dd^cG^+, \quad \frac{1}{d}f^*(dd^cG) = dd^cG.$$

If $\deg(f) > 1$, we suspect that

$$J^+ = \operatorname{supp}(dd^cG^+), \quad J(f) = \operatorname{supp}(dd^cG).$$

Heinemann [105] studied a family $\mathcal{S}_d(\mathbb{C}^m)$ of polynomial self-mappings on $\mathbb{C}^m$. An entire mapping $f : \mathbb{C}^m \longrightarrow \mathbb{C}^m$ is in $\mathcal{S}_d(\mathbb{C}^m)$ iff one can find constants $m_f, M_f > 0$, $r_0 \in \mathbb{R}_+$ such that

$$m_f\|z\|^d \le \|f(z)\| \le M_f\|z\|^d \tag{7.10}$$

holds for $\|z\| > r_0$. If $f \in \mathcal{S}_d(\mathbb{C}^m)$, then f is a polynomial self-mappings on $\mathbb{C}^m$ with $\deg(f) = d$ (cf. [71]). The inequality (7.10) implies that the analytic set $f^{-1}(0)$ is a bounded closed subset of $\mathbb{C}^m$, and hence is finite by Lemma B.6. Thus for $f \in \mathcal{L}_d^m(\mathbb{C}^m)$, $f \in \mathcal{S}_d(\mathbb{C}^m)$ iff $f^{-1}(0) = \{0\}$.

Take $\hat{f} \in \mathcal{L}_d^m(\mathbb{C}^m)$ such that $\deg(f - \hat{f}) < d$. Then

$$\|f(z) - \hat{f}(z)\| = O(\|z\|^{d-1}).$$

Thus $f \in \mathcal{S}_d(\mathbb{C}^m)$ iff $\hat{f} \in \mathcal{S}_d(\mathbb{C}^m)$ iff $\hat{f}^{-1}(0) = \{0\}$ iff $\tilde{f}^{-1}(0) = \{0\}$. Thus $\tilde{f}$ is the reduced lifted mapping of the holomorphic mapping $\mathbf{f} = \mathbb{P} \circ \tilde{f}$ on $\mathbb{P}^m$. Note that $\tilde{f} : \mathbb{C}^{m+1} \longrightarrow \mathbb{C}^{m+1}$ is surjective and d^{m+1} to 1 mapping. It is easy to prove the following result:

Lemma 7.6 ([105]) *Each* $f \in \mathcal{S}_d(\mathbb{C}^m)$ *is* d^m *to 1 surjective mapping.*

Corollary 7.1 *Each* $f \in \mathcal{S}_d(\mathbb{C}^m)$ *has* d^{mp} *periodic points of order* $p \in \mathbb{N}$ *(counted with multiplicity).*

Remark. W. J. Zhang [284] announced that if p is prime and large enough, then each $f \in S_d(\mathbb{C}^m)$ $(d \geq 2)$ has a p-cycle at least.

Note that $f \in S_d(\mathbb{C}^m)$ implies $f^n \in S_{d^n}(\mathbb{C}^m)$, and hence

$$d(f) = d[f] = d.$$

If $d \geq 2$, letting $R_1 = M_f^{-\frac{1}{d-1}}, R_2 = m_f^{-\frac{1}{d-1}}$, we have

$$R_2 \left(\frac{\|z\|}{R_2} \right)^{d^n} \leq \|f^n(z)\| \leq R_1 \left(\frac{\|z\|}{R_1} \right)^{d^n},$$

for $\|z\| > r_0$. Then $F_{dc}(f)$ is just the basin of attraction for 'infinity',

$$\bigcup_{n=0}^{\infty} f^{-n}(\mathbb{C}^m - \mathbb{C}^m[R_f]) = F_{dc}(f) \subset F(f),$$

where $R_f = \max\{r_0, R_2\}$, as it is the complement of the compact set K^+ with

$$K^+ = \bigcap_{n=0}^{\infty} f^{-n}(\mathbb{C}^m[R_f]).$$

Hence

$$K^+ = J(f) \cup F_{uc}(f) \subset \mathbb{C}^m[R_f].$$

Note that K^+ is not empty. We obtain

$$J(f) = J^+ \neq \emptyset.$$

Remark. One can show that $J(f)$ is backward invariant by Theorem 1.11 and the following open mapping theorem (cf. [97], p.108):

Theorem 7.9 *A holomorphic mapping $f : \mathbb{C}^m \longrightarrow \mathbb{C}^m$ with discrete fibers is open.*

For m polynomials $f_i : \mathbb{C} \longrightarrow \mathbb{C}$ in one variable with $\deg(f_i) \geq 2$, define a polynomial vector $f : \mathbb{C}^m \longrightarrow \mathbb{C}^m$ by setting

$$f(z_1, ..., z_m) = (f_1(z_1), ..., f_m(z_m)).$$

Heinemann [105] derived

$$J(f) = J(f_1) \times \cdots \times J(f_m).$$

An polynomial mapping $f : \mathbb{C}^m \longrightarrow \mathbb{C}^m$ is in $\mathcal{R}_\alpha(\mathbb{C}^m)$ if

$$\liminf_{\|z\| \to \infty} \frac{\|f(z)\|}{\|z\|^\alpha} > 0$$

holds for some $\alpha > 0$. Obviously, $f \in \mathcal{R}_\alpha(\mathbb{C}^m)$ iff $f : \mathbb{C}^m \longrightarrow \mathbb{C}^m$ is an polynomial mapping such that one can find constants $m_f, M_f > 0$, $r_0 \in \mathbb{R}_+$ satisfying

$$m_f \|z\|^\alpha \leq \|f(z)\| \leq M_f \|z\|^d$$

for $\|z\| > r_0$, where $d = \deg(f)$. Let $f \in \mathcal{R}_\alpha(\mathbb{C}^m)$ with $\alpha > 1$. Then f is proper (cf. Heinemann [105]) with

$$K^+ = J(f) \cup F_{uc}(f), \quad J(f) = J^+,$$

and hence f admits a continuation to $\overline{\mathbb{C}^m} = \mathbb{C}^m \cup \{\infty\}$ by setting $f(\infty) = \infty$. Also the 'infinity' is an attractive fixed point of f. Klimek [140] proved that K^+ is completely invariant, compact and L-regular, where a compact set E in $\mathbb{C}^m$ is said to be L-regular if the L-extremal function corresponding to E

$$G_E = \sup\{u \mid u(z) \leq \beta + \log^+ \|z\| \text{ for } z \in \mathbb{C}^m \text{ and } u \leq 0 \text{ on } E\}$$

is continuous, and Kosek [148] showed that the pluricomplex Green function G_{K^+} of K^+ has a Hölder continuity property, i.e., there exist constants $c > 0$ and $\delta \in (0,1]$ such that

$$G_{K^+}(z) \leq c\{\text{dist}(z, K^+)\}^\delta, \quad \text{if dist}(z, K^+) \leq 1.$$

The proof depends on Klimek's inequality (see [141]):

$$\frac{1}{d} G_E \circ f \leq G_{f^{-1}(E)} \leq \frac{1}{\alpha} G_E \circ f.$$

7.3 The group $\text{Aut}(\mathbb{C}^2)$

Let $\mathcal{G} \subset \text{Aut}(\mathbb{C}^2)$ be the group of all polynomial automorphisms of $\mathbb{C}^2$. Note that the Jacobian determinant of any transformation in $\mathcal{G}$ must be a non-zero constant. The famous "Jacobian conjecture" is the converse claim that any polynomial mapping with non-zero constant Jacobian determinant must necessarily be a polynomial automorphism (cf. [29]).

Let $\mathcal{A} \subset \mathcal{G}$ be the six-dimensional group consisting of all affine automorphisms of $\mathbb{C}^2$, i.e., $a \in \mathcal{A}$ iff

$$a(z, w) = (\alpha_{11} z + \alpha_{12} w + \beta_1, \alpha_{21} z + \alpha_{22} w + \beta_2)$$

with $\mathcal{J}a = \det(\alpha_{ij}) \neq 0$. Note that $g \in \mathcal{G}$ has degree one iff $g \in \mathcal{A}$. Write

$$a(z, w) = (z, w)A + b, \quad A = (\alpha_{ij}), \quad b = (\beta_1, \beta_2).$$

Take $H \in GL(2; \mathbb{C})$. Then $h(z, w) = (z, w)H \in \mathcal{A}$ with

$$h \circ a \circ h^{-1}(z, w) = (z, w)H^{-1}AH + bH.$$

Thus every $a \in \mathcal{A}$ is conjugate to the automorphism

$$(z, w) \mapsto (\alpha z + \delta, \beta w + \gamma), \quad \alpha \neq \beta, \quad \text{or } (z, w) \mapsto (\beta z + w + \delta, \beta w + \gamma).$$

Let $\mathcal{E} \subset \mathcal{G}$ be the group consisting of all polynomial automorphisms which are *elementary* in the sense that they carry each line of the form $w = c(= \text{constant})$ to a line of the form $w = c'(= \text{constant})$. It is not difficult to check that a polynomial automorphism e is elementary iff it can be written as

$$e(z,w) = (\alpha z + p(w), \beta w + \gamma)$$

where $\alpha, \beta, \gamma \in \mathbb{C}, \alpha\beta \neq 0$, and p is a polynomial. Obviously, $\mathcal{J}e = \alpha\beta$, and

$$e^{-1}(z,w) = (\alpha^{-1}z - \alpha^{-1}p(\beta^{-1}w - \beta^{-1}\gamma), \beta^{-1}w - \beta^{-1}\gamma)$$

with $\deg(e^{-1}) = \deg(e)$. A *Hénon mapping* is an automorphism which has the form

$$g(z,w) = (p(z) - \alpha w, z) \tag{7.11}$$

with $\deg(p) \geq 2, \alpha \neq 0$. Obviously, g is conjugate to the automorphism

$$f(z,w) = (w, p(w) - \alpha z), \tag{7.12}$$

i.e., $f = h \circ g \circ h^{-1}$, where $h = h^{-1} : (z,w) \mapsto (w,z)$, also called a *generalized Hénon mapping*. Its inverse is given by

$$f^{-1}(z,w) = (\frac{1}{\alpha}(p(z) - w), z), \tag{7.13}$$

and the derivative is

$$f'(z,w) = \begin{pmatrix} 0 & 1 \\ -\alpha & p'(w) \end{pmatrix}. \tag{7.14}$$

Here we introduce the following classification of elementary mappings, due to Friedland and Milnor [91].

Theorem 7.10 *Every element of $\mathcal{E}$ is $\mathcal{E}$-conjugate to one of the following types of automorphisms.*

1) $(z,w) \mapsto (\alpha z, \beta w)$;

2) $(z,w) \mapsto (\alpha z, w + 1)$ or $(z,w) \mapsto (z + 1, \beta w)$;

3) $(z,w) \mapsto (\beta^d(z + w^d), \beta w)$ for some integer $d \geq 1$;

4) $(z,w) \mapsto (\beta^d(z + w^d q(w^r)), \beta w)$ for some $d \geq 0$, where β is a primitive r-th root of unity, and where q is a nonconstant polynomial.

Thus every elementary transformation which is not $\mathcal{E}$-conjugate to an affine transformation is $\mathcal{E}$-conjugate to a transformation of the form

$$e(z,w) = (\alpha z + \alpha h(w), \beta w)$$

where $h(w)$ is a polynomial which satisfies the identity

$$\alpha h(w) = h(\beta w).$$

One very useful consequence of the equation is the following simple expression for the n-fold iterate of e.

$$e^n(z,w) = (\alpha^n(z + nh(w)), \beta^n w).$$

Hence the dynamics of the elementary transformations are quite simple.

The following theorem is due to Jung [131]. Its proofs can be found in [103],[131] or [171].

Theorem 7.11 *The group $\mathcal{G}$ is generated by subgroups $\mathcal{A}$ and $\mathcal{E}$ in the sense that every element of $\mathcal{G}$ is a finite composition of elements from $\mathcal{A}$ and $\mathcal{E}$.*

A more precise result is due to Van der Kulk [261] which states that the group $\mathcal{G}$ is an amalgamated free product of the subgroups $\mathcal{A}$ and $\mathcal{E}$; see also Rentschler [210] and Friedland and Milnor [91] (It appears that Van der Kulk was not aware of Jung's work [131]). Friedland and Milnor [91] introduced the following concept:

Definition 7.1 *A group element $g_n \circ g_{n-1} \circ \cdots \circ g_1$ of $\mathcal{G}$ is called a reduced word of length n if each factor g_i belongs to either $\mathcal{A}$ or $\mathcal{E}$ but not to the intersection $\mathcal{A} \cap \mathcal{E}$, and if no two consecutive factors belong to the same subgroups $\mathcal{A}$ or $\mathcal{E}$.*

It follows immediately from Jung's theorem that every element g of $\mathcal{G}$ can be expressed as a reduced word

$$g = g_n \circ g_{n-1} \circ \cdots \circ g_1, \tag{7.15}$$

unless $g \in \mathcal{A} \cap \mathcal{E}$. Moreover, such a representation of g is unique in the sense: For any $s \in \mathcal{A} \cap \mathcal{E}$ and any $i > 1$ we can replace g_i by the product $g_i \circ s$, and simultaneously replace g_{i-1} by $s^{-1} \circ g_{i-1}$, where s should be treated as units in this representation. The crucial step in the proof of the uniqueness, given in [91] (see also [216]), is to show that any reduced word (7.15) satisfies

$$\deg(g) = \deg(g_n) \cdot \deg(g_{n-1}) \cdots \deg(g_1).$$

As immediate corollaries, we see that any reduced word, unless it consists of a single affine factor, must have degree two or more, and that $\deg(g^{-1}) = \deg(g)$. Another important consequence is that no reduced word is equal to the identity element of $\mathcal{G}$. By using this fact to start the induction, the uniqueness follows from a straightforward induction on the length n.

It follows immediately from the uniqueness that the length $n \geq 1$ of such a reduced word is an invariant of the group element g. Furthermore, the sequence of degrees of the factors g_i also forms an invariant. If $g \in \mathcal{G} - \mathcal{A}$, the *polydegree* $(d_1, ..., d_m)$ of the reduced word g is defined to be the sequence of integers $d_j \geq 2$ which is obtained from the sequence $(\deg(g_1), ..., \deg(g_n))$ by crossing out all of the ones. The set $\mathcal{G}[d_1, ..., d_m] \subset \mathcal{G}$ consisting of all group elements with polydegree $(d_1, ..., d_m)$ forms a smooth analytic manifold of dimension $d_1 + ... + d_m + 6$ (cf. [91]). It follows that $\mathcal{G}$ can be decomposed as a disjoint union

$$\mathcal{G} = \mathcal{A} \cup \mathcal{G}[2] \cup \mathcal{G}[3] \cup \mathcal{G}[2,2] \cup \mathcal{G}[4] \cup \mathcal{G}[5] \cup \cdots$$

of smooth Zariski open subsets of algebraic varieties.

From the representation (7.15) of polynomial automorphisms it follows that every $g \in \mathcal{G}$ is either conjugate in $\mathcal{G}$ to an element of $\mathcal{E}$, or else g is conjugate to a reduced word

$$a_m \circ e_m \circ \cdots \circ a_1 \circ e_1$$

of even length $2m$ with $a_i \in \mathcal{A}$, $e_i \in \mathcal{E}$ ($1 \leq i \leq m$). In the second case, the group element g is called *cyclically reduced*. The length of a cyclically reduced word are invariant under conjugation, and the polydegree is invariant up to cyclic permutation. Obviously, a generalized Hénon mapping (7.12) is cyclically reduced since $f = a \circ e$ with

$$a(z,w) = (w,z), \quad e(z,w) = (-\alpha z + p(w), w).$$

Theorem 7.12 ([91]) *Every cyclically reduced element of $\mathcal{G}$ is conjugate to a composition*

$$(z_1, z_0) \overset{g_1}{\longmapsto} (z_2, z_1) \overset{g_2}{\longmapsto} \cdots \overset{g_m}{\longmapsto} (z_{m+1}, z_m)$$

of Hénon mappings, where $g_i(z,w) = (p_i(z) - \alpha_i w, z)$. Furthermore, this composition can be chosen so that the coefficient of the highest degree in each polynomial p_i is ± 1, and so that the next highest coefficient is zero.

Thus any cyclically reduced conjugacy class of polydegree $(d_1, ..., d_m)$ can be represented by a mapping of the form $g(z_1, z_0) = (z_{m+1}, z_m)$ where the sequence $z_2, ..., z_{m+1}$ is defined inductively by the formula $z_{i+1} = p_i(z_i) - \alpha_i z_{i-1}$, with

$$p_i(z_i) = \pm z_i^{d_i} + \sum_{k=0}^{d_i-2} \alpha_{ik} z_i^k.$$

This normal form depends on just $d_1 + \cdots + d_m$ parameters.

One special case of Theorem 7.12 is the following: Every automorphism $g \in \mathcal{G}$ of prime degree is affinely conjugate either to a Hénon mapping or to an elementary transformation. In the simplest case, if $\deg(g) = 2$ and if g is not affinely conjugate to an elementary transformation, one obtains an unique normal form

$$(z, w) \mapsto (z^2 + c - \alpha w, z)$$

depending on two parameters c and α, which is equivalent to the normal form studied in [108].

Theorem 7.13 ([91]) *If $g \in \mathcal{G}$ is cyclically reduced of degree d, then the number of fixed points of g is precisely equal to d counting multiplicity.*

Proof. According to Theorem 7.12, the mapping g is conjugate to a composition of Hénon mappings. Evidently the number of fixed points will not be changed if we replace g by a conjugate $h \circ g \circ h^{-1}$. Hence we may assume for example that $g = g_m \circ \cdots \circ g_1$ is the composition of Hénon mappings

$$(z_1, z_0) \overset{g_1}{\longmapsto} (z_2, z_1) \overset{g_2}{\longmapsto} \cdots \overset{g_m}{\longmapsto} (z_{m+1}, z_m)$$

where

$$z_2 = p_1(z_1) - \alpha_1 z_0, ..., z_{m+1} = p_m(z_m) - \alpha_m z_{m-1}. \tag{7.16}$$

For a fixed point (z_1, z_0) of g, we must have $z_1 = z_{m+1}, z_0 = z_m$; and the equations (7.16) reduce to a system of m polynomial equations in m complex variables as follows:

$$p_1(z_1) - \alpha_1 z_m - z_2 = \cdots = p_m(z_m) - \alpha_m z_{m-1} - z_1 = 0. \tag{7.17}$$

Let $d_i \geq 2$ be the degree of the polynomial p_i, so that $d = d_1 \cdots d_m$ is the total degree of g. The theorem follows from Theorem 7.12 and Bézout theorem. □

As an immediate corollary, it follows that

$$\#\text{Fix}(g^n) = d^n$$

counting multiplicity, since g^n is cyclically reduced of degree d^n. If g is an elementary or affine transformation, it can have at most one isolated fixed point with a multiplicity equal to 1 (see [91]). Hence no elementary or affine transformation can be topologically conjugate to a cyclically reduced element of the group $\mathcal{G}$. Friedland and Milnor [91] also proved the following result:

Theorem 7.14 *If $g \in \mathcal{G}$ is cyclically reduced, then g has periodic points of period p for every sufficiently large prime p.*

P. Ahern and F. Forstneric [8] exhibited the following result from G. Buzzard:

Lemma 7.7 *If a flow $\{f^t\} \subset \text{Aut}(\mathbb{C}^2)$ is of $f^1 \in \mathcal{G}$, then f^1 is conjugate in $\mathcal{G}$ to an elementary mapping $e \in \mathcal{E}$.*

Proof. Assume, to the contrary, that $f^1 \in \mathcal{G}$ is conjugate to a cyclically reduced word

$$g = a_m \circ e_m \circ \cdots \circ a_1 \circ e_1$$

of even length $2m$, w.l.o.g., we may assume $f^1 = g$. Let z_0 be a periodic point of period n of g which is not a fixed point of g. Then

$$g^n(f^t(z_0)) = f^n(f^t(z_0)) = f^t(f^n(z_0)) = f^t(g^n(z_0)) = f^t(z_0),$$

i.e., $O(z_0) = \{f^t(z_0)\}_{t \in \mathbb{R}} \subset \text{Fix}(g^n)$. This is impossible since $\text{Fix}(g^n)$ is finite, and $O(z_0)$ is a nontrivial closed curve. Therefore f^1 must be conjugate to an elementary mapping. □

Thus by using this result and according to the classification of elementary mappings in Theorem 7.10, the flows on $\mathbb{C}^2$ whose time one mapping is polynomial are classified in [8]. The classification theorem for polynomial flows $\{f^t\}_{t \in \mathbb{R}} \subset \mathcal{G}$ on $\mathbb{C}^2$ were given in 1977 by Suzuki [252], [253]. In 1985, this result was rediscovered by Bass and Meisters [30].

Theorem 7.15 (Ahern-Forstneric [8]) *If a flow $\{f^t\} \subset \text{Aut}(\mathbb{C}^2)$ is such that $f^1 = e \in \mathcal{E}$ is one of the normal forms in Theorem 7.10, then f^t is conjugate to one of the following types of flows, respectively.*

1) $(z, w) \mapsto (\exp(\mu t)z, \exp(\lambda t)w)$, $\exp(\mu) = \alpha$, $\exp(\lambda) = \beta$, *here it is assumed that* $e(z, w) = (\alpha z, \beta w)$ *is aperiodic, i.e.,* $e^n \neq id$ $(n \geq 1)$.

2) $(z, w) \mapsto (z + t, \exp(\lambda t)w)$, $\exp(\lambda) = \beta$, *here it is assumed that the infinitesimal generator is a polynomial vector field on $\mathbb{C}^2$.*

3) $(z, w) \mapsto (\exp(d\lambda t)(z + tw^d), \exp(\lambda t)w)$, $\exp(\lambda) = \beta$.

4) $(z, w) \mapsto (z + tw^d q(w^r), w)$ *if* $\beta = 1$, *otherwise e does not belong to any flow* $\{f^t\} \subset \text{Aut}(\mathbb{C}^2)$.

7.4 Dynamics in $\mathbb{C}^2$

Lemma 7.8 ([91]) *For every generalized Hénon mapping*

$$f : (z, w) \mapsto (w, y) = (w, p(w) - \alpha z),$$

and for $r > 0$, there exists a constant $R > 0$ so that $|w| > R$ implies that either $|y| > r|w|$ or $|z| > r|w|$ or both.

Proof. Since $p(w)$ has degree two or more, we can choose $R > 0$ so that $|p(w)| > r(|w| + |\alpha w|)$ whenever $|w| > R$. It then follows that either $|z| > r|w|$ or

$$|y| = |p(w) - \alpha z| \geq |p(w)| - r|\alpha w| > r|w|,$$

whenever $|w| > R$. $\qquad\qquad\qquad\qquad\qquad\qquad\qquad\qquad\qquad\qquad\qquad\qquad\qquad$ $\square$

Define

$$
\begin{align}
V_r^- &= \{(z, w) \mid |w| > R, |w| > r|z|\} \tag{7.18}\\
V_r^+ &= \{(z, w) \mid |z| > R, |z| > r|w|\} \tag{7.19}\\
V^- &= V_1^-, \quad V^+ = V_1^+ \tag{7.20}\\
V &= \{(z, w) \mid |z| \leq R, |w| \leq R\}, \tag{7.21}
\end{align}
$$

and consider the space $\mathcal{F}$ of finite compositions of generalized Hénon mappings:

$$\mathcal{F} = \{f = f_k \circ f_{k-1} \circ \cdots \circ f_1 \mid f_j(z, w) = (w, p_j(w) - \alpha_j z), \alpha_j \in \mathbb{C}_*\}. \tag{7.22}$$

Take $f = f_k \circ f_{k-1} \circ \cdots \circ f_1 \in \mathcal{F}$. Fix $r \geq 1$. Choose R large enough so that Lemma 7.8 holds for each f_j.

If $(z, w) \in V_{r-1}^-$, by Lemma 7.8, we have

$$|y_j| > r|w| \quad (y_j = p_j(w) - \alpha_j z),$$

which implies $f_j(z, w) \in V_r^-$ for $j = 1, ..., k$, and since $|w| > R$, hence

$$f_j(V_{r-1}^-) \subset V_r^- \subset V_{r-1}^-. \tag{7.23}$$

If $f_j(z, w) = (w, y_j) \in V_{r-1}^+$, Lemma 7.8 gives $|z| > r|w|$ since $|w| > R$ and $|y_j| < r|w|$. This implies $(z, w) \in V_r^+$. Hence

$$f_j^{-1}(V_{r-1}^+) \subset V_r^+ \subset V_{r-1}^+. \tag{7.24}$$

Thus

$$f(V_{r-1}^-) \subset V_r^-, \quad f^{-1}(V_{r-1}^+) \subset V_r^+. \tag{7.25}$$

Assume $(z, w) \in V$. We consider two cases. If $|y_j| \leq R$, then $(w, y_j) \in V$ since $|w| \leq R$. If $|y_j| > R$, then, since $|w| \leq R$, we have $|y_j| > |w|$ so $(w, y_j) \in V^-$. Thus

$$f(V^- \cup V) \subset V^- \cup V. \tag{7.26}$$

Assume $(w, y_j) \in V$. If $|z| \leq R$, then, since $|w| \leq R$ we have $(z, w) \in V$. If $|z| > R$, then we have $|z| > |w|$ since $|w| \leq R$, and so $(z, w) \in V^+$. Hence

$$f^{-1}(V^+ \cup V) \subset V^+ \cup V. \tag{7.27}$$

Assume that $p = (x_0, x_1) \notin V_r^+ \cup V$. Then we have $|x_1| > R, r|x_1| \geq |x_0|$. Thus $s|x_1| > |x_0|$ for $s > r$, i.e., $(x_0, x_1) \in V_{s-1}^-$. Write $(x_j, x_{j+1}) = f_j(x_0, x_1)$. By induction, (7.24) implies $(x_j, x_{j+1}) \in V_s^-$ for $j \geq 1$. In particular, $|x_{j+1}| > s|x_j|$ so that $|x_{j+1}| > s^j|x_1|$. Therefore $\{f^n(x_0, x_1) = (x_{kn}, x_{kn+1})\}$ is not bounded so that $p \notin K^+$. Thus one has

$$K^+ \subset V_r^+ \cup V. \tag{7.28}$$

Similarly,

$$K^- \subset V_r^- \cup V, \quad K \subset V. \tag{7.29}$$

We let π_1 and π_2 denote projection to the first and second coordinates, respectively.

Proposition 7.2 *For $f \in \mathcal{F}$, the following relationships hold:*
1) $V^- \subset f^{-1}(V^-) \subset f^{-2}(V^-) \subset \cdots$; $\bigcup_{n=0}^{\infty} f^{-n}(V^-) = \mathbb{C}^2 - K^+$;
2) $V^+ \subset f(V^+) \subset f^2(V^+) \subset \cdots$; $\bigcup_{n=0}^{\infty} f^n(V^+) = \mathbb{C}^2 - K^-$;
3) $V_1 \supset V_2 \supset \cdots$; $\bigcap_{n=1}^{\infty} V_n = K$, where $V_n = f^n(V) \cap f^{-n}(V)$;
4) $W^s(K) = K^+, W^u(K) = K^-$;
5) $(K^+)^\circ \subset F(f), (K^-)^\circ \subset F(f^{-1})$.

Proof. 1) The first part of the assertions is obvious. If $q \notin K^+$, then we must show that $f^n(q) \in V^-$ for some large n. If this is not the case, then the sequence $\{f^n(q)\}$ is a subset of $V \cup V^+$, which has a subsequence that tends to infinity. Thus we must have a subsequence of $\{|\pi_1 \circ f^n(q)|\}$ tending to infinity. Let us "expand" this sequence by putting the terms

$$\{f_1 \circ f^n(q), f_2 \circ f_1 \circ f^n(q), ..., f_{k-1} \circ \cdots \circ f_2 \circ f_1 \circ f^n(q)\}$$

between $f^n(q)$ and $f^{n+1}(q)$. Thus a subsequence of this expanded sequence must tend to infinity. In particular, we have

$$|\pi_1(q')| < |\pi_1 \circ f_j(q')| \quad \text{for some } 1 \leq j \leq k.$$

This is not possible, however, since $q' \in V^+$, and $\pi_1 \circ f_j(z, w) = w$. We prove 2) by applying the same argument to f^{-1}.

For 3) we show that $V_n \subset V_{n-1}$. We consider the inclusions

$$f^n(V) \subset f^{n-1}(V) \cup V^-, \quad f^{-n}(V) \subset f^{n-1}(V) \cup V^+.$$

For $n = 1$, this is a consequence of (7.26) and (7.27), and the result follows by induction. Since $K \subset V$, we see that $K \subset \cap V_n$. On the other hand, the orbits of points in $\cap V_n$ are clearly bounded in forward time, so the two sets coincide.

For 4), we note that $W^s(K) \subset K^+$ is obvious. To prove the reverse inequality, it suffices to show that if $U \subset K^+$ is any neighborhood of K in K^+ and $q \in K^+$, then there is an R' such that $f^n(q) \in U$ for $n > R'$. For any $p \in V \cap (K^+ - U)$ there is an m such that

$f^{-m}(p) \notin V$. The set $V \cap (K^+ - U)$ is compact, so there is a fixed number N such that this holds for $m > N$. Now choose N_0 such that $f^{N_0}(q) \in V$. It follows that if $n > N + N_0$ then $f^n(q) \in U$. The another statement follows if we replace f by f^{-1}.

For 5), fix a point $p \in (K^+)^o$. Take a bounded neighborhood $W \subset\subset (K^+)^o$ of p. It follows from 4) that for any neighborhood U of K in K^+, there an n_0 such that

$$f^n(W) \subset U \quad \text{for } n \geq n_0.$$

In particular, $f^n|_W$ is bounded. Thus the Montel Theorem implies that $\{f^n\}$ is normal in W, so that $p \in F(f)$. $\square$

Proposition 7.3 ([91]) *For $f \in \mathcal{F}$, the nonwandering set $\Omega(f)$ is contained in K, and hence is compact.*

Proof. Let $f = f_k \circ f_{k-1} \circ \cdots \circ f_1$. It is convenient to extend periodically by setting $f_{j+k} = f_j$ so that the notation

$$f_j(x_{j-1}, x_j) = (x_j, x_{j+1}),$$

makes sense for all integer values of j. We will show that the total orbit

$$O((x_0, x_1)) = \{(x_{ik}, x_{ik+1})\}_{i=-\infty}^{\infty}$$

under the composition of f is bounded iff $|x_j| \leq R$ for all j. Suppose, to the contrary, that $|x_n| > R$ for all n. Then according to Lemma 7.8 we must have either $|x_{n+1}| > r|x_n| (r \geq 1)$, which implies inductively that

$$|x_{n+j+1}| > r|x_{n+j}|, \quad j = 0, 1, \cdots,$$

or $|x_{n-1}| > r|x_n|$ which implies that

$$|x_{n-j-1}| > r|x_{n-j}|, \quad j = 0, 1, \cdots.$$

In either case the resulting monotonic sequence of numbers must diverge to infinity. It follows immediately that $\Omega(f) \subset K$. In fact, every point which does not belong to K has either an open neighborhood whose forward images diverge monotonically to infinity or an open neighborhood whose backward images diverge monotonically to infinity. $\square$

If $f \in \mathcal{G}$ is conjugate either to an affine transformation or to an elementary transformation, then $\Omega(f) = K(f)$. If we choose a suitable representative g with the conjugacy class, then the set $\Omega(g) = K(g)$, if non-vacuous, is either a linear space on which g acts linearly, or a union of finitely many 'horizontal' lines $w = \text{const}$, consisting entirely of periodic points, on which g acts by an isometry (see [91]).

Proposition 7.4 ([34]) *For $f \in \mathcal{F}$, the set $R(f)$ of chain recurrent points is contained in K.*

Proof. Let $f = f_k \circ f_{k-1} \circ \cdots \circ f_1$. For given $\delta_j > 0$ we may choose R such that for $(z, w) \in V^-$,

$$(1 - \delta_j)|w|^{d_j} < |\pi_2 \circ f_j(z, w)| < (1 + \delta_j)|w|^{d_j} \quad (d_j = \deg(f_j)).$$

Applying this once, we get

$$
\begin{aligned}
|\pi_2 \circ f_2 \circ f_1(z, w)| &= |\pi_2 \circ f_2(\pi_1 \circ f_1(z, w), \pi_2 \circ f_1(z, w))| \\
&< (1 + \delta_2)|\pi_2 \circ f_1(z, w)|^{d_2} \\
&< (1 + \delta_2)(1 + \delta_1)^{d_2}|w|^{d_1 d_2}.
\end{aligned}
$$

By induction, then, we have

$$|\pi_2 \circ f(z, w)| < (1 + \delta_k)(1 + \delta_{k-1})^{d_k}(1 + \delta_{k-2})^{d_k d_{k-1}} \cdots (1 + \delta_1)^{d_k \cdots d_1}|w|^d.$$

Thus by taking $\delta_k, ..., \delta_1$ sufficiently small, one has

$$(1 - \delta)|w|^d < |\pi_2 \circ f(z, w)| < (1 + \delta)|w|^d \quad (d = \deg(f)),$$

for $\delta > 0$ sufficiently small. Thus if R is sufficiently large, for $(z, w) \in V^-$,

$$
\begin{aligned}
(1 - \delta)^{d^{n-1}/d_k}|w|^{d^n/d_k} &< |z_n| < (1 + \delta)^{d^{n-1}/d_k}|w|^{d^n/d_k} \\
(1 - \delta)^{d^{n-1}}|w|^{d^n} &< |w_n| < (1 + \delta)^{d^{n-1}}|w|^{d^n}.
\end{aligned}
$$

If $p \notin K^+$, then $f^T(p) \in V^-$ for some large T. Thus $\varepsilon > 0$ may be chosen small enough so that any ε-trajectory enters V^- after T steps. By the inequalities above, it is clear that an ε-trajectory that enters V^- will remain there. Thus p cannot be chain recurrent. A similar argument holds if $p \notin K^-$. $\qquad\square$

Theorem 7.16 ([35]) *Let f be a polynomial diffeomorphism of $\mathbb{C}^2$ satisfying $d(f) > 1$. Let p be a saddle point of f. Then J^+ is the closure of the stable manifold $W^s(p)$ and J^- is the closure of $W^u(p)$.*

The theorem was conjectured by J. Hubbard. This result was proved in [34], [35]. Here p is a saddle point of f iff p is a periodic point of period k with eigenvalues λ_1 and λ_2 of $(f^k)'(p)$ labeled so that $|\lambda_1| < 1 < |\lambda_2|$. If p is saddle, then it follows from the Stable Manifold Theorem that the stable and unstable sets of p are immersed complex manifolds biholomorphically equivalent to $\mathbb{C}$. Bedford and Smillie [35] also proved that the boundary of any basin of attraction of a polynomial diffeomorphism of $\mathbb{C}^2$ satisfying $d(f) > 1$ is J^+.

It was shown in [34] that $G^\pm$ is continuous on $\mathbb{C}^2$ and is a plurisubharmonic Green function of $K^\pm$. Since the function $-G^+$ is a plurisubharmonic exhaustion of the complement of K^+, the set K^+ is pseudoconvex. Also the support of the current $dd^c G^+$ is all of J^+, and K^+, K^-, J^+, and J^- are connected which was conjectured by Friedland and Milnor [91]. Further, if J is a hyperbolic set for $f \in \mathcal{F}$, then the interior of K^+ consists of the basins of finitely many attractive cycles, and the periodic points are dense in J which answers the question of J. H. Hubbard in the hyperbolic case. For all of these results, see [34].

According to [34] and [37], the equilibrium measure μ of K is defined as

$$\mu = 4dd^c G^+ \wedge dd^c G^-.$$

It was proved in [37] that μ is an f-invariant Borel probability measure on K supported on J and f is ergodic with respect to μ. He Wu [271] proved that for μ almost every $p \in K$, the stable and unstable manifolds $W^s(p) \subset J^+$ and $W^u(p) \subset J^-$, are immersed complex manifolds and biholomorphically equivalent to $\mathbb{C}$. The same results were also obtained independently by Bedford, Lyubich and Smillie [33].

The topological entropy $h_{top}(f)$ is usually defined for a continuous mapping f from a compact metrizable space into itself. It will be convenient to extend this definition to the case of a proper mapping from a suitable locally compact space M into itself by defining $h_{top}(f)$ to be the *topological entropy* of the canonical extension of f over the one-point compactification M^* (cf. [91]). With this understanding, it follows that every polynomial automorphism g of complex plane has a well defined topological entropy $h_{top}(g) \geq 0$. If g is an elementary transformation or an affine transformation, then it follows easily that

$$\limsup_{n \to \infty} \frac{\log^+ P_n(g)}{n} = h_{top}(g) = 0, \tag{7.30}$$

where $P_n(g)$ is the number of periodic points of g with period exactly equal to n, not necessarily the smallest possible period. If g is cyclically reduced of degree d, Friedland and Milnor [91] proved

$$0 \leq h_{top}(g) \leq \limsup_{n \to \infty} \frac{\log^+ P_n(g)}{n} \leq \log d. \tag{7.31}$$

The inequality $0 \leq h_{top}(g)$ is true by definition, and the inequality

$$\limsup_{n \to \infty} \frac{\log^+ P_n(g)}{n} \leq \log d$$

is an immediate consequence of Theorem 7.13. Actually Friedland and Milnor showed that Katok's argument of Theorem 3.8 applies also to the mapping g, and conjectured that

$$\limsup_{n \to \infty} \frac{\log^+ P_n(g)}{n} = h_{top}(g)$$

for every polynomial automorphism g. Smillie [237] showed that the equality

$$\limsup_{n \to \infty} \frac{\log^+ P_n(g)}{n} = h_{top}(g) = \log d, \tag{7.32}$$

always holds in the cyclically reduced case.

Take $f \in \mathrm{Hol}(\mathbb{C}^2, \mathbb{C}^2)$ such that $f(0) = 0$. Let $A = f'(0)$ and let λ_1, λ_2 be the eigenvalues of A. If $(\lambda_1, \lambda_2) \in \mathrm{P}_{\mathrm{II}}^2$ is not algebraically resonant, Poincaré theorem shows that f can be linearized at the point $z = 0$. More precisely, there exists a unique germ ϕ of a biholomorphism with $\phi(0) = 0$ and $\phi'(0) = I$ such that in a neighborhood of 0 the identity

$$\phi \circ f = A\phi$$

holds. If $|\lambda_1|^2 < |\lambda_2| \leq |\lambda_1|$ and if $f \in \mathrm{Aut}(\mathbb{C}^2)$, then ϕ extends to a biholomorphic mapping $\phi : \mathrm{Att}(0) \longrightarrow \mathbb{C}^2$.

Assume that A is diagonal with eigenvalues $\lambda_1 = e^{2\pi i\theta}$, $|\lambda_2| < 1$. The matrix A satisfies the Brjuno condition if and only if

$$\sum_m 2^{-m} \log \left(\inf_{1 \leq k \leq 2^m - 1} |1 - e^{2k\pi i\theta}| \right)^{-1} < \infty, \tag{7.33}$$

which is true if there are positive constants c, ν such that

$$\left| \theta - \frac{p}{q} \right| \geq \frac{c}{|q|^\nu}$$

for all $p, q \in \mathbb{Z}$. Siegel and Brjuno proved that if A satisfies the Brjuno condition, then f can be linearized at the point $z = 0$. Further, if $f \in \mathrm{Aut}(\mathbb{C}^2)$, then either f is conjugate to A in the group $\mathcal{E}$ or there exists a a biholomorphic mapping $\phi : D \longrightarrow \Delta \times \mathbb{C}$, where D is the connected component of $(K^+)^o$ containing 0 such that $\phi \circ f = A\phi$ (see [86]).

7.5 Borel's theorem on $\mathbb{C}^m$

Let $z = (z_1, \cdots, z_m)$ be the natural coordinate system in $\mathbb{C}^m$ and set

$$\tau(z) = \|z\|^2 = \sum_{j=1}^m |z_j|^2,$$

$$\upsilon = dd^c\tau, \quad \sigma = d^c \log\tau \wedge (dd^c \log\tau)^{m-1}.$$

Let f be a meromorphic function on $\mathbb{C}^m$ and denote by D_f^a the divisor $\{f = a\}$ for a point $a \in \mathbb{P}^1$. Define the *valence function* by

$$\int_{r_0}^r t^{1-2m} \int_{(D_f^a)_{\sqrt{\tau}}[t]} \upsilon^{m-1} dt = \begin{cases} N(r, f) & : \quad a = \infty \\ N(r, \frac{1}{f-a}) & : \quad a \neq \infty \end{cases} \tag{7.34}$$

Then one has the well-known *Jensen Formula* (cf. [245]):

$$N\left(r, \frac{1}{f}\right) - N(r, f) = m(r; |f|) - m(r_0; |f|), \tag{7.35}$$

where

$$m(r; h) = \int_{\mathbb{C}^m_{\sqrt{\tau}}(r)} \log h\sigma.$$

Write $\log^+ x = \max\{\log x, 0\}$ for $x \geq 0$ and define the *compensation function* by

$$m(r, f) = \int_{\mathbb{C}^m_{\sqrt{\tau}}(r)} \log^+ |f|\sigma, \tag{7.36}$$

and define the *Nevanlinna characteristic function* by

$$T(r, f) = N(r, f) + m(r, f) \tag{7.37}$$

which is an increasing function of r. The following equation is the so called *first main theorem*

$$T\left(r, \frac{1}{f-a}\right) = T(r, f) + O(1) \quad (a \in \mathbb{P}^1). \tag{7.38}$$

It has the following relation with the Ahlfors-Shmizu characteristic $T_f(r)$ defined in § B.7:

$$T(r, f) = T_f(r) + O(1). \tag{7.39}$$

Also it is well-known that f is non-constant if and only if $T(r, f) \to \infty$ as $r \to \infty$, and rational if and only if

$$\lim_{r \to \infty} \frac{T(r, f)}{\log r} < \infty.$$

The following result belongs to Nevanlinna[183] and Vitter[262] (also see [41]):

Lemma 7.9 (The Lemma on the Logarithmic Derivative) *Let f be a meromorphic function on $\mathbb{C}^m$. Then for $j = 1, \cdots, m$,*

$$\| \ m\left(r, \frac{1}{f}\frac{\partial f}{\partial z_j}\right) \leq 17 \log^+(rT(r, f)),$$

where $\|$ indicates that the inequality holds only outside a set of finite measure on $\mathbb{R}^+$.

The following lemma is a generalization of a result of Nevanlinna[184] (in one variable version).

Lemma 7.10 *Let $f_j(j = 1, \cdots, n)$ be linearly independent meromorphic functions on $\mathbb{C}^m$ such that*

$$f_1 + \cdots + f_n = 1. \tag{7.40}$$

Define

$$T(r) = \max_k \{T(r, f_k)\}.$$

Then

$$\| \ T(r, f_j) \ < \ \sum_{k=1}^{n} N\left(r, \frac{1}{f_k}\right) - \sum_{k \neq j} N(r, f_k)$$

$$+ \ N(r, W) - N\left(r, \frac{1}{W}\right) + O\{\log(rT(r))\}, \quad 1 \leq j \leq n \tag{7.41}$$

where $W = \det(\frac{\partial^{\alpha_i} f_j}{\partial z^{\alpha_i}})$ is the Wronskian of $f_1, ..., f_n$ with $\alpha_i \in \mathbb{Z}_+^m$, and

$$0 = |\alpha_0| < |\alpha_1| \leq \cdots \leq |\alpha_{n-1}| \leq n - 1. \tag{7.42}$$

Proof. Since $f_1, \cdots, f_n$ are linearly independent, then there exists $\alpha_i \in \mathbb{Z}_+^m$ with the property (7.42) such that the Wronskian $W \not\equiv 0$. By (7.40) and

$$\frac{\partial^{\alpha_i} f_1}{\partial z^{\alpha_i}} + \cdots + \frac{\partial^{\alpha_i} f_n}{\partial z^{\alpha_i}} = 0 \quad i = 1, \cdots, n - 1,$$

we have $f_j = \frac{g_j}{g}, j = 1, \cdots, n$, where

$$g = \frac{W}{f_1 \cdots f_n} = \det\left(\frac{1}{f_j}\frac{\partial^{\alpha_i} f_j}{\partial z^{\alpha_i}}\right),$$

and g_j is the minor of the j-th term on the first row of g. Then

$$m(r, f_j) \leq m(r, g_j) + m\left(r, \frac{1}{g}\right)$$
$$= m(r, g_j) + m(r, g) + N(r, g) - N\left(r, \frac{1}{g}\right) + O(1), \qquad (7.43)$$

where we use Jensen's formula which also gives

$$N(r, g) - N\left(r, \frac{1}{g}\right) = -m(r; |g|) + O(1)$$
$$= -m(r; |W|) + \sum_{k=1}^{n} m(r; |f_k|) + O(1)$$
$$= N(r, W) - N\left(r, \frac{1}{W}\right) + \sum_{k=1}^{n}\left\{N\left(r, \frac{1}{f_k}\right) - N(r, f_k)\right\} + O(1). \qquad (7.44)$$

By using Lemma 7.9, we obtain

$$\| \quad m(r, g_j) + m(r, g) = O\{\log(rT(r))\}. \qquad (7.45)$$

Hence (7.41) follows from (7.43), (7.44), and (7.45). □

Corollary 7.2 *Assume that the conditions of Lemma 7.10 hold. Then*

$$\| \quad T(r, f_j) < \sum_{k=1}^{n} N\left(r, \frac{1}{f_k}\right) + (n-1)\sum_{k \neq j} N(r, f_k) + O\{\log(rT(r))\}, \quad j = 1, \cdots, n. \quad (7.46)$$

As an application of Lemma 7.10 and Corollary 7.2, we can prove the following result which is a generalization of the Borel theorem[48] in one variable case.

Theorem 7.17 ([119]) *Let $f_j \not\equiv 0, a_j (j = 0, \cdots, n)$ be meromorphic functions on $\mathbb{C}^m$ such that*

$$a_0 f_0 + a_1 f_1 + \cdots + a_n f_n = 0, \qquad (7.47)$$

and such that

$$\| \quad \sum_{k=0}^{n}\left\{N\left(r, \frac{1}{f_k}\right) + N(r, f_k)\right\} = o(\lambda(r)), \qquad (7.48)$$

$$\| \quad T(r, a_j) = o(\lambda(r)), \quad j = 0, 1, \cdots, n, \qquad (7.49)$$

for a continuous, increasing non-negative function $\lambda(r)$ of $r \in \mathbb{R}^+$. Assume that

$$\lim_{r \to \infty}\frac{\lambda(r)}{\log r} = +\infty. \qquad (7.50)$$

Then

1) There exist constants $c_j (0 \leq j \leq n, j \neq k)$ not all zero such that

$$\sum_{j \neq k} c_j a_j f_j = 0,$$

when

$$\lambda(r) = \max_{0 \leq j \leq n, j \neq k} \left\{ T\left(r, \frac{f_j}{f_k}\right) \right\}.$$

2) $a_j = 0, j = 0, 1, 2, \cdots, n$, when

$$\lambda(r) = \min_{0 \leq j,k \leq n, j \neq k} \left\{ T\left(r, \frac{f_j}{f_k}\right) \right\}.$$

Proof. To prove 1), without loss of generality, let $k = 0$. Then 1) is obvious if $a_0 \equiv 0$. Now assume $a_0 \not\equiv 0$. Set

$$g_j = -\frac{a_j f_j}{a_0 f_0}, \quad j = 1, \cdots, n. \tag{7.51}$$

Then $g_1 + \cdots + g_n = 1$. This is sufficient to prove that $g_1, \cdots, g_n$ are linearly dependent. If $g_1, \cdots, g_n$ are linearly independent, then Corollary 7.2 implies

$$\| \ T(r, g_j) < \sum_{k=1}^{n} N\left(r, \frac{1}{g_k}\right) + (n-1) \sum_{k \neq j} N(r, g_k) + O\{\log(r\hat{T}(r))\}, \tag{7.52}$$

where $\hat{T}(r) = \max_j \{T(r, g_j)\}$. Note that

$$\| \ T(r, g_j) = T\left(r, \frac{f_j}{f_0}\right) + o(\lambda(r)), \quad j = 1, \cdots, n \tag{7.53}$$

which gives

$$\| \ \hat{T}(r) = \lambda(r) + o(\lambda(r)). \tag{7.54}$$

It follows that

$$\begin{aligned}
\| \ T\left(r, \frac{f_j}{f_0}\right) \ &< \ \sum_{k=1}^{n} \left\{ N(r, f_0) + N\left(r, \frac{1}{f_k}\right) \right\} \\
&\quad + (n-1) \sum_{k \neq j} \left\{ N(r, f_k) + N\left(r, \frac{1}{f_0}\right) \right\} + o(\lambda(r)) \\
&= \ o(\lambda(r)).
\end{aligned}$$

Therefore

$$\| \ \lambda(r) = o(\lambda(r)).$$

This is impossible. Hence $g_1, \cdots, g_n$ are linearly dependent.

For 2), first we consider the case $n = 1$. Since $f_j \not\equiv 0 (j = 0, 1)$, then $a_0 \not\equiv 0$ and $a_1 \not\equiv 0$ if one of a_0 and a_1 is not identically zero. Hence

$$\| \ \lambda(r) \leq T\left(r, \frac{f_0}{f_1}\right) = T\left(r, \frac{a_1}{a_0}\right) \leq T(r, a_1) + T(r, a_0) + O(1) = o(\lambda(r))$$

which is impossible. Hence $a_0 = a_1 = 0$.

Assume that 2) holds for any integer $\leq n - 1$. Now we use induction to prove the assertion for case n. If one of $a_0, \cdots, a_n$ is identically zero, say $a_n \equiv 0$, then we have $a_0 f_0 + \cdots + a_{n-1} f_{n-1} = 0$. By the assumption of induction, we have $a_0 = \cdots = a_{n-1} = 0$. Hence we may assume $a_j \neq 0$ for $j = 0, 1, \cdots, n$. Set g_j by (7.51). If there exist constants $c_j (j = 1, \cdots, n)$ such that

$$c_1 g_1 + \cdots + c_n g_n = 0$$

which gives

$$c_1 a_1 f_1 + \cdots + c_n a_n f_n = 0,$$

by induction, then $c_j a_j = 0$ $(j = 1, \cdots, n)$. Thus $c_j = 0$ $(j = 1, \cdots, n)$. Hence $g_1, \cdots, g_n$ are linearly independent. Then Corollary 7.2 gives (7.52). Note that

$$\lambda(r) \leq \lambda_0(r) = \max_{1 \leq j \leq n} \left\{ T\left(r, \frac{f_j}{f_0}\right) \right\}.$$

Then (7.53) and (7.54) also hold for $\lambda(r) = \lambda_0(r)$. Thus we obtain

$$\| \lambda_0(r) = o(\lambda_0(r));$$

which is impossible.   □

Note that for an entire function g, $\mathrm{ord}(f) < \mathrm{ord}\,(e^g)$ implies

$$\| T(r, f) = o(T(r, e^g)).$$

See the proof of Lemma 3.4 in [38]. Hence Theorem 7.17 immediately implies the following $\mathbb{C}^m$ version of the Borel's theorem given by Berenstein-Chang-Li [38]:

Corollary 7.3 *Let $a_j(j = 0, \cdots, n)$ be meromorphic functions on $\mathbb{C}^n$ with $\mathrm{ord}(a_j) \leq \rho$ and $g_j(j = 0, \cdots, n)$ be entire functions on $\mathbb{C}^n$ such that $g_j - g_k(j \neq k)$ are transcendental functions or polynomials of degree higher than ρ. Then*

$$\sum_{k=0}^{n} a_k e^{g_k} = 0$$

holds only when $a_0 = \cdots = a_n = 0$.

For the meromorphic function $f_j \not\equiv 0$ $(j = 0, \cdots, n)$ on $\mathbb{C}^m$, set

$$\lambda(r) = \min_{j,k,j \neq k} \left\{ T\left(r, \frac{f_j}{f_k}\right) \right\}.$$

Assume that (7.50) and (7.48) hold. Then the set of all meromorphic functions a on $\mathbb{C}^m$ with

$$\| T(r, a) = o(\lambda(r))$$

forming a field, denote it by $\mathcal{M}(f_0, \cdots, f_n)$. Theorem 7.17 shows that $f_0, \cdots, f_n$ are linearly independent in $\mathcal{M}(f_0, \cdots, f_n)$. Note that the field $\mathbb{C}(z_1, \cdots, z_m)$ of rational functions in $\mathbb{C}^m$ is a subfield of $\mathcal{M}(f_0, \cdots, f_n)$. Thus, Theorem 7.17 yields the following corollary by Narasimhan [182]:

Corollary 7.4 *Let $g_0, \cdots, g_n$ be entire functions in $\mathbb{C}^m$ and none of $g_j - g_k(j \neq k)$ can be a constant. Then $e^{g_0}, \cdots, e^{g_n}$ are linearly independent in $\mathbb{C}(z_1, \cdots, z_m)$.*

7.6   Dynamics of composite mappings

For $f = (f_1, ..., f_m) \in \mathrm{Hol}(\mathbb{C}^n, \mathbb{C}^m)$, we denote the set of zero points of f by

$$Z(f) = \{z \in \mathbb{C}^n | f(z) = 0\}, \tag{7.55}$$

and write

$$e^f = (e^{f_1}, ..., e^{f_m}).$$

For another $g = (g_1, ..., g_m) \in \mathrm{Hol}(\mathbb{C}^n, \mathbb{C}^m)$, we define

$$f \odot g = (f_1 g_1, ..., f_m g_m),$$

and denote

$$\mathcal{A} = \{Q \odot e^\alpha \mid Q \in \mathrm{Pol}(\mathbb{C}^n, \mathbb{C}^m), \alpha \in \mathrm{Hol}(\mathbb{C}^n, \mathbb{C}^m)\}, \tag{7.56}$$

where $\mathrm{Pol}(\mathbb{C}^n, \mathbb{C}^m) \subset \mathrm{Hol}(\mathbb{C}^n, \mathbb{C}^m)$ is the set of polynomial mappings. Obviously,

$$\mathrm{Pol}(\mathbb{C}^n, \mathbb{C}^m) \subset \mathcal{A}.$$

For $P \in \mathrm{Hol}(\mathbb{C}^n, \mathbb{C}^m)$, write

$$\mathcal{A}_P = P + \mathcal{A} = \{P + a \mid a \in \mathcal{A}\}.$$

The set $Z(f)$ is said to be affine algebraic if $Z(f) = Z(P)$ for some $P \in \mathrm{Pol}(\mathbb{C}^n, \mathbb{C}^m)$, and will be called transcendental if $Z(f)(\neq \emptyset)$ is not affine algebraic. If $f \in \mathcal{A}$, then $Z(f)$ is affine algebraic. A theorem of Stoll [245] implies that $Z(f)$ is infinite if $Z(f)$ is transcendental. Obviously the set $\mathrm{Fix}(f)$ of fixed points of f is just the analytic set $Z(f - id)$, where $id : \mathbb{C}^m \to \mathbb{C}^m$ is the identity. If $f \in \mathcal{A}_{id}$, then $\mathrm{Fix}(f)$ is affine algebraic.

Definition 7.2 *If $f \in \mathrm{Hol}(\mathbb{C}^m, \mathbb{C}^m) - \mathcal{A}$, then $Z(f)$ is said to be quasitranscendental .*

Let $f : \mathbb{C}^m \longrightarrow \mathbb{C}^m$ be a holomorphic mapping such that $f^{-1}(a)$ is discrete for some $a \in \mathbb{C}^m$. Let $n(r, f = a)$ denote the algebraic number of points in $f^{-1}(a)$ and define

$$N(r, f = a) = \int_{r_0}^r n(t, f = a) \frac{dt}{t} \qquad (r > r_0 > 0) \tag{7.57}$$

$$p(z) = \left(1 - \frac{1}{(1 + \|z\|)^{m-1}}\right), \quad q_a(z) = \frac{1}{\|z - a\|^{2m-2}} \quad (z \in \mathbb{C}^m) \tag{7.58}$$

$$\lambda_a = \frac{1}{4\pi(m-1)}(p + q_a)(dd^c \tau)^{m-1} \tag{7.59}$$

where τ is defined by $\tau(z) = \|z\|^2$. Also define

$$m_f(r, a) = \frac{1}{2} \int_{\mathbb{C}^m_{\sqrt{\tau}}\langle r \rangle} f^* \lambda_a \wedge d^c \log \tau \tag{7.60}$$

$$D_f(r, a) = \frac{1}{2} \int_{\mathbb{C}^m_{\sqrt{\tau}}[r]} f^* \lambda_a \wedge dd^c \log \tau \tag{7.61}$$

Let $l : \mathbb{C}^m \longrightarrow \mathbb{P}^m$ be the inclusion, defined by

$$l(z_1, \cdots, z_m) = [1, z_1, \cdots, z_m],$$

and let Ω be the Fubini-Study form on $\mathbb{P}^m$. Then

$$\omega = l^*\Omega = dd^c \log(1 + \tau). \tag{7.62}$$

H. Wu[273] proved the first main theorem as follows:

$$T_{m,f}(r, \omega) = N(r, f = a) + m_f(r, a) - m_f(r_0, a) - D_f(r, a) + D_f(r_0, a), \tag{7.63}$$

which yields

$$N(r, f = a) \leq T_{m,f}(r, \omega) + D_f(r, a) + O(1). \tag{7.64}$$

By Theorem III of Carlson-Griffiths[58], we have

$$T_{m,f}(r, \omega) \leq c_\theta T_{l \circ f}(\theta r)^m, \tag{7.65}$$

for constants $\theta > 1$ and c_θ. Note that

$$T_{l \circ f}(r) = \int_{\mathbb{C}^m_{\sqrt{\tau}}(r)} \log\sqrt{1 + \|f\|^2} d^c \log \tau \wedge (dd^c \log \tau)^{m-1} + O(1). \tag{7.66}$$

Hence we obtain

$$N(r, f = a) \leq c_\theta (\log M(\theta r, f))^m + D_f(r, a) + O(1). \tag{7.67}$$

However, in general, we are unable to estimate the growth of $D_f(r, a)$ in terms of $\log M(r, f)$. Cornalba and Shiffman[65] constructed a holomorphic mapping $f : \mathbb{C}^2 \longrightarrow \mathbb{C}^2$ such that $\log M(r, f) = O(r^\varepsilon)$ and $n(r, f = 0) \geq e^r$. Stoll[249] obtained a Bézout estimate of the average rate of growth of the volumes of certain complex analytic varieties in terms of the growth of the defining functions.

Now let $P : \mathbb{C}^m \longrightarrow \mathbb{C}^m$ be a polynomial mapping and let $f \in \mathrm{Hol}(\mathbb{C}^m, \mathbb{C}^m)$. Assume that $Z(f - P)$ is discrete. Then (7.67) yields

$$N(r, f = P) \leq c_\theta (\log M(\theta r, f) + \log r)^m + D_{f-P}(r, 0) + O(1). \tag{7.68}$$

For $f, g \in \mathrm{Hol}(\mathbb{C}^m, \mathbb{C}^m)\text{-}\mathcal{A}_P$, we conjectured [120]

$$\limsup_{r \to \infty} \frac{N(r, f \circ g = P)}{\log M(r, g)} = +\infty. \tag{7.69}$$

For the case $m = 1$, see Rosenbloom[217] and Yang-Zheng[275].

Given $f, g \in \mathrm{Hol}(\mathbb{C}^m, \mathbb{C}^m)$. Let z_0 be a fixed point of $f \circ g$. Then $g \circ f(g(z_0)) = g(z_0)$ implies that $g(z_0)$ is a fixed point of $g \circ f$. If z_1, z_2 are distinct fixed points of $f \circ g$, then $g(z_1) \neq g(z_2)$. Hence we have injective mappings

$$g : \mathrm{Fix}(f \circ g) \longrightarrow \mathrm{Fix}(g \circ f) \quad \text{and} \quad f : \mathrm{Fix}(g \circ f) \longrightarrow \mathrm{Fix}(f \circ g). \tag{7.70}$$

A mapping $f = (f_1, \cdots, f_m) \in \mathrm{Hol}(\mathbb{C}^m, \mathbb{C}^m)$ is said to be *algebraically degenerate* on a ring κ if there exists a polynomial $P \in \kappa[x_1, \cdots, x_m]$ such that $P(f_1, \cdots, f_m) \equiv 0$. Otherwise f is said to be *algebraically non-degenerate* on κ.

Theorem 7.18 ([120]) *Take $f \in \text{Hol}(\mathbb{C}^m, \mathbb{C}^m)$ and $P = (P_1, \cdots, P_m) \in \text{Hol}(\mathbb{C}^m, \mathbb{C}^m)$ such that each P_k is a nonlinear polynomial and such that $f \circ P$ is algebraically non-degenerate on $\mathbb{C}[z_1, \cdots, z_m]$. Then either $\text{Fix}(P \circ f)$ or $\text{Fix}(f \circ P)$ is quasitranscendental.*

Proof. Assume that $\text{Fix}(P \circ f)$ and $\text{Fix}(f \circ P)$ are, to the contrary, not quasitranscendental. Then $P \circ f$ is of the following form

$$P(f(z)) = (z_1 + Q_1 e^{\alpha_1}, \cdots, z_m + Q_m e^{\alpha_m}), \tag{7.71}$$

where $Q_k \in \mathbb{C}[z_1, \cdots, z_m], \alpha_k \in \text{Hol}(\mathbb{C}^m, \mathbb{C}), k = 1, \cdots, m$. If there exists some k such that $Q_k \equiv 0$, then we have $P_k \circ f(z) = z_k$. Thus $P_k(f \circ P) = P_k$. This is impossible since $f \circ P$ is algebraically non-degenerate on $\mathbb{C}[z_1, \cdots, z_m]$. Hence $Q_k \not\equiv 0$ for $k = 1, \cdots, m$. Similarly we have

$$\alpha_k \not\equiv \text{const.}, \quad k = 1, \cdots, m.$$

Let $f \circ P$ assume the following form:

$$f(P(z)) = (z_1 + S_1 e^{\beta_1}, \cdots, z_m + S_m e^{\beta_m}), \tag{7.72}$$

where $S_k \in \mathbb{C}[z_1, \cdots, z_m], \beta_k \in \text{Hol}(\mathbb{C}^m, \mathbb{C}), k = 1, \cdots, m$. Since $f \circ P$ is algebraically non-degenerate on $\mathbb{C}[z_1, \cdots, z_m]$, we have

$$S_k \not\equiv 0, \quad \beta_k \not\equiv \text{const.}, \quad k = 1, \cdots, m.$$

Write

$$P_k(z) = \sum_n a_{k,n} z_1^{n_1} \cdots z_m^{n_m}, \quad a_{k,n} \in \mathbb{C} - \{0\}, n = (n_1, \cdots, n_m) \in \mathbb{Z}_+^m.$$

Then we have

$$
\begin{aligned}
P_k(f(P(z))) &= \sum_n a_{k,n}(z_1 + S_1 e^{\beta_1})^{n_1} \cdots (z_m + S_m e^{\beta_m})^{n_m} \\
&= P_k(z) + \sum_n \sum_{1 \leq i_l \leq n_l} a_{k,n,i} \exp(i_1 \beta_1 + \cdots + i_m \beta_m),
\end{aligned} \tag{7.73}
$$

where $i = (i_1, \cdots, i_m) \in \mathbb{Z}_+^m$ and

$$a_{k,n,i} = a_{k,n} \binom{n_1}{i_1} \cdots \binom{n_m}{i_m} z_1^{n_1 - i_1} \cdots z_m^{n_m - i_m} S_1^{i_1} \cdots S_m^{i_m} \not\equiv 0.$$

By (7.71) and (7.73), we obtain

$$\sum_n \sum_{1 \leq i_l \leq n_l} a_{k,n,i} \exp(i_1 \beta_1 + \cdots + i_m \beta_m) = Q_k \circ P \exp(\alpha_k \circ P), \tag{7.74}$$

If there exists some $(j_1, \cdots, j_m) \in \mathbb{Z}^m - \{0\}$ such that

$$j_1 \beta_1 + \cdots + j_m \beta_m = c = \text{const},$$

then

$$e^c = (e^{\beta_1})^{j_1} \cdots (e^{\beta_m})^{j_m} = \left(\frac{f_1 \circ P - z_1}{S_1}\right)^{j_1} \cdots \left(\frac{f_m \circ P - z_m}{S_m}\right)^{j_m},$$

that is, $f \circ P$ is algebraically degenerate on $\mathbb{C}[z_1, \cdots, z_m]$. This is a contradiction. Hence for any $(j_1, \cdots, j_m) \in \mathbb{Z}^m - \{0\}$, we have

$$j_1 \beta_1 + \cdots + j_m \beta_m \not\equiv \text{const.}$$

If $i_1 \beta_1 + \cdots + i_m \beta_m - \alpha_k \circ P = c = \text{const.}$, we have

$$\left(\frac{f_1 \circ P - z_1}{S_1} \right)^{i_1} \cdots \left(\frac{f_m \circ P - z_m}{S_m} \right)^{i_m} = e^c (P_k(f \circ P) - P_k)/Q_k \circ P$$

i.e., $f \circ P$ is algebraically degenerate on $\mathbb{C}[z_1, \cdots, z_m]$, which is a contradiction. Hence

$$i_1 \beta_1 + \cdots + i_m \beta_m - \alpha_k \circ P \not\equiv \text{const.}$$

By Theorem 7.17, (7.74) yields $Q_k \circ P \equiv 0$, a contradiction. $\qquad \square$

Remark. In one complex variable case, the transcendency of f implies that $f \circ P$ is algebraically non-degenerate on $\mathbb{C}[z]$, and the corresponding result of Theorem 7.18 was obtained by Rosenbloom[217].

We can say more about (7.70). Let M be a smooth manifold and take $f, g \in C(M, M)$. Set $h = f \circ g$ and $k = g \circ f$. Note that

$$g \circ h^n = k^n \circ g, \quad f \circ k^n = h^n \circ f, \tag{7.75}$$

for all $n \in \mathbb{Z}_+$. Similarly we can obtain injective mappings

$$g : \text{Fix}(h^n) \longrightarrow \text{Fix}(k^n) \quad \text{and} \quad f : \text{Fix}(k^n) \longrightarrow \text{Fix}(h^n), \tag{7.76}$$

for all $n \in \mathbb{Z}^+$, and hence

$$g : \text{Per}(h) \longrightarrow \text{Per}(k) \quad \text{and} \quad f : \text{Per}(k) \longrightarrow \text{Per}(h). \tag{7.77}$$

If f and g are open mappings, by using (7.75) it is easy to prove the following fact:

$$g : F_{uc}(h) \longrightarrow F_{uc}(k) \quad \text{and} \quad f : F_{uc}(k) \longrightarrow F_{uc}(h). \tag{7.78}$$

Conversely, if $g(x) \in F_{uc}(k)$ for some $x \in M$, then we see $h(x) = f(g(x)) \in F_{uc}(h)$, and hence $x \in F_{uc}(h)$ since $F_{uc}(h)$ is backward invariant by Theorem 1.23. Thus we obtain the following:

Proposition 7.5 *If f and g are continuous open self-mappings on a smooth manifold M, then $x \in F_{uc}(f \circ g)$ if and only if $g(x) \in F_{uc}(g \circ f)$.*

Further, we have the following:

Proposition 7.6 *Let f and g be continuous open self-mappings on a smooth manifold M. Let U_0 be a component of $F_{uc}(f \circ g)$ and let V_0 be the component of $F_{uc}(g \circ f)$ that contains $g(U_0)$. Then U_0 is wandering if and only if V_0 is wandering.*

Proof. Set $h = f \circ g$ and $k = g \circ f$. Let U_n be the component of $F_{uc}(h)$ containing $h^n(U_0)$ and let V_n be the component of $F_{uc}(k)$ containing $k^n(V_0)$. Then (7.75) imply

$$g(h^n(U_0)) = k^n(g(U_0)), \quad f(k^n(V_0)) = h^n(f(V_0)), \quad n = 1, 2, \ldots,$$

which yield

$$g(U_n) \subset V_n, \quad f(V_n) \subset U_{n+1}.$$

Therefore $U_n = U_m$ implies $V_n = V_m$, and $V_n = V_m$ gives $U_{n+1} = U_{m+1}$.   □

If f and g are nonlinear entire functions on $\mathbb{C}$, Baker and Singh [25], Bergweiler and Wang [40], Qiao[203], Poon and Yang [200] proved that the propositions also are true for Fatou sets $F(f \circ g)$ and $F(g \circ f)$. Generally, we don't know whether these are true. If f and g satisfy some Lipschitz condition, we can prove that Proposition 7.5 and 7.6 are true for sets $F_{equ}(f \circ g)$ and $F_{equ}(g \circ f)$.

Proposition 7.7 *Take $f, g \in \text{Diff}^\infty(M, M)$ and $p > 0$. Suppose that M is compact, orientable and that f, f^{-1}, g, g^{-1} are orientation preserving. Let μ be the measure induced by a volume form Ω of M. If f and g are measure preserving, then $x \in F_\mu^p(f \circ g)$ if and only if $g(x) \in F_\mu^p(g \circ f)$.*

Proof. Set $h = f \circ g$ and $k = g \circ f$. Take $x \in F_\mu^p(h)$. Then there exists a neighborhood U of x such that

$$\lim_{n \to \infty} \left\| \frac{1}{n} \sum_{j=0}^{n-1} \psi \circ h^j - \overline{\psi} \right\|_{p, U} = 0,$$

for every $\psi \in C_0(M)$. Since f is orientation preserving and since M is compact, then there is a positive number c such that $0 \leq f^*\Omega/\Omega \leq c$. Thus ,

$$c^{-1} \left\| \frac{1}{n} \sum_{j=0}^{n-1} \phi \circ k^j - \overline{\phi} \right\|_{p, g(U)}^p \leq \left\| \frac{1}{n} \sum_{j=0}^{n-1} \phi \circ k^j \circ g - \overline{\phi} \right\|_{p, U}^p$$

$$\leq \left\| \frac{1}{n} \sum_{j=0}^{n-1} \psi \circ h^j - \overline{\psi} \right\|_{p, U}^p$$

$$\to 0 \quad \text{as } n \to \infty,$$

where $\psi = \phi \circ g$ and $\overline{\psi} = \overline{\phi}$ since g is measure preserving. Therefore $g(x) \in F_\mu^p(k)$, i.e., $g(F_\mu^p(h)) \subset F_\mu^p(k)$. Similarly, we also have $f(F_\mu^p(k)) \subset F_\mu^p(h)$.

Conversely, if $g(x) \in F_\mu^p(k)$ for some $x \in M$, then we see $h(x) = f(g(x)) \in F_\mu^p(h)$, and hence $x \in F_\mu^p(h)$ since $F_\mu^p(h)$ is backward invariant by Theorem 2.18.   □

Consequently we can prove the following result:

Proposition 7.8 *Take $f, g \in \text{Diff}^\infty(M, M)$ and $p > 0$. Suppose that M is compact, orientable and that f, f^{-1}, g, g^{-1} are orientation preserving. Let μ be the measure induced by a volume form Ω of M. If f and g satisfy some Lipschitz condition, then $x \in F_p(f \circ g)$ if and only if $g(x) \in F_p(g \circ f)$.*

Related to the results of this section, see Hu and Yang [126].

Appendix A

Foundations of differentiable dynamics

In this appendix, we will introduce some notations, terminologies and basic facts used in dynamics.

A.1 Basic notations in differentiable geometry

Let $\mathbb{N}, \mathbb{Z}, \mathbb{Q}, \mathbb{R}$, and $\mathbb{C}$ denote the positive integer set, integer ring, rational number field, real number field, and complex number field, respectively. If κ is a set, define

$$\begin{aligned} \kappa^n &= \kappa \times \cdots \times \kappa \quad (n - \text{times}) \\ &= \{x = (x_1, ..., x_n) \mid x_i \in \kappa, 1 \le i \le n\}. \end{aligned}$$

Given a function $\tau : \kappa \longrightarrow \mathbb{R} \cup \{-\infty\}$, define the *open pseudoball*, the *closed pseudoball* and the *pseudosphere* of radius r respectively by

$$\kappa(r) = \kappa_\tau(r) = \{x \in \kappa \mid \tau(x) < r\},$$

$$\kappa[r] = \kappa_\tau[r] = \{x \in \kappa \mid \tau(x) \le r\},$$

$$\kappa\langle r\rangle = \kappa_\tau\langle r\rangle = \{x \in \kappa \mid \tau(x) = r\},$$

and define the intervals

$$\kappa(s, r) = \kappa_\tau(s, r) = \{x \in \kappa \mid s < \tau(x) < r\},$$

$$\kappa(s, r] = \kappa_\tau(s, r] = \{x \in \kappa \mid s < \tau(x) \le r\},$$

$$\kappa[s, r) = \kappa_\tau[s, r) = \{x \in \kappa \mid s \le \tau(x) < r\},$$

$$\kappa[s, r] = \kappa_\tau[s, r] = \{x \in \kappa \mid s \le \tau(x) \le r\}.$$

For instance, if $\tau = id : \kappa \longrightarrow \kappa$ is the identity mapping, we have

$$\mathbb{Z}^+ = \mathbb{N} = \mathbb{Z}(0, \infty), \quad \mathbb{Z}_+ = \mathbb{Z}[0, \infty), \quad \mathbb{R}^+ = \mathbb{R}(0, \infty), \quad \mathbb{R}_+ = \mathbb{R}[0, \infty).$$

We also often consider functions $d : \kappa \times \kappa \longrightarrow \mathbb{R} \cup \{-\infty\}$, say, distance functions. For this case, fix a point $x_0 \in \kappa$ and set $\tau(x) = d(x, x_0)$. Write

$$\kappa(x_0; r) = \kappa_\tau(r), \quad \kappa(x_0; s, r) = \kappa_\tau(s, r),$$

and so on.

Let M be a topological space. If A, B are subsets of M, A^o denotes interior points of A, $\overline{A}$ closure of A, $A^c = M - A$ the complement of A, $\partial A = \overline{A} - A^o$ boundary of A, and $A \subset\subset B$ denotes that A is relatively compact in B, i.e., $\overline{A}$ is compact in B. If N also is a topological space, let $C(M, N)$ be the space of continuous mappings from M into N. The *compact-open topology* on $C(M, N)$ is defined as follows: let K and U be compact and open sets in M and N respectively, and let

$$\mathcal{W}(K, U) = \{f \in C(M, N) | f(K) \subset U\}.$$

The open sets of the compact-open topology are then the unions of finite intersections of such $\mathcal{W}(K, U)$, i.e. $\{\mathcal{W}(K, U)\}$ is a subbase of $C(M, N)$. A basic fact is that, if M, N are both second countable, so is $C(M, N)$ in this topology.

Define a function $| \, | : \mathbb{Z}_+^m = (\mathbb{Z}_+)^m \longrightarrow \mathbb{Z}_+$ by

$$|\alpha| = \alpha_1 + \cdots + \alpha_m,$$

for $\alpha = (\alpha_1, \cdots, \alpha_m) \in \mathbb{Z}_+^m$ so that the sets $\mathbb{Z}_+^m[r], \mathbb{Z}_+^m\langle r \rangle$, and so on, are well-defined for $\tau = | \, |$. Then

$$\#\mathbb{Z}_+^m\langle r \rangle = \binom{m - 1 + r}{r},$$

$$\#\mathbb{Z}_+^m[r] = \sum_{k=0}^{r} \binom{m - 1 + k}{k} = \binom{m + r}{r},$$

where $\#S$ denotes the cardinal number of the set S. For coordinates $x = (x_1, \cdots, x_m)$ of $\mathbb{R}^m$, also define $\| \, \| : \mathbb{R}^m \longrightarrow \mathbb{R}_+$ by

$$\|x\| = (x_1^2 + \cdots + x_m^2)^{1/2},$$

and we will write

$$x^\alpha = x_1^{\alpha_1} \cdots x_m^{\alpha_m}.$$

Set

$$dx = (dx_1, ..., dx_m), \quad \frac{d}{dx} = \left(\frac{\partial}{\partial x_1}, ..., \frac{\partial}{\partial x_m} \right).$$

Then

$$dx^\alpha = dx_1^{\alpha_1} \cdots dx_m^{\alpha_m}, \quad \frac{d^\alpha}{dx^\alpha} = \left(\frac{d}{dx} \right)^\alpha = \left(\frac{\partial}{\partial x_1} \right)^{\alpha_1} \cdots \left(\frac{\partial}{\partial x_m} \right)^{\alpha_m} = \frac{\partial^{|\alpha|}}{\partial x_1^{\alpha_1} \cdots \partial x_m^{\alpha_m}}.$$

Let U be an open set of $\mathbb{R}^m$. Take $f \in C(U, \mathbb{R}^n)$. The mapping f is said to be of C^r, and write $f \in C^r(U, \mathbb{R}^n)$ if f is continuous for the case $r = 0$; or f is differentiable of order r, i.e., $\frac{d^\alpha f}{dx^\alpha} \in C(U, \mathbb{R}^n)$ for all $\alpha \in \mathbb{Z}_+^m[r]$ for the case $r \in \mathbb{Z}^+$; or f is smooth, i.e., $\frac{d^\alpha f}{dx^\alpha} \in C(U, \mathbb{R}^n)$ for all $\alpha \in \mathbb{Z}_+^m$ for the case $r = \infty$; or f is analytic, i.e., f can be expressed by a convergent power series in a neighborhood of each point of U for the case $r = \omega$.

Definition A.1 *A Hausdorff topological space M is called a C^s differential manifold if there exists a family $\mathcal{A} = \{(U_\alpha, \varphi_\alpha)\}_{\alpha \in \Lambda}$ which satisfies the following conditions:*

1) $\{U_\alpha\}$ is an open covering of M;

2) $\varphi_\alpha : U_\alpha \longrightarrow \varphi_\alpha(U_\alpha) \subset \mathbb{R}^m$ is a homeomorphism onto an open subset $\varphi_\alpha(U_\alpha)$ of $\mathbb{R}^m$ for each $\alpha \in \Lambda$;

3) $\varphi_\beta \circ \varphi_\alpha^{-1} \in C^s(\varphi_\alpha(U_\alpha \cap U_\beta), \mathbb{R}^m)$ if $U_\alpha \cap U_\beta \neq \emptyset$;

4) $\mathcal{A}$ is maximal, i.e., if $\varphi : U \longrightarrow \varphi(U) \subset \mathbb{R}^m$ is a homeomorphism from an open subset U of M onto an open subset $\varphi(U)$ of $\mathbb{R}^m$ such that if $U_\alpha \cap U \neq \emptyset$ for $\alpha \in \Lambda$, $\varphi \circ \varphi_\alpha^{-1} \in C^s(\varphi_\alpha(U_\alpha \cap U), \mathbb{R}^m)$ and $\varphi_\alpha \circ \varphi^{-1} \in C^s(\varphi(U_\alpha \cap U), \mathbb{R}^m)$, then $(U, \varphi) \in \mathcal{A}$.

A C^s differential manifold also is called a *topological manifold* if $s = 0$, a *smooth manifold* if $s = \infty$, and an *analytic manifold* if $s = \omega$. The family $\mathcal{A}$ in Definition A.1 is called a C^s differential structure (or atlas) of M. The pair $(U, \varphi) \in \mathcal{A}$ is called a *local coordinate system* (or chart) of M, $\varphi = (x_1, ..., x_m)$ is said to be a *local coordinate* on U, and U is said to be a *coordinate neighborhood*. In this book, C^s differential manifolds are always Hausdorff, with a countable basis, and, unless otherwise stated, connected and without boundary.

Assume that M and N are C^s differential manifolds with C^s differential structures $\mathcal{A} = \{(U_\alpha, \varphi_\alpha)\}$ and $\mathcal{B} = \{(V_\mu, \psi_\mu)\}$, respectively. A mapping $f : M \longrightarrow N$ is called C^r differentiable $(0 \leq r \leq s)$ at a point $p \in M$ if there exist $(U, \varphi) \in \mathcal{A}$ and $(V, \psi) \in \mathcal{B}$ such that $p \in U, f(p) \in V$, and such that $\psi \circ f \circ \varphi^{-1} : \varphi(U) \longrightarrow \psi(V)$ is C^r differentiable at the point $\varphi(p)$. The mapping f is called C^r differentiable on M if f is C^r differentiable at each point of M. Let $C^r(M, N)$ be the space of C^r differentiable mappings from M into N. A C^r diffeomorphism of a C^s differential manifold M onto another C^s differential manifold N is a homeomorphism f such that both f and f^{-1} are C^r differentiable. Denote the space of C^r diffeomorphisms of M onto N by $\text{Diff}^r(M, N)$. Also write $\text{Diff}^0(M, N) = \text{Hom}(M, N)$ for the space of homeomorphisms of M onto N. A diffeomorphism of M onto itself also is called a *differentiable transformation* of M (or, simply, a transformation). Usually, a C^∞ property also is said to be a *smoothness*.

Assume that M is a smooth manifold. We shall now define a *tangent vector* (or simply a *vector*) at a point p of M. Let $\mathcal{F}_p$ be the algebra of germs of differentiable functions of class C^∞ defined in a neighborhood of p. Let γ be a curve of class C^∞ which passes through the point p, i.e., there is a positive δ such that $\gamma : (-\delta, \delta) \longrightarrow M$ is C^∞ differentiable, and $\gamma(0) = p$. The vector tangent to the curve γ at p is a mapping $X_p : \mathcal{F}_p \longrightarrow \mathbb{R}$ defined by

$$X_p f = \frac{df(\gamma(t))}{dt}\Big|_{t=0}.$$

In other words, $X_p f$ is the derivative of f in the direction of the curve $\gamma(t)$ at $t = 0$. We often write the tangent vector as follows:

$$X_p = \frac{d\gamma(t)}{dt}\Big|_{t=0} = \dot{\gamma}(0).$$

The vector X_p satisfies the following conditions:

(i) $X_p(af + bg) = a \cdot X_p f + b \cdot X_p g$, for $f, g \in \mathcal{F}_p, a, b \in \mathbb{R}$;

(ii) $X_p(fg) = f(p) \cdot X_p g + g(p) \cdot X_p f$, for $f, g \in \mathcal{F}_p$.

The set of tangent vectors at p, denoted by $T(M)_p$, is called the *tangent space* of M at p. It is easy to prove that $T(M)_p$ is a vector space of dimension m, where m is the dimension of M. If $x_1, ..., x_m$ is a local coordinate system in a coordinate neighborhood U of p, then $(\frac{\partial}{\partial x_1})_p, ..., (\frac{\partial}{\partial x_m})_p$ is a basis of $T(M)_p$.

A *vector field* X on a manifold M is an assignment of a vector $X_p \in T(M)_p$ to each point p of M. The mapping $p \mapsto X_p$ forms a section of the *tangent bundle*

$$T(M) = \bigcup_{x \in M} T(M)_x.$$

If f is a differentiable function on M, then Xf is a function on M defined by

$$Xf(p) = X_p f.$$

A vector field X is called differentiable if Xf is differentiable for every differentiable function f. In terms of a local coordinate system $x_1, ..., x_m$, a vector field X may be expressed by

$$X = \sum_{i=1}^{m} \xi_i \frac{\partial}{\partial x_i},$$

where ξ_i are functions defined in the coordinate neighborhood, called the *components* of X with respect to $x_1, ..., x_m$. Then X is differentiable if and only if (hereafter iff) its components ξ_i are differentiable. If X, Y are vector fields on M, define the *Poisson bracket* $[X, Y]$ as a mapping from the ring of functions on M into itself by

$$[X, Y]f = X(Yf) - Y(Xf).$$

Then $[X, Y]$ is a vector field satisfying Jacobi's identity:

$$[[X, Y], Z] + [[Y, Z], X] + [[Z, X], Y] = 0,$$

for another vector field Z on M.

For a point p of M, the dual vector space $T^*(M)_p$ of the tangent space $T(M)_p$ is called the space of *covectors* or *cotangent space* at p. For $f \in \mathcal{F}_p$, the *total differential* $(df)_p$ of f at p is defined by

$$\langle X_p, (df)_p \rangle = X_p f \quad \text{for } X_p \in T(M)_p,$$

where $\langle , \rangle$ denotes the value of the second entry on the first entry as a linear functional on $T(M)_p$. Obviously, $(df)_p \in T^*(M)_p$. In fact, we have

$$T^*(M)_p = \{(df)_p \mid f \in \mathcal{F}_p\}.$$

If $x_1, ..., x_m$ is a local coordinate system in a neighborhood of p, then the total differentials $(dx_1)_p, ..., (dx_m)_p$ form a basis for $T^*(M)_p$. In fact, they form the dual basis of the basis $(\frac{\partial}{\partial x_1})_p, ..., (\frac{\partial}{\partial x_m})_p$ for $T(M)_p$.

An assignment of a covector at each point p is called a 1-*form* (*differential form of degree* 1), which forms a section of the *cotangent bundle*

$$T^*(M) = \bigcup_{x \in M} T^*(M)_x.$$

If f is a differentiable function on M, the total differential df of f is the section of $T^*(M)$ defined by $p \mapsto (df)_p$. Now we can extend $\langle,\rangle$ to be a linear relation between sections of $T(M)$ and $T^*(M)$ by setting

$$\langle X, df \rangle = Xf,$$

for vector fields X on M. Let

$$\bigwedge T^*(M)_p = \sum_{r=0}^{m} \bigwedge_r T^*(M)_p$$

be the exterior algebra over $T^*(M)_p$. An r-form ω is an assignment of an element of degree r in $\wedge T^*(M)_p$ to each point p of M, which forms a section of the r-th exterior power of the cotangent bundle:

$$\bigwedge_r T^*(M) = \bigcup_{x \in M} \bigwedge_r T^*(M)_x.$$

In a coordinate neighborhood of p, every r-form ω can be uniquely written as

$$\omega = \sum_{i_1 < \cdots < i_r} a_{i_1 \ldots i_r} dx_{i_1} \wedge \cdots \wedge dx_{i_r},$$

where $a_{i_1 \ldots i_r}$ are functions defined in the coordinate neighborhood of p and are called the *components* of ω with respect to $x_1, \ldots, x_m$. The r-form ω is called differentiable if $a_{i_1 \ldots i_r}$ are differentiable (this condition is independent of the choice of a local coordinate system). An r-form ω can be defined also as a skew-symmetric r-linear mapping

$$\omega : T(M)_p \times \cdots \times T(M)_p \longrightarrow \mathbb{R}$$

at each point p. The two definitions are related as follows. If $\omega_1, \ldots, \omega_r$ are 1-forms and $X_1, \ldots, X_r$ are vector fields, then

$$(\omega_1 \wedge \cdots \wedge \omega_r)(X_1, \ldots, X_r) = \frac{1}{r!} \det(\langle X_k, \omega_j \rangle). \tag{A.1}$$

Let

$$T_s^r(p) = (\bigotimes_r T(M)_p) \otimes (\bigotimes_s T^*(M)_p)$$

be the *tensor space of type* (r,s) at a point $p \in M$. An element of $T_s^r(p)$ is called a *tensor of type* (r,s), or *tensor of contravariant degree r and covariant degree s*. A *tensor field* K *of type* (r,s) on M is an assignment of a tensor $K(p) \in T_s^r(p)$ to each point p of M. The mapping $p \mapsto K(p)$ forms a section of the *tensor bundle of type* (r,s)

$$T_s^r = \bigcup_{x \in M} T_s^r(x).$$

In a coordinate neighborhood of p, every tensor field K of type (r,s) can be uniquely written as

$$K = \sum K_{j_1 \cdots j_s}^{i_1 \cdots i_r} \frac{\partial}{\partial x_{i_1}} \otimes \cdots \otimes \frac{\partial}{\partial x_{i_r}} \otimes dx_{j_1} \otimes \cdots \otimes dx_{j_s},$$

where $K_{j_1\cdots j_s}^{i_1\cdots i_r}$ are functions defined in the coordinate neighborhood of p and are called the *components* of K with respect to $x_1, ..., x_m$. The tensor K is called differentiable if $K_{j_1\cdots j_s}^{i_1\cdots i_r}$ are differentiable (this condition is independent of the choice of a local coordinate system). An r-form ω is nothing but a covariant tensor field of degree r which is skew-symmetric.

Generally, we denote the totality of smooth sections of a vector bundle $\pi : \mathbf{E} \longrightarrow M$ by $\Gamma(\mathbf{E})$ which is the subset of elements s of $C^\infty(M, \mathbf{E})$ with $\pi \circ s = id$. In particular,

$$A^r(M) = \Gamma(\bigwedge_r T^*(M))$$

is the totality of differentiable r-forms on M for each $r = 0, 1, ..., m$ with $A^0(M) = C^\infty(M, \mathbb{R})$. We set

$$A(M) = \sum_{r=0}^m A^r(M).$$

With respect to the exterior product, $A(M)$ forms an algebra over $\mathbb{R}$. *Exterior differentiation* d can be characterized as follows:
   (i) $d : A(M) \longrightarrow A(M)$ is an $\mathbb{R}$-linear mapping such that $d(A^r) \subset A^{r+1}$;
   (ii) df is the total differential for each $f \in A^0(M)$;
   (iii) $d(\omega \wedge \sigma) = d\omega \wedge \sigma + (-1)^r \omega \wedge d\sigma$ for $\omega \in A^r(M), \sigma \in A^s(M)$;
   (iv) $d^2 = 0$.
In terms of the local coordinate system, then

$$d\omega = \sum_{i_1 < \cdots < i_r} da_{i_1\ldots i_r} \wedge dx_{i_1} \wedge \cdots \wedge dx_{i_r}.$$

A *connection* in a vector bundle $\pi : \mathbf{E} \longrightarrow M$ is a mapping

$$\nabla : \Gamma(\mathbf{E}) \longrightarrow \Gamma(T^*(M) \otimes \mathbf{E}),$$

satisfying the following conditions:
   1) $\nabla(s_1 + s_2) = \nabla s_1 + \nabla s_2, \quad s_1, s_2 \in \Gamma(\mathbf{E})$;
   2) $\nabla(as) = da \otimes s + a\nabla s, \quad s \in \Gamma(\mathbf{E}), a \in C^\infty(M, \mathbb{R})$.
If $X \in \Gamma(T(M))$, i.e., X is a smooth vector field on M, and if $s \in \Gamma(\mathbf{E})$, define

$$\nabla_X s = \langle X, \nabla s \rangle,$$

where $\langle , \rangle$ denotes to take the value between $T(M)$ and $T^*(M)$. Then $\nabla_X s \in \Gamma(\mathbf{E})$, which called the *covariant derivative of s in the direction (or with respect to)* X. A section $s \in \Gamma(\mathbf{E})$ is said to be *parallel* if $\nabla s = 0$, and *parallel in the direction of a curve* γ if

$$\frac{\nabla s}{dt} = \nabla_{\dot\gamma(t)} s = 0.$$

A basic fact is that there exist connections on any vector bundle. The connections on the tangent bundle $T(M)$ are called *affine connections* on M. If $\dot\gamma(t)$ is parallel in the direction of γ, then γ is said to be a *geodesic*. The *torsion tensor field* T of an affine connection ∇ can be expressed as follows:

$$T(X,Y) = \nabla_X Y - \nabla_Y X - [X,Y],$$

where X and Y are smooth vector fields on M.

The connection ∇ on $\mathbf{E}$ induces a connection on the dual bundle $\mathbf{E}^*$, also denoted by ∇, defined by

$$d\langle s, s^* \rangle = \langle \nabla s, s^* \rangle + \langle s, \nabla s^* \rangle, \tag{A.2}$$

for $s \in \Gamma(\mathbf{E}), s^* \in \Gamma(\mathbf{E}^*)$, where $\langle, \rangle$ is the inner product between $\mathbf{E}$ and $\mathbf{E}^*$. If ∇_1, ∇_2 are connections on vector bundles $\mathbf{E}_1, \mathbf{E}_2$, respectively, they induce a connection ∇ on vector bundles $\mathbf{E} = \mathbf{E}_1 \otimes \mathbf{E}_2$ as follows:

$$\nabla(s_1 \otimes s_2) = \nabla_1 s_1 \otimes s_2 + s_1 \otimes \nabla_2 s_2, \tag{A.3}$$

for $s_1 \in \Gamma(\mathbf{E}_1), s_2 \in \Gamma(\mathbf{E}_2)$. Thus an affine connection ∇ on M induces the connection on the cotangent bundle $T^*(M)$, and further, induces the connection on the tensor bundle T_s^r of type (r,s), all denoted by ∇.

Let $(U;x)$ be a local coordinate system and $\{e_1, ..., e_q\}$ a local frame of $\mathbf{E}$ on U. We can write

$$\nabla e_\alpha = \sum_{\beta=1}^q \omega_\alpha^\beta \otimes e_\beta, \quad \omega_\alpha^\beta = \sum_{i=1}^m \Gamma_{\alpha i}^\beta dx_i, \quad \Omega_\alpha^\beta = d\omega_\alpha^\beta - \sum_{\nu=1}^q \omega_\alpha^\nu \wedge \omega_\nu^\beta, \tag{A.4}$$

where $\Gamma_{\alpha i}^\beta$ are smooth functions on U, called the *coefficients of the connection* ∇ with respect to the local coordinates x_i. Define

$$\omega = \begin{pmatrix} \omega_1^1 & \omega_1^2 & \cdots & \omega_1^q \\ \omega_2^1 & \omega_2^2 & \cdots & \omega_2^q \\ \cdots & \cdots & \cdots & \cdots \\ \omega_q^1 & \omega_q^2 & \cdots & \omega_q^q \end{pmatrix}, \quad \Omega = \begin{pmatrix} \Omega_1^1 & \Omega_1^2 & \cdots & \Omega_1^q \\ \Omega_2^1 & \Omega_2^2 & \cdots & \Omega_2^q \\ \cdots & \cdots & \cdots & \cdots \\ \Omega_q^1 & \Omega_q^2 & \cdots & \Omega_q^q \end{pmatrix}, \quad e = \begin{pmatrix} e_1 \\ \vdots \\ e_q \end{pmatrix}.$$

Then we have

$$\nabla e = \omega \otimes e, \quad \nabla e^* = - {}^t\omega \otimes e^*, \quad \Omega = d\omega - \omega \wedge \omega,$$

where e^* is the dual frame of e for $\mathbf{E}^*$ on U and where ${}^t\omega$ denotes the transpose of the matrix ω. Here ω and Ω are called the *connection matrix* and the *curvature matrix* of ∇ on U, respectively. Take vectors $X_p, Y_p \in T(M)_p, p \in U$. The curvature matrix Ω induces a linear mapping $R(X_p, Y_p)$ on the fibre $\pi^{-1}(p)$ as follows:

$$R(X_p, Y_p)s_p = \sum_{\alpha,\beta=1}^q a_\alpha \langle X_p \wedge Y_p, \Omega_\alpha^\beta|_p \rangle e_\beta|_p$$

for $s_p = \sum_{\alpha=1}^q a_\alpha e_\alpha|_p \in \pi^{-1}(p)$. If X and Y are smooth vector fields on M, then a linear mapping

$$R(X,Y) : \Gamma(\mathbf{E}) \longrightarrow \Gamma(\mathbf{E})$$

is defined by

$$(R(X,Y)s)(p) = R(X_p, Y_p)s_p, \quad s \in \Gamma(\mathbf{E}), \quad p \in M$$

which is called the *curvature operator* of ∇ and satisfies the following relation

$$R(X,Y) = \nabla_X \nabla_Y - \nabla_Y \nabla_X - \nabla_{[X,Y]}.$$

Write

$$\Omega_\alpha^\beta = \frac{1}{2} \sum_{k,l=1}^m R_{\alpha kl}^\beta dx_k \wedge dx_l,$$

with $R_{\alpha kl}^\beta = -R_{\alpha lk}^\beta$. It follows that

$$R\left(\frac{\partial}{\partial x_k}, \frac{\partial}{\partial x_l}\right) e_\alpha = \sum_{\beta=1}^q R_{\alpha kl}^\beta e_\beta.$$

Given a curve $\gamma(t) = (x_1(t), ..., x_m(t))$ on U and a section $s \in \Gamma(\mathbf{E})$, and wrote

$$\dot{\gamma}(t) = \sum_{i=1}^m \frac{dx_i}{dt} \frac{\partial}{\partial x_i}, \qquad s = \sum_{\alpha=1}^q \lambda_\alpha e_\alpha,$$

then s is parallel in the direction of a curve γ iff it satisfies the equation

$$\langle \dot{\gamma}(t), \nabla s \rangle = \sum_{\alpha=1}^q \left(\frac{d\lambda_\alpha}{dt} + \sum_{\beta,i} \Gamma_{\beta i}^\alpha \frac{dx_i}{dt} \lambda_\beta\right) s_\alpha = 0,$$

that is,

$$\frac{d\lambda_\alpha}{dt} + \sum_{\beta,i} \Gamma_{\beta i}^\alpha \frac{dx_i}{dt} \lambda_\beta = 0, \quad 1 \le \alpha \le q. \tag{A.5}$$

If ∇ is an affine connection on M, then $\{e_i = \frac{\partial}{\partial x_i} \mid 1 \le i \le m\}$ is a local frame of the tangent bundle $T(M)$ on U. Thus γ is a geodesic iff

$$\frac{d^2 x_i}{dt^2} + \sum_{j,k=1}^m \Gamma_{jk}^i \frac{dx_j}{dt} \frac{dx_k}{dt} = 0. \tag{A.6}$$

There is a well-defined *exponential mapping* $\exp_p$ from a small ball in $T(M)_p$ to a neighborhood of $p \in M$ defined by sending a tangent vector v to the point $\gamma(1)$, where γ is the geodesic with $\gamma(0) = p$ and $\dot{\gamma}(0) = v$.

A mapping $f \in C^1(M, N)$ induces a linear mapping

$$f^* : T^*(N)_{f(p)} \longrightarrow T^*(M)_p$$

as follows:

$$f^*((dg)_{f(p)}) = (d(g \circ f))_p,$$

for $(dg)_{f(p)} \in T^*(N)_{f(p)}$, and extends a linear mapping

$$f^* : A(N) \longrightarrow A(M)$$

as follows: in terms of a local coordinates $u_1, ..., u_n$ $(n = \dim N)$ on N, if we write a s-form $\eta \in A^s(N)$ as

$$\eta = \sum_{j_1 < \cdots < j_s} b_{j_1 \ldots j_s} du_{j_1} \wedge \cdots \wedge du_{j_s},$$

then

$$f^*\eta = \sum_{j_1 < \cdots < j_s} b_{j_1 \ldots j_s} \circ f \, df_{j_1} \wedge \cdots \wedge df_{j_s},$$

where $f_j = u_j \circ f$.

The conjugate mapping

$$f_* : T(M)_p \longrightarrow T(N)_{f(p)}$$

of f^*, i.e.,

$$\langle f_* X, \alpha \rangle = \langle X, f^* \alpha \rangle,$$

for $X \in T(M)_p, \alpha \in T^*(N)_{f(p)}$, is called the differential induced by f. If $g \in C^1(N, L)$, we have the following facts

$$(g \circ f)_* = g_* \circ f_*, \quad (g \circ f)^* = f^* \circ g^*. \tag{A.7}$$

Also we have

$$f^*\eta(X_1, ..., X_s) = \eta(f_* X_1, ..., f_* X_s)$$

for vector fields $X_1, ..., X_s$ on M and for a s-form $\eta \in A^s(N)$. Thus f determines an element of $T(N)_{f(p)} \otimes T^*(M)_p$, denoted by $(df)_p$, such that $f_* X = \langle X, (df)_p \rangle$ and $f^*\alpha = \langle (df)_p, \alpha \rangle$ for $X \in T(M)_p, \alpha \in T^*(N)_{f(p)}$.

Let $x = (x_1, ..., x_m)$ $(m = \dim M)$ be local coordinates at p and let $u = (u_1, ..., u_n)$ be local coordinates at $f(p)$. Then

$$\left\langle \left(\frac{\partial}{\partial x_i} \right)_p, (df)_p \right\rangle = f_* \left(\frac{\partial}{\partial x_i} \right)_p = \sum_{j=1}^n \frac{\partial f_j}{\partial x_i}(p) \left(\frac{\partial}{\partial u_j} \right)_{f(p)},$$

$$\left\langle (df)_p, (du_j)_{f(p)} \right\rangle = f^*((du_j)_{f(p)}) = (df_j)_p = \sum_{i=1}^m \frac{\partial f_j}{\partial x_i}(p)(dx_i)_p,$$

and hence

$$(df)_p = \sum_{i=1}^m \sum_{j=1}^n \frac{\partial f_j}{\partial x_i}(p) \left(\frac{\partial}{\partial u_j} \right)_{f(p)} \otimes (dx_i)_p.$$

Thus there is a matrix, called the *Jacobi's matrix* of f at p and denoted by $\frac{df}{dx}(p)$, such that

$$\left(f_* \left(\frac{\partial}{\partial x_1} \right)_p, ..., f_* \left(\frac{\partial}{\partial x_m} \right)_p \right) = \left(\left(\frac{\partial}{\partial u_1} \right)_{f(p)}, ..., \left(\frac{\partial}{\partial u_n} \right)_{f(p)} \right) \frac{df}{dx}(p),$$

or simply write

$$f_* \left(\frac{d}{dx} \right)_p = \left(\frac{d}{du} \right)_{f(p)} \frac{df}{dx}(p).$$

In the sequel, we will identify df with f_* or $\frac{df}{dx}(p)$.

A point $p \in M$ is called a *critical point* of f if

$$\text{rank}((df)_p) < \min\{\dim M, \dim N\}.$$

The set of critical points of f is denoted by $C_f = Cri(f)$, and the range $f(C_f)$ of f on the critical points is called the *wrinkle*. Moreover $w \in N$ is called a *regular value* if for all $x \in f^{-1}(w)$ the differential $(df)_x$ has maximal rank. Otherwise w is called a *singular value*. If $w \in N$ is a regular value, then $f^{-1}(w)$ is a submanifold of M whose dimension is $\dim M - \dim N$.

Theorem A.1 (Sard) *If $\dim M = \dim N$, then the Lebesgue measure of $f(C_f)$ in N is zero. The Lebesgue measure of $f(C_f)$ in N is also zero if $M, N \in C^\infty$ and $f \in C^\infty(M, N)$.*

A mapping $f \in C^\infty(M, N)$ is said to be an *immersion* if the differential

$$(df)_p : T(M)_p \longrightarrow T(N)_{f(p)}$$

is injective for all $p \in M$. The mapping f is said to be an *embedding* if f is an injective immersion which has a continuous inverse $f^{-1} : f(M) \subset N \longrightarrow M$. The classic Whitney's Theorem (see [111]) states that if $\dim M = m$ then there exists an embedding

$$w : M \longrightarrow \mathbb{R}^{2m+1}.$$

A.2   Dynamical systems

Definition A.2 *Let M be a set and let κ be an additive semigroup. Given a mapping $f : M \times \kappa \longrightarrow M$, write $f^t(x) = f(x, t)$. If f^t satisfies the following axiom:*
    1) identity axiom: $f^0 = id$, where 0 is the unit of κ;
    2) additive axiom: $f^t \circ f^s = f^{t+s}$, for all $t, s \in \kappa$,
then (M, κ, f) or simply $\mathcal{F} = \{f^t\}_{t \in \kappa}$ is called a dynamical system (or simply DS) defined on M, or an action of the semigroup κ on M, or a κ-flow, and M is called the phase space of the action.

Let $\text{Map}(M, M)$ be the set of self-mappings on the set M. Then $\text{Map}(M, M)$ is a semigroup under the composition:

$$(f, g) \in \text{Map}(M, M) \times \text{Map}(M, M) \longmapsto f \circ g \in \text{Map}(M, M).$$

Thus a dynamical system (M, κ, f) is a semigroup homomorphism

$$f : \kappa \longrightarrow \text{Map}(M, M)$$

with $f^0 = id$.

Lemma A.1 *Let G be a group. Then*
    1) every semigroup homomorphism $\sigma : \mathbb{R}_+ \longrightarrow G$ can be extended in a unique way to a group homomorphism of $\mathbb{R}$ into G;
    2) if G is finite, every semigroup homomorphism $\sigma : \mathbb{R}_+ \longrightarrow G$ is trivial.

Proof. 1) follows by noting that for $t < 0$, $\sigma(t) = \sigma(-t)^{-1}$ is the unique extension. To prove 2), extend σ to a group homomorphism $\sigma : \mathbb{R} \longrightarrow G$, and let K be the kernel of σ. Let k be the order of G. Then $kt \in K$ for all $t \in \mathbb{R}$, and hence $\mathbb{R} = k\mathbb{R} \subset K \subset \mathbb{R}$, that is, σ is trivial. $\qquad\square$

Given a dynamical system $\mathcal{F} = \{f^t\}_{t\in\kappa}$ defined on M. The mapping $t \mapsto f^t(x)$ is called the *motion* of the point x (any such mapping will also be called a motion of the DS). The *(phase) trajectory* of a point x or a motion $t \mapsto f^t(x)$ is the set $\mathcal{F}(x) = \{f^t(x)\}_{t\in\kappa}$. Any such set is called a trajectory of the DS.

In order to analyze a system mathematically, one needs to have some structure on M and restrictions on the DS f. This will either be the structure of a topological space or of a manifold. There are four major cases:

    (i) M is a measure space and f is measure-preserving, i.e., the case of *ergodic theory*;

    (ii) M is a topological space and f is continuous, i.e., the case of *topological dynamics*;

    (iii) M is a differentiable manifold and f is differentiable, i.e., the case of *differentiable dynamics*; and

    (iv) M is a complex manifold and f is holomorphic, i.e., the case of *complex dynamics*.

In this section, we explain the case (ii) and (iii) more clearly. Other cases will be discussed in the following sections. Let M be a C^s differential manifold and let $0 \le r \le s \le \infty$. First we let $\kappa = \mathbb{R}_+$. Now C^r differentiability of f and Definition A.2 imply that

$$f : M \times \mathbb{R}_+ \longrightarrow M$$

is a C^r differentiable mapping. Then for every $t \in \mathbb{R}_+$, $f^t \in C^r(M, M)$. Thus we obtain a semigroup homomorphism

$$f : \mathbb{R}_+ \longrightarrow C^r(M, M).$$

The actions of $\mathbb{R}_+$ are usually called *semiflows*.

If the image of the homomorphism contains in the group $\text{Diff}^r(M, M)$. By Lemma A.1, it is extended to a group homomorphism

$$f : \mathbb{R} \longrightarrow \text{Diff}^r(M, M).$$

Thus we obtain a $\mathbb{R}$-flow $\{f^t\}$. Usually a $\mathbb{R}$-flow also is said to be a *(phase) flow* or a 1-*parameter group of differentiable transformations* of M. Flows and semiflows are known as *dynamical systems with continuous time*.

If $\kappa = \mathbb{Z}_+$, C^r differentiability of f means that for every $t \in \mathbb{Z}_+$, $f^t \in C^r(M, M)$. This reduces to the fact that f is C^r differentiable if and only if $f^1 \in C^r(M, M)$. Thus we obtain a semigroup homomorphism

$$f : \mathbb{Z}_+ \longrightarrow C^r(M, M).$$

The actions of $\mathbb{Z}_+$ are usually called a *cascade* which is used to contrast it to a "semiflow"(or "flow"). In this case, we often use n and nearby letters of the alphabet instead of $t, s, \ldots$. Obviously, a cascade $\{f^n\}_{n\in\mathbb{Z}_+}$ is given by the *iterate* of the mapping f^1, i.e.,

$$f^0 = id, \quad f^n = f^{n-1} \circ f^1 = f^1 \circ f^{n-1}(n > 0).$$

For convenience, we will identify f with f^1, and say that the cascade is generated by f.

If $f = f^1 \in \mathrm{Diff}^r(M, M)$, let the notation f^n denotes the n-th iterate of the map f when $n > 0$ and the $(-n)$-th iterate of f^{-1} when $n < 0$. Then we obtain a group homomorphism

$$f : \mathbb{Z} \longrightarrow \mathrm{Diff}^r(M, M).$$

The groups and semigroups formed by iterates are also known as *dynamical systems with discrete time*.

Let us consider flows more carefully. Each phase flow $\{f^t\}$ induces a smooth vector field X as follows: For every point $p \in M$, $X(p)$ is the vector tangent to the curve $x(t) = f^t(p)$ at $t = 0$, i.e.,

$$X(p) = \frac{dx(t)}{dt}\Big|_{t=0} = \dot{x}(0).$$

The vector field X is called the *phase velocity field of a flow* $\{f^t\}$, or the *generating field* (or *infinitesimal generator*) of the 1-parameter group $\{f^t\}$. The latter completely determines the flow: for a fixed p and variable t, $x(t) = f^t(p)$ satisfies

$$\dot{x}(t) = \frac{dx(t)}{dt} = X(x(t)). \tag{A.8}$$

so that $f^t(p)$ is a solution of the differential equation

$$\dot{x} = X(x) \tag{A.9}$$

with initial condition p. Hence a trajectory $\mathcal{F}(x) = \{f^t(x)\}_{t \in \kappa}$ of a point x or a motion $t \mapsto f^t(x)$ is the same as the solution of the differential equation (A.9). The equation (A.9) is called an *autonomous equation*. Generally, an equation whose right-hand side also depends on time, e.g.,

$$\dot{x} = X(t, x), \tag{A.10}$$

is called a *nonautonomous equation*.

Let $U \subset M$ be a coordinate neighborhood with coordinates $(x_1, ..., x_m)$ and write

$$X = \sum_{i=1}^{m} \xi_i \frac{\partial}{\partial x_i}. \tag{A.11}$$

If our initial point p is represented by coordinates $(x_1^0, ..., x_m^0)$, then the differential equation (A.9) with initial condition p is the system of first-order ordinary differential equations

$$\frac{dx_i}{dt} = \xi_i(x_1, ..., x_m) \tag{A.12}$$

with initial conditions

$$x_i(0) = x_i^0, \quad i = 1, ..., m. \tag{A.13}$$

Conversely, let X be a smooth vector field on a manifold M. A curve $x(t)$ in M is called an *integral curve* of X if, for every parameter value t, the vector $X(x(t))$ is tangent to the curve at $x(t)$. We know from the standard theory of ordinary differential equations that for any point $p \in M$, there is an unique integral curve $x(t)$ of X, defined for $|t| < \varepsilon$ for some $\varepsilon > 0$, such that $x(0) = p$. More generally, one has the following properties:

Proposition A.1 *Let X be a smooth vector field on a manifold M. For any point $p \in M$, there exist a neighborhood U of p, a positive number ε and a mapping $f : U \times (-\varepsilon, \varepsilon) \longrightarrow M$ such that*

1) for each $t \in (-\varepsilon, \varepsilon)$, $f^t : x \rightarrow f(x, t)$ is a diffeomorphism of U onto the open set $f^t(U)$ of M;

2) if $t, s, t + s \in (-\varepsilon, \varepsilon)$, and if $x, f^s(x) \in U$, then $f^{t+s}(x) = f^t(f^s(x))$.

Proof ([145]). Let $x_1, ..., x_m$ be a local coordinate system in a neighborhood W of p such that $x_i(p) = 0$. Given X in W by (A.11) and consider the system (A.12) of first-order ordinary differential equations. By the fundamental theorem for systems of ordinary differential equations, there exists a unique set of functions $f_1(x, t), ..., f_m(x, t)$, defined for $x = (x_1, ..., x_m)$ with $|x_i| < \delta_1$ and for $|t| < \varepsilon_1$, which forms a solution of the differential equations for each fixed x and satisfy the initial conditions:

$$f_i(x, 0) = x_i, \quad i = 1, ..., m.$$

For $|t| < \varepsilon_1$, and $x \in U_1 = \{x \mid |x_i| < \delta_1\}$, set

$$f^t(x) = (f_1(x, t), ..., f_m(x, t)).$$

If $|t|, |s|, |t + s| < \varepsilon_1; x, f^s(x) \in U_1$, then the functions $g_i(t) = f_i(x, t + s)$ are easily seen to be a solution of the differential equations for the initial conditions $g_i(0) = f_i(x, s)$. By the uniqueness of the solution, we have $g_i(t) = f_i(f^s(x), t)$. This proves that

$$f^{t+s}(x) = f^t(f^s(x)).$$

Since f^0 is the identity transformation of U_1, there exist $\delta > 0$ and $\varepsilon > 0$ such that, for $U = \{x \mid |x_i| < \delta\}$, then $f^t(U) \subset U_1$ if $|t| < \varepsilon$. Hence

$$f^{-t}(f^t(x)) = f^t(f^{-t}(x)) = f^0(x) = x \quad \text{for every } x \in U, |t| < \varepsilon.$$

This proves that f^t is a diffeomorphism of U for $|t| < \varepsilon$. From the construction of f^t, it is obvious that f^t induces the given vector field X in U. $\square$

The family $\{f^t\}$ in Proposition A.1 is called a *local 1-parameter group of local transformations* defined on $U \times (-\varepsilon, \varepsilon)$, which also is called a local 1-parameter group generated by X in a neighborhood of p. If there exists a global 1-parameter group of transformations of M which induces X, then we say that X is *complete*. We also say that the field X defines a dynamical system. If $f^t(x)$ is defined on $M \times (-\varepsilon, \varepsilon)$ for some ε, then X is complete. On a compact manifold M, every vector field X is complete.

Definition A.3 *For $f \in C^r(M, M), g \in C^r(N, N)$, f and g are C^k-conjugate $(k \leq r)$ if there exist a C^k diffeomorphism $h : M \longrightarrow N$ such that*

$$g = h \circ f \circ h^{-1}.$$

If h is a homeomorphism, the mappings f and g are topologically conjugate. If M and N are smooth, and if h is a smooth diffeomorphism, then we speak of smooth conjugacy.

Usually C^k conjugacy also is called C^k equivalence or C^k isomorphism, in other words, which means that f differs from g by a C^k coordinate change. This certainly looks like a natural equivalence relation in differentiable dynamics. Obviously, we have

$$g^n = h \circ f^n \circ h^{-1}, \quad n \in \mathbb{Z}_+.$$

Definition A.4 *Two dynamical systems $\{f^t\}_{t \in \kappa}$ and $\{g^t\}_{t \in \kappa}$ with phase spaces M and N are isomorphic or conjugate if there exist a bijection $h : M \longrightarrow N$ such that*

$$h \circ f^t = g^t \circ h \quad \text{for all } t \in \kappa.$$

If $\{f^t\}_{t \in \kappa}$ and $\{g^t\}_{t \in \kappa}$ are topological DSs, and if h is a homeomorphism, then h is called an isomorphism or conjugacy (of the DSs), and the DSs are said to be (topologically) isomorphic or (topologically) conjugate. If the DSs are C^r, and if h is a C^k ($k \leq r$) diffeomorphism, then we speak of C^k isomorphism or C^k conjugacy, and smooth isomorphism or smooth conjugacy if $k = \infty$.

Proposition A.2 *Let $\{f^t\}$ and $\{g^t\}$ be the 1-parameter group of transformations generated by a vector field X on M and a vector field Y on N, respectively. Then $\{f^t\}$ and $\{g^t\}$ are C^r-conjugate iff there is an $h \in \mathrm{Diff}^r(M, N)$ such that $Y = h_* X$, i.e., there being a r-times differentiable change of coordinates $y = h(x)$ which transforms the differential equation $\dot{x} = X(x)$ of f^t into that $\dot{y} = Y(y)$ of g^t.*

Proof. ($\Leftarrow$) It is clear that $h \circ f^t \circ h^{-1}$ is a 1-parameter group of transformations. To show that it induces the vector field $h_* X$, let q be an arbitrary point of N and $p = h^{-1}(q)$. Since f^t induces X, the vector $X_p \in T(M)_p$ is tangent to the curve $x(t) = f^t(p)$ at $p = x(0)$. It follows that the vector

$$(h_* X)_q = h_*(X_p) \in T(N)_q$$

is tangent to the curve $y(t) = h \circ f^t(p) = h \circ f^t \circ h^{-1}(q)$ at $q = y(0)$.

($\Rightarrow$) Recall C^r-conjugacy of $\{f^t\}$ and $\{g^t\}$ means that there is $h \in \mathrm{Diff}^r(M, N)$ such that $h(f^t(x)) = g^t(h(x))$. Differentiate this equation with respect to t to obtain

$$\{(dh)_{f^t(x)} \frac{d}{dt} f^t(x)\}|_{t=0} = \frac{d}{dt} g^t(h(x))|_{t=0} \tag{A.14}$$

or

$$(dh)_x X(x) = Y(h(x)), \tag{A.15}$$

since $f^0 = \mathrm{id}$. Now consider the change of coordinates $y = h(x)$ applied to $\dot{x} = X(x)$. We find

$$\dot{y} = (dh)_x \dot{x} = (dh)_x X(x) = Y(h(x)) = Y(y) \tag{A.16}$$

as required. Thus when h exhibits the conjugacy of f^t and g^t, the differential mapping h_* transforms the vector field $X(x)$ into $Y(y)$ with $y = h(x)$. $\square$

Let X be a vector field on M, and f^t a local 1-parameter group of local transformations generated by X in a neighborhood U of p. We obtain a linear isomorphism

$$\tilde{f}^t = (\bigotimes_r (df^t)_{f^{-t}(p)}) \otimes (\bigotimes_k (f^{-t})_p^*) : T_k^r(f^{-t}(p)) \longrightarrow T_k^r(p)$$

between the tensor spaces of type (r, k). Given a tensor field K of type (r, k) on M, we define a tensor $\tilde{f}^t K(p)$ by

$$\tilde{f}^t K(p) = \tilde{f}^t(K(f^{-t}(p))).$$

The *Lie derivative* $L_X K$ of K with respect to X at p is defined by

$$L_X K(p) = \lim_{t \to 0} \frac{1}{t}\{K(p) - \tilde{f}^t K(p)\} = -\frac{d}{dt}\tilde{f}^t K(p)|_{t=0}.$$

It is linear and satisfies (see [145])

$$\frac{d}{dt}\tilde{f}^t K = \lim_{s \to 0} \frac{1}{s}\{\tilde{f}^{t+s}K - \tilde{f}^t K\} = -\tilde{f}^t L_X K, \tag{A.17}$$

$$L_X(K \otimes K') = (L_X K) \otimes K' + K \otimes (L_X K'), \tag{A.18}$$

$$L_X L_Y - L_Y L_X = L_{[X,Y]}, \tag{A.19}$$

$$d \circ L_X = L_X \circ d, \tag{A.20}$$

for all tensor fields K, K' and for all vector fields X, Y.

If $K = Y$ is a vector field on M, then $\tilde{f}^t = f_*^t$, and (see [145])

$$\frac{d}{dt}f_*^t Y = \lim_{s \to 0} \frac{1}{s}\{f_*^{t+s}Y - f_*^t Y\} = -f_*^t[X, Y]. \tag{A.21}$$

Thus the *Lie derivative* $L_X Y$ of Y with respect to X is just $[X, Y]$.

If $K = \omega$ is a r-form, then $\tilde{f}^t \omega = (f^{-t})^* \omega$. Hence (A.17) gives

$$\frac{d}{dt}f^{t*}\omega = f^{t*}L_X \omega. \tag{A.22}$$

Here $L_X \omega$ satisfies the following formula:

$$(L_X \omega)(Y_1, ..., Y_r) = X(\omega(Y_1, ..., Y_r)) - \sum_j \omega(Y_1, ..., [X, Y_j], ..., Y_r). \tag{A.23}$$

If $K = \phi$ is a function on M, then

$$L_X \phi = \lim_{t \to 0} \frac{1}{t}\{\phi - \phi \circ f^{-t}\} = X\phi.$$

Remark. Let M be a set and let κ be a multiplicative semigroup. Given a mapping $f : M \times \kappa \longrightarrow M$, write $f_t(x) = f(x, t)$. If f_t satisfies the following axiom:

1) identity axiom: $f_1 = id$, where 1 is the unit of κ;

2) multiplicative axiom: $f_t \circ f_s = f_{ts}$, for all $t, s \in \kappa$,

then (M, κ, f) or simply $\mathcal{F} = \{f_t\}_{t \in \kappa}$ is called an action of the semigroup κ on M. If κ is a topological group, then (M, κ, f) is called a *topological transformation group* . Here we omit the discussion for topological transformation groups.

Remark. Let M and κ be sets. For a mapping $f : M \times \kappa \longrightarrow M$, we also write $f_t(x) = f(x, t)$ and denote $\mathcal{F} = \{f_t \mid t \in \kappa\}$. Set

$$\kappa^{\mathbb{N}} = \{(t_1, t_2, ...) \mid t_n \in \kappa, \ n \in \mathbb{N}\},$$

and define the *shift transformation* $\sigma : \kappa^{\mathbb{N}} \longrightarrow \kappa^{\mathbb{N}}$ by $\sigma(t)_n = t_{n+1}$. The dynamical system $(\kappa^{\mathbb{N}}, \sigma)$ is called the *one-sided shift on κ* or the *full one-sided κ-shift*. The two-sided sequence space is

$$\kappa^{\mathbb{Z}} = \{(..., t_{-2}, t_{-1}, t_0, t_1, t_2, ...) \mid t_n \in \kappa, \ n \in \mathbb{Z}\},$$

and the shift transformation $\sigma : \kappa^{\mathbb{Z}} \longrightarrow \kappa^{\mathbb{Z}}$ can be defined similarly by $\sigma(t)_n = t_{n+1}$. The dynamical system $(\kappa^{\mathbb{Z}}, \sigma)$ is called the *two-sided shift on κ* or the *full κ-shift*. Take $t \in \kappa^{\mathbb{N}}$ and write

$$t = (t_1, t_2, ...) = (t^n, \sigma^n(t)), \ t^n = (t_1, t_2, ..., t_n), \ ^n t = (t_n, t_{n-1}, ..., t_1).$$

Then t induces two dynamical systems on M as follows

$$\mathbf{R}_t(\mathcal{F}) = \{f_{t^n} = f_{t_1} \circ f_{t_2} \circ \cdots \circ f_{t_n} \mid n \in \mathbb{N}\},$$

$$\mathbf{L}_t(\mathcal{F}) = \{f_{^n t} = f_{t_n} \circ f_{t_{n-1}} \circ \cdots \circ f_{t_1} \mid n \in \mathbb{N}\}.$$

A.3   Fixed points and cycles

Definition A.5 *A fixed point of a self-mapping f of a set M is a point $x \in M$ such that $f(x) = x$. Denote the set of fixed points of f by $\mathrm{Fix}(f)$.*

Take a given dynamical system $\mathcal{F} = \{f^t\}_{t\in\kappa}$ defined on M, where $\kappa = \mathbb{R}, \mathbb{R}_+, \mathbb{Z}$, or $\mathbb{Z}_+$. If $\mathcal{F}(x) = \{f^t(x)\}_{t\in\kappa} = \{x\}$, the point x is called a *singular point, equilibrium point , rest point*, and so on. Denote the set of equilibrium points of the DS by $\mathrm{Fix}(\mathcal{F})$. Obviously, we have

$$\mathrm{Fix}(\mathcal{F}) = \bigcap_{t\in\kappa} \mathrm{Fix}(f^t).$$

If $\kappa = \mathbb{Z}$, or $\mathbb{Z}_+$, and if the cascade $\mathcal{F}$ is generated by f, then

$$\mathrm{Fix}(\mathcal{F}) = \mathrm{Fix}(f).$$

If $r|k$, i.e., r is a factor dividing k, then $\mathrm{Fix}(f^r) \subset \mathrm{Fix}(f^k)$.

Definition A.6 *A k-cycle (or cycle of order k) of a mapping $f : M \longrightarrow M$ is a k-tuple of pairwise different elements $x_0, ..., x_{k-1}$ of M such that*

$$f(x_i) = x_{i+1} \ (0 \leq i < k - 1); \ f(x_{k-1}) = x_0.$$

For a k-cycle $\{x_0, ..., x_{k-1}\}$, obviously x_i is of $f^k(x_i) = x_i$, i.e.,

$$\{x_0, ..., x_{k-1}\} \subset \mathrm{Fix}(f^k).$$

A 1-cycle of f is just a fixed point of f. Set

$$\mathrm{Per}(f) = \bigcup_{k=1}^{\infty} \mathrm{Fix}(f^k).$$

The points in $\text{Per}(f)$ are called *periodic points* of the mapping f. Obviously, $x \in \text{Per}(f)$ iff $f^k(x) = x$ for some $k \in \mathbb{Z}^+$. All such k are called *periods* of the point x. Often the word "period" is understood to be the minimal period. If k is the minimal period of x, then $k \geq 1$, and $\{x, f(x), ..., f^{k-1}(x)\}$ is a k-cycle of f. Hence $\text{Per}(f)$ just is the union of cycles of f. Thus the elements of a k-cycle are also called *periodic points of order* k. If $\mathcal{F} = \{f^n\}$ is the cascade generated by f, a trajectory $\mathcal{F}(x)$ is called *cyclic* or *acyclic* according to whether it contains a cycle or not. Obviously, finite trajectories are cyclic.

For a flow or semiflow $\mathcal{F} = \{f^t\}_{t\in\kappa}$ defined on M, where $\kappa = \mathbb{R}$, or $\mathbb{R}_+$, a trajectory $\{f^t(x)\}_{t\in\kappa}$ is said to be *periodic* if it is the trajectory of a *periodic motion*, i.e., if

$$f^{t+T}(x) = f^t(x) \quad \text{for all } t \text{ and some } T > 0.$$

All such T are called *periods* of the motion. This point x is then called a *periodic point* of the system. Denote the set of periodic points by $\text{Per}(f)$ (or $\text{Per}(\{f^t\}_{t\in\kappa})$). Note that each periodic trajectory of the period T is contained in $\text{Fix}(f^T)$. Then

$$\text{Per}(f) = \bigcup_{0<T\in\kappa} \text{Fix}(f^T).$$

According to the definition, periodicity also includes the case where $f^t(x)$ does not depend on t. But, in speaking of a periodic trajectory, one often tacitly excludes equilibrium points. A periodic trajectory of $\mathcal{F}$ is called a *closed trajectory*, because it is a closed curve if $f^t(x)$ depends continuously on t. Also all the periods are integral multiples of a certain *minimal period* $T > 0$. A periodic trajectory of the minimal period T is sometimes called a T-*cycle*.

A point $x \in M$ is *preperiodic* under f if it is not periodic but if some image, say $f^p(x)$, is: in this case there exist positive integers p and k with

$$x, f(x), f^2(x), ..., f^p(x), ..., f^{p+k-1}(x)$$

distinct but with $f^{p+k}(x) = f^p(x)$.

A self-mapping f of a set M is *pointwise periodic* if every element x lies in a $k(x)$-cycle; f is *periodic* if there exists a positive integer k such that $f^k(x) = x$ for all x in M. We list some main properties:

(a) A self-mapping f of an arbitrary set M is pointwise periodic if and only if for every subset $A \subset M$, $f(A) \subset A$ implies $f(A) = A$.

(b) Every pointwise periodic (in particular, every periodic) self-mapping of a set is a bijection.

(c) Every continuous periodic self-mapping of a topological space is a homeomorphism of the space onto itself [Whyburn 1942].

Let M be a differential manifold and consider a continuous mapping $f : M \longrightarrow M$. Firstly, we need to know when $\text{Fix}(f) \neq \emptyset$.

Definition A.7 *Let* (M, d) *be a metric space. A mapping* $f : M \longrightarrow M$ *is called contracting if there exists* $0 \leq \lambda < 1$ *such that for any* $x, y \in M$,

$$d(f(x), f(y)) \leq \lambda d(x, y). \tag{A.24}$$

Lemma A.2 (Hille [110]) *Let f be a self-mapping of a complete metric space (M, d). Assume that for some $k \in \mathbb{N}$, f^k is contracting. Then f has a unique fixed point.*

Proof. Since f^k is contracting, there exists $0 \leq \lambda < 1$ such that for any $x, y \in M$,

$$d(f^k(x), f^k(y)) \leq \lambda d(x, y). \tag{A.25}$$

By iterating (A.25), one sees that for any positive integer n

$$d(f^{nk}(x), f^{nk}(y)) \leq \lambda^n d(x, y), \tag{A.26}$$

which implies

$$\lim_{n \to \infty} d(f^{nk}(x), f^{nk}(y)) = 0.$$

On the other hand, for any $x \in M$ the sequence $\{f^{nk}(x)\}_{n \in \mathbb{N}}$ is a Cauchy sequence because for $m \geq n$,

$$
\begin{aligned}
d(f^{mk}(x), f^{nk}(x)) &\leq \sum_{i=0}^{m-n-1} d(f^{(n+i+1)k}(x), f^{(n+i)k}(x)) \\
&\leq \sum_{i=0}^{m-n-1} \lambda^{n+i} d(f^k(x), x) \leq \frac{\lambda^n}{1-\lambda} d(f^k(x), x) \to 0, \quad \text{(A.27)}
\end{aligned}
$$

as $n \to \infty$. Thus there exists a fixed limit

$$\lim_{n \to \infty} f^{nk}(x) = p,$$

for all $x \in M$. Note that for any $x \in M$ and any integer n one has

$$
\begin{aligned}
d(p, f^k(p)) &\leq d(p, f^{nk}(x)) + d(f^{nk}(x), f^{(n+1)k}(x)) + d(f^{(n+1)k}(x), f^k(p)) \\
&\leq (1+\lambda) d(p, f^{nk}(x)) + \lambda^n d(x, f^k(x)) \to 0.
\end{aligned}
$$

We have $f^k(p) = p$. Since $\mathrm{Fix}(f^k) = \{p\}$, we can prove $\mathrm{Fix}(f) = \{p\}$ easily. $\qquad \square$

If the condition (A.24) is weakened, one has the following result:

Lemma A.3 (Edelstein [74]) *Let f be a self-mapping of a complete metric space (M, d) with*

$$d(f(x), f(y)) < d(x, y) \quad (x \neq y).$$

If an iterative sequence $\{f^n(x)\}$ contains a convergent subsequence, then the limit of this subsequence is the unique fixed point of f.

Remark. If M is a complete Riemannian manifold, and if $f \in C^1(M, M)$, then f is a contraction iff the norm of the differential is bounded by a constant $\lambda < 1$. Expanding mappings can be defined similarly to contracting mappings. Let (M, d) be a metric space. A continuous mapping $f : M \longrightarrow M$ is called *expanding* if for some $\gamma > 1, \varepsilon_0 > 0$ and every $x, y \in M$ with $x \neq y$ and $d(x, y) < \varepsilon_0$ one has

$$d(f(x), f(y)) > \gamma d(x, y).$$

It follows immediately that if M is a Riemannian manifold and if $f : M \longrightarrow M$ is a differentiable expanding mapping, then for any $x \in M$, $(df)_x : T(M)_x \longrightarrow T(M)_{f(x)}$ is expanding with respect to the norm generated by the Riemannian metric: There exists $\gamma > 1$ such that for every $X \in T(M)_x - \{0\}$, we have

$$\|(df)_x X\| > \gamma \|X\|.$$

On compact manifolds this is sufficient (cf. [134], p.72).

Now let $f : M \longrightarrow M$ be a C^1 map. We say that a fixed point $p \in M$ of f is *nondegenerate* if it is isolated and satisfies

$$\det((df)_p - id) \neq 0;$$

under these circumstances, we define the *index* $\iota_f(p)$ of f at p to be

$$\iota_f(p) = \operatorname{sign} \det((df)_p - id).$$

The index $\iota_f(p)$ can be calculated in terms of eigenvalues:

$$\iota_f(p) = (-1)^s = (-1)^{m+u},$$

$$s = \#\{i | \lambda_i < 1\}, \quad u = \#\{i | \lambda_i > 1\},$$

where $\#$ denotes the cardinal number of the set, and where $\{\lambda_i\}_{i=1}^m$ ($m = \dim M$) are the eigenvalues of $(df)_p$. In fact, one has

$$
\begin{aligned}
\operatorname{sign} \det((df)_p - id) &= \operatorname{sign} \prod_{i=1}^m (\lambda_i - 1) \\
&= \operatorname{sign} \prod_{\lambda_i < 1} (\lambda_i - 1) \cdot \operatorname{sign} \prod_{\lambda_i > 1} (\lambda_i - 1) \cdot \operatorname{sign} \prod_{\lambda_i \in \mathbf{C} - \mathbf{R}} (\lambda_i - 1) \\
&= (-1)^s
\end{aligned}
$$

since pairing complex conjugate eigenvalues shows $\operatorname{sign} \prod_{\lambda_i \in \mathbf{C} - \mathbf{R}} (\lambda_i - 1) > 0$.

Let M be compact and $H_{DR}^r(M, \mathbb{R})$ the r-th de Rham cohomology group of M, $0 \leq r \leq m$. Then

$$b_r = \dim H_{DR}^r(M, \mathbb{R})$$

is just the r-th Betti number. Note that f^* induces homomorphisms

$$f^* : H_{DR}^r(M, \mathbb{R}) \longrightarrow H_{DR}^r(M, \mathbb{R})$$

so that $f^*|_{H_{DR}^r(M, \mathbb{R})}$ has a matrix representation. Then the *Lefschetz number* of f is defined by

$$\operatorname{Lef}(f) = \sum_{r=0}^m (-1)^r \operatorname{tr}(f^*|_{H_{DR}^r(M, \mathbb{R})}).$$

If f has only nondegenerate fixed points, the *Lefschetz Fixed-Point Formula* (see [101])

$$\sum_{p \in \operatorname{Fix}(f)} \iota_f(p) = \operatorname{Lef}(f)$$

holds. Obviously, we have

$$\#\text{Fix}(f) \geq |\text{Lef}(f)|,$$

and in particular,

$$\text{Lef}(f) \neq 0 \Longrightarrow \text{Fix}(f) \neq \emptyset.$$

Assume that f is homotopic to a C^1 mapping $g : M \longrightarrow M$. Then $\text{Lef}(f) = \text{Lef}(g)$ since f, g induce same homomorphisms of cohomology groups. If g is constant, the mappings $g^*|_{H^r_{DR}(M,\mathbb{R})}$ are zero for $r > 0$, and id if $r = 0$ so that

$$\text{Lef}(f) = \text{Lef}(g) = b_0 \neq 0, \quad \text{Fix}(f) \neq \emptyset.$$

If $g = id$, then $g^* = id$ and hence

$$\text{tr}(f^*|_{H^r_{DR}(M,\mathbb{R})}) = b_r.$$

In particular,

$$\text{Lef}(f) = \text{Lef}(id) = \sum_{r=0}^{m}(-1)^r b_r = \chi(M),$$

where $\chi(M)$ is the Euler-Poincaré characteristic of M. Since f^n is also homotopic to the identity, then $\text{Lef}(f^n) = \chi(M)$. Consequently for maps homotopic to the identity with only isolated fixed points,

$$\sum_{p\in\text{Fix}(f^n)} \iota_{f^n}(p) = \chi(M). \qquad (A.28)$$

As an immediate corollary to the Lefschetz Fixed-Point Formula one obtain:

Lemma A.4 (Brouwer) *Let $f : M \longrightarrow M$ be a continuous mapping of a compact convex subset M of $\mathbb{R}^m$ into itself. Then f has a fixed point.*

As another corollary to the Lefschetz Fixed-Point Formula one will prove the Hopf Index Theorem. To do this, the following fact will be needed:

Lemma A.5 *Let $\sigma(t)$ be a continuous group of linear transformations of $\mathbb{R}^m$, i.e., $\sigma(t) : \mathbb{R}^m \longrightarrow \mathbb{R}^m$ for any $t \in \mathbb{R}$, $\det \sigma(t) \neq 0$, $\sigma(t + s) = \sigma(t)\sigma(s)$, $\lim_{t_n \to t} \sigma(t_n) = \sigma(t)$. Then there exists a linear transformation $\mathbf{A} : \mathbb{R}^m \longrightarrow \mathbb{R}^m$ such that $\sigma(t) = \exp(\mathbf{A}t)$ for all $t \in \mathbb{R}$.*

This is a special case of a general theorem on the representation of a continuous semi-group of operators in a Banach space (see [73]). Note that the *exponential matrix*

$$\exp(\mathbf{A}t) = \sum_{k=0}^{\infty} \frac{1}{k!}(\mathbf{A}t)^k$$

which obviously satisfies

$$\exp(\mathbf{A}(t + s)) = \exp(\mathbf{A}t)\exp(\mathbf{A}s),$$

is just the flow generated by $\mathbf{A}x$.

Let X be a smooth vector field on a manifold M and let f^t be the 1-parameter group of transformations generated by X. Let $p \in M$ be a zero of X. Then $f^t(p) = p$ for all $t \in \mathbb{R}$. Since $f^{t+s} = f^t \circ f^s$ for $t, s \in \mathbb{R}$, we have at the point p

$$(df^{t+s})_p = (df^t)_p (df^s)_p.$$

It follows that there exists a linear transformation $\mathbf{A} : T(M)_p \longrightarrow T(M)_p$ such that for all $t \in \mathbb{R}$,

$$(df^t)_p = \exp(\mathbf{A}t). \tag{A.29}$$

In terms of local coordinates $x = (x_1, ..., x_m)$ centered at p, writing

$$X = \sum_{i=1}^m \xi_i \frac{\partial}{\partial x_i},$$

then one can prove that $\mathbf{A}$ is given by

$${}^t(\xi_1, ..., \xi_m) = \mathbf{A} \, {}^t x + O(\|x\|^2),$$

where ${}^t x$ denotes the transpose of the vector x.

Definition A.8 *The operator $\mathbf{A}$ in formula (A.29) is called the Hessian of the vector field X at p, denoted by $\dot{X}(p)$. A zero point p of X is said to be nondegenerate if $\mathbf{A}$ is nonsingular, and be hyperbolic if $\mathbf{A}$ does not have purely imaginary complex numbers as eigenvalues.*

By the definition, if p is a hyperbolic zero point of X, then p is nondegenerate. By the inverse function theorem, one can prove that if p is a nondegenerate zero point of X, then p is an isolated point of the set of all zero points of X. In this case we define the *index* $\iota_X(p)$ of X at p by

$$\iota_X(p) = \operatorname{sign} \det \mathbf{A}.$$

Thus if X is given as above near a zero p, then in terms of the coordinates x,

$$(df^t)_p - id = \mathbf{A}t + O(t^2),$$

and for $t > 0$ yet sufficiently small,

$$\iota_{f^t}(p) = \operatorname{sign} \det((df^t)_p - id) = \operatorname{sign} \det \mathbf{A} = \iota_X(p).$$

Since f^t is homotopic to the identity, then

$$\operatorname{Lef}(f^t) = \chi(M);$$

and we have

Theorem A.2 (Hopf index theorem) *For a smooth vector field X on a compact manifold M with isolated zero points we have*

$$\sum_{X(p)=0} \iota_X(p) = \chi(M).$$

A.4   Orbits and invariant sets

If f is a mapping of a set M into itself, a subset E of M is:
  (a) *forward invariant* if $f(E) = E$;
  (b) *backward invariant* if $f^{-1}(E) = E$;
  (c) *completely invariant* if $f^{-1}(E) = E = f(E)$.
If f is injective, forward invariance imply backward invariance and completely invariance. Generally, we have the following relations:

Lemma A.6 *1) If E is backward invariant, then*

$$f(E) = E \cap f(M) \subset E.$$

2) If $f^{-1}(E) \subset E, f(E) \subset E$, then E is backward invariant.

Proof. If $f^{-1}(E) = E$, then $f(f^{-1}(E)) = f(E)$ so that we obtain

$$E \cap f(M) = f(E).$$

Now we prove 2). If $E - f^{-1}(E) \neq \emptyset$, there exist $x \in E - f^{-1}(E)$ and $y \in E$ such that $f(x) = y$. Thus $x \in f^{-1}(E)$. This is impossible. Hence, E is backward invariant.   □

Thus if E is backward invariant and if $E \subset f(M)$, then E is forward invariant and completely invariant.

Corollary A.1 *If f is surjective (that is, if $f(M) = M$), then the following are equivalent:*
  1) E is completely invariant;
  2) E is backward invariant;
  3) $f^{-1}(E) \subset E, f(E) \subset E$.
Further, if f also is injective, these also are equivalent to
  4) E is forward invariant.

The requirement of surjectivity here is crucial and without it, there is a difference between forward and complete invariance: for example, $\mathbb{C}$ is backward invariant, but not completely invariant, under the mapping $z \mapsto \exp(z)$ since $\exp(\mathbb{C}) = \mathbb{C} - \{0\}$.

If E is completely invariant under f, and if h is a bijection of M onto itself, then $h(E)$ is completely invariant under the conjugate mapping $g = h \circ f \circ h^{-1}$.

The operator f^{-1} commutes with the intersection operator, that is, for any collection $\{E_\alpha\}$ of sets,

$$f^{-1}(\cap E_\alpha) = \cap f^{-1}(E_\alpha),$$

and, because of this, the intersection of a family of backward invariant sets is itself backward invariant. This means that we can take any subset E_0 and form the intersection, say E, of all those backward invariant sets which contain E_0: obviously, E is then the smallest backward invariant set that contains E_0 and we say that E_0 *generates* E.

Given a dynamical system $\{f^t\}_{t \in \kappa}$ defined on M, where $\kappa = \mathbb{R}, \mathbb{R}_+, \mathbb{Z}$, or $\mathbb{Z}_+$. For a subset $E \subset M$ and for $0 \leq t \in \kappa$, we always write

$$f^{-t}(E) := (f^t)^{-1}(E).$$

Of course, $f^{-t} = (f^t)^{-1}$ if $\kappa = \mathbb{R}, \mathbb{Z}$. We often use the following facts:

$$f^{-t}(f^{-s}(E)) = f^{-s}(f^{-t}(E)) = f^{-t-s}(E) \text{ for } t \geq 0, s \geq 0,$$

$$f^t(f^{-s}(E)) = \begin{cases} f^{t-s}(E \cap f^s(M)) & : \quad \text{if } t \geq s \geq 0, \\ f^{t-s}(E) \cap f^t(M) & : \quad \text{if } s \geq t \geq 0. \end{cases}$$

For the dynamical system $\{f^t\}_{t \in \kappa}$, a subset E of M is:

(d) *forward invariant* if $f^t(E) = E$ for all $t \geq 0$;

(e) *backward invariant* if $f^{-t}(E) = E$ for all $t \geq 0$;

(f) *completely invariant* if $f^{-t}(E) = E = f^t(E)$ for all $t \geq 0$.

Obviously, if the system is a cascade generated by a mapping f, these notions are equivalent to those of f.

Denote the *forward orbit* of x by

$$O^+(x) = \bigcup_{t \geq 0} \{f^t(x)\},$$

which also is called the *positive orbit* or *positive trajectory* through x. The elements in $O^+(x)$ are called *successors* of x. For the case $\kappa = \mathbb{Z}$, or $\mathbb{Z}_+$, a *negative trajectory* through x is a sequence $\{x_{-n}, n = 0, 1, 2, ...\}$ such that $x_0 = x, f(x_{-n-1}) = x_{-n}$ for all n. For the case $\kappa = \mathbb{R}$, or $\mathbb{R}_+$, a *negative trajectory* through x is a curve $\gamma : \mathbb{R}(-\infty, 0] \longrightarrow X$ such that $\gamma(0) = x$, and

$$f^t(\gamma(-s)) = \gamma(t - s) \quad \text{for } 0 \leq t \leq s.$$

Define the *backward orbit* of x by

$$O^-(x) = \bigcup_{t \geq 0} \{f^{-t}(x)\}.$$

Obviously, $O^-(x)$ is the union of all negative trajectories through x. The elements in $O^-(x)$ are called *predecessors* (or *preimages* under $f^t, t \in \kappa$) of x. Define the *total orbit*:

$$O(x) = O^+(x) \cup O^-(x).$$

Generally, for a subset $E \subset M$, we also define the forward, backward and total orbits of E respectively by

$$O^+(E) = \bigcup_{t \geq 0} f^t(E), \quad O^-(E) = \bigcup_{t \geq 0} f^{-t}(E), \quad O(E) = O^+(E) \cup O^-(E).$$

Obviously, we have

$$O^+(E) = \bigcup_{x \in E} O^+(x), \quad O^-(E) = \bigcup_{x \in E} O^-(x).$$

The set E is:

(g) *plus invariant* if $O^+(E) = E$;

(h) *minus invariant* if $O^-(E) = E$;

(i) *invariant* if $O^+(E) = E$ and if $O^-(E) = E$.

Obviously, $O^+(x)$ is plus invariant. If $\kappa = \mathbb{R}$, or $\mathbb{Z}$, $O^-(x)$ is minus invariant and $O(x)$ is invariant.

Proposition A.3 *A subset $E \subset M$ is invariant iff E is backward invariant.*

Proof. If E is invariant, then

$$O^+(E) = \bigcup_{t \geq 0} f^t(E) = E, \quad O^-(E) = \bigcup_{t \geq 0} f^{-t}(E) = E$$

imply $f^t(E) \subset E, f^{-t}(E) \subset E$ for all $t \geq 0$ so that E is backward invariant under f^t for all $t \geq 0$ by Lemma A.6.

Conversely, if E is backward invariant under f^t for all $t \geq 0$, then we have

$$O^-(E) = \bigcup_{t \geq 0} f^{-t}(E) = E,$$

$$O^+(E) = \bigcup_{t \geq 0} f^t(E) = \bigcup_{t \geq 0} \{E \cap f^t(M)\} = E,$$

i.e., E is invariant.   □

In Proposition A.3, the backward invariance of sets E under f^t for all $t \geq 0$ is equivalent to the following: there exists some $R > 0$ such that $f^{-r}(E) = E$ for all $r \in \kappa(0, R]$. If so, we specially have

$$f^{-R}(E) = E.$$

By induction, we obtain

$$f^{-nR}(E) = E, \quad n = 0, 1, \ldots$$

Since each $t \geq 0$ is of the form $t = nR + r$ with $n \in \mathbb{Z}_+, r \in \kappa[0, R)$, we obtain

$$f^{-t}(E) = f^{-nR-r}(E) = f^{-nR}(f^{-r}(E)) = E.$$

Thus for a dynamical system with discrete time formed by f, a subset E is invariant iff E is backward invariant under f.

Corollary A.2 *If $\kappa = \mathbb{Z}$ or $\mathbb{R}$, then the following are equivalent:*
1) E is invariant;
2) E is backward invariant;
3) E is forward invariant;
4) E is completely invariant;
5) $f^{-t}(E) \subset E, f^t(E) \subset E$ for all $t \geq 0$.

For any x and y in M, we define the relation $\sim$ on M by $x \sim y$ if and only if there exist non-negative numbers t and s with

$$f^t(x) = f^s(y),$$

that is, x and y have a common successor. Obviously, the relation $\sim$ is symmetric and reflexive, and it is also transitive since

$$f^t(x) = f^s(y), f^p(y) = f^q(z) \implies f^{t+p}(x) = f^{s+q}(z),$$

thus $\sim$ is an equivalence relation on M. We denote the equivalence class containing x by $[x]$, and we call this the *(grand) orbit* of x. Since $\sim$ is an equivalence relation, the two orbits are either identical or disjoint. Obviously, an orbit consists precisely of all successors and all predecessors of all successors of any one of its elements, that is,

$$[x] = O^+(x) \cup \left\{ \bigcup_{t \geq 0} O^-(f^t(x)) \right\} = \bigcup_{t \geq 0} O(f^t(x)).$$

Thus if $\kappa = \mathbb{R}$, or $\mathbb{Z}$, we have $[x] = O(x)$.

Definition A.9 *A invariant set E is said to be a minimal set if for any invariant set E', $E' \subset E$ implies $E' = E$.*

Theorem A.3 *The orbits are precisely the minimal sets which are invariant.*

Proof. First we prove that an orbit $[x]$ is backward invariant under f^t for all $t \geq 0$. By definition, $y \in [x]$ implies $f^t(y) \in [x]$ and $f^{-t}(y) \subset [x]$; thus $f^t([x]) \subset [x]$ and $f^{-t}([x]) \subset [x]$. Then $f^{-t}([x]) = [x]$ by Lemma A.6.

Next letting $E \subset M$ have the property $f^{-t}(E) = E$ for all $t \geq 0$, we claim that $x \in E$ implies $[x] \subset E$. By Lemma A.6, we have $f^t(E) = E \cap f^t(X) \subset E$. Let y be an arbitrary element of $[x]$. Then $f^t(x) = f^s(y)$ for some $t, s \geq 0$. Hence $f^s(y) = f^t(x) \in f^t(E) \subset E$, and $y \in f^{-s}(E) = E$. Thus E contains, along with x, the entire orbit on which x lies; this was to be proved.

Now every orbit $[x]$ has the property $f^{-t}([x]) = [x]$ for all $t \geq 0$ and no proper subset of $[x]$ has this property. Let $E \subset [x]$. Then $[x] \subset E$ by our claim and hence $E = [x]$, so $[x]$ is minimal with respect to the property. Conversely, every E with the property $f^{-t}(E) = E$ for all $t \geq 0$ is a union of orbits by our claim and this union is disjoint. Thus E is minimal if and only if it consists of exactly one orbit. Now Theorem A.3 follows from Proposition A.3. $\square$

Corollary A.3 *Given a dynamical system $\{f^t\}_{t \in \kappa}$ defined on M, a subset E of M is invariant iff it is a union of equivalence classes $[x]$. If this is the case, then its complement $M - E$ must also be a union of equivalence classes and, therefore, also invariant.*

For a subset E of M, define

$$[E] = \bigcup_{x \in E} [x].$$

Then $[E]$ is invariant. Obviously, if E is invariant, then $[E] = O(E) = E$. The invariant set $[E]$ is the minimal element in all invariant sets containing E.

Theorem A.4 *Assume that for each $t \in \kappa$, f^t is a continuous, open mapping of a topological space M into itself and suppose that E is invariant. Then so are the interior E^o, the boundary ∂E, and the closure $\overline{E}$, of E.*

Proof. As f^t is continuous on M, $f^{-t}(E^o)$ is an open subset of $f^{-t}(E)$, and hence (by invariance) of E. Thus $f^{-t}(E^o) \subset E^o$. Similarly, as f^t is an open mapping, $f^t(E^o)$ is an open subset of $f^t(E) = E \cap f^t(M) \subset E$ and so $f^t(E^o) \subset E^o$. Thus E^o is backward invariant under f^t for all $t \geq 0$. Hence E^o is invariant.

We now know that the invariance of E implies that of $M - E$ and E^o, and the usual topological arguments guarantee that the closure and the boundary of E are also invariant. This completes the proof.   □

Definition A.10 *A point x is said to be exceptional for a dynamical system $\{f^t\}_{t \in \kappa}$ defined on M when $[x]$ is finite, and the set of such points is denoted by* $\mathrm{Exc}(f)$.

Note that the set $\mathrm{Exc}(f)$ is invariant. If E is a finite invariant set, then $E \subset \mathrm{Exc}(f)$.

Theorem A.5 *If $x \in \cap_{t \geq 0} f^t(M)$, the backward orbit $O^-(x)$ of x is finite iff $[x]$ is finite. If so, $[x] = O^-(x)$.*

Proof. As $O^-(x) \subset [x]$, thus $O^-(x)$ is finite if so does $[x]$. Now we show that the converse is also true. To do this, define the non-empty sets

$$B_r = \bigcup_{s \geq r} f^{-s}(x) \text{ for } s, r \geq 0$$

so that $f^{-t}(B_r) = B_{t+r}$ and

$$[x] \supset O^-(x) = B_0 \supset B_r \supset B_s$$

for $s, r \in \kappa$ with $0 < r < s$. We now assume that $O^-(x)$ is finite. Then each B_r is finite, so there is some $s \geq 0$ with $B_s = B_{s+t}$ for all $t \geq 0$; this means that $f^{-t}(B_s) = B_s$ and so B_s is backward invariant under f^t for all $t \geq 0$. It follows that B_s contains some orbit $[y]$, and as it is a subset of $[x]$, it must be $[x]$: thus $[x]$ is finite as required. In fact, we have obtained $[x] = O^-(x)$.   □

Note that two cycles are either identical or disjoint. In the latter case, two elements, one from each cycle, have no common successor, and thus cannot belong to a same (grand) orbit. Hence an orbit contains at most one cycle. An orbit is called *cyclic* or *acyclic* according to whether it contains a cycle or not. Finite orbits are cyclic. Also an orbit is acyclic iff it contains an infinite forward orbit.

If a closed (or compact) invariant set E is such that if for any closed (or compact) invariant set E', $E' \subset E$ implies $E' = E$, then E also is said to be a *minimal orbit closure* (or minimal compact set).

Proposition A.4 *Assume that for each $t \in \kappa$, f^t is a continuous, open mapping of a topological space M into itself. Then a set E is minimal orbit closure iff E is the closure $\overline{[x]}$ of the orbit $[x]$ of a point $x \in M$.*

Proof. If each f^t is open mapping, then by Theorem A.3 and Theorem A.4, the closure $\overline{[x]}$ of the orbit of every point $x \in M$ is a minimal orbit closure. Conversely, if E is a minimal orbit closure, then $x \in E$ implies $[x] \subset E$ since E is invariant, and hence $\overline{[x]} \subset E$ since E is closed, and further $\overline{[x]} = E$ since E is minimal, i.e., the orbit of every point $x \in E$ is dense in E.   □

Proposition A.5 *Assume that for each* $t \in \kappa$, f^t *is a continuous, open mapping of a connected topological space* M *into itself. Then a minimal orbit closure* E *is the entire* M *if* $E^o \neq \emptyset$.

Proof. Assume that $E^o \neq \emptyset$. By theorem A.4, the boundary $\partial E = E - E^o$ of E is invariant. Clearly, this set is closed and contained in E, and hence it is empty by the minimality of E. Consequently, E is a closed-open set in M and thus we have $E = M$ since M is connected. □

A. J. Schwartz[226] proved that if E is a minimal orbit closure of a flow on a two-dimensional connected manifold M, then E is either an equilibrium point or a closed trajectory or the entire manifold M. Further, if M is compact, and if $E = M$, then M is the torus $\mathbb{T}^2$ (Kneser[142]).

Definition A.11 *A topological dynamical system* $\{f^t\}_{t \in \kappa}$ *on* M *is called topological transitive if there exists a point* $x \in M$ *such that its orbit* $[x]$ *is dense in* M. *If the orbit of every point is dense in* M, *then the DS is said to be minimal.*

Proposition A.6 *Every invariant closed (or compact) set* A *contain minimal orbit closures.*

Proof. Let $\mathcal{I}_f(A)$ be the collection of all closed invariant subsets contained in A, partially ordered by inclusion. Since the intersection of any number of closed invariant subsets is still closed and invariant, any totally ordered subsets of $\mathcal{I}_f(A)$ has a lower bound. Then by Zorn's Lemma, $\mathcal{I}_f(A)$ contains a minimal element, that is, a closed invariant set E that contains no closed invariant subsets. □

Generally, a component U of a subset E of M is a:

(j) *fixed component* if $f(U) = U$;

(k) *periodic component* if $f^k(U) = U$ for some $k \geq 1$. The minimal k is the *period* of the component;

(l) *preperiodic component* if $f^p(U)$ is periodic for some $p \geq 1$;

(m) *wandering component* if all $\{f^n(U)\}$ are distinct.

(n) *recurrent component* if there exists $x_0 \in U$ such that $\{f^{n_j}(x_0)\}$ is relatively compact in U for some subsequence $\{n_j\}$.

A.5 The isometric ℝ-actions

We will need some facts in the Riemannian geometry. Let M be a smooth manifold of dimension m and g a smooth symmetric covariant tensor field of degree 2 on M. If $(U; x)$ is a local coordinate system on M, the tensor field g can be given by

$$g = \sum_{i,j=1}^{m} g_{ij} dx_i \otimes dx_j,$$

where $g_{ij} = g_{ji}$ are smooth functions on U. For each $p \in M$, g defines a bilinear function on $T(M)_p$ by

$$g(X,Y) = \sum_{i,j=1}^{m} g_{ij}\xi_i\eta_j, \quad X = \sum_{i=1}^{m} \xi_i \frac{\partial}{\partial x_i}, Y = \sum_{j=1}^{m} \eta_j \frac{\partial}{\partial x_j}.$$

If there exists a vector $X \in T(M)_p$ such that $g(X,Y) = 0$ for all $Y \in T(M)_p$ which implies $X = 0$, then g is said to be *nondegenerate* at p. Obviously, g is nondegenerate at p iff $\det(g_{ij}(p)) \neq 0$. If $g(X,X) \geq 0$ for all $X \in T(M)_p$, and $g(X,X) = 0$ iff $X = 0$, then g is said to be *positive definite* at p. Obviously, g is positive definite at p iff the matrix $(g_{ij}(p))$ is positive definite. A positive definite g is nondegenerate. If there exists a smooth symmetric covariant tensor field g of degree 2 on M such that g is positive definite at each point of M, then M is said to be a *Riemannian manifold*, and g is called a *Riemannian metric*. There exist Riemannian metrics on smooth manifolds.

On a Riemannian manifold M,the *arc length* of a differentiable curve $\gamma = \gamma(t), a \leq t \leq b$, of class C^1 is defined by

$$\text{Length}(\gamma) = \int_a^b g(\dot{\gamma}(t), \dot{\gamma}(t))^{\frac{1}{2}} dt.$$

This definition can be generalized to a piecewise differentiable curve of class C^1 in an obvious manner. If M is connected, the *distance* $d_g(x,y)$ between two points x and y on M is, by definition, the infinimum of the lengths of all piecewise differentiable curves of class C^1 joining x and y. The topology defined by d_g is the same as the manifold topology of M.

Theorem A.6 *If (M,g) is a Riemannian manifold, then there exists a unique affine connection ∇ on M such that $\nabla g = 0$ and $T(X,Y) = 0$, that is, $\nabla_X Y - \nabla_Y X = [X,Y]$, for all vector fields X, Y on M. The connection is called the Levi-Civita connection or the Riemannian connection.*

This is a basic theorem in the Riemannian geometry. For a proof of the theorem, see [145]. Let $\omega = (\omega_i^j)$ be the connection matrix of ∇ with respect to the local coordinates x_i. Then $\nabla g = 0$ holds iff

$$dg_{ij} = \omega_i^k g_{kj} + \omega_j^k g_{ik}.$$

Generally the Levi-Civita connection ∇ satisfies the following relation

$$dg(X,Y) = g(\nabla X, Y) + g(X, \nabla Y), \tag{A.30}$$

for all vector fields X and Y (also see [145]).

Let $R(X,Y)$ denote the curvature operator of $T(M)_x$ determined by $X, Y \in T(M)_x$. The *Riemannian curvature tensor* of M, denoted also by R, is the tensor field of covariant degree 4 defined by

$$R(X,Y,Z,W) = g(R(Z,W)X,Y),$$

for $X, Y, Z, W \in T(M)_x$. Its components R_{ijkl} with respect to a local coordinate system $x_1, ..., x_m$ in M are given by

$$R_{ijkl} = R\left(\frac{\partial}{\partial x_i}, \frac{\partial}{\partial x_j}, \frac{\partial}{\partial x_k}, \frac{\partial}{\partial x_l}\right) = \sum_{n=1}^{m} R_{ikl}^n g_{nj}.$$

The Riemannian curvature tensor, considered as a quadrilinear mapping

$$R : T(M)_x \times T(M)_x \times T(M)_x \times T(M)_x \longrightarrow \mathbb{R}$$

at each $x \in M$, possesses the following properties:
    (a) $R(X, Y, Z, W) = -R(Y, X, Z, W)$;
    (b) $R(X, Y, Z, W) = -R(X, Y, W, Z)$;
    (c) $R(X, Y, Z, W) + R(X, Z, W, Y) + R(X, W, Y, Z) = 0$;
    (d) $R(X, Y, Z, W) = R(Z, W, X, Y)$.
For each plane p in $T(M)_x$, the *sectional curvature* $K(p)$ for p is defined by

$$K(p) = -\frac{R(X, Y, X, Y)}{g(X, X)g(Y, Y) - g(X, Y)^2},$$

where X, Y is a basis of p.

If $f : M \longrightarrow N$ is a C^k-differentiable mapping of the Riemannian manifold (M, g) into a Riemannian manifold (N, h) such that $f^*h = g$, i.e.,

$$h(f_*X, f_*Y) = g(X, Y),$$

for all $X, Y \in T(M)_p$ and for all $p \in M$, then f is called C^k-*isometric*. In particular, if $(M, g) = (N, h)$, and if f is a C^k-isometric diffeomorphism, it is also called a C^k-*isometry*. The set of C^k-isometries of (M, g) forms a subgroup of the group $\mathrm{Diff}^k(M, M)$ of C^k-diffeomorphisms. It is called the *isometry group* $\mathrm{Iso}^k(M) = \mathrm{Iso}^k(M, g)$. The following lemma is due to van Dantzig and van der Waerden [260]. For the proof, see also Kobayashi and Nomizu [145].

Lemma A.7 *If M is a connected, locally compact metric space, the group $\mathrm{Iso}^k(M)$ is locally compact with respect to the compact-open topology, and its isotropy subgroup*

$$\mathrm{Iso}_p^k(M) = \{f \in \mathrm{Iso}^k(M) \mid f(p) = p\}$$

is compact for each $p \in M$. If M is moreover compact, then $\mathrm{Iso}^k(M)$ is compact.

Remark 1. An isometric mapping $f : M \longrightarrow N$ is necessarily an immersion since $X \in T(M)_p$ and $f_*X = 0$ implies

$$0 = h(f_*X, f_*X) = g(X, X), \quad \text{i.e., } X = 0.$$

Remark 2. Any distance preserving surjective mapping $f : M \longrightarrow N$ is an isometric diffeomorphism. Here the distances are induced by the Riemannian metrics.

Remark 3. Let M be compact. Then one can introduce a topology on the set $\mathcal{G}(M)$ of Riemannian metrics of M and show that there is an open dense subset in $\mathcal{G}(M)$ such that for an element g in this subset, $\mathrm{Iso}^\infty(M) = \{id\}$, this means, generically $\mathrm{Iso}^\infty(M)$ is finite.

If (M, g) is Riemannian, and if f^t is an isometry for all $t \in \mathbb{R}$, then the flow $\{f^t\}$ is called 1-*parameter group of isometries* or *isometric $\mathbb{R}$-action* . The associated vector field X_f is called a *Killing vector field*.

Proposition A.7 *For a Killing vector field X_f on (M, g), the identity*

$$g(\nabla_Y X_f, Z) + g(Y, \nabla_Z X_f) = 0 \tag{A.31}$$

holds for all vector fields Y and Z. Conversely, if a vector field X_f satisfies (A.31), the local 1-parameter group generated by X_f consists of isometries.

Proof. f^t being isometric means

$$g(f_*^t Y, f_*^t Z) = g(Y, Z) \tag{A.32}$$

for all $Y, Z \in T(M)_p$ and for all $p \in M$. Differentiating this with respect to t at $t = 0$ and using (A.30) yields

$$
\begin{aligned}
0 &= g\left(\frac{\nabla}{dt} f_*^t Y, f_*^t Z\right)|_{t=0} + g\left(f_*^t Y, \frac{\nabla}{dt} f_*^t Z\right)|_{t=0} \\
&= g(\nabla_Y X_f, Z) + g(Y, \nabla_Z X_f).
\end{aligned}
$$

Conversely, if a vector field X_f satisfies (A.31), then this implies that the left hand side of (A.32) is independent of t. Thus the mapping f^t is an isometry.   □

Proposition A.8 *Let $\{f^t\}$ be an isometric $\mathbb{R}$-action on M. Let X_f be the associated Killing vector field. Consider the function*

$$\phi : M \longrightarrow \mathbb{R}; \quad \phi = g(X_f, X_f).$$

The orbit $O(p) = \{f^t(p) \mid t \in \mathbb{R}\}$ is a non-constant geodesic iff

$$\phi(p) \neq 0, \quad (d\phi)_p = 0.$$

In particular, if the $\mathbb{R}$-action induces an S^1-action, the orbit $O(p)$ will be a closed geodesic, i.e., there exists $T > 0$ with

$$\dot{x}(T) = \dot{x}(0) \quad (x(t) = f^t(p)).$$

Proof. Choose $X_0 \in T(M)_p$ and a curve $\gamma : [-\varepsilon, \varepsilon] \longrightarrow M$ with $\dot{\gamma}(0) = X_0$. Then $\phi(f^t(\gamma(s)))$ is independent of t. Note that

$$
\begin{aligned}
g\left(\dot{x}(t), \frac{\partial}{\partial s} f^t(\gamma(s))\right)|_{s=0} &= g(f_*^t X_f(p), f_*^t X_0) \\
&= g(X_f(p), X_0) = \text{ constant .}
\end{aligned}
$$

We find

$$
\begin{aligned}
\frac{1}{2}\langle X_0, (d\phi)_p\rangle &= \frac{1}{2}\frac{d\phi(\gamma(s))}{ds}|_{s=0} = \frac{1}{2}\frac{d\phi(f^t(\gamma(s)))}{ds}|_{s=0} \\
&= \frac{1}{2}\frac{d}{ds}g\left(\frac{\partial}{\partial t}f^t(\gamma(s)), \frac{\partial}{\partial t}f^t(\gamma(s))\right)|_{s=0} \\
&= g\left(\dot{x}(t), \frac{\nabla}{ds}\frac{\partial}{\partial t}f^t(\gamma(s))\right)|_{s=0} \\
&= g\left(\dot{x}(t), \frac{\nabla}{dt}\frac{\partial}{\partial s}f^t(\gamma(s))\right)|_{s=0} \\
&= -g\left(\frac{\nabla\dot{x}(t)}{dt}, f_*^t X_0\right),
\end{aligned}
$$

with $X_0 \in T(M)_p$, and $f_*^t X_0 \in T(M)_{x(t)}$ can also be prescribed arbitrarily. Thus $(d\phi)_p = 0$ is equivalent to $\frac{\nabla \dot{x}(t)}{dt} = 0$. □

Definition A.12 *An isometric immersion* $f : (M, g) \longrightarrow (N, h)$ *is called totally geodesic at* $p \in M$ *if every geodesic* $\gamma(t)$ *with initial vector* $X \in T(M)_p$ *is carried under* f *into a geodesic of* N *for all* t *in some neighborhood of* 0. *The mapping* f *is called a totally geodesic immersion if it is totally geodesic for every* $p \in M$.

Theorem A.7 *Let* $f : (M, g) \longrightarrow (M, g)$ *be an isometry. Then every connected component of* Fix(f) *is a totally geodesic submanifold.*

Theorem A.8 *The set of Killing vector fields on a Riemannian manifold* M *is a subalgebra of the Lie algebra of all vector fields on* M.

The proof of theorems can be found in [139] and [179].

A.6 Stability

Take a topological dynamical system $\{f^t\}_{t \in \kappa}$ defined on M, where $\kappa = \mathbb{R}, \mathbb{R}_+, \mathbb{Z}$, or $\mathbb{Z}_+$. For any set $A \subset M$, define the ω-*limit* set $L^+(A)$ of A as

$$L^+(A) = \bigcap_{s \geq 0} \overline{\bigcup_{t \geq s} f^t(A)}.$$

It can be characterized by saying that $y \in L^+(A)$ iff there are a sequence $x_j \in A$ and a sequence $t_j \to +\infty$ such that $f^{t_j}(x_j) \to y$ as $j \to +\infty$. Hence for a point x, the ω-limit set $L^+(x)$ of x can be characterized by saying that $y \in L^+(x)$ iff there is a sequence $t_j \to +\infty$ such that $f^{t_j}(x) \to y$ as $j \to +\infty$. Therefore we have

$$L^+(A) = \overline{\bigcup_{x \in A} L^+(x)}.$$

Obviously, $L^+(x)$ is closed. If $L^+(x) \neq \emptyset$, then $L^+(x)$ is plus invariant, and

$$\overline{O^+(x)} = O^+(x) \cup L^+(x).$$

In particular, if $L^+(x) = \{y\}$, then y is a singular point such that

$$\lim_{t \to \infty} f^t(x) = y.$$

Further, if $\kappa = \mathbb{R}$, or $\mathbb{Z}$, $L^+(x)$ is also minus invariant so that it is invariant. If M is compact, the set $L^+(x)$ is nonempty.

Similarly, for any set $A \subset M$, define the α-*limit* set $L^-(A)$ of A as

$$L^-(A) = \bigcap_{s \geq 0} \overline{\bigcup_{t \geq s} f^{-t}(A)}.$$

Similar characterizations hold for $L^-(x)$ and $L^-(A)$. If $\kappa = \mathbb{R}$, or $\mathbb{Z}$, the set $L^-(x)$ is also invariant.

Remark. Let H be a real-valued function on M. If M is a compact smooth Riemannian manifold, the gradient vector field $\mathrm{grad}(H)$ generates a flow f^t, called the *gradient flow* of H. Suppose that in local coordinates $(x_1, ..., x_m)$ the Riemannian metric has the form $g = \sum g_{ij} dx_i \otimes dx_j$. Then

$$\mathrm{grad}(H) = G^{-1}\left(\frac{\partial H}{\partial x_1}, ..., \frac{\partial H}{\partial x_m}\right),$$

where $G = (g_{ij})$ and G^{-1} is the inverse matrix. Thus $\mathrm{grad}(H)$ is a smooth vector field on M. For this flow, $L^+(x)$ and $L^-(x)$ consist of critical points of H, that is, fixed points of the gradient flow, and $L^+(x)$ is either a single point or an infinite set for any $x \in M$ (see [134], p. 37). Hence if the function H has only isolated critical points, every trajectory of the gradient flow of H converges to a critical point of H as $t \to \infty$.

Definition A.13 *A point $x_0 \in M$ is said to be plus (or minus) Poisson stable if $x_0 \in L^+(x_0)$ (or $x_0 \in L^-(x_0)$), and called Poisson stable if both $x_0 \in L^+(x_0)$ and $x_0 \in L^-(x_0)$.*

Let $P^+(f)$, $P^-(f)$ and $P(f)$ be the closure of the set of plus Poisson stable, minus Poisson stable and Poisson stable, respectively. Obviously, if $x_0 \in L^+(x_0)$, then $O^+(x_0) \subset L^+(x_0)$ and $O^+(x_0) \subset P^+(f)$. Hence

$$\overline{O^+(x_0)} = L^+(x_0) \subset P^+(f).$$

Obviously we also have

$$\overline{\mathrm{Per}(f)} \subset P^+(f) \subset L^+(M).$$

If $\kappa = \mathbb{Z}$ or $\mathbb{R}$, then $x_0 \in L^+(x_0)$ also implies

$$L^-(x_0) \subset \overline{O(x_0)} = L^+(x_0) \subset P^+(f).$$

Therefore if x_0 is Poisson stable, then

$$L^-(x_0) = \overline{O(x_0)} = L^+(x_0).$$

Now we have

$$\overline{\mathrm{Per}(f)} \subset P(f) \subset P^+(f) \cup P^-(f) \subset L^+(M) \cup L^-(M).$$

For a subset A of M, we denote by $\mathrm{Att}(A)$ its *basin of attraction*, that is the set

$$\mathrm{Att}(A) = \{x \in M \mid L^+(x) \subset A\}.$$

Definition A.14 *A set A is called an attractor of the dynamical system $\{f^t\}_{t \in \kappa}$ if A is a closed forward invariant set such that $\mathrm{Att}(A)$ contains a residual subset of an open subset of M.*

Here we say that S is a *residual set* if it is the countable intersection of open dense sets. In the sequel, we consider some kind of attractors.

Definition A.15 *A set A is asymptotically stable if there exists a neighborhood U of A satisfying $U \subset \mathrm{Att}(A)$.*

Lemma A.8 *If A is asymptotically stable, $\mathrm{Att}(A)$ is open.*

Proof. Fix $x \in \mathrm{Att}(A)$. We can find $t > 0$ such that $f^t(x) \in U$, here U is a neighborhood of A described in Definition A.15. Then there exists a neighborhood V of x such that $f^t(V) \subset U$. Evidently $V \subset \mathrm{Att}(A)$. $\qquad\square$

Definition A.16 *A plus (or minus) invariant set A is said to be plus (or minus) stable if, for any neighborhood U of A there exists a neighborhood V of A such that $O^+(V) \subset U$ (or $O^-(V) \subset U$). A set A is said to be plus asymptotically stable if A is plus stable and asymptotically stable.*

In dynamics, a plus (resp., minus) stable equilibrium point is usually said to be stable for $t > 0$ or future stable (resp., stable for $t < 0$ or past stable). Similarly one can define stability for all t.

Definition A.17 *A set A is said to separate invariant sets if there exists a neighborhood U of A such that any invariant set in U contains in A. An invariant set A is called locally maximal (or isolated) if there exists a neighborhood U of A such that for any invariant set A' with $A \subset A' \subset U$, A' coincides with A.*

Obviously, if A is an invariant set, then A separate invariant sets if and only if A is locally maximal.

Theorem A.9 (cf. [259]) *Take $\kappa = \mathbb{Z}$ or $\mathbb{R}$. Assume that M is locally compact and that a subset $A \subset M$ is plus invariant and plus stable. Then A is asymptotically stable iff A separate invariant sets.*

Proof. Here we give a rough proof. Assume that A is asymptotically stable. Then there exists a neighborhood U of A such that for any point $x \in U$ the set $L^+(x) \subset A$. Let A' be an invariant set in U. Then we have $A' = f^t(A') \subset f^t(U)$ for $t \geq 0$, and hence

$$A' \subset \bigcap_{t \geq 0} f^t(U) \subset A,$$

i.e., A separate invariant sets.

Conversely, assume that A separate invariant sets. By the definition, there exists a neighborhood U of A such that any invariant set in U contains in A. Construct U carefully such that there is a neighborhood U' of A with $\overline{U'} \subset U$. Since A is plus stable, there exists a neighborhood V of A such that $O^+(V) \subset U'$. Define

$$A' = \bigcup_{x \in V} L^+(x).$$

Note that $O^+(V) \subset U'$ implies $A' \subset \overline{U'} \subset U$. Since A' is invariant, we have $A' \subset A$, i.e., A is asymptotically stable. $\qquad\square$

Definition A.18 *A domain U of the phase space is said to be absorbing if*

$$\overline{f^t(U)} \subset U \quad for \ t > 0.$$

Proposition A.9 *If $\kappa = \mathbb{Z}$ or $\mathbb{R}$ and if U is an absorbing domain, then the following set*

$$A = \bigcap_{t \geq 0} f^t(U) = \bigcap_{t \geq 0} f^t(\overline{U})$$

is closed, locallly maximal and invariant.

Proof. Let U be a absorbing neighborhood of A such that $A = \bigcap_{t\geq 0} f^t(U)$. Note that $f^t(U) \subset \overline{f^t(U)} \subset U$ for $t > 0$ imply $U \subset f^{-t}(U)$. Then

$$f^{-s}(A) = \bigcap_{t \geq 0} f^{-s}(f^t(U)) = \bigcap_{t \geq 0} f^{t-s}(U) = A,$$

for $s \geq 0$, and hence A is invariant by Proposition A.3.

Note that

$$f^s(f^t(U)) \subset f^s(\overline{f^t(U)}) \subset f^s(U) \quad (t > 0, \quad s \in \kappa).$$

We obtain

$$A = \bigcap_{s \geq -t} f^s(f^t(U)) \subset \bigcap_{s \geq -t} f^s(\overline{f^t(U)}) \subset \bigcap_{s \geq -t} f^s(U) = A,$$

which implies that A is closed.

If A' is an invariant set with $A \subset A' \subset U$, then

$$A = f^t(A) \subset f^t(A')(= A') \subset f^t(U)$$

for all $t \in \kappa$. Thus we obtain

$$A \subset A' \subset \bigcap_{t \geq 0} f^t(U) = A.$$

Since A is invariant, for any $t > 0$ we have

$$A \subset \bigcap_{s \geq 0} f^s(\overline{U}) = \bigcap_{s \geq 0} f^{s-t}(f^t(\overline{U})) = \bigcap_{s \geq 0} f^{s-t}(\overline{f^t(U)}) \subset \bigcap_{s \geq 0} f^{s-t}(U) = A.$$

$\square$

Remark. Ya. B. Pesin[230], V. I. Arnold and Yu. S. Il'yashenko defined the set A in Proposition A.9 as an attractor. If $\kappa = \mathbb{Z}$ or $\mathbb{R}$, A. Katok and B. Hasselblatt [134] define an attractor as a compact subset A such that there exists a plus invariant neighborhood U of A with $A = \bigcap_{t\geq 0} f^t(U)$. Pilyugin [195] defines an attractor to be a plus asymptotically stable compact subset. A number of basic generic properties of attractors of homeomorphisms were described in [130]. For the topology of attractors, see [136] and [174].

Assume that M is a compact smooth manifold. Let N be another smooth manifold with a distance function d. For $f, g \in C(M, N)$, define

$$\rho_0(f, g) = \max_{x \in M}\{d(f(x), g(x))\}.$$

It is easy to see that ρ_0 is a metric on $C(M,N)$. The topology generated by the metric ρ_0 is also called the C^0-*topology* of $C(M,N)$. Obviously, $C(M,N)$ is complete with the metric ρ_0. If $M = N$, and if $f,g \in \mathrm{Hom}(M,M)$, define a metric on $\mathrm{Hom}(M,M)$ by

$$\hat{\rho}_0(f,g) = \rho_0(f,g) + \rho_0(f^{-1}, g^{-1}).$$

It is easy to see that $\rho_0, \hat{\rho}_0$ generate the same topology in $\mathrm{Hom}(M,M)$, since by compactness, all the mappings in question are uniformly continuous. Under the C^0-topology, $\mathrm{Hom}(M,M)$ is a closed subset of $C(M,M)$, and also is complete and so it is a Baire space, which means that any countable intersection of open dense subsets of $\mathrm{Hom}(M,M)$ is dense in $\mathrm{Hom}(M,M)$; such an intersection is called a *residual* subset of $\mathrm{Hom}(M,M)$. A property of homeomorphisms is said to be *generic* if the set of $f \in \mathrm{Hom}(M,M)$ satisfying the property contains a residual set.

Take $f,g \in C^r(M,N)$. Fix a finite covering of M by open sets $V_1, ..., V_l$ such that each $\overline{V_i}$ is contained in the domain of a coordinate atlas $(U_\alpha, \varphi_\alpha)$ of M. Define

$$\rho^{(r)}(f,g) = \max_{1 \leq i \leq l} \sup_{x \in \varphi_\alpha(V_i)} \sum_{\mu \in Z_+^m[r]} \left\| \frac{d^\mu w \circ f \circ \varphi_\alpha^{-1}}{dx^\mu}(x) - \frac{d^\mu w \circ g \circ \varphi_\alpha^{-1}}{dx^\mu}(x) \right\|,$$

where $w : N \longrightarrow \mathbb{R}^{2n+1}$ ($n = \dim N$) is an embedding. Also define

$$\rho^{(\infty)}(f,g) = \sum_{k=1}^{\infty} 2^{-k} \frac{\rho^{(k)}(f,g)}{1 + \rho^{(k)}(f,g)}.$$

It is easy to see that $\rho^{(r)}$ is a metric on $C^r(M,N)$. This metric induces the C^r-*topology* (as M is compact the topology is independent on the choice of $V_1, ..., V_l$). If $f,g \in \mathrm{Diff}^r(M,M)$, it is usual to use the following metric

$$\hat{\rho}_r(f,g) = \rho^{(r)}(f,g) + \rho^{(r)}(f^{-1}, g^{-1}).$$

Definition A.19 *A $f \in C^r(M,M)$ is C^k structurally stable $(0 \leq k \leq r)$ if there exists a neighborhood $\mathcal{U}$ of f in the C^k topology such that every mapping $g \in \mathcal{U}$ is topologically conjugate to f. A $f \in C(M,M)$ is topologically stable if it is C^0 structurally stable.*

In many references, a somewhat stronger version of structural stability often appears, which is both natural and practical.

Definition A.20 *A $f \in C^r(M,M)$ is C^k strongly structurally stable $(0 \leq k \leq r)$ if it is C^k structurally stable and in addition for any $g \in \mathcal{U}$ one can choose a conjugating homeomorphism h_g in such a way that both h_g and h_g^{-1} uniformly converge to the identity as g converges to f in the C^k topology.*

Definition A.21 *Two C^r flows $\{f^t\}$ on M and $\{g^t\}$ on N are said to be C^k orbit equivalent $(k \leq r)$ if there exists a $h \in \mathrm{Diff}^k(M,N)$ such that the flow $\{h \circ f^t \circ h^{-1}\}$ is a time change of the flow $\{g^t\}$, i.e., for each $x \in N$, $\{h \circ f^t \circ h^{-1}(x)\}$ and $\{g^t(x)\}$ coincide and the orientations given by the change of t in the positive direction are the same. If $k = 0$, $\{f^t\}$ and $\{g^t\}$ are also called topologically orbit equivalent.*

By definition, there is an increasing function σ_x of t for every x such that

$$g^{\sigma_x(t)}(x) = h \circ f^t \circ h^{-1}(x).$$

If two flows $\{f^t\}$ and $\{g^t\}$ are C^k (orbit) equivalent, then their vectors fields X and Y are also said to be C^k (orbit) equivalent.

Similarly, there are the notion of structural stability for flows.

Definition A.22 *A C^r flow $\{f^t\}$ on a manifold M is said to be C^k structurally stable $(0 \leq k \leq r)$ if there is a C^k-neighborhood $\mathcal{U}$ of the vector field generating f^t such that the f^t is topologically orbitally equivalent to any flow g^t generated by a vector field in $\mathcal{U}$.*

Definition A.23 *A C^r flow $\{f^t\}$ on a manifold M is said to be C^k strongly structurally stable) if, for any neighborhood $\mathcal{V}$ of the homeomorphism id in $\mathrm{Hom}(M, M)$ with the C^0-topology, there is a C^k-neighborhood $\mathcal{U}$ of the vector field generating f^t such that the f^t is topologically orbitally equivalent to any flow g^t generated by a vector field in $\mathcal{U}$ by means of some $h \in \mathcal{V}$.*

A.7   Measure and integral

Definition A.24 *Let M be a nonempty set. An algebra $\mathcal{B}$ for M is a nonempty collection of subsets of M such that*
   1) $B \in \mathcal{B} \Rightarrow B^c = M - B \in \mathcal{B}$;
   2) $B_1, B_2 \in \mathcal{B} \Rightarrow B_1 \cup B_2 \in \mathcal{B}$.
If 2) is replaced by
   2') $B_1, B_2, \cdots \in \mathcal{B} \Rightarrow \bigcup_{i=1}^{\infty} B_i \in \mathcal{B}$,
then $\mathcal{B}$ is called a σ-algebra on M, the pair $(M, \mathcal{B})$ is called a measurable space, and the elements of $\mathcal{B}$ are called its measurable sets.

Obviously, a σ-algebra is an algebra. If $\mathcal{B}$ is an algebra on M, then
   (i) $\emptyset, M \in \mathcal{B}$;
   (ii) $B_1, B_2 \in \mathcal{B} \Rightarrow B_1 \cap B_2 \in \mathcal{B}, B_1 - B_2 \in \mathcal{B}$;
   (iii) $B_1, B_2, \cdots, B_n \in \mathcal{B} \Rightarrow \bigcup_{i=1}^{n} B_i \in \mathcal{B}, \bigcap_{i=1}^{n} B_i \in \mathcal{B}$.
The conditions 1) and 2') in Definition A.24 imply

$$\bigcap_{i=1}^{\infty} B_i = \left(\bigcup_{i=1}^{\infty} B_i^c \right)^c \in \mathcal{B}.$$

Let $\mathcal{O}$ be a nonempty collection of subsets of M. The intersection of all σ-algebra which contain $\mathcal{O}$ also is a σ-algebra, called the σ-algebra generated by $\mathcal{O}$, denoted by $\mathcal{B}(\mathcal{O})$. If M is a topological space, a topology $\mathcal{O}$ on M can generate an associated σ-algebra $\mathcal{B} = \mathcal{B}(\mathcal{O})$ by simply adding in all countable unions and complements of existing sets (and also of those we generate), and is called the *Borel σ-algebra*. The elements of $\mathcal{B}(\mathcal{O})$ are called the Borel sets or Borel measurable.

Definition A.25 *Given a measurable space* $(M, \mathcal{B})$. *A measure* μ *on* $(M, \mathcal{B})$ *is a mapping* $\mu : \mathcal{B} \longrightarrow \mathbb{R}_+ \cup \{\infty\}$ *such that*

    1) $\mu(\emptyset) = 0$;

    2) $B_1, B_2, \cdots \in \mathcal{B}(B_i \cap B_j = \emptyset, i \neq j) \Rightarrow \mu(\bigcup_{i=1}^{\infty} B_i) = \sum_{i=1}^{\infty} \mu(B_i)$.

The triple $(M, \mathcal{B}, \mu)$ *is called a measure space . If* $\mu(M) < \infty$, *then* μ *is called a finite measure. In the case* $\mu(M) = 1$, μ *is called a probability measure, and* $(M, \mathcal{B}, \mu)$ *a probability space.*

Example A.1 (Dirac measure) *If we fix any point* $p \in M$ *we can define a probability measure* $\delta_p : \mathcal{B} \longrightarrow \mathbb{R}_+$ *by*

$$\delta_p(A) = \begin{cases} 1 & : \quad p \in A \\ 0 & : \quad p \notin A \, . \end{cases}$$

Rather than specifying the measure on all sets in the σ-algebra $\mathcal{B}$ it is usually enough to specify it on somewhat smaller families. The basic result in this direction is the following.

Theorem A.10 (Extension Theorem) *Assume that* $\mathcal{O}$ *is a nonempty collection of subsets of* M *such that*

    1) $\emptyset, M \in \mathcal{O}$;

    2) $B_1, B_2, \cdots, B_n \in \mathcal{O} \Rightarrow \bigcup_{i=1}^{n} B_i \in \mathcal{O}, \bigcap_{i=1}^{n} B_i \in \mathcal{O}$.

Let $\mathcal{B} = \mathcal{B}(\mathcal{O})$ *be the* σ-algebra generated by $\mathcal{O}$. Then any mapping $\mu : \mathcal{O} \longrightarrow \mathbb{R}_+$ which satisfies:

    3) $\mu(\emptyset) = 0$;

    4) $B_1, B_2, \cdots, B_n \in \mathcal{B}(B_i \cap B_j = \emptyset, i \neq j) \Rightarrow \mu(\bigcup_{i=1}^{n} B_i) = \sum_{i=1}^{n} \mu(B_i)$,

extends uniquely to a measure on $(M, \mathcal{B})$.

Here we exhibit some basic properties of measures. Let $\mathcal{B}$ be an algebra for M and let μ be a measure. Then

    (a) $\mu(A) \leq \mu(B)$ if $A, B \in \mathcal{B}$ with $A \subset B$;

    (b) $\mu(B - A) = \mu(B) - \mu(A)$ if $A, B \in \mathcal{B}$ with $A \subset B$ and $\mu(B) < \infty$;

    (c) $\mu\left(\bigcup_{n=1}^{\infty} A_n\right) \leq \sum_{n=1}^{\infty} \mu(A_n)$ if $\{A_n\}_{n \geq 1} \subset \mathcal{B}$ with $\bigcup_{n=1}^{\infty} A_n \in \mathcal{B}$.

Further if $\mathcal{B}$ is a σ-algebra, then

    (d) $\mu\left(\bigcup_{n=1}^{\infty} A_n\right) = \lim_{n \to \infty} \mu(A_n)$ if $\{A_n\}_{n \geq 1} \subset \mathcal{B}$ with $A_n \subset A_{n+1}$ for $n \geq 1$.

Definition A.26 *Let* M *be a separable locally compact Hausdorff space and* $\mathcal{B}$ *the Borel* σ-algebra. Then a Borel measure is a measure μ defined on $\mathcal{B}$ such that $\mu(K) < \infty$ when K is compact.

A measurable subset A of M is σ-finite, or μ is σ-finite on A if A is a countable union of sets of finite measure. Also μ is σ-finite if it is σ-finite on M. Let $\mathcal{B}$ be a σ-algebra containing the Borel sets. A measure μ on $\mathcal{B}$ is said to be regular if $\mu(K)$ is finite for all compact K, and if

$$\begin{aligned} \mu(B) &= \inf\{\mu(U) \mid B \subset U, \ U \text{ is open }\}, \\ \mu(B) &= \sup\{\mu(K) \mid K \subset B, \ K \text{ is compact }\}, \end{aligned}$$

for all $B \in \mathcal{B}$.

Definition A.27 *Given measurable spaces* $(M, \mathcal{B})$ *and* $(N, \mathcal{R})$, *mapping* $f : M \longrightarrow N$ *is measurable if* $f^{-1}(A) \in \mathcal{B}$ *whenever* $A \in \mathcal{R}$.

By the definition, if $f : M \longrightarrow N$ is measurable, and $g : N \longrightarrow L$ is measurable, then the composite $g \circ f$ is measurable. If N is a topological space, and $\mathcal{R}$ is the Borel σ-algebra, then f is measurable if and only if the inverse image of every open set is measurable. If $M = N$, we let $\mathcal{B} = \mathcal{R}$. From now on, our mappings will be defined in a topological space, and have values in a topological space, with the Borel σ-algebras. Thus any $f \in C(M, N)$ is automatically measurable. Let $\mathcal{L}^0(M, N) = \mathcal{L}^0(M, N; \mathcal{B}, \mathcal{R})$ be the set of measurable mappings from M into N.

Now we give the definition of the integral induced by a measure. Let $(M, \mathcal{B}, \mu)$ be a measure space. A measurable function ϕ on M, i.e., $\phi \in \mathcal{L}^0(M, \mathbb{R})$, is called a *simple function* if $\phi(M)$ is finite, which is equivalent to satisfy that there exist $E_1, ..., E_s \in \mathcal{B}, a_1, ..., a_s \in \mathbb{R}$ such that $E_i \cap E_j = \emptyset (i \neq j), M = \bigcup_{i=1}^{s} E_i$, and

$$\phi(x) = \sum_{i=1}^{s} a_i \chi_{E_i}(x),$$

where

$$\chi_{E_i}(x) = \begin{cases} 1 & : & x \in E_i \\ 0 & : & x \notin E_i . \end{cases}$$

Further if $\phi(M) \subset \mathbb{R}_+$ and if $A \in \mathcal{B}$, define the *integral*

$$\int_A \phi d\mu = \sum_{i=1}^{s} a_i \mu(E_i \cap A),$$

and ϕ is said to be *integrable* if

$$\int_A \phi d\mu < \infty.$$

If ϕ is a non-negative measurable function and if $A \in \mathcal{B}$, define

$$\int_A \phi d\mu = \sup \left\{ \int_A \psi d\mu \mid \psi \text{ simple functions with } 0 \leq \psi \leq \phi \text{ on } A \right\}.$$

Generally, define

$$\int_A \phi d\mu = \int_A \phi^+ d\mu - \int_A \phi^- d\mu,$$

where

$$\phi^+(x) = \max\{\phi(x), 0\}, \quad \phi^-(x) = \max\{-\phi(x), 0\}.$$

Define

$$\mathcal{L}^p(M) = \mathcal{L}^p(M, \mathcal{B}, \mu) = \left\{ \phi \in \mathcal{L}^0(M, \mathbb{R}) \mid \int_M |\phi|^p d\mu < +\infty \right\} \quad (0 < p < \infty),$$

$$\mathcal{L}^\infty(M) = \mathcal{L}^\infty(M, \mathcal{B}, \mu) = \{ \phi \in \mathcal{L}^0(M, \mathbb{R}) \mid \phi \text{ is bounded a.e. } (\mu) \},$$

where a.e. means "almost everywhere". Define the *norm*

$$\|\phi\|_p = \|\phi\|_{p,M} = \left\{ \int_M |\phi|^p d\mu \right\}^{1/p}, \quad \phi \in \mathcal{L}^p(M),$$

$$\|\phi\|_\infty = \|\phi\|_{\infty,M} = \inf\{C \mid |\phi(x)| \le C \text{ on } M \text{ a.e. } (\mu)\}, \quad \phi \in \mathcal{L}^\infty(M).$$

To make $\mathcal{L}^p(M)$ into a Banach space we need to identify mappings equivalent under the equivalence relation: $\phi \sim \psi$ if ϕ and ψ differ only on a set of zero measure. We can write

$$L^p = L^p(M, \mathcal{B}, \mu) = \mathcal{L}^p(M)/\sim \quad (0 < p \le \infty).$$

Then $L^p(1 \le p \le \infty)$ is complete with the distance $\|\phi - \psi\|_p$. We also have the following property:

Proposition A.10 *If $\mu(M) < \infty$, and if $\phi \in L^\infty$, then $\phi \in L^p(1 \le p < \infty)$ with*

$$\lim_{p \to \infty} \|\phi\|_p = \|\phi\|_\infty.$$

Let $\Sigma(M)$ be the non-empty set of all Borel probability measures on a measurable space $(M, \mathcal{B})$. If M is compact metrizable, then $\Sigma(M)$ possesses a natural convex structure, i.e., if $\mu_1, \mu_2 \in \Sigma(M)$, then $t\mu_1 + (1 - t)\mu_2 \in \Sigma(M), 0 \le t \le 1$, where

$$(t\mu_1 + (1 - t)\mu_2)(A) = t\mu_1(A) + (1 - t)\mu_2(A), \quad \text{for } A \in \mathcal{B},$$

and a natural topology, called the *weak* topology*. Namely, $\mu_n \to \mu$ if

$$\int_M \phi d\mu_n \to \int_M \phi d\mu$$

for every continuous function ϕ. Then $\Sigma(M)$ is compact with respect to this topology (Alaoglu); this can be easily seen from a usual diagonal argument.

Let $(M, \mathcal{B})$ be a measurable space and suppose μ, ν are two probability measures on $(M, \mathcal{B})$. We say μ is *absolutely continuous* with respect to ν if $\mu(B) = 0$ whenever $\nu(B) = 0$. The following theorem characterises absolute continuity.

Theorem A.11 (Radon-Nikodym) *Let μ, ν be two probability measures on a measurable space $(M, \mathcal{B})$. Then μ is absolutely continuous with respect to ν iff there exists $\phi \in L^1(M, \mathcal{B}, \nu)$ with $\phi \ge 0$ and $\int_M \phi d\nu = 1$ such that*

$$\mu(B) = \int_B \phi d\nu, \quad B \in \mathcal{B}.$$

The function ϕ is unique, and is called the Radon-Nikodym derivative of μ with respect to ν and denoted by $d\mu/d\nu$.

Next we introduce the integral induced by a Riemannian metric. Let M be a Riemannian manifold of dimension m. In terms of a local coordinate system $\{x_i\}$ on a coordinate neighborhood U, the Riemannian metric g can be given by

$$g = \sum_{i,j=1}^{m} g_{ij} dx_i \otimes dx_j.$$

Set $G = \det(g_{ij})$. Then $G > 0$ and is C^∞ on U. For $\phi \in C(M, \mathbb{R})$, denote the support of ϕ by

$$\text{supp}(\phi) = \overline{\{x \in M \mid \phi(x) \neq 0\}}.$$

Let $C_0(M)$ be the set of continuous functions with compact supports. If $\phi \in C_0(M)$ with $\text{supp}(\phi) \subset U$, define the *integral* of ϕ on M by the following usual integral

$$\int_M \phi = \int_U \phi \sqrt{G} dx_1 \cdots dx_m, \tag{A.33}$$

which is well-defined.

Let $\{U_\alpha\}_{\alpha \in \Lambda}$ be a locally finite coordinate covering of M, i.e., every point of M has a neighborhood which intersects only finitely many U_α's. Let $\{\rho_\alpha\}$ be a *partition of unity subordinate to the covering* $\{U_\alpha\}$, i.e., a family of differentiable functions $\{\rho_\alpha\}$ on M satisfying the following conditions:

    i) $0 \leq \rho_\alpha \leq 1$ on M for every $\alpha \in \Lambda$;
    ii) $\text{supp}(\rho_\alpha) \subset U_\alpha$ for every $\alpha \in \Lambda$;
    iii) $\sum_{\alpha \in \Lambda} \rho_\alpha(x) = 1$.

For $\phi \in C_0(M)$, define

$$\int_M \phi = \sum_\alpha \int_{U_\alpha} \rho_\alpha \phi, \tag{A.34}$$

which is independent of the choice of the covering $\{U_\alpha\}$ and the partition $\{\rho_\alpha\}$ of unity. Obviously, if $\phi \in C_0(M)$ with $\phi \geq 0$, then $\int_M \phi \geq 0$. Hence we obtain a positive linear functional $\int_M : C_0(M) \longrightarrow \mathbb{R}$. By Riesz's representation theorem, there exists a unique Borel measure $V = Vol_M$ on M called the *volume measure* such that for all $\phi \in C_0(M)$,

$$\int_M \phi = \int_M \phi dV. \tag{A.35}$$

By the standard measure theory, we obtain the linear functional $\int_M : L^p(M) \longrightarrow \mathbb{R}$ for $0 < p \leq \infty$.

Finally we define the integral of differential forms. We begin from the following notation.

Definition A.28 *A smooth m-dimensional manifold M is called orientable if there exists a continuous m-form Ω on M such that Ω vanishes nowhere. Such a form Ω is called to define an orientation of M. The two forms Ω and Ω' are called to define a same orientation of M if Ω'/Ω is a positive function on M.*

A connected orientable manifold has just two orientations. In fact, if two m-forms Ω and Ω' define orientations of M, the function

$$h = \Omega'/\Omega \quad (\text{ i.e., } \Omega' = h\Omega)$$

vanishes nowhere on M. Since M is connected, h is positive or negative on M. Hence either Ω' and Ω define a same orientation of M, or Ω' and $-\Omega$ define a same orientation of M.

Assume that the manifold M has an orientation defined by an m-form Ω. Let $\{(U_\alpha, \varphi_\alpha)\}$ be a coordinate covering of M and set $\varphi_\alpha = (x_1^\alpha, ..., x_m^\alpha)$. Write

$$\Omega = h_\alpha dx_1^\alpha \wedge \cdots \wedge dx_m^\alpha.$$

If $h_\alpha > 0$ on U_α, the coordinate system x_i^α is said to be *compatible* with the orientation of M, which is equivalent to the following condition

$$\Omega\left(\frac{\partial}{\partial x_1^\alpha}, ..., \frac{\partial}{\partial x_m^\alpha}\right) > 0.$$

Obviously, we can choose a coordinate covering $\{(U_\alpha, \varphi_\alpha)\}$ of M such that each coordinate system x_i^α is compatible with the orientation of M. Thus for $\phi \in C_0(M)$, we can define the *integral*

$$\int_M \phi\Omega = \sum_\alpha \int_{U_\alpha} \rho_\alpha \phi h_\alpha dx_1^\alpha \cdots dx_m^\alpha, \tag{A.36}$$

where $\{\rho_\alpha\}$ is a partition of unity subordinate to the covering $\{U_\alpha\}$. The definition is independent of the choice of the covering $\{U_\alpha\}$ and the partition $\{\rho_\alpha\}$ of unity. Note that every continuous m-form Φ with a compact support can be written as $\Phi = \phi\Omega$ for a $\phi \in C_0(M)$ so that $\int_M \Phi$ is well-defined. Also note that $\phi \mapsto \int_M \phi\Omega$ is a positive functional on $C_0(M)$. There exists an unique Borel measure μ_Ω such that

$$\int_M \phi\Omega = \int_M \phi d\mu_\Omega. \tag{A.37}$$

Definition A.29 *Assume that the manifold M has an orientation defined by an m-form Ω. A frame $\{e_1, ..., e_m\}$ on M is said to be positively oriented if $\Omega(e_1, ..., e_m) > 0$. A continuous m-form Φ is said to be a (positive) volume element on M if Φ is positive on positively oriented frames. It is said to be normalized if*

$$\int_M \Phi = 1.$$

Thus if M has an orientation defined by an m-form Ω, then Ω is a positive volume element on M. Note that

$$h_\beta = h_\alpha \text{Jacobi}(\varphi_\alpha \circ \varphi_\beta^{-1}),$$

if $U_\alpha \cap U_\beta \neq \emptyset$, where

$$\text{Jacobi}(\varphi_\alpha \circ \varphi_\beta^{-1}) = \det\left(\frac{\partial x_i^\alpha}{\partial x_j^\beta}\right)$$

is the Jacobi determinant of the coordinate transformation. Hence if covering $\{(U_\alpha, \varphi_\alpha)\}$ of M is compatible with the orientation of M, one has

$$\text{Jacobi}(\varphi_\alpha \circ \varphi_\beta^{-1}) > 0. \tag{A.38}$$

Conversely, one can show that if there exists a covering $\{(U_\alpha, \varphi_\alpha)\}$ of M such that (A.38) holds, then M is orientable.

Here we compare the integral of differential forms with the integral induced by a Riemannian metric. Assume that M is orientable, $\varphi_\alpha = (x_1^\alpha, ..., x_m^\alpha)$ is a compatible coordinate system with the orientation of M, write

$$g = \sum_{i,j=1}^{m} g_{ij}^\alpha dx_i^\alpha \otimes dx_j^\alpha.$$

and set $G_\alpha = \det(g_{ij}^\alpha)$. Note that

$$g_{ij}^\beta = g_{kl}^\alpha \frac{\partial x_k^\alpha}{\partial x_i^\beta} \frac{\partial x_l^\alpha}{\partial x_j^\beta}, \quad \sqrt{G_\beta} = \sqrt{G_\alpha} \text{Jacobi}(\varphi_\alpha \circ \varphi_\beta^{-1}).$$

Then a positive volume element Θ is defined on M by

$$\Theta|_{U_\alpha} = \sqrt{G_\alpha} dx_1^\alpha \wedge \cdots \wedge dx_m^\alpha.$$

If $\{e_1, ..., e_m\}$ is any orthonormal positively oriented frame fields, then

$$\langle e_1 \wedge \cdots \wedge e_m, \Theta \rangle = m! \Theta(e_1, ..., e_m) = 1,$$

i.e., if $\{\theta_1, ..., \theta_m\}$ is the dual frame fields of $\{e_1, ..., e_m\}$, then

$$\Theta = \theta_1 \wedge \cdots \wedge \theta_m.$$

The form Θ is unique, called the *associated volume element* of the metric g. For $\phi \in C_0(M)$, we have

$$\int_M \phi d\mu_\Theta = \int_M \phi \Theta = \int_M \phi = \int_M \phi dV. \tag{A.39}$$

Hence we have

$$\Theta = dV, \quad \mu_\Theta = Vol_M. \tag{A.40}$$

Theorem A.12 (Moser) *Let M be a smooth compact orientable manifold and Ω_0, Ω_1 two volume elements on M with the same total volume:*

$$\int_M \Omega_0 = \int_M \Omega_1.$$

Then there exists a diffeomorphism f such that $f^* \Omega_1 = \Omega_0$.

For the proof, see [134].

Appendix B

Foundations of complex dynamics

In this appendix, we introduce the foundations of complex dynamics.

B.1 Complex structures

Let κ be real field $\mathbb{R}$ or complex field $\mathbb{C}$ and let V be a vector space of dimension n over κ, which will be called a real (or complex) vector space if $\kappa = \mathbb{R}$ (or $\kappa = \mathbb{C}$). Let V^* be the *dual vector space* of V which consists of all κ-valued linear functions on V. The *inner product* between $X \in V$ and $\alpha \in V^*$ is defined by

$$\langle X, \alpha \rangle = \alpha(X).$$

Indentify $V^{**} = V$ by $\langle X, \alpha \rangle = \langle \alpha, X \rangle$ and $(\bigwedge_p V)^* = \bigwedge_p V^*$ by

$$\langle X_1 \wedge ... \wedge X_p, \alpha_1 \wedge ... \wedge \alpha_p \rangle = \det(\langle X_i, \alpha_j \rangle).$$

Take $p, q \in \mathbb{Z}[1, n]$ and take $\xi \in \bigwedge_p V$ and $\alpha \in \bigwedge_q V^*$. If $q \leq p$, the *interior product* $\xi \angle \alpha \in \bigwedge_{p-q} V$ is uniquely defined by

$$\langle \xi \angle \alpha, \beta \rangle = \langle \xi, \alpha \wedge \beta \rangle,$$

for all $\beta \in \bigwedge_{p-q} V^*$. If $p = q$, then

$$\xi \angle \alpha = \langle \xi, \alpha \rangle \in \mathbb{C} = \bigwedge_0 V$$

by definition. On the other hand, if $q > p$, we define $\xi \angle \alpha \in \bigwedge_{q-p} V^*$ such that if $\eta \in \bigwedge_{q-p} V$,

$$\langle \eta, \xi \angle \alpha \rangle = \langle \xi \wedge \eta, \alpha \rangle.$$

Let $\bigotimes_p V$ be the p-fold tensor product of V and let $\mathcal{S}_p$ be the permutation group of $\{1, ..., p\}$. For each $\nu \in \mathcal{S}_p$ a linear isomorphism $\nu : \bigotimes_p V \longrightarrow \bigotimes_p V$ is uniquely defined by

$$\nu(\xi_1 \otimes \cdots \otimes \xi_p) = \xi_{\nu^{-1}(1)} \otimes \cdots \otimes \xi_{\nu^{-1}(p)}$$

for $\xi_j \in V$. A vector $\xi \in \bigotimes_p V$ is said to be *symmetric* iff $\nu(\xi) = \xi$ for all $\nu \in S_p$. The set of all symmetric vectors in $\bigotimes_p V$ is a linear subspace $\coprod_p V$ of $\bigotimes_p V$ with

$$\dim \coprod_p V = \binom{n-1+p}{p}.$$

Here $\coprod_p V$ is called the *p-fold symmetric tensor product* of V. The linear mapping

$$S_p = \frac{1}{p!} \sum_{\nu \in S_p} \nu : \bigotimes_p V \longrightarrow \bigotimes_p V$$

is called the *symmetrizer* of $\bigotimes_p V$ with $S_p\left(\bigotimes_p V\right) = \coprod_p V$. If $\xi \in \coprod_p V$ and $\eta \in \coprod_q V$, the symmetric tensor product

$$\xi \amalg \eta = \eta \amalg \xi = S_{p+q}(\xi \otimes \eta) \in \coprod_{p+q} V$$

is defined. Let $\xi^d \in \coprod_{dp} V$ be the d-th symmetric tensor power of $\xi \in \coprod_p V$. The inner product between $\bigotimes_p V$ and $\left(\bigotimes_p V\right)^* = \bigotimes_p V^*$ restricts the inner product between $\coprod_p V$ and $\coprod_p V^*$ given by

$$\langle \xi_1 \amalg \cdots \amalg \xi_p, \alpha_1 \amalg \cdots \amalg \alpha_p \rangle = \frac{1}{p!} \sum_{\nu \in S_p} \langle \xi_1, \alpha_{\nu(1)} \rangle \cdots \langle \xi_p, \alpha_{\nu(p)} \rangle$$

for all $\xi_j \in V$ and $\alpha_j \in V^*$, so that we can identify $\coprod_p V^* = \left(\coprod_p V\right)^*$. Hence a linear isomorphism

$$\sim : \coprod_p V^* \longrightarrow \mathcal{L}_p(V)$$

is defined by

$$\tilde{\alpha}(\xi) = \langle \xi^p, \alpha \rangle, \quad \xi \in V, \alpha \in \coprod_p V^*,$$

where $\mathcal{L}_p(V)$ is the vector space of all homogeneous polynomials of degree p on V.

Now let V be a real vector space. Take a basis $\{e_1, ..., e_n\}$ in V and let $\{e^{*1}, ..., e^{*n}\}$ be the dual basis of $\{e_1, ..., e_n\}$, i.e.,

$$\langle e_j, e^{*l} \rangle = \delta_j^l = \begin{cases} 1 & : \ j = l \\ 0 & : \ j \neq l \,. \end{cases}$$

Let $\mathbf{J}$ be a linear endomorphism of V and let $\mathbb{J}_e = (J_l^k)$ be the matrix of $\mathbf{J}$ under the basis $\{e_1, ..., e_n\}$, i.e.,

$${}^t(\mathbf{J}e_1, ..., \mathbf{J}e_n) = \mathbb{J}_e \, {}^t(e_1, ..., e_n), \tag{B.1}$$

where $^t e$ denotes the transpose of the vector $e = (e_1, ..., e_n)$. Then the endomorphism on V induces that of V^*, also denoted by $\mathbf{J}$, defined by

$$\langle X, \mathbf{J} X^* \rangle = \langle \mathbf{J} X, X^* \rangle, \tag{B.2}$$

for $X \in V, X^* \in V^*$. Thus we have

$$(\mathbf{J} e^{*1}, ..., \mathbf{J} e^{*n}) = (e^{*1}, ..., e^{*n}) \mathbb{J}_e. \tag{B.3}$$

The endomorphism $\mathbf{J}$ also defines a 2-tensor, denoted also by $\mathbf{J}$, as follows

$$\mathbf{J} = \sum_{k,l=1}^{n} J_l^k e_k \otimes e^{*l}. \tag{B.4}$$

Definition B.1 *Let V be a real vector space. A complex structure on V is a linear endomorphism $\mathbf{J}$ of V such that*

$$\mathbf{J}^2 = -id : V \longrightarrow V$$

where id stands for the identity transformation of V.

Lemma B.1 *A real vector space V has a complex structure iff $\dim V$ is even.*

Proof. If λ is an eigenvalue of $\mathbf{J}$ and if X is an eigenvector of $\mathbf{J}$ for λ, then $X \neq 0$ and

$$\mathbf{J} X = \lambda X, \quad -X = \mathbf{J}^2 X = \lambda \mathbf{J} X = \lambda^2 X$$

which gives $\lambda = \pm i$. Hence the eigenpolynomial of $\mathbf{J}$ is

$$|\lambda I_n - \mathbb{J}_e| = (\lambda^2 + 1)^{n/2}.$$

Therefore n is even.

Conversely, assume that $\dim V = 2m$. Take a basis $\{e_1, ..., e_{2m}\}$ in V and define a linear endomorphism $\mathbf{J}$ of V such that

$$\mathbf{J} e_j = e_{m+j}, \quad \mathbf{J} e_{m+j} = -e_j \quad (j = 1, ..., m). \tag{B.5}$$

Obviously $\mathbf{J}$ is a complex structure on V. $\qquad \square$

Assume that a real vector space V has a complex structure $\mathbf{J}$ so that we can assume

$$\dim V = 2m.$$

Since the eigenvectors of $\mathbf{J}$ are complex, we have to discuss the complexification of V, i.e.,

$$V_{\mathbb{C}} = V \otimes_{\mathbb{R}} \mathbb{C} = \mathbb{C}\{e_1, ..., e_{2m}\},$$

which is a complex $2m$-dimensional vector space. Then

$$V = \mathbb{R}\{e_1, ..., e_{2m}\}$$

is a real subspace of V_C in a natural manner. The complex structure of V extends in a natural manner to that of V_C defined by

$$\mathbf{J}(X + iY) = \mathbf{J}X + i\mathbf{J}Y, \quad X, Y \in V.$$

For $Z = X + iY \in V_C$, we denote the conjugate of Z by $\overline{Z} = X - iY$. Then $\mathbf{J}\overline{Z} = \overline{\mathbf{J}Z}$. Define

$$\mathbf{V} = \{Z \in V_C \mid \mathbf{J}Z = iZ\},$$
$$\overline{\mathbf{V}} = \{Z \in V_C \mid \mathbf{J}Z = -iZ\}.$$

If $Z = X + iY \in \mathbf{V}$, then $\mathbf{J}X = -Y, \mathbf{J}Y = X$ so that $\mathbf{J}\overline{Z} = -i\overline{Z}$. The following proposition is evident:

Proposition B.1 *Let V be a real $2m$-dimensional vector space. Then*
   1) $\mathbf{V} = \{X - i\mathbf{J}X \mid X \in V\}, \overline{\mathbf{V}} = \{X + i\mathbf{J}X \mid X \in V\};$
   2) $V_C = \mathbf{V} \oplus \overline{\mathbf{V}}, \quad \overset{\wedge}{_r}V_C = \underset{p+q=r}{\bigoplus}(\overset{\wedge}{_p}\mathbf{V}) \wedge (\overset{\wedge}{_q}\overline{\mathbf{V}});$
   3) The mapping $Z \in \mathbf{V} \mapsto \overline{Z} \in \overline{\mathbf{V}}$ is a real linear isomorphism. Hence $\mathbf{V}$ and $\overline{\mathbf{V}}$ are
complex m-dimensional vector spaces.

From now on, we fix a basis $\{\xi_1, ..., \xi_m\}$ of $\mathbf{V}$. Then

$$\overline{\mathbf{V}} = \mathbb{C}\{\overline{\xi}_1, ..., \overline{\xi}_m\},$$

$$V_C = \mathbb{C}\{\xi_1, ..., \xi_m, \overline{\xi}_1, ..., \overline{\xi}_m\}.$$

Thus we can write

$$\xi_k = \frac{1}{2}(e_k - i\mathbf{J}e_k), \quad \overline{\xi}_k = \frac{1}{2}(e_k + i\mathbf{J}e_k), \quad e_k \in V, 1 \le k \le m. \tag{B.6}$$

Obviously,

$$(e_1, ..., e_m, \mathbf{J}e_1, ..., \mathbf{J}e_m) = (\xi_1, ..., \xi_m, \overline{\xi}_1, ..., \overline{\xi}_m)P, \tag{B.7}$$

where

$$P = \begin{pmatrix} I_m & iI_m \\ I_m & -iI_m \end{pmatrix}. \tag{B.8}$$

Hence $(e_1, ..., e_m, \mathbf{J}e_1, ..., \mathbf{J}e_m)$ is a basis of V. Under the basis, the matrix $\mathbf{J}_e$ of $\mathbf{J}$ is

$$\mathbf{J}_e = \begin{pmatrix} 0 & I_m \\ -I_m & 0 \end{pmatrix}. \tag{B.9}$$

The matrix $\mathbf{J}_e$ has the following property:

$$\begin{pmatrix} A & C \\ B & D \end{pmatrix}\mathbf{J}_e = \mathbf{J}_e\begin{pmatrix} A & C \\ B & D \end{pmatrix} \iff A = D, C = -B,$$

for $A, B, C, D \in GL(m; \mathbb{R})$. Hence the complex general linear group $GL(m; \mathbb{C})$ of degree m can be identified with the subgroup of $GL(2m; \mathbb{R})$ consisting of matrices which commute with $\mathbf{J}_e$. Thus we obtain an embedding

$$\mathcal{P} : GL(m; \mathbb{C}) \longrightarrow GL(2m; \mathbb{R})$$

called the *real representation* of $GL(m; \mathbb{C})$, which is given by

$$\mathcal{P}(A + iB) = \begin{pmatrix} A & -B \\ B & A \end{pmatrix} \quad \text{for } A, B \in GL(m; \mathbb{R}). \tag{B.10}$$

Obviously, we have

$$\mathcal{P}(\mathbf{H}) = {}^t P \begin{pmatrix} 0 & \frac{1}{2}\overline{\mathbf{H}} \\ \frac{1}{2}\mathbf{H} & 0 \end{pmatrix} P \quad \text{for } \mathbf{H} \in GL(m; \mathbb{C}). \tag{B.11}$$

For $\mathbf{H}, \mathbf{H}_1, \mathbf{H}_2 \in GL(m; \mathbb{C})$, we also have

$$\mathcal{P}(\mathbf{H}_1 \mathbf{H}_2) = \mathcal{P}(\mathbf{H}_1)\mathcal{P}(\mathbf{H}_2), \quad \mathcal{P}(\mathbf{H}^{-1}) = \mathcal{P}(\mathbf{H})^{-1}, \tag{B.12}$$

$$\mathcal{P}(\mathbf{H}_1) = \mathcal{P}(\mathbf{H}_2) \Longleftrightarrow \mathbf{H}_1 = \mathbf{H}_2. \tag{B.13}$$

For $\mathbf{H} \in GL(m; \mathbb{C})$, $\lambda \in \mathbb{C}$, it is easy to prove that there exists a constant $c \neq 0$ such that

$$\det(\mathcal{P}(\mathbf{H}) - \lambda I_{2m}) = c \det(\mathbf{H} - \lambda I_m)\overline{\det(\mathbf{H} - \overline{\lambda} I_m)}.$$

Hence eigenvalues of $\mathbf{H}$ also are eigenvalues of $\mathcal{P}(\mathbf{H})$.

We now define another real representation

$$\mathfrak{R} : \mathbf{V} \longrightarrow V,$$

by setting $\mathfrak{R}(Z) = \sqrt{2}\mathrm{Re}(Z)$. For $Z \in \mathbf{V}$, write

$$Z = \sum_{k=1}^{m} z_k \xi_k, \quad z_k = x_k + iy_k.$$

Then

$$\mathfrak{R}(Z) = \sqrt{2}\mathrm{Re}(Z) = \frac{1}{\sqrt{2}}(Z + \overline{Z}) = \frac{1}{\sqrt{2}}\sum_{k=1}^{m}(x_k e_k + y_k \mathbf{J} e_k).$$

Note that

$${}^t(z_1, ..., z_m, \overline{z}_1, ..., \overline{z}_m) = P \, {}^t(x_1, ..., x_m, y_1, ..., y_m).$$

For $Z \in \mathbf{V}$, $A, B \in GL(m; \mathbb{R})$, we have

$$\mathfrak{R}((A + iB)Z) = \mathcal{P}(A + iB)\mathfrak{R}(Z).$$

Now we consider the transformation of bases in V. Let $(s_1, ..., s_m, \mathbf{J}s_1, ..., \mathbf{J}s_m)$ be another basis of V and set

$$(s_1, ..., s_m, \mathbf{J}s_1, ..., \mathbf{J}s_m) = (e_1, ..., e_m, \mathbf{J}e_1, ..., \mathbf{J}e_m)G. \tag{B.14}$$

Then we obtain $G\mathbf{J}_e = \mathbf{J}_e G$ so that G is of the form

$$G = \begin{pmatrix} A & -B \\ B & A \end{pmatrix} = \mathcal{P}(A + iB) \quad \text{for } A, B \in GL(m; \mathbb{R}). \tag{B.15}$$

If we set

$$\zeta_k = \frac{1}{2}(s_k - i\mathbf{J}s_k), \quad k = 1, ..., m,$$

then

$$(\zeta_1, ..., \zeta_m) = (\xi_1, ..., \xi_m)(A + iB). \tag{B.16}$$

Obviously,

$$\det(G) = |\det(A + iB)|^2 > 0.$$

Thus we have proved the following result:

Theorem B.1 *Let $\mathbf{J}$ be a complex structure on a $2m$-dimensional real vector space V. Then there exist elements $\{e_1, ..., e_m\}$ of V such that $\{e_1, ..., e_m, \mathbf{J}e_1, ..., \mathbf{J}e_m\}$ is a basis for V. Such bases all define the same orientation of V.*

Let V^* be the dual space of V. Its complexification $V^*_{\mathbf{C}}$ is the dual space of $V_{\mathbf{C}}$. Let $\{e^{*1}, ..., e^{*2m}\}$ be the dual basis of $\{e_1, ..., e_m, \mathbf{J}e_1, ..., \mathbf{J}e_m\}$. Then (B.3) and (B.9) imply

$$\mathbf{J}e^{*k} = -e^{*m+k}, \quad \mathbf{J}e^{*m+k} = e^{*k}. \tag{B.17}$$

Define

$$\lambda^k = e^{*k} - i\mathbf{J}e^{*k}, \quad k = 1, ..., m. \tag{B.18}$$

Then we have

$$\begin{aligned} \langle \xi_l, \lambda^k \rangle &= \langle \overline{\xi}_l, \overline{\lambda}^k \rangle = \delta^k_l \\ \langle \xi_l, \overline{\lambda}^k \rangle &= \langle \overline{\xi}_l, \lambda^k \rangle = 0. \end{aligned} \tag{B.19}$$

Hence $\{\lambda^1, ..., \lambda^m\}$ and $\{\overline{\lambda}^1, ..., \overline{\lambda}^m\}$ are the dual bases of $\{\xi_1, ..., \xi_m\}$ and $\{\overline{\xi}_1, ..., \overline{\xi}_m\}$, respectively. Thus we have a direct sum decomposition as above:

$$V^*_{\mathbf{C}} = \mathbf{V}^* \oplus \overline{\mathbf{V}}^*, \quad \bigwedge V^*_{\mathbf{C}} = \bigoplus_r \bigoplus_{p+q=r} \left(\bigwedge_p \mathbf{V}^* \right) \wedge \left(\bigwedge_q \overline{\mathbf{V}}^* \right).$$

Obviously, we have

$$\begin{aligned} \mathbf{V}^* &= \{\lambda \in V^*_{\mathbf{C}} \mid \mathbf{J}\lambda = i\lambda\} \\ &= \{\alpha - i\mathbf{J}\alpha \mid \alpha \in V^*\} \\ &= \{\lambda \in V^*_{\mathbf{C}} \mid \langle Z, \lambda \rangle = 0, Z \in \overline{\mathbf{V}}\}, \\ \overline{\mathbf{V}}^* &= \{\lambda \in V^*_{\mathbf{C}} \mid \mathbf{J}\lambda = -i\lambda\} \\ &= \{\alpha + i\mathbf{J}\alpha \mid \alpha \in V^*\} \\ &= \{\lambda \in V^*_{\mathbf{C}} \mid \langle Z, \lambda \rangle = 0, Z \in \mathbf{V}\}. \end{aligned} \tag{B.20}$$

Finally we consider the transformation of bases in V^*. From (B.7), we easily prove

$$^t(e^{*1}, ..., e^{*m}, -\mathbf{J}e^{*1}, ..., -\mathbf{J}e^{*m}) = P^{-1} \, {}^t(\lambda^1, ..., \lambda^m, \overline{\lambda}^1, ..., \overline{\lambda}^m). \tag{B.21}$$

Let $(s_1, ..., s_m, \mathbf{J}s_1, ..., \mathbf{J}s_m)$ be another basis of V satisfying (B.14) and define

$$\gamma^k = s^{*k} - i\mathbf{J}s^{*k}, \quad k = 1, ..., m,$$

where $\{s^{*1}, ..., s^{*m}, -\mathbf{J}s^{*1}, ..., -\mathbf{J}s^{*m}\}$ is the dual basis of $\{s_1, ..., s_m, \mathbf{J}s_1, ..., \mathbf{J}s_m\}$. Then

$$^t(s^{*1}, ..., s^{*m}, -\mathbf{J}s^{*1}, ..., -\mathbf{J}s^{*m}) = G \, {}^t(e^{*1}, ..., e^{*m}, -\mathbf{J}e^{*1}, ..., -\mathbf{J}e^{*m}), \tag{B.22}$$

$$^t(\gamma^1, ..., \gamma^m) = (A + iB) \, {}^t(\lambda^1, ..., \lambda^m). \tag{B.23}$$

B.2   Hermitian structures

Definition B.2 *The index of a bilinear function φ on a vector space V, is defined to be the maximal dimension of a subspace of V on which φ is negative definite; the nullity is the dimension of the null-space, i.e., the subspace consisting of all $X \in V$ such that $\varphi(X, Y) = 0$ for every $Y \in V$.*

Let V be a real vector space of dimension n. Take a basis $\{e_1, ..., e_n\}$ in V and let $\{e^{*1}, ..., e^{*n}\}$ be the dual basis of $\{e_1, ..., e_n\}$. Let $\varphi : V \times V \longrightarrow \mathbb{R}$ be a bilinear function which can be expressed by a 2-tensor

$$\begin{aligned}
\varphi &= \sum_{k,l=1}^{n} \varphi_{kl} e^{*k} \otimes e^{*l} \\
&= (e^{*1}, ..., e^{*n}) \Phi \, {}^t(e^{*1}, ..., e^{*n}), \tag{B.24} \\
\Phi &= (\varphi_{kl}), \quad \varphi_{kl} = \varphi(e_k, e_l), \quad 1 \le k, l \le n. \tag{B.25}
\end{aligned}$$

For each $X \in V$, we can define an element $\varphi_X \in V^*$ by setting $\varphi_X(Y) = \varphi(X, Y)$ for all $Y \in V$. By using the basis, we can write

$$\varphi_X = \sum_{k,l=1}^{n} \varphi_{kl} \langle X, e^{*k} \rangle e^{*l}.$$

If the mapping $X \mapsto \varphi_X$ is an isomorphism from V to V^*, then φ is said to be *nondegenerate*. Obviously, φ is nondegenerate iff

$$\det(\Phi) \ne 0,$$

or iff φ has nullity equal to 0.

Assume that φ is skew-symmetric, i.e.,

$$\varphi(Y, X) = -\varphi(X, Y), \quad \text{for all } X, Y \in V.$$

Then φ can be expressed by a 2-form

$$\varphi = \sum_{k,l=1}^{n} \varphi_{kl} e^{*k} \wedge e^{*l}. \tag{B.26}$$

Now φ_X is given by

$$\varphi_X = \frac{1}{2} X \lrcorner \varphi.$$

Note that Φ is a skew-symmetric, i.e.,

$$^t\Phi = -\Phi.$$

Hence the nondegeneracy of φ implies that $\dim V$ is even. Assume that $\dim V = 2m$. Since φ is skew-symmetric, we can choose the base $\{e_1, ..., e_{2m}\}$ of V such that φ is of the following form:

$$\varphi = 2\sum_{k=1}^{m} \lambda_k e^{*2k-1} \wedge e^{*2k}, \tag{B.27}$$

where $-\lambda_k^2$ is the eigenvalue of Φ. Note that

$$\det(\Phi) = (\lambda_1 \cdots \lambda_m)^2, \quad \varphi^m = 2^m m!(\lambda_1 \cdots \lambda_m)e^{*1} \wedge \cdots \wedge e^{*2m}.$$

Therefore

$$\det(\Phi) \neq 0 \iff \varphi^m \neq 0.$$

Thus we obtain

Lemma B.2 *Let φ be a skew-symmetric 2-form on a real vector space V. Then φ is nondegenerate iff $\dim V = 2m$ is even and the m-th exterior power φ^m of φ is not zero.*

Lemma B.3 *Let φ be a skew-symmetric 2-form on a real vector space V. Then φ is nondegenerate iff $\dim V = 2m$ for some $m \in \mathbb{N}$ and there is a symplectic basis $\{e_1, ..., e_{2m}\}$ of V such that*

$$\begin{aligned}
\varphi(e_k, e_{m+k}) &= 1, \quad k = 1, ..., m \\
\varphi(e_k, e_l) &= 0, \quad |k - l| \neq m.
\end{aligned} \tag{B.28}$$

Proof. ($\Leftarrow$) With respect to the basis, $\Phi = \mathbb{J}_e$ so that $\det(\Phi) \neq 0$. Hence φ is nondegenerate.

($\Rightarrow$) Since φ is nondegenerate there exist e_1, e_{m+1} such that $\varphi(e_1, e_{m+1}) \neq 0$, and without loss of generality we may assume $\varphi(e_1, e_{m+1}) = 1$. By skew-symmetry

$$\varphi(e_1, e_1) = \varphi(e_{m+1}, e_{m+1}) = 0, \quad \varphi(e_{m+1}, e_1) = -1,$$

so the matrix of $\varphi|_{V_1}$, where $V_1 = \mathbb{R}\{e_1, e_{m+1}\}$, with respect to (e_1, e_{m+1}) is

$$\Phi_1 = \begin{pmatrix} 0 & 1 \\ -1 & 0 \end{pmatrix}. \tag{B.29}$$

Now we use induction. Define

$$V_2 = \{X \in V \mid \varphi(X, Y) = 0 \text{ for all } Y \in V_1\}.$$

Since for $X \in V$, we have

$$X - \varphi(X, e_{m+1})e_1 + \varphi(X, e_1)e_{m+1} \in V_2,$$

then

$$V_1 \cap V_2 = \{0\}, \quad V_1 \oplus V_2 = V.$$

This claim is obtained inductively. □

Using the basis in Lemma B.3 and using the complex structure defined by (B.5), we can prove that a non-degenerate skew-symmetric 2-form φ on V satisfies

$$\varphi(\mathbf{J}X, \mathbf{J}Y) = \varphi(X, Y).$$

Definition B.3 *A non-degenerate skew-symmetric 2-form φ is called a symplectic form. A real vector space V with a symplectic form φ is called a symplectic vector space. A subspace E of the symplectic vector space (V, φ) is said to be isotropic if $\varphi|_E = 0$, and to be non-degenerate if $(E, \varphi|_E)$ is a symplectic vector space. An isotropic subspace of dimension $m = \frac{1}{2} \dim V$ is said to be Lagrangian.*

Thus the basis in Lemma B.3 gives a decomposition of V as a direct sum of two Lagrangian subspaces. Notice that by nondegeneracy of φ, an isotropic subspace has a dimension of at most $m = \frac{1}{2} \dim V$, so Lagrangian subspaces are maximal isotropic subspaces.

If (V, φ) and (W, ψ) are symplectic vector spaces, an invertible mapping $\mathbf{A} : V \longrightarrow W$ is called *symplectic* if $\mathbf{A}^*\psi = \varphi$, where

$$\mathbf{A}^*\psi(X, Y) = \psi(\mathbf{A}X, \mathbf{A}Y) \quad \text{for } X, Y \in V.$$

An immediate observation from the preceding results is that symplectic mappings preserve volume and orientation.

Proposition B.2 *A symplectic $\mathbf{A} : V \longrightarrow V$ is an automorphism. If λ is an eigenvalue of $\mathbf{A}$, so are $\bar{\lambda}, \lambda^{-1}$, and $\bar{\lambda}^{-1}$.*

Proof. If $\mathbf{A}X = 0$ for $X \in V$, then

$$0 = \varphi(\mathbf{A}X, \mathbf{A}Y) = \mathbf{A}^*\varphi(X, Y) = \varphi(X, Y) \quad \text{for all } Y \in V,$$

implies $X = 0$. Thus $\mathbf{A} : V \longrightarrow V$ is an automorphism.

Since φ is non-degenerate, we have the isomorphism

$$\hat{\varphi} : V \longrightarrow V^*; \quad X \mapsto \varphi_X.$$

Note that

$$\varphi(\mathbf{A}X, Y) = \varphi(X, \mathbf{A}^{-1}Y),$$

i.e., $\hat{\varphi} \circ \mathbf{A} = {}^t\mathbf{A}^{-1} \circ \hat{\varphi}$, where ${}^t\mathbf{A}$ denotes the transpose of $\mathbf{A}$. Thus ${}^t\mathbf{A}^{-1}$ is conjugate to $\mathbf{A}$.

Let λ be an eigenvalue of $\mathbf{A}$. Since $\mathbf{A}$ is real, $\bar{\lambda}$ is also an eigenvalue of $\mathbf{A}$. From what we just saw, λ^{-1} and $\bar{\lambda}^{-1}$ are also eigenvalues of $\mathbf{A}$. □

Remark. The proof also shows that $\det(\mathbf{A}) = \pm 1$.

Definition B.4 *A complex value function* $h : V \times V \longrightarrow \mathbb{C}$ *is called a Hermitian structure on the real vector space V with a complex structure $\mathbf{J}$ if*

1) $h(a_1 X_1 + a_2 X_2, Y) = a_1 h(X_1, Y) + a_2 h(X_2, Y)$;

2) $h(Y, X) = \overline{h(X, Y)}$;

3) $h(\mathbf{J}X, Y) = i h(X, Y)$,

for any $X, X_1, X_2, Y \in V; a_1, a_2 \in \mathbb{R}$.

Theorem B.2 *There exists a Hermitian structure on the real vector space V with a complex structure $\mathbf{J}$ iff there exists a real skew-symmetric bilinear function* $\varphi : V \times V \longrightarrow \mathbb{R}$ *such that*

$$\varphi(\mathbf{J}X, \mathbf{J}Y) = \varphi(X, Y).$$

Proof. ($\Longrightarrow$) For $X, Y \in V$, write

$$2h(X, Y) = g(X, Y) + i\varphi(X, Y), \tag{B.30}$$

where $g, \varphi : V \times V \longrightarrow \mathbb{R}$ are real bilinear functions. Then the property 2) in Definition B.4 implies that g is symmetric, and φ skew-symmetric, i.e.,

$$g(Y, X) = g(X, Y), \quad \varphi(Y, X) = -\varphi(X, Y). \tag{B.31}$$

By the property 3) of Definition B.4, we have

$$g(\mathbf{J}X, Y) = -\varphi(X, Y), \quad \varphi(\mathbf{J}X, Y) = g(X, Y). \tag{B.32}$$

Therefore

$$\begin{aligned} g(\mathbf{J}X, \mathbf{J}Y) &= -\varphi(X, \mathbf{J}Y) = \varphi(\mathbf{J}Y, X) = g(Y, X) = g(X, Y), \\ \varphi(\mathbf{J}X, \mathbf{J}Y) &= \varphi(X, Y). \end{aligned} \tag{B.33}$$

($\Longleftarrow$) Set $g(X, Y) = \varphi(\mathbf{J}X, Y)$ and define $h(X, Y)$ by (B.30). It is easy to prove that h is a Hermitian structure. $\square$

From the proof of the theorem, we see that the Hermitian structure satisfies

$$h(\mathbf{J}X, \mathbf{J}Y) = h(X, Y). \tag{B.34}$$

Similarly, we have

Theorem B.3 *There exists a Hermitian structure on the real vector space V with a complex structure $\mathbf{J}$ iff there exists a real symmetric bilinear function* $g : V \times V \longrightarrow \mathbb{R}$ *such that*

$$g(\mathbf{J}X, \mathbf{J}Y) = g(X, Y). \tag{B.35}$$

The symmetric 2-tensor g in this theorem is called a *Hermitian inner product*. If $g(X, X) > 0$ for all $X \in V$ with $X \neq 0$, then g is said to be *positive definite*. If so, define the norm

$$\|X\| := \sqrt{g(X, X)}.$$

Further, if $Z \in \mathbf{V}$, define

$$\|Z\| = \|\Re(Z)\|.$$

If we write

$$g = \sum_{k,l=1}^{2m} g_{kl} e^{*k} \otimes e^{*l}$$

$$= (e^{*1}, ..., e^{*2m}) G \, {}^t(e^{*1}, ..., e^{*2m}), \tag{B.36}$$

$$G = (g_{kl}), \quad g_{kl} = g(e_k, e_l), \quad 1 \leq k, l \leq 2m, \tag{B.37}$$

then g is positive definite iff the matrix G is positive definite. Hence the positive definite g is nondegenerate. The Hermitian structure h is called *positive definite* if g is positive definite.

Given a Hermitian structure h on the real vector space V with a complex structure $\mathbf{J}$ such that they define g, φ by (B.30). Let $\{e_1, ..., e_{2m}\}$ be a basis of V and let $\{e^{*1}, ..., e^{*2m}\}$ be its dual basis. We can write

$$h = \sum_{k,l=1}^{2m} h_{kl} e^{*k} \otimes e^{*l}, \tag{B.38}$$

$$H = (h_{kl}), \quad h_{kl} = h(e_k, e_l), \quad 1 \leq k, l \leq 2m. \tag{B.39}$$

Now assume that $\{e_1, ..., e_{2m}\}$ and $\{e^{*1}, ..., e^{*2m}\}$ satisfy (B.5) and (B.17) and write

$$A = (g_{kl}), \quad B = (g_{k,m+l}), \quad \mathbf{H} = (h_{kl}), \quad 1 \leq k, l \leq m.$$

Then $\mathbf{H} = \frac{1}{2}(A + iB)$, and we obtain

$$G = \begin{pmatrix} A & -B \\ B & A \end{pmatrix} = \mathcal{P}(2\mathbf{H}), \quad \Phi = \begin{pmatrix} B & A \\ -A & B \end{pmatrix} = \mathcal{P}(-2i\mathbf{H}), \quad H = \begin{pmatrix} \mathbf{H} & i\mathbf{H} \\ -i\mathbf{H} & \mathbf{H} \end{pmatrix}.$$

In terms of the basis $\{\lambda^1, ..., \lambda^m, \overline{\lambda}^1, ..., \overline{\lambda}^m\}$ of $V_{\mathbb{C}}^*$, we have

$$h = (e^{*1}, ..., e^{*2m}) H \, {}^t(e^{*1}, ..., e^{*2m})$$

$$= (\lambda^1, ..., \lambda^m, \overline{\lambda}^1, ..., \overline{\lambda}^m) \, {}^t P^{-1} H P^{-1} \, {}^t(\lambda^1, ..., \lambda^m, \overline{\lambda}^1, ..., \overline{\lambda}^m)$$

$$= (\overline{\lambda}^1, ..., \overline{\lambda}^m) \mathbf{H} \, {}^t(\lambda^1, ..., \lambda^m),$$

i.e., h can be given by

$$h = \sum_{k,l=1}^{m} h_{kl} \lambda^k \otimes \overline{\lambda}^l. \tag{B.40}$$

Then we have

$$g = 2\mathrm{Re}(h) = \sum_{k,l=1}^{m} h_{kl} (\lambda^k \otimes \overline{\lambda}^l + \overline{\lambda}^l \otimes \lambda^k), \tag{B.41}$$

$$\varphi = -2i \sum_{k,l=1}^{m} h_{kl} \lambda^k \wedge \overline{\lambda}^l. \tag{B.42}$$

We also have

$$\det(G) = \det(\Phi) = |\det(2\mathbf{H})|^2. \tag{B.43}$$

Thus we have proved

Theorem B.4 *There exists a symplectic form on a real vector space V iff there exists a nondegenerate Hermitian inner product.*

Definition B.5 *Let E be a complex vector space. A complex value function $F : E \times E \longrightarrow \mathbb{C}$ is called a Hermition structure on E if*
1) $F(b_1 Z_1 + b_2 Z_2, W) = b_1 F(Z_1, W) + b_2 F(Z_2, W)$;
2) $F(W, Z) = \overline{F(Z, W)}$,
for any $Z, Z_1, Z_2, W \in E; b_1, b_2 \in \mathbb{C}$. If $F(Z, Z) > 0$ for all $Z \in E$ with $Z \neq 0$, then F is said to be positive definite.

Now we continue to discuss the Hermitian structure h on the real vector space V. Let F be one of the functions h, g or φ. For $X_j, Y_j \in V$ $(j = 1, 2)$, define

$$
\begin{aligned}
F(X_1 + iY_1, X_2 + iY_2) \quad &:= \quad F(X_1, X_2) + F(Y_1, Y_2) \\
&\quad + i(F(Y_1, X_2) - F(X_1, Y_2)).
\end{aligned} \tag{B.44}
$$

A simple calculation shows that h and g are extended uniquely to Hermitian structures on $V_{\mathbb{C}}$, and φ satisfies the property 1) in Definition B.5 such that

$$
\begin{aligned}
\varphi(W, Z) &= -\overline{\varphi(Z, W)} = -\varphi(\overline{Z}, \overline{W}), \tag{B.45} \\
2h(Z, W) &= g(Z, W) + i\varphi(Z, W) \tag{B.46}
\end{aligned}
$$

for $Z, W \in V_{\mathbb{C}}$. For the basis $\{\xi_1, ..., \xi_m\}$ of $\mathbf{V}$, we have

$$
\begin{aligned}
h(\xi_k, \xi_l) &= h_{kl}, \ h(\xi_k, \overline{\xi}_l) = h(\overline{\xi}_k, \overline{\xi}_l) = 0, \tag{B.47} \\
g(\xi_k, \xi_l) &= h_{kl}, \ g(\xi_k, \overline{\xi}_l) = 0, g(\overline{\xi}_k, \overline{\xi}_l) = h_{lk}, \tag{B.48} \\
\varphi(\xi_k, \xi_l) &= -ih_{kl}, \ \varphi(\xi_k, \overline{\xi}_l) = 0, \varphi(\overline{\xi}_k, \overline{\xi}_l) = ih_{lk}. \tag{B.49}
\end{aligned}
$$

Also we have

$$
\begin{aligned}
2h(X + iY, X + iY) &= g(X, X) + g(Y, Y) + 2\varphi(X, Y), \tag{B.50} \\
g(X + iY, X + iY) &= g(X, X) + g(Y, Y), \tag{B.51} \\
\overline{g(Z, W)} &= g(\overline{Z}, \overline{W}), \quad Z, W \in V_{\mathbb{C}}, \tag{B.52} \\
h(Z, W) &= \sum_{k,l=1}^{m} h_{kl} \lambda^k(Z) \overline{\lambda}^l(\overline{W}), \quad Z, W \in \mathbf{V}, \tag{B.53} \\
g(Z, W) &= h(Z, W) + h(\overline{Z}, \overline{W}), \quad Z, W \in V_{\mathbb{C}}, \tag{B.54} \\
\varphi(Z, W) &= -2i \sum_{k,l=1}^{m} h_{kl} \lambda^k \wedge \overline{\lambda}^l(Z, \overline{W}), \quad Z, W \in V_{\mathbb{C}}. \tag{B.55}
\end{aligned}
$$

Lemma B.4 *Let* $A : V \longrightarrow V$ *be a symplectic transformation. Consider the spaces*

$$V(\lambda) = \{X \in V \mid (A\lambda^{-1} - id)^k X = 0 \text{ for some } k \in \mathbb{Z}_+\}.$$

Then $V_{\mathbb{C}}$ *is the* φ*-orthogonal sum of non-degenerate subspaces of the form*

$$\begin{cases} V(\lambda) \oplus V(\overline{\lambda}^{-1}) & : & \lambda\overline{\lambda} \neq 1 \\ V(\lambda) & : & \lambda\overline{\lambda} = 1. \end{cases}$$

Here λ *runs through the eigenvalues of* A. *In particular,* $\det(A) = 1$.

Proof. Let $\lambda\overline{\sigma} \neq 1$. We claim

$$\varphi(V(\lambda), V(\sigma)) = 0.$$

To see this we denote by $V_k(\lambda)$ the subspace of $V(\lambda)$ which is annihilated by $(A\lambda^{-1} - id)^k$. For $X \in V_1(\lambda), Y \in V_1(\sigma)$ we have

$$\varphi(X, Y) = \varphi(AX, AY) = \lambda\overline{\sigma}\varphi(X, Y) = 0.$$

Assume we already know that

$$\varphi(V_{r-1}(\lambda), V_s(\sigma)) = \varphi(V_r(\lambda), V_{s-1}(\sigma)) = 0.$$

For $(r, s) = (1, 2)$ or $(2, 1)$, we have just proved this. Then we find for $X \in V_r(\lambda), Y \in V_s(\sigma)$, that

$$\varphi((A\lambda^{-1} - id)X, AY) = \varphi(X, (A\sigma^{-1} - id)Y) = 0.$$

Hence

$$0 = \varphi(X, Y) - \varphi(AX, AY) = (1 - \lambda\overline{\sigma})\varphi(X, Y).$$

Finally, for $0 \neq X \in V(\lambda)$, there exists a X^* with $\varphi(X, X^*) \neq 0$. We may assume $X^* \in V(\overline{\lambda}^{-1})$. If $\lambda\overline{\lambda} \neq 1$, then $V(\overline{\lambda}^{-1}) \neq V(\lambda)$ and $V(\lambda) \oplus V(\overline{\lambda}^{-1})$ is non-degenerate. If $\lambda\overline{\lambda} = 1$, $V(\lambda)$ is non-degenerate. $\square$

Corollary B.1 *Let* $A : V \longrightarrow V$ *be a symplectic transformation. Then we have for* V *the decomposition*

$$V = V^s \oplus V^u \oplus V^c,$$

into A*-invariant subspaces.* V^s *is called the stable subspace of* V. *It is generated by the* $V(\lambda)$ *with* $|\lambda| < 1$, V^u *the unstable subspace of* V *is generated by the* $V(\lambda)$ *with* $|\lambda| > 1$, V^c *the center subspace of* V *is generated by the* $V(\lambda)$ *with* $|\lambda| = 1$. V^s *and* V^u *are isotropic.* $V^s \oplus V^u$ *as well as* V^c *are non-degenerate.*

B.3   Complex manifolds

We identify $\mathbb{C}^m$ with $\mathbb{R}^{2m}$ by setting

$$z_j = x_j + iy_j \ (i = \sqrt{-1}); \quad x_j, y_j \in \mathbb{R}, \quad j = 1, \cdots, m.$$

The cotangent space to a point in $\mathbb{C}^m \cong \mathbb{R}^{2m}$ is spanned by $\{dx_j, dy_j\}$; it will often be more convenient, however, to work with the complex basis

$$dz_j = dx_j + idy_j, \quad d\bar{z}_j = dx_j - idy_j$$

and the dual basis in the tangent space

$$\frac{\partial}{\partial z_j} = \frac{1}{2}\left(\frac{\partial}{\partial x_j} - i\frac{\partial}{\partial y_j}\right), \quad \frac{\partial}{\partial \bar{z}_j} = \frac{1}{2}\left(\frac{\partial}{\partial x_j} + i\frac{\partial}{\partial y_j}\right).$$

Also write

$$\frac{d}{dz} = \left(\frac{\partial}{\partial z_1}, ..., \frac{\partial}{\partial z_m}\right), \quad \frac{d}{d\bar{z}} = \left(\frac{\partial}{\partial \bar{z}_1}, ..., \frac{\partial}{\partial \bar{z}_m}\right).$$

With this notation, the formula for the total differential becomes

$$df = \sum_{j=1}^{m} \frac{\partial f}{\partial z_j}dz_j + \sum_{j=1}^{m} \frac{\partial f}{\partial \bar{z}_j}d\bar{z}_j,$$

where $f : D \longrightarrow \mathbb{C}^n$ is differentiable on an open subset D of $\mathbb{C}^m$. We denote the first term ∂f and the second term $\bar{\partial}f$; ∂ and $\bar{\partial}$ are differential operators invarint under a complex linear change of coordinates.

A C^∞ mapping $f : D \longrightarrow \mathbb{C}^n$ is called *holomorphic* if $\bar{\partial}f = 0$; this is equivalent to $f(z_1, ..., z_m)$ being holomorphic in each variable z_j separately. Then by definition, $\mathrm{Hol}(D, \mathbb{C}^n)$ is the set of all *holomorphic mappings* from D to $\mathbb{C}^n$.

Definition B.6 *A Hausdorff topological space M with a countable basis is called a complex manifold of dimension m if there exists a family $\mathcal{A} = \{(U_\alpha, \varphi_\alpha)\}$ which satisfies the following conditions:*

1) $\{U_\alpha\}$ is an open covering of M;
2) $\varphi_\alpha : U_\alpha \longrightarrow \varphi_\alpha(U_\alpha) \subset \mathbb{C}^m$ is a homeomorphism onto an open subset $\varphi_\alpha(U_\alpha)$ of $\mathbb{C}^m$;
3) $\varphi_\beta \circ \varphi_\alpha^{-1} \in \mathrm{Hol}(\varphi_\alpha(U_\alpha \cap U_\beta), \mathbb{C}^m)$ if $U_\alpha \cap U_\beta \neq \emptyset$.

The pair $(U_\alpha, \varphi_\alpha)$ is called a holomorphic coordinate atlas of M, and $\varphi_\alpha = (z_1, ..., z_m)$ is said to be a local holomorphic coordinate system on U_α.

Given another complex manifold N, mapping $f \in C(M, N)$ is called *holomorphic* if for every $p \in M$, there exist a holomorphic coordinate atlas (U, φ) around p and a holomorphic coordinate atlas (V, ψ) around $f(p)$ such that $f(U) \subset V$, and such that $\psi \circ f \circ \varphi^{-1} \in \mathrm{Hol}(\varphi(U), \psi(V))$. Let $\mathrm{Hol}(M, N)$ be the set of all *holomorphic mappings* from M to N. The Cauchy integral formula implies in a standard way that $\mathrm{Hol}(M, N)$ is closed in $C(M, N)$.

If $f \in \mathrm{Hom}(M, N)$, and if f, f^{-1} are holomorphic, then f is said to be *biholomorphic*. If $f \in \mathrm{Hom}(M, M)$ is biholomorphic, then f is said to be an *automorphism* on M. Let $\mathrm{Aut}(M)$ be the *group of automorphisms* on M. The group operation is composition.

Let M be a complex manifold of dimension m, $p \in M$ any point, and $z = (z_1, ..., z_m)$ a holomorphic coordinate system around p. There are three different notions of a tangent space to M at p, which we now describe:

(i) $T(M)_p$ is the usual *real tangent space* to M at p, where we consider M as a real manifold of dimension $2m$. $T(M)_p$ can be realized as the space of $\mathbb{R}$-linear derivations on the ring of real-valued C^∞ functions in a neighborhood of p; if we write $z_j = x_j + iy_j$,

$$T(M)_p = \mathbb{R}\left\{ \frac{\partial}{\partial x_j}, \frac{\partial}{\partial y_j} \right\}.$$

(ii) $T_\mathbb{C}(M)_p = T(M)_p \otimes_\mathbb{R} \mathbb{C}$ is called the *complexified tangent space* to M at p. It can be realized as the space of $\mathbb{C}$-linear derivations on the ring of complex-valued C^∞ functions in a neighborhood of p. We can write

$$T_\mathbb{C}(M)_p = \mathbb{C}\left\{ \frac{\partial}{\partial x_j}, \frac{\partial}{\partial y_j} \right\} = \mathbb{C}\left\{ \frac{\partial}{\partial z_j}, \frac{\partial}{\partial \bar{z}_j} \right\},$$

so that we can obtain the *complex tangent bundle*

$$T_\mathbb{C}(M) = \bigcup_{p \in M} T_\mathbb{C}(M)_p.$$

(iii) $\mathbf{T}(M)_p = \mathbb{C}\{\frac{\partial}{\partial z_j}\} \subset T_\mathbb{C}(M)_p$ is called the *holomorphic tangent space* to M at p. It can be realized as the subspace of $T_\mathbb{C}(M)_p$ consisting of derivations that vanish on antiholomorphic functions (i.e., f such that $\bar{f}$ is holomorphic), and so it is independent of the holomorphic coordinate system $(z_1, ..., z_m)$ chosen. The subspace $\overline{\mathbf{T}}(M)_p = \mathbb{C}\{\frac{\partial}{\partial \bar{z}_j}\}$ is called the *antiholomorphic tangent space* to M at p; clearly

$$T_\mathbb{C}(M)_p = \mathbf{T}(M)_p \oplus \overline{\mathbf{T}}(M)_p.$$

The subspaces $\{\mathbf{T}(M)_p \subset T_\mathbb{C}(M)_p\}$ form a subbundle $\mathbf{T}(M) \subset T_\mathbb{C}(M)$, called the *holomorphic tangent bundle*.

Observe that for M, N complex manifolds, any $f \in C^\infty(M, N)$ induces the linear mapping

$$(df)_{p,\mathbb{R}} = (df)_p : T(M)_p \longrightarrow T(N)_{f(p)}$$

and hence a mapping

$$(df)_{p,\mathbb{C}} : T_\mathbb{C}(M)_p \longrightarrow T_\mathbb{C}(N)_{f(p)},$$

but do not in general induce a linear mapping from $\mathbf{T}(M)_p$ to $\mathbf{T}(N)_{f(p)}$. In fact, a mapping $f : M \longrightarrow N$ is holomorphic iff

$$(df)_{p,\mathbb{C}}(\mathbf{T}(M)_p) \subset \mathbf{T}(N)_{f(p)}$$

for all $p \in M$. If so, we denote the induced mapping by

$$f'(p) : \mathbf{T}(M)_p \longrightarrow \mathbf{T}(N)_{f(p)},$$

which is called the *holomorphic differential*.

Now let M, N be complex manifolds, $z = (z_1, ..., z_m)$ be holomorphic coordinates centered at $p \in M$, $w = (w_1, ..., w_n)$ holomorphic coordinates centered at $q \in N$ and $f : M \longrightarrow N$ a holomorphic mapping with $f(p) = q$. Corresponding to the various tangent spaces to M and N at p and q, we have different notions of the *Jacobian* of f, as follows. If we write

$$z_j = x_j + i y_j, \quad w_\alpha = u_\alpha + i v_\alpha,$$

then in terms of the bases $\{\frac{\partial}{\partial x_j}, \frac{\partial}{\partial y_j}\}$ and $\{\frac{\partial}{\partial u_\alpha}, \frac{\partial}{\partial v_\alpha}\}$ for $T(M)_p$ and $T(N)_q$, the linear mapping $(df)_{p,\mathbb{R}}$ is given by the $2m \times 2n$ Jacobi's matrix

$$\frac{df}{d(x,y)}(p) = \begin{pmatrix} \frac{\partial u}{\partial x}(p) & \frac{\partial u}{\partial y}(p) \\ \frac{\partial v}{\partial x}(p) & \frac{\partial v}{\partial y}(p) \end{pmatrix} = \mathcal{P}\left(\frac{\partial u}{\partial x}(p) + i\frac{\partial v}{\partial x}(p)\right)$$

such that

$$f_*\left(\frac{\partial}{\partial x}, \frac{\partial}{\partial y}\right) = \left(\frac{\partial}{\partial u}, \frac{\partial}{\partial v}\right)\frac{df}{d(x,y)},$$

where $u = {}^t(u_1, ..., u_n)$, $v = {}^t(v_1, ..., v_n)$,

$$\frac{\partial}{\partial x} = \left(\frac{\partial}{\partial x_1}, ..., \frac{\partial}{\partial x_m}\right), \quad \frac{\partial}{\partial y} = \left(\frac{\partial}{\partial y_1}, ..., \frac{\partial}{\partial y_m}\right),$$

and where we used the Cauchy-Riemann equations

$$\frac{\partial u}{\partial x} = \frac{\partial v}{\partial y}, \quad \frac{\partial u}{\partial y} = -\frac{\partial v}{\partial x},$$

In terms of the bases $\{\frac{\partial}{\partial z_j}, \frac{\partial}{\partial \bar{z}_j}\}$ and $\{\frac{\partial}{\partial w_\alpha}, \frac{\partial}{\partial \bar{w}_\alpha}\}$ for $T_{\mathbb{C}}(M)_p$ and $T_{\mathbb{C}}(N)_q$, $(df)_{p,\mathbb{R}}$ is given by

$$
\begin{aligned}
(df)_{p,\mathbb{R}} &= \left(\left(\frac{\partial}{\partial u}\right)_p, \left(\frac{\partial}{\partial v}\right)_p\right)\frac{df}{d(x,y)}(p) \, {}^t((dx)_p, (dy)_p) \\
&= \left(\left(\frac{d}{dw}\right)_p, \left(\frac{d}{d\bar{w}}\right)_p\right)\frac{df}{d(z,\bar{z})}(p) \, {}^t((dz)_p, (d\bar{z})_p) \\
&= \left(\frac{d}{dw}\right)_p\frac{df}{dz}(p) \, {}^t(dz)_p + \left(\frac{d}{d\bar{w}}\right)_p\frac{\overline{df}}{dz}(p) \, {}^t(d\bar{z})_p,
\end{aligned}
$$

where

$$\frac{df}{d(z,\bar{z})}(p) = \begin{pmatrix} \frac{df}{dz}(p) & 0 \\ 0 & \frac{\overline{df}}{dz}(p) \end{pmatrix},$$

which is just the Jacobi's matrix of $(df)_{p,\mathbb{C}}$, and where

$$\frac{df}{dz}(p) = \frac{\partial u}{\partial x}(p) + i\frac{\partial v}{\partial x}(p) = \left(\frac{\partial w_\alpha}{\partial z_j}(p)\right)$$

is just the Jacobi's matrix of $f'(p)$ in terms of the bases $\frac{\partial}{\partial z}$ and $\frac{\partial}{\partial w}$ for $\mathbf{T}(M)_p$ and $\mathbf{T}(N)_q$, respectively, such that

$$f_*\left(\frac{d}{dz}\right) = \left(\frac{d}{dw}\right)\frac{df}{dz}.$$

In particular, one has

$$\mathrm{rank}((df)_{p,\mathbf{R}}) = 2 \cdot \mathrm{rank}(f'(p)).$$

Further if $m = n$, then

$$\det((df)_{p,\mathbf{R}}) = |\mathcal{J}f(p)|^2 \geq 0,$$

where

$$\mathcal{J}f(p) = \det(f'(p)).$$

Hence *holomorphic mappings are orientation preserving.*

We take the *natural orientation* on $\mathbb{C}^m$ to be given by the $2m$-form

$$\left(\frac{\sqrt{-1}}{2}\right)^m dz_1 \wedge d\bar{z}_1 \wedge \cdots \wedge dz_m \wedge d\bar{z}_m = dx_1 \wedge dy_1 \wedge \cdots \wedge dx_m \wedge dy_m.$$

It is clear that if $\varphi_\alpha : U_\alpha \longrightarrow \mathbb{C}^m$ and $\varphi_\beta : U_\beta \longrightarrow \mathbb{C}^m$ are holomorphic coordinate mappings on the complex manifold M, the pullbacks via φ_α and φ_β of the natural orientation on $\mathbb{C}^m$ agree on $U_\alpha \cap U_\beta$. Thus any complex manifold has a natural orientation which is preserved under holomorphic mappings.

Take $f \in \mathrm{Hol}(M, N)$ and let $A \subset M$ be an analytic subset and denote by $f|_A$ the restriction of f on A. Then for an arbitrary $p \in A$, $(f|_A)^{-1}((f|_A)(p))$ is an analytic subset of M called the *fiber* of $f|_A$ at $f(p)$. We define the *rank* of $f|_A$ at p by

$$\mathrm{rank}_p(f|_A) = \dim_p A - \dim_p(f|_A)^{-1}((f|_A)(p)).$$

If A is irreducible, then the *rank* of $f|_A$ is defined by

$$\mathrm{rank}(f|_A) = \sup_{p \in A} \mathrm{rank}_p(f|_A).$$

In general, let $A = \bigcup_{\lambda \in \Lambda} A_\lambda$ be the irreducible decomposition of A. Then set

$$\mathrm{rank}(f|_A) = \sup_{\lambda \in \Lambda} \mathrm{rank}(f|_{A_\lambda}).$$

We have the following fact:

$$\mathrm{rank}(f|_A) = \sup_{p \in A_{reg}} \mathrm{rank}((f|_{A_{reg}})'(p)),$$

where A_{reg} is the set of all regular points of A. If $(f|_A)(A) = N$, then $\mathrm{rank}(f|_A) = \dim N$.

Let M be a complex manifold and let $T_\mathbb{C}^*(M)_z = T^*(M)_z \otimes_\mathbb{R} \mathbb{C}$ be the *complexified cotangent space* to M at z. Then we have the decomposition

$$T_\mathbb{C}^*(M)_z = \mathbf{T}^*(M)_z \oplus \overline{\mathbf{T}}^*(M)_z, \quad \bigwedge_r T_\mathbb{C}^*(M)_z = \bigoplus_{p+q=r} \left(\bigwedge_p \mathbf{T}^*(M)_z\right) \wedge \left(\bigwedge_q \overline{\mathbf{T}}^*(M)_z\right).$$

Correspondingly, we can write

$$A^r(M) = \bigoplus_{p+q=r} A^{p,q}(M),$$

where

$$A^{p,q}(M) = \left\{ \omega \in A^r(M) \mid \omega(z) \in \left(\bigwedge_p \mathbf{T}^*(M)_z \right) \wedge \left(\bigwedge_q \overline{\mathbf{T}}^*(M)_z \right) \text{ for all } z \in M \right\}.$$

The form $\omega \in A^{p,q}(M)$ is said to be of type (p,q). We denote by $\pi^{p,q}$ the projection mappings

$$\pi^{p,q} : A^*(M) \longrightarrow A^{p,q}(M).$$

Then the exterior differentiation

$$d : A^{p,q}(M) \longrightarrow A^{p+1,q}(M) \oplus A^{p,q+1}(M)$$

has the decomposition

$$d = \partial + \overline{\partial},$$

where

$$\partial = \pi^{p+1,q} \circ d : A^{p,q}(M) \longrightarrow A^{p+1,q}(M),$$
$$\overline{\partial} = \pi^{p,q+1} \circ d : A^{p,q}(M) \longrightarrow A^{p,q+1}(M).$$

We will use the operator

$$d^c = \frac{i}{4\pi}(\overline{\partial} - \partial).$$

Remark. As usual, we have the *complex cotangent bundle*

$$T_{\mathbf{C}}^*(M) = T_{\mathbf{C}}(M)^* = \bigcup_{z \in M} T_{\mathbf{C}}^*(M)_z,$$

and the *holomorphic cotangent bundle*

$$\mathbf{T}^*(M) = \mathbf{T}(M)^* = \bigcup_{z \in M} \mathbf{T}^*(M)_z.$$

Note that if $f \in \mathrm{Hol}(M, N)$, the linear mapping

$$f^* : T^*(N)_{f(p)} \longrightarrow T^*(M)_p$$

induces a linear mapping

$$f^* : \mathbf{T}^*(N)_{f(p)} \longrightarrow \mathbf{T}^*(M)_p,$$

further, if $M = \mathbb{C}^m$, which induces a linear mapping

$$f^{(k)*} : \mathbf{T}^*(N)_{f(p)} \longrightarrow \coprod_k \mathbf{T}^*(\mathbb{C}^m)_p$$

given by

$$f^{(k)*}((dg)_{f(p)}) = (d^k(g \circ f))_p = \sum_{\alpha \in \mathbb{Z}_+^m\langle k \rangle} \frac{k!}{\alpha!} \frac{d^\alpha g \circ f}{dz^\alpha}(p)dz^\alpha$$

for $(dg)_{f(p)} \in \mathbf{T}^*(N)_{f(p)}$, where $\alpha! = \alpha_1! \cdots \alpha_m!$ for $\alpha = (\alpha_1, ..., \alpha_m) \in \mathbb{Z}_+^m\langle k \rangle$. Thus f determines an element of $\mathbf{T}(N)_{f(p)} \otimes \coprod_k \mathbf{T}^*(\mathbb{C}^m)_p$, denoted by $(d^k f)_p$, such that $f^{(k)*}\alpha = \langle (d^k f)_p, \alpha \rangle$ for $\alpha \in \mathbf{T}^*(N)_{f(p)}$. Let $w = (w_1, ..., w_n)$ be local holomorphic coordinates at $f(p)$. Then

$$(d^k f)_p = \sum_{\alpha \in \mathbb{Z}_+^m\langle k \rangle} \sum_{j=1}^n \frac{k!}{\alpha!} \frac{d^\alpha f_j}{dz^\alpha}(p) \left(\frac{\partial}{\partial w_j} \right)_{f(p)} \otimes dz^\alpha,$$

where $f_j = w_j \circ f$. Thus there is a matrix, called the k-th Jacobi's matrix of f at p and denoted by $\frac{d^k f}{dz^k}(p)$, such that

$$f_*^{(k)} \left(\frac{d^k}{dz^k} \right)_p = \left(\frac{d}{dw} \right)_{f(p)} \frac{d^k f}{dz^k}(p),$$

where

$$f_*^{(k)} : \coprod_k \mathbf{T}(\mathbb{C}^m)_p \longrightarrow \mathbf{T}(N)_{f(p)}$$

is the conjugate mapping of $f^{(k)*}$, and where

$$\frac{d^k}{dz^k} = \left(\frac{d^\alpha}{dz^\alpha} \mid \alpha \in \mathbb{Z}_+^m\langle k \rangle \right).$$

If $X = \sum_{i=1}^m \xi_i \frac{\partial}{\partial z_i}$, then

$$\langle X^k, (d^k f)_p \rangle = \left(\frac{d}{dw} \right)_{f(p)} \frac{d^k f}{dz^k}(p)\xi^k,$$

where $\xi = (\xi_1, ..., \xi_m)$, and

$$\xi^k = {}^t \left(\frac{k!}{\alpha!} \xi^\alpha \mid \alpha \in \mathbb{Z}_+^m\langle k \rangle \right).$$

Let M be a complex manifold of dimension m. A *Hermitian metric* h on M is given by a positive definite Hermitian structure

$$h_p : \mathbf{T}(M)_p \times \mathbf{T}(M)_p \longrightarrow \mathbb{C}$$

on the holomorphic tangent space at p for each $p \in M$, depending smoothly on p, that is, such that for local coordinates z on M the functions

$$h_{kl}(z) = h_z \left(\frac{\partial}{\partial z_k}, \frac{\partial}{\partial z_l} \right)$$

are C^∞. In terms of the local coordinates z, the Hermitian metric is given by

$$h = \sum_{k,l} h_{kl}(z)dz_k \otimes d\bar{z}_l.$$

A complex manifold with a Hermitian metric is called a *Hermitian manifold*.

A *coframe* for the Hermitian metric is an m-tuple $(\theta_1, ..., \theta_m)$ of forms of type $(1,0)$ such that

$$h = \sum_k \theta_k \otimes \bar\theta_k,$$

i.e., such that, in terms of the Hermitian structure induced on $\mathbf{T}^*(M)_z$ by h_z on $\mathbf{T}(M)_z$, $(\theta_1(z), ..., \theta_m(z))$ is an orthonormal basis for $\mathbf{T}^*(M)_z$. From this description it is clear that coframes always exit locally. The dual of a coframe is called a frame.

The real and imaginary parts of a Hermitian inner product on a complex vector space give an ordinary inner product and an alternating quadratic form, respectively, on the underlying real vector space. We see that for a Hermitian metric h on M,

$$g = 2\mathrm{Re}(h) : T(M)_p \otimes T(M)_p \longrightarrow \mathbb{R}$$

is a Riemannian metric on M, called the induced Riemannian metric of the Hermitian metric. When we speak of distance, area, or volume on a complex manifold with a Hermitian metric, we always refer to the induced Riemannian metric. For example, one has

Lemma B.5 (Hopf-Rinow) *A Hermitian manifold is a complete metric space iff every closed and bounded subset is compact.*

In the usual method, the Riemannian metric g is extended uniquely to a positive definite Hermitian structure

$$g : T_{\mathbb{C}}(M)_p \times T_{\mathbb{C}}(M)_p \longrightarrow \mathbb{C}$$

on the complexified tangent space at p for each $p \in M$. For local coordinates z on M, we have

$$g\left(\frac{\partial}{\partial z_k}, \frac{\partial}{\partial z_l}\right) = g\left(\frac{\partial}{\partial \bar z_k}, \frac{\partial}{\partial \bar z_l}\right) = h_{kl}, \quad g\left(\frac{\partial}{\partial z_k}, \frac{\partial}{\partial \bar z_l}\right) = 0.$$

It is then customary to write

$$ds_M^2 = g = 2\sum_{k,l} h_{kl}(z)dz_k d\bar z_l,$$

for the metric g, where

$$dz_k d\bar z_l = \frac{1}{2}(dz_k \otimes d\bar z_l + d\bar z_l \otimes dz_k).$$

We also see that since the quadratic form

$$\varphi = 2\mathrm{Im}(h) : T(M)_p \otimes T(M)_p \longrightarrow \mathbb{R}$$

is alternating, it represents a real differential form of degree 2. In terms of the local coordinates z, the form is given by

$$\varphi = -2i\sum_{k,l} h_{kl}(z)dz_k \wedge d\bar z_l.$$

Usually we use the form

$$\omega = -\frac{1}{2}\varphi,$$

which is called the *associated (1,1)-form* of the metric. Generally, we say that a (1,1)-form η on M is *positive* at $p \in M$ if it locally is wrote as

$$\eta = i \sum_{k,l} a_{kl}(z) dz_k \wedge d\bar{z}_l$$

such that $(a_{kl}(p))$ is a positive definite Hermitian matrix. We write $\eta > 0$ if η is positive everywhere, $\eta \geq 0$ if η is positive semidefinite, and $\eta > \eta'$ if $\eta - \eta' > 0$. Thus the associated (1,1)-form ω is positive everywhere. A (r,r)-form η on M is *positive* (resp., *non-negative*) at $p \in M$, denoted by $\eta(p) > 0$ (resp., $\eta \geq 0$), if

$$\langle Z_1 \wedge \cdots \wedge Z_r \wedge \overline{iZ_1} \wedge \cdots \wedge \overline{iZ_r}, \eta(p) \rangle > 0 \text{ (resp., } \geq 0)$$

for any linearly independent vectors $Z_1, ..., Z_r$ in $\mathbf{T}(M)_p$.

Explicitly, if $(\theta_1, ..., \theta_m)$ is a coframe for h, we write

$$\theta_k = \alpha_k + i\beta_k,$$

where α_k, β_k are real differential forms. The induced Riemannian metric is given by

$$g = 2 \sum_k (\alpha_k \otimes \alpha_k + \beta_k \otimes \beta_k),$$

and the associated (1,1)-form of the metric is given by

$$\omega = 2 \sum_k \alpha_k \wedge \beta_k = i \sum_k \theta_k \wedge \bar{\theta}_k,$$

and the volume element associated to g is given by

$$\Theta = 2^m \alpha_1 \wedge \beta_1 \wedge \cdots \wedge \alpha_m \wedge \beta_m,$$

so that the m-th exterior power is

$$\omega^m = m!\Theta.$$

Next let A be an analytic subset of pure dimension k of M. The form $\frac{1}{k!}\omega^k$ is a positive (k,k)-form, which induces a volume measure on A. Let $\text{Vol}(A)$ be the real $2k$-dimensional volume of A with respect to the volume measure associated with $\frac{1}{k!}\omega^k$, restricted to A. Then we have

Theorem B.5 (Wirtinger theorem) *Let $A \subset M$ be a complex submanifold of dimension k. Then*

$$\text{Vol}(A) = \frac{1}{k!} \int_A \omega^k.$$

If $M = \mathbb{C}^m$ and if $A \subset \mathbb{C}^m$ is compact, then A can be characterized by the following result:

Lemma B.6 ([218]) *Every compact analytic subvariety of $\mathbb{C}^m$ is a finite set of points.*

Note that the standard Hermitian metric h on $\mathbb{C}^m$ is given by

$$h = \frac{1}{2} \sum_{j=1}^{m} dz_j \otimes d\bar{z}_j.$$

With respect to the metric h, the induced Riemannian metric

$$g = 2\mathrm{Re}(h) = \sum_{j=1}^{m} dz_j d\bar{z}_j$$

coincides with the Euclidean metric on $\mathbb{C}^m$. Now the associated (1,1)-form of the Euclidean metric is given by

$$\omega = \frac{i}{2} \sum_{j=1}^{m} dz_j \wedge d\bar{z}_j.$$

Lemma B.7 ([250]) *Let A be an analytic subset of pure dimension k of $\mathbb{C}^m(r)$ such that $0 \in A$. Then we have*

$$\mathrm{Vol}(A(t)) \geq \frac{\pi^k}{k!} t^{2k}, \quad 0 < t < r,$$

where the volume measure of A is induced by the Euclidean form ω.

B.4   Hamiltonian manifolds

Definition B.7 *An almost complex structure on a real differentiable manifold M is a smooth tensor field $\mathbf{J}$ of type $(1,1)$ which is an endomorphism of the tangent space $T(M)_x$ at every point $x \in M$ such that $\mathbf{J}^2 = -\mathrm{id}$. A manifold with a fixed almost complex structure is called an almost complex manifold.*

Here the smoothness of $\mathbf{J}$ means that if X is a smooth vector field on M, then $\mathbf{J}X$ also is a smooth vector field on M. By Lemma B.1 and Theorem B.1, we have

Theorem B.6 *Every almost complex manifold is of even dimensions and is orientable.*

Ehresmann and Hopf [240], p. 217, proved that a 4-dimensional unit sphere

$$S^4 = \{(x_1, ..., x_5) \in \mathbb{R}^5 \mid x_1^2 + ... + x_5^2 = 1\}$$

has no almost complex structures. Hence the inverse theorem is not true.

Let $\dim M = 2m$. In each tangent space $T(M)_x$ we fix a basis $\{e_1, ..., e_m, \mathbf{J}e_1, ..., \mathbf{J}e_m\}$. To give an orientation to M, we consider the family of all local coordinate systems $x_1, ..., x_{2m}$ of M such that, at each point x where the coordinate system $x_1, ..., x_{2m}$ is valid, the basis $(\partial/\partial x_1)_x, ..., (\partial/\partial x_{2m})_x$ of $T(M)_x$ differs from the above chosen basis by a linear transformation with positive determinant. It is a simple matter to verify that the family of local coordinate systems defines an orientation of M, called the *natural orientation*.

Theorem B.7 *Complex manifolds must be almost complex manifolds.*

Proof. Let M be a complex manifold of dimension m and $z_k = x_k + iy_k$ a complex local coordinate system of M. We identify M with an underlying differentiable manifold. Then x_k, y_k is a local coordinate system of the underlying real manifold M so that $\partial/\partial x_1, ..., \partial/\partial x_m, \partial/\partial y_1, ..., \partial/\partial y_m$ is the natural basis of $T(M)_x$ on the coordinate neighborhood. Define a linear transformation

$$\mathbf{J}_x : T(M)_x \longrightarrow T(M)_x$$

by setting

$$\mathbf{J}_x\left(\frac{\partial}{\partial x_k}\right) = \frac{\partial}{\partial y_k}, \quad \mathbf{J}_x\left(\frac{\partial}{\partial y_k}\right) = -\frac{\partial}{\partial x_k}, \quad (1 \le k \le m). \tag{B.56}$$

Obviously, $\mathbf{J}_x^2 = -id : T(M)_x \longrightarrow T(M)_x$.

Next we prove that the linear transformation $\mathbf{J}_x$ is independent of the choice of complex local coordinates z_k. Let $w_k = u_k + iv_k$ be another complex local coordinate system on a neighborhood of the point x so that we have the Cauchy-Riemann equations

$$\frac{\partial x_j}{\partial u_k} = \frac{\partial y_j}{\partial v_k}, \quad \frac{\partial x_j}{\partial v_k} = -\frac{\partial y_j}{\partial u_k}.$$

Thus we have

$$\mathbf{J}_x\left(\frac{\partial}{\partial u_k}\right) = \mathbf{J}_x\left(\sum_{j=1}^m \frac{\partial x_j}{\partial u_k}\frac{\partial}{\partial x_j} + \sum_{j=1}^m \frac{\partial y_j}{\partial u_k}\frac{\partial}{\partial y_j}\right) = \frac{\partial}{\partial v_k},$$

$$\mathbf{J}_x\left(\frac{\partial}{\partial v_k}\right) = \mathbf{J}_x\left(\sum_{j=1}^m \frac{\partial x_j}{\partial v_k}\frac{\partial}{\partial x_j} + \sum_{j=1}^m \frac{\partial y_j}{\partial v_k}\frac{\partial}{\partial y_j}\right) = -\frac{\partial}{\partial u_k}.$$

Therefore $\mathbf{J}_x$ is well-defined. □

The almost complex structure defined by (B.56) is called a *canonical almost complex structure* of the complex manifold M. It also gives

$$\mathbf{J}_x(dx_k) = -dy_k, \quad \mathbf{J}_x(dy_k) = dx_k. \tag{B.57}$$

An almost complex structure $\mathbf{J}$ on a manifold M is called a *complex structure* if M is an underlying differentiable manifold of a complex manifold which induces $\mathbf{J}$ in the way just described in the proof of the theorem. Let M and M' be almost complex manifolds with almost complex structures $\mathbf{J}$ and $\mathbf{J}'$, respectively. A mapping $f : M \longrightarrow M'$ is said to be *almost complex* if $\mathbf{J}' \circ f_* = f_* \circ \mathbf{J}$. It is easy to show that a mapping $f : M \longrightarrow M'$ between complex manifolds M and M' is holomorphic iff f is almost complex with respect to the complex structures of M and M'.

Theorem B.8 (Newlander-Nirenberg [187]) *An almost complex structure is a complex structure iff it has no torsion.*

Nijenhuis, Woolf, Kohn and Hörmander also proved the theorem if $\mathbf{J}$ satisfies weak differentiable conditions. Here the *torsion* of an almost complex structure $\mathbf{J}$ is defined to be the tensor field N of type $(1,2)$ given by

$$\frac{1}{2}N(X,Y) = [\mathbf{J}X, \mathbf{J}Y] - [X,Y] - \mathbf{J}[X, \mathbf{J}Y] - \mathbf{J}[\mathbf{J}X, Y] \tag{B.58}$$

for any vector fields X and Y. An almost complex structure also is said to be *integrable* if it has no torsion. An affine connection ∇ on M is said to be *almost complex* if the almost complex structure $\mathbf{J}$ of M is parallel with respect to ∇. From the general theory of connections, we know that any almost complex manifold admits an almost complex affine connection (provided it is paracompact) such that its torsion T is given by $N = 8T$. Thus an almost complex manifold admits a torsion-free almost complex affine connection iff the almost complex structure has no torsion.

Let ∇ be an almost complex connection on M with the almost complex structure $\mathbf{J}$. Take a local frame $\{e_1, ..., e_{2m}\}$ of $T(M)$ on an open subset U of M such that $e_{m+k} = \mathbf{J}e_k$ for $1 \le k \le m$. Thus we can write

$$\mathbf{J} = \sum_{k=1}^{m} \{e_{m+k} \otimes e^{*k} - e_k \otimes e^{*m+k}\}$$

on U, where $\{e^{*1}, ..., e^{*2m}\}$ is the dual frame for $T^*(M)$ over U. By a simple calculation, we can prove that $\nabla\mathbf{J} = 0$ holds on U iff

$$\omega_{m+k}^{m+l} = \omega_k^l, \quad \omega_{m+k}^l = -\omega_k^{m+l}, \quad 1 \le k, l \le m,$$

where $\omega = (\omega_\alpha^\beta)$ is the connection matrix. Define

$$\psi_k^l = \omega_k^l - i\omega_{m+k}^l, \quad k, l = 1, ..., m$$

and set $\psi = (\psi_k^l)$. Then ω is a real representation of $\overline{\psi}$, that is, $\omega = \mathcal{P}(\overline{\psi})$. If we set

$$\Psi = (\Psi_k^l), \quad \Psi_k^l = \Omega_k^l - i\Omega_{m+k}^l, \quad k, l = 1, ..., m,$$

where $\Omega = (\Omega_\alpha^\beta)$ is the curvature matrix, then

$$\Psi_k^l = d\psi_k^l - \sum_{j=1}^{m} \psi_k^j \wedge \psi_j^l, \quad k, l = 1, ..., m$$

or $\Psi = d\psi - \psi \wedge \psi$ and hence $\Omega = \mathcal{P}(\overline{\Psi})$. Thus we can obtain

$$R(e_\alpha, e_\beta)\mathbf{J}e_\gamma = \mathbf{J}R(e_\alpha, e_\beta)e_\gamma.$$

More generally, the curvature operator R of an almost complex affine connection satisfies the following property (see [145]):

$$R(X, Y) \circ \mathbf{J} = \mathbf{J} \circ R(X, Y),$$

for all vector fields X and Y. Define

$$\xi_k = \frac{1}{2}(e_k - i\mathbf{J}e_k), \quad \overline{\xi}_k = \frac{1}{2}(e_k + i\mathbf{J}e_k), \quad 1 \le k \le m,$$

and extend ∇ linearly to $\Gamma(T_\mathbb{C}(M))$. We have

$$\nabla\xi_k = \sum_{l=1}^{m} \psi_k^l \otimes \xi_l, \quad \nabla\overline{\xi}_k = \sum_{l=1}^{m} \overline{\psi}_k^l \otimes \overline{\xi}_l, \quad 1 \le k \le m,$$

that is, the almost complex connection ∇ induces connections in the subbundles $\mathbf{T}(M)$ and $\overline{\mathbf{T}}(M)$. The converse also is true. Note that the curvature operator $R(Z, W)$ of ∇ also is well-defined for vectors $Z, W \in \Gamma(T_{\mathbb{C}}(M))$. If ∇ is almost complex, $R(Z, W)$ is expressed on U by

$$R(Z, W)s = \sum_{k,l=1}^{m} \lambda_k \langle Z \wedge W, \Psi_k^l \rangle \xi_l + \sum_{k,l=1}^{m} \gamma_k \langle Z \wedge W, \overline{\Psi}_k^l \rangle \overline{\xi}_l$$

for $s = \sum_{k=1}^{m} \{\lambda_k \xi_k + \gamma_k \overline{\xi}_k\}$. If ∇ also has no torsion, then the coeffcients of the connection ∇ satisfy

$$\Gamma_{m+k,m+j}^{l} = -\Gamma_{kj}^{l}, \quad \Gamma_{k,m+j}^{l} = \Gamma_{m+k,j}^{l}, \quad 1 \le j, k, l < m,$$

and hence $\psi_k^l = \sum_{j=1}^{m} \Gamma_{kj}^{l} dz_j$ are $(1, 0)$-forms, where

$$\Gamma_{kj}^{l} = \Gamma_{kj}^{l} - i\Gamma_{m+k,j}^{l}, \quad \Gamma_{kj}^{l} = \Gamma_{jk}^{l}.$$

Definition B.8 *An almost complex manifold with a Riemannian metric g invariant by the almost complex structure $\mathbf{J}$, i.e.,*

$$g(\mathbf{J}X, \mathbf{J}Y) = g(X, Y),$$

for any vector fields X and Y, is called an almost Hermitian manifold.

Every almost complex manifold admits a $\mathbf{J}$-invariant Riemannian metric provided it is paracompact. In fact, given an almost complex manifold M, take any Riemannian metric g_0 which exists provided M is paracompact. We obtain a $\mathbf{J}$-invariant Riemannian metric g by setting

$$g(X, Y) = g_0(X, Y) + g_0(\mathbf{J}X, \mathbf{J}Y)$$

for any vector fields X and Y.

A $\mathbf{J}$-invariant Riemannian metric g of an almost Hermitian manifold M defines a positive definite Hermitian inner product on each tangent space $T(M)_x$. The *fundamental 2-form* φ of M is defined by

$$\varphi(X, Y) = -g(\mathbf{J}X, Y)$$

with the almost complex structure $\mathbf{J}$ and the metric g for any vector fields X and . Obviously, φ is non-degenerate at each point of M. By Theorem B.3, there also exists a positive definite Hermitian structure h on each tangent space $T(M)_x$ given by

$$2h(X, Y) = g(X, Y) + i\varphi(X, Y),$$

which is called a *Hermitian metric* of almost Hermitian manifold M. If M is a complex manifold, the Hermitian structure h can be extended uniquely to a Hermitian structure on each complexified tangent space $T_{\mathbb{C}}(M)_x$, which defines a Hermitian metric on the complex manifold M.

Definition B.9 *A Hermitian metric on an almost con.   manifold is called a Kähler metric if the fundamental 2-form is closed. An almost co...plex manifold (resp. a complex manifold) with a Kähler metric is called an almost Kähler manifold (resp. a Kähler manifold).*

Let M be an almost Hermitian manifold with almost complex structure $\mathbf{J}$ and the $\mathbf{J}$-invariant Riemannian metric g. Let φ be the fundamental 2-form, N the torsion of $\mathbf{J}$ and ∇ the Levi-Civita connection defined by g, which also is said to be induced by the Hermitian metric h. One has the following formula (see [145])

$$2g((\nabla_X \mathbf{J})Y, Z) = 6d\varphi(X, \mathbf{J}Y, \mathbf{J}Z) - 6d\varphi(X, Y, Z) + g(N(Y, Z), \mathbf{J}X).$$

As an application, we know that the Levi-Civita connection ∇ defined by g is almost complex iff $\mathbf{J}$ has no torsion and φ is closed. In particular, if M is a Hermitian manifold, the Levi-Civita connection induced by the Hermitian metric is almost complex iff the fundamental 2-form is closed.

Now further assume that M is a Kähler manifold. Let $\xi_1, ..., \xi_m$ be a frame for $\mathbf{T}(M)$ over U and set

$$h_{kl} = h(\xi_k, \xi_l), \quad 1 \le k, l \le m, \quad \mathbf{H} = (h_{kl}).$$

A simple calculation shows

$$dh_{kl} = \sum_{j=1}^{m} \{\psi_k^j h_{jl} + \overline{\psi}_l^j h_{kj}\},$$

or generally

$$dh(Z, W) = h(\nabla Z, W) + h(Z, \nabla W)$$

holds for $Z, W \in \Gamma(\mathbf{T}(M))$. Note that ψ has type $(1, 0)$. Comparing types, we see

$$\partial \mathbf{H} = \psi \mathbf{H}, \quad \overline{\partial} \mathbf{H} = \mathbf{H} \, {}^t\overline{\psi},$$

and hence the curvature matrix

$$\Psi = -\partial\overline{\partial}\mathbf{H} \cdot \mathbf{H}^{-1} + \partial\mathbf{H} \cdot \mathbf{H}^{-1} \wedge \overline{\partial}\mathbf{H} \cdot \mathbf{H}^{-1}$$

has type $(1, 1)$. A global closed $(1, 1)$-form

$$\mathrm{Ric}_M = -2i \sum_{k=1}^{m} \Psi_k^k = 2i\partial\overline{\partial} \log(\det \mathbf{H})$$

is well-defined, called the *Ricci form* of M for the metric h. Now the curvature matrix Ω also has type $(1, 1)$. By using the components $R_{\alpha kl}^\beta$ of the curvature operator with respect to a local coordinate system $x_1, ..., x_m, y_1, ..., y_m$ in M, then Ω_α^β and Ψ_n^j are given by

$$\Omega_\alpha^\beta = \frac{1}{2} \sum_{k,l=1}^{m} \mathbf{R}_{\alpha kl}^\beta dz_k \wedge d\overline{z}_l, \quad 1 \le \alpha, \beta \le 2m,$$

$$\Psi_n^j = \sum_{k,l=1}^{m} K_{nkl}^j dz_k \wedge d\overline{z}_l, \quad 1 \le n, j \le m,$$

where $z_k = x_k + iy_k$, and where

$$\mathbf{R}_{\alpha kl}^\beta = R_{\alpha kl}^\beta - iR_{\alpha, m+k, l}^\beta, \quad K_{nkl}^j = \frac{1}{2} \left(\mathbf{R}_{nkl}^j - i\mathbf{R}_{m+n, kl}^j \right).$$

Thus for the frame $\{\xi_k = \frac{\partial}{\partial z_k} \mid 1 \le k \le m\}$ of $\mathbf{T}(M)$ on U, we have

$$R(\xi_k, \bar{\xi}_l)\xi_n = \sum_{j=1}^{m} K_{nkl}^j \xi_j, \quad R(\bar{\xi}_k, \xi_l)\bar{\xi}_n = \sum_{j=1}^{m} \overline{K}_{nkl}^j \bar{\xi}_j.$$

Define

$$K_{njkl} = h(R(\xi_k, \bar{\xi}_l)\xi_n, \xi_j) = \sum_{\beta=1} K_{nkl}^\beta h_{\beta j}.$$

Then

$$\Gamma_{nk}^j = \sum_{\beta=1}^{m} \frac{\partial h_{n\beta}}{\partial z_k} h^{\beta j}, \quad K_{nkl}^j = -\frac{\partial \Gamma_{nk}^j}{\partial \bar{z}_l}, \quad K_{njkl} = -\frac{\partial^2 h_{nj}}{\partial z_k \partial \bar{z}_l} + \sum_{p,q} h^{pq} \frac{\partial h_{np}}{\partial z_k} \frac{\partial h_{qj}}{\partial \bar{z}_l},$$

where (h^{kl}) is the inverse matrix of (h_{kl}). The curvature operator R of M also possess the following properties (see [145]):

$$R(\mathbf{J}X, \mathbf{J}Y) = R(X, Y),$$

for all vector fields X and Y.

Recall that for each plane p in $T(M)_x$, the sectional curvature $K(p)$ was defined by $K(p) = -R(X, Y, X, Y)$, where X, Y is an orthonormal basis for p, and where R is the Riemannian curvature tensor of M with the metric g. If p is invariant by the complex structure $\mathbf{J}$, then $K(p)$ is called the *holomorphic sectional curvature* by p. If p is invariant by $\mathbf{J}$ and X is a unit vector in p, then X, $\mathbf{J}X$ is an orthonormal basis for p and hence

$$K(p) = -R(X, \mathbf{J}X, X, \mathbf{J}X).$$

If $\dim M = 1$, the holomorphic sectional curvature is nothing but the Gaussian curvature.

Given two $\mathbf{J}$-invariant planes p and p' in $T(M)_x$, the *holomorphic bisectional curvature* $K(p, p')$ is defined by

$$K(p, p') = -R(X, \mathbf{J}X, Y, \mathbf{J}Y),$$

where X (resp. Y) is a unit vector in p (resp. p'). Obviously, $K(p, p) = K(p)$ and Bianchi's identity implies

$$K(p, p') = -R(X, Y, X, Y) - R(X, \mathbf{J}Y, X, \mathbf{J}Y).$$

Write

$$Z = X - i\mathbf{J}X = \sum_k \lambda_k \xi_k, \quad W = Y - i\mathbf{J}Y = \sum_k \eta_k \xi_k.$$

Then

$$-4R(X, \mathbf{J}X, Y, \mathbf{J}Y) = h(R(W, \overline{W})Z, Z) = \sum K_{njkl} \lambda_n \bar{\lambda}_j \eta_k \bar{\eta}_l,$$

and hence

$$K(p, p') = \frac{\sum K_{njkl} \lambda_n \bar{\lambda}_j \eta_k \bar{\eta}_l}{(\sum h_{nj} \lambda_n \bar{\lambda}_j)(\sum h_{kl} \eta_k \bar{\eta}_l)}.$$

Definition B.10 *A smooth manifold M with a non-degenerate 2-form (resp. a closed 2-form) φ, i.e., φ is non-degenerate at each point of M, is called an almost symplectic or almost Hamiltonian manifold (resp. a symplectic or Hamiltonian manifold). A non-degenerate closed 2-form φ is called a symplectic form.*

Hence an almost Hermitian manifold is an almost Hamiltonian manifold. An almost Kähler manifold is a Hamiltonian manifold. Obviously, if (M, φ) is an almost Hamiltonian manifold, then M is even-dimensional and φ^m is a volume form $(m = \frac{1}{2} \dim M)$. In particular, M is orientable. If (M, φ) is a Hamiltonian manifold, then a subbundle of the tangent bundle $T(M)$ of M is said to be *isotropic* if at every point $x \in M$, it defines an isotropic subspace of $T(M)_x$, and *Lagrangian* if at every point $x \in M$, it defines an Lagrangian subspace of $T(M)_x$. A smooth submanifold of M is said to be *isotropic* if its tangent bundle is an isotropic subbundle, and *Lagrangian* if its tangent bundle is a Lagrangian subbundle of $T(M)$. A diffeomorphism $f : (M, \varphi) \longrightarrow (N, \psi)$ between Hamiltonian manifolds such that $f^*\psi = \varphi$ is called a *symplectic diffeomorphism*. If $(M, \varphi) = (N, \psi)$, it is also called a *canonical transformation*.

Proposition B.3 *Let $N = T^*(M)$ be the total space of the cotangent bundle $\pi : T^*(M) \longrightarrow M$ of a differentiable manifold M. Then N carries a canonical symplectic structure given by the 2-form $\varphi^* = -d\theta$. Here the 1-form θ is defined by*

$$\theta : X \in T(N)_\omega \mapsto \langle \pi_* X, \omega \rangle.$$

Note. In the definition of θ we use the following commutative diagram

$$
\begin{array}{ccccccc}
T(N) & \overset{\pi_N}{\longrightarrow} & N & \quad & (x, y, \dot{x}, \dot{y}) & \longmapsto & (x, y) \\
\downarrow \pi_* & & \downarrow \pi & & \downarrow & & \downarrow \\
T(M) & \overset{\pi_M}{\longrightarrow} & M & \quad & (x, \dot{x}) & \longmapsto & (x)
\end{array}
$$

Proof. Clearly, φ^* is closed. To see that it is non-degenerate we write down the local representation of θ. Take a local coordinate system $(U; x)$ on M. These induce coordinates (x, y) on $T^*(U) = \pi^{-1}(U) \subset N$. There we have written x instead of $x \circ \pi$ for the first set of coordinates on $T^*(U)$. Then $\theta(X)$ is given by $\langle \dot{x}, y \rangle$, i.e.,

$$\theta = \sum_k y_k dx_k.$$

Hence

$$\varphi^* = -d\theta = \sum_k dx_k \wedge dy_k,$$

which shows that φ^* is non-degenerate.   □

We see even more. All charts of the canonical atlas of $T^*(M)$ derived from an atlas of M are symplectic charts. By this we mean that the representation of the symplectic 2-form φ^* with respect to the chart has a diagonal form $\sum dx_k \wedge dy_k$. There exists a symplectic atlas (i.e., an atlas the charts of which are all symplectic) for every symplectic manifold. This is the content of the so-called Darboux theorem:

Theorem B.9 (Darboux theorem [134]) *Let (M, φ) be a Hamiltonian manifold. Then for every point $x \in M$, there exists coordinates $\{x_k, y_k\}$ on a neighborhood U of x such that on U,*

$$\varphi = \sum_{k=1}^{m} dx_k \wedge dy_k.$$

B.5 Complex Euclidean spaces $\mathbb{C}^m$

Here we collect some basic notations and results on $\mathbb{C}^m$ which are often used in complex dynamics. If $z = (z_1, ..., z_m) \in \mathbb{C}^m$, let $\| \ \ \|$ denote the ordinary Euclidean norm

$$\|z\| = \left(\sum_{j=1}^{m} |z_j|^2 \right)^{1/2},$$

of $\mathbb{C}^m$, induced by the canonical Hermitian structure

$$(z, w) = \sum_{j=1}^{m} z_j \overline{w_j}.$$

Then

$$\mathbb{C}^m(a; r) = \{z \mid \|z - a\| < r\} \subset \mathbb{C}^m,$$

be the r-ball about $a \in \mathbb{C}^m$. Also write

$$\mathbb{C}^m(r) = \mathbb{C}^m(0; r), \quad \Delta^m = \mathbb{C}^m(1), \quad \Delta = \Delta^1.$$

For a sequence $r = (r_1, ..., r_m) \in (\mathbb{R}^+)^m$ of positive numbers, denote the polydisc

$$\Delta^m(a, r) = \mathbb{C}(a_1; r_1) \times \cdots \times \mathbb{C}(a_m; r_m) \subset \mathbb{C}^m.$$

Abbreviate as $\Delta^m(r) = \Delta^m(0, r)$. If $r = (r_1, ..., r_m) = (s, ..., s) \in (\mathbb{R}^+)^m$, write $\Delta_s^m = \Delta^m(r)$, and abbreviate $\Delta^m = \Delta_1^m$.

For $\alpha = (\alpha_1, ..., \alpha_m) \in \mathbb{Z}^m, z = (z_1, ..., z_m) \in \mathbb{C}^m$, define

$$z^\alpha = z_1^{\alpha_1} \cdots z_m^{\alpha_m}.$$

For $\beta = (\beta_1, ..., \beta_m) \in \mathbb{Z}^m$, $\alpha \le \beta$ means $\alpha_j \le \beta_j, j = 1, \cdots, m$; $\alpha < \beta$ means $\alpha \le \beta$ and $\alpha \ne \beta$. For $j \in \mathbb{Z}$, we also write

$$\alpha + j = (\alpha_1 + j, \cdots, \alpha_m + j)$$

so that

$$z^{\alpha+j} = z_1^{\alpha_1+j} \cdots z_m^{\alpha_m+j}, \quad z^{-j} = z_1^{-j} \cdots z_m^{-j}.$$

For $\alpha = (\alpha_1, \cdots, \alpha_m) \in (\mathbb{Z}_+)^m$, write

$$\alpha! = \alpha_1! \cdots \alpha_m!, \quad \frac{d^\alpha}{dz^\alpha} = \frac{\partial^{|\alpha|}}{\partial z_1^{\alpha_1} \cdots \partial z_m^{\alpha_m}}.$$

The classical Cauchy formula of one complex variable imples the following result:

Proposition B.4 (Cauchy formula) *Let* $f \in \mathrm{Hol}(D, \mathbb{C}^n)$, *where* D *is an open subset of* $\mathbb{C}^m$. *Take* $a \in D, r \in (\mathbb{R}^+)^m$ *such that* $\overline{\Delta^m(a, r)} \subset D$. *Then for* $z \in \Delta^m(a, r)$,

$$f(z) = \frac{1}{(2\pi i)^m} \int_\Gamma f(\zeta)(\zeta - z)^{-1} d\zeta,$$

$$\frac{d^\alpha f}{dz^\alpha}(z) = \frac{\alpha!}{(2\pi i)^m} \int_\Gamma f(\zeta)(\zeta - z)^{-\alpha - 1} d\zeta,$$

where $\Gamma = \partial\mathbb{C}(a_1; r_1) \times \cdots \times \partial\mathbb{C}(a_m; r_m)$ *is the skeleton of* $\Delta^m(a, r)$*, and where* $d\zeta = d\zeta_1 \wedge \cdots \wedge d\zeta_m$*.*

As consequence, a mapping $f : D \longrightarrow \mathbb{C}^n$ is holomorphic iff it has local power series expansions in the variables z_j, i.e., for any $a \in D$,

$$f(z) = \sum_{\alpha \in \mathbb{Z}_+^m} c_\alpha(f, a)(z - a)^\alpha$$

holds on a neighborhood of a. If so, we have

$$c_\alpha(f, a) = \frac{1}{\alpha!} \frac{d^\alpha f}{dz^\alpha}(a),$$

and hence

$$f(z) = \sum_{k=0}^\infty \frac{1}{k!} \frac{d^k f}{dz^k}(a)(z - a)^k.$$

In addition, the *Cauchy inequality*

$$\left\| \frac{d^\alpha f}{dz^\alpha}(a) \right\| \le \sqrt{n}\alpha! r^{-\alpha} \max_{\zeta \in \Gamma} \|f(\zeta)\|$$

follows.

Proposition B.5 (Hartogs theorem) *Let D be an open subset of $\mathbb{C}^m$, $m > 1$, and let K be a compact subset of D such that $D - K$ is connected. For every $f \in \text{Hol}(D - K, \mathbb{C}^n)$, one can find $F \in \text{Hol}(D, \mathbb{C}^n)$ so that $F|_{D-K} = f$.*

Proposition B.6 (Weierstrass theorem) *Let D be an open subset of $\mathbb{C}^m$ and let $f_k \in \text{Hol}(D, \mathbb{C}^n)(k = 1, 2, \ldots)$. If f_k converges uniformly on compact subsets of D, then $\lim f_k = f \in \text{Hol}(D, \mathbb{C}^n)$, and $\{\frac{d^\alpha f_k}{dz^\alpha}\}$ converges to $\{\frac{d^\alpha f}{dz^\alpha}\}$ on compact subsets of D for $\alpha \in \mathbb{Z}_+^m$.*

A family $\mathcal{F} \subset \text{Hol}(D, \mathbb{C}^n)$ is called *uniformly bounded* on a subset E of D iff there exists a constant $K = K_E > 0$ such that

$$\|f(z)\| < K \quad \text{for all } z \in E, f \in \mathcal{F},$$

and is said to be *locally uniformly bounded* on D if D is a union of open sets in each of which the family is uniformly bounded. Obviously, a family is locally uniformly bounded on D if and only if it is uniformly bounded on any compact subset of D.

Proposition B.7 (Vitali-Porter theorem) *If a sequence $\{f_n\} \subset \text{Hol}(D, \mathbb{C})$ is locally uniformly bounded on a domain $D \subset \mathbb{C}$ and if it converges for a set of points $z \in D$ having a limit point in D, then $\{f_n\}$ converges locally uniformly on D to some $f \in \text{Hol}(D, \mathbb{C})$.*

Proposition B.8 (Montel theorem) *If a family $\mathcal{F} \subset \text{Hol}(D, \mathbb{C}^n)$ is uniformly bounded on any compact subset of D, then $\mathcal{F}$ is equicontinuous on D.*

Proofs can be completed in the fashion of Narasimhan[181] and Hörmander[115] by using the Cauchy formula. The Weierstrass theorem shows that $\mathrm{Hol}(D, \mathbb{C}^n)$ is closed in $C(D, \mathbb{C}^n)$. If a family $\mathcal{F} \subset \mathrm{Hol}(D, \mathbb{C}^n)$ is uniformly bounded on any compact subset of D, then the Montel theorem and Arzela-Ascoli theorem show that $\mathcal{F}$ is relatively compact in $\mathrm{Hol}(D, \mathbb{C}^n)$, and hence each sequence of $\mathcal{F}$ contains a subsequence which converges uniformly on compact subsets of D. Therefore the family $\mathcal{F}$ is normal on D.

For $z = (z_1, ..., z_m) \in \mathbb{C}^m$, abbreviate

$$\omega(z, \bar{z}) = \frac{(m-1)!}{(2\pi i)^m \|z\|^{2m}} \sum_{j=1}^m (-1)^{j-1} \bar{z}_j d\bar{z}_1 \wedge \cdots \wedge \hat{d\bar{z}_j} \wedge \cdots \wedge d\bar{z}_m \wedge dz_1 \wedge \cdots \wedge dz_m,$$

where $\hat{d\bar{z}_j}$ means that the term is omitted.

Theorem B.10 *Let D be a bounded domain in $\mathbb{C}^n$, and its boundary ∂D is piecewise smooth. Let $f = (f_1, \cdots, f_m) : \overline{D} \longrightarrow \mathbb{C}^m$ be a holomorphic mapping such that ∂D does not contain zeros of f Then f has only isolated zeros in D, and their number, with each zero counted as many times as its multiplicity, is expressed by the formula*

$$n(D, f) = \int_{\partial D} \omega(f(\zeta), \overline{f(\zeta)}). \tag{B.59}$$

A proof can be found in [10]. In Theorem B.10, if we only assume that $f : D \longrightarrow \mathbb{C}^n$ is holomorphic and continuous on $\overline{D}$, the *Martinelli-Bochner formula*

$$\int_{\partial D} f(\zeta)\omega(\zeta - z, \bar{\zeta} - \bar{z}) = \begin{cases} f(z) & : \quad z \in D \\ 0 & : \quad z \notin \overline{D} \end{cases}$$

holds. Further, if f and D satisfy the conditions of Theorem B.10, then $f(\partial D)$ is homologous to the cycle $N \cdot \partial \mathbb{C}^m(r)$, where N is an integer, so that the Martinelli-Bochner formula implies that the integral

$$\int_{\partial D} f^* \omega(\zeta, \bar{\zeta}) = \int_{f(\partial D)} \omega(\zeta, \bar{\zeta}) = N \int_{\partial \mathbb{C}^m(r)} \omega(\zeta, \bar{\zeta}) = N$$

is an integer, and Theorem B.10 shows that the integer N is just $n(D, f)$.

Theorem B.11 (Rouché principle) *Suppose that the mapping f and the domain D satisfy the conditions of Theorem B.10, and let $g = (g_1, \cdots, g_m) : \overline{D} \longrightarrow \mathbb{C}^m$ be a holomorphic mapping such that for each point $z \in \partial D$ there is at least one index $j(j = 1, ..., m)$ such that*

$$|g_j(z)| < |f_j(z)|. \tag{B.60}$$

Then f and $f + g$ have the same number of zeros in D, i.e., $n(D, f) = n(D, f + g)$.

Condition (B.60) holds, for example, if $\|g\| < \|f\|$ on ∂D. This condition can be relaxed. In fact, one can require only that

$$f(z) + tg(z) \neq 0 \quad \text{for } z \in \partial D \text{ and } 0 \leq t \leq 1. \tag{B.61}$$

It suffices that at each point $z \in \partial D$ there is at least one index j such that

$$\mathrm{Re}[g_j(z)] < \mathrm{Re}[f_j(z)] \text{ or } \mathrm{Im}[g_j(z)] < \mathrm{Im}[f_j(z)].$$

If (B.61) holds, the integral

$$n(D, f + tg) = \int_{\partial D} \omega(f + tg, \overline{f + tg})$$

is a continuous function of t on $\mathbb{R}[0,1]$, and its values are integers. Therefore $n(D, f + tg)$ is constant, and so $n(D, f) = n(D, f + g)$. The proof of Theorem B.11 follows.

The following theorem is classic:

Theorem B.12 (Inverse function theorem) *Let D be open set in $\mathbb{C}^m$ with $0 \in D$ and $f \in \mathrm{Hol}(D, \mathbb{C}^m)$ such that the Jacobian determinant*

$$\mathcal{J}f(0) = \det\left(\frac{\partial f_j}{\partial z_i}(0)\right) \neq 0.$$

Then f is one-to-one in a neighborhood of 0, and f^{-1} is holomorphic at $f(0)$.

Definition B.11 *An open set $D \subset \mathbb{C}^m$ is called pseudoconvex if there exists a continuous plurisubharmonic function τ in D such that*

$$D(r) = \{z \in D \mid \tau(z) < r\} \subset\subset D$$

for every $r \in \mathbb{R}$.

Here a plurisubharmonic function τ in D means a semicontinuous function from above, i.e., $D(r)$ is open for every $r \in \mathbb{R}$, with values in $\mathbb{R} \cup \{-\infty\}$ such that for arbitrary $z, w \in \mathbb{C}^m$, the function $t \mapsto \tau(z + tw)$ is subharmonic in the part of $\mathbb{C}$ where it is defined. A function $\tau \in C^2(D, \mathbb{R})$ is plurisubharmonic if and only if $dd^c\tau \geq 0$. Further, if $dd^c\tau > 0$, then τ is said to be strictly plurisubharmonic. For other definitions of pseudoconvexity and their equivalence with this definition, see [53], [158] and [191].

A set D in $\mathbb{C}^m$ is called *convex* if $z, w \in D$ implies $tz + (1 - t)w \in D$ for $0 \leq t \leq 1$. If $D \neq \emptyset$, there exists the smallest convex set containing D, which is called the *convex hull* of D. An open set D in $\mathbb{C}^m$ is called a *tube* if there is an open set D' in $\mathbb{R}^m$, called the *base* of D, such that

$$D = \{z \mid \mathrm{Re}(z) \in D'\}$$

(see [115]).

A domain $D' \subset \mathbb{R}^n$ has C^r *boundary* (or is a C^r *domain*) if there is a C^r function $\rho : \mathbb{R}^n \longrightarrow \mathbb{R}$, called a C^r *defining function* for D', such that:
(i) $D' = \{x \in \mathbb{R}^n \mid \rho(x) < 0\}$,
(ii) $\partial D' = \{x \in \mathbb{R}^n \mid \rho(x) = 0\}$, and
(iii) $\mathrm{grad}(\rho)$ is not vanishing on $\partial D'$.
If ρ_1 is another C^r defining function for D', it is easy to check that there exists a never vanishing C^r function $h : \mathbb{R}^n \longrightarrow \mathbb{R}^+$ such that $\rho_1 = h\rho$. The C^r boundary $\partial D'$ also is a C^r manifold embedded in $\mathbb{R}^n$. In particular, for every $x \in \partial D'$ the tangent space $T(\partial D')_x$

can be identified with the kernel of $(d\rho)_x$. Let $\rho : \mathbb{R}^n \longrightarrow \mathbb{R}$ be a C^2 defining function for D'. According to Krantz [149], p.102, a C^2 domain $D' \subset \mathbb{R}^n$ is convex iff the Hessian $\nabla^2 \rho$ of ρ at every $x \in \partial D'$ is positive semidefinite on $T(\partial D')_x$. A C^2 domain $D' \subset \mathbb{R}^n$ is said to be *strongly* (or *strictly*) *convex* at $x \in \partial D'$ if the Hessian $\nabla^2 \rho$ of ρ at x is positive definite on $T(\partial D')_x$; and *strongly* (or *strictly*) *convex* if it is so at each point of $\partial D'$. The definition is independent of the chosen defining function. Every strongly convex domain has a C^2 defining function such that its Hessian is positive definite on the whole of $T(\mathbb{R}^n)_x$ for every $x \in \partial D'$ (see [149]). A convex domain which is not strongly convex is sometimes called *weakly convex*.

Now we move on to the complex case. Let $D \subset \mathbb{C}^m$ be a bounded domain with C^2 boundary and defining function $\rho : \mathbb{C}^m \longrightarrow \mathbb{R}$. The holomorphic tangent space $\mathbf{T}(\partial D)_x$ of ∂D at $x \in \partial D$ is the kernel of $\partial \rho(x)$, that is

$$\mathbf{T}(\partial D)_x = \left\{ v \in \mathbf{T}(\mathbb{C}^m)_x \mid \sum_{k=1}^{m} \frac{\partial \rho}{\partial z_k}(x) v_k = 0 \right\}.$$

Then a bounded C^2 domain $D \subset \mathbb{C}^m$ is *strongly* (or *strictly*) *pseudoconvex* at a point $x \in \partial D$ if for some (and hence all) C^2 defining function ρ for D, the *Levi form* $dd^c \rho$ of ρ is positive definite on $\mathbf{T}(\partial D)_x$; D is *strongly* (or *strictly*) *pseudoconvex* if it is so at each point of ∂D. Every strongly pseudoconvex domain has a C^2 defining function such that its Levi form is positive definite on $\mathbf{T}(\mathbb{C}^m)_x$ for every $x \in \partial D$ (see [149]).

In the following theorem, we collect some results concerning the range of non-degenerate holomorphic mappings from $\mathbb{C}^m$ into itself. These results either are well known or follow easily from standard results of the theory of analytic functions of several complex variables.

Theorem B.13 *If $f \in \mathrm{Hol}(\mathbb{C}^m, \mathbb{C}^m)$ is non-degenerate, i.e., $\mathcal{J}f \not\equiv 0$, then*

1) The convex hull of $f(\mathbb{C}^m)$ equals $\mathbb{C}^m$, and for every $z \in \mathbb{C}^m$ there exist $a_1, ..., a_{2m} \in \mathbb{C}^m$ and $\lambda_1 \geq 0, ..., \lambda_{2m} \geq 0$ such that

$$z = \sum_{k=1}^{2m} \lambda_k f(a_k), \quad \sum_{k=1}^{2m} \lambda_k = 1, \quad \mathcal{J}f(a_k) \neq 0 (k = 1, ..., 2m).$$

2) $f(\mathbb{C}^m)$ omits at most m distinct hyperplanes.
If, further, f is ono-to-one, then the following properties hold:
3) $f(\mathbb{C}^m)$ is pseudoconvex.
4) The volume of $f(\mathbb{C}^m)$ is infinite.
5) If $\mathbb{C}^m - f(\mathbb{C}^m)$ is bounded, then f is onto.
6) If $f(\mathbb{C}^m)$ contains an open connected tube whose covex hull equals $\mathbb{C}^m$, then f is onto.

For the proof, see [70]. Also we know that holomorphic mappings with dense range are dense in $\mathrm{Hol}(\mathbb{C}^m, \mathbb{C}^n)(m \geq 1, n \geq 1)$, and that surjective holomorphic mappings are dense in $\mathrm{Hol}(\mathbb{C}^m, \mathbb{C}^m)(m \geq 1)$ (see [93] or [70]).

A well known theorem ([46], Chapter 8, Theorem 9) shows that if D is a connected open subset of $\mathbb{C}^m$, and if a sequence $f_n \in \mathrm{Hol}(D, \mathbb{C}^m)$ converges uniformly to f on compact subsets of D, and if f_n are one-to-one, then either f is one-to-one or f is degenerate. Also

H. Cartan [59] showed in the case where $m = 2$ that if f_n converge uniformly to id on compact subsets of D, then $f_n|_K$ is eventually one-to-one for every compact subset K of D and that K' is eventually contained in $f_n(D)$ for every compact subset K' of D. Dixon and Esterle [70] further showed that if $f_n \in \text{Hol}(D, \mathbb{C}^m)$ are one-to-one, and converge uniformly to f on compact subsets of D, and if f is non-degenerate, then f_n^{-1} converge uniformly to f^{-1} on compact subsets of $f(D)$.

Definition B.12 *An open set $D \subset \mathbb{C}^m$ is called a domain of holomorphy if there are no open sets D_1 and D_2 in $\mathbb{C}^m$ with the following properties:*

1) $\emptyset \neq D_1 \subset D_2 \cap D$;

2) D_2 is connected and not contained in D;

3) For every $f \in \text{Hol}(D, \mathbb{C})$ there is $f_2 \in \text{Hol}(D_2, \mathbb{C})$ (necessarily uniquely determined) such that $f = f_2$ in D_1.

Roughly speaking, the definition means that there is no part of the boundary across which every element in $\text{Hol}(D, \mathbb{C})$ can be continued analytically. We know that for every D there is a largest $\tilde{D}$ to which all functions in $\text{Hol}(D, \mathbb{C})$ can be continued analytically, and $\tilde{D}$ is then a domain of holomorphy. One of the main developments in the theory of several complex variables is the solution of the Levi problem (see [115]):

Theorem B.14 *An open set in $\mathbb{C}^m$ is a domain of holomorphy if and only if it is pseudo-convex.*

Definition B.13 *A domain of holomorphy $D \subset \mathbb{C}^m$ is called a Runge domain if polynomials are dense in $\text{Hol}(D, \mathbb{C})$, that is, if every $f \in \text{Hol}(D, \mathbb{C})$ can be uniformly approximated on an arbitrary compact set in D by analytic polynomials.*

Corollary B.2 ([70]) *Let $\text{Aut}(\mathbb{C}^m)$ be the group consisting of automorphisms of $\mathbb{C}^m$. If $\{f_n\}_{n \geq 1} \subset \text{Aut}(\mathbb{C}^m)$ converges uniformly to $f \in \text{Hol}(\mathbb{C}^m, \mathbb{C}^m)$ on compact subsets of $\mathbb{C}^m$, and if f is non-degenerate, then*

1) f is one-to-one;

2) $f(\mathbb{C}^m)$ is a Runge domain;

3) f_n^{-1} converge uniformly to f^{-1} on compact subsets of $f(\mathbb{C}^m)$;

4) $\|f_n^{-1}\| \to \infty$ uniformly over $\mathbb{C}^m - f(\mathbb{C}^m)$ as $n \to \infty$.

It is well known that with the topology of uniform convergence on compact subsets $\text{Aut}(\mathbb{C}^m)$ becomes a topological group. We see in particular that the mapping $f \mapsto f^{-1}$ is continuous on $\text{Aut}(\mathbb{C}^m)$ with the topology of uniform convergence on compact subsets. When $m = 1$, this group is quite easy to describe: its members are the functions that send z to $az + b(a, b \in \mathbb{C}, a \neq 0)$. But when $m \geq 2$, $\text{Aut}(\mathbb{C}^m)$ is infinite dimensional, for it contains mappings of form

$$z \mapsto z + \phi_1(z)e_j \quad (1 \leq j \leq m) \tag{B.62}$$

or more generally,

$$z \mapsto z + (\phi_1(z) + \phi_2(z)z_j - z_j)e_j \quad (1 \leq j \leq m) \tag{B.63}$$

with ϕ_1, ϕ_2 holomorphic functions on $\mathbb{C}^n$ independent of z_j, ϕ_2 nowhere 0, where $\{e_1, ..., e_m\}$ is the standard basis for $\mathbb{C}^m$. Following [215] automorphisms of form (B.62) will be called *shears*, and those of form (B.63) will be called *overshears*.

Let $\mathrm{Aut}_1(\mathbb{C}^m)$ be the group consisting of automorphisms of $\mathbb{C}^m$ with Jacobian determinant 1. Jung's theorem [131] shows that the group of polynomial automorphisms of $\mathbb{C}^2$ with Jacobian determinant 1 is generated by polynomial shears. Andersén [13] proved that for $m \geq 2$ shears generate a dense subgroup of the subgroup $\mathrm{Aut}_1(\mathbb{C}^m)$ in the topology of uniform convergence on compact subsets. Andersén and Lempert [14] obtained that for $m \geq 2$ the subgroup of $\mathrm{Aut}(\mathbb{C}^m)$ generated by overshears is dense.

B.6   Complex projective spaces $\mathbb{P}^m$ and torus $\mathbb{C}^m/\Gamma$

In this section, we introduce two important complex spaces. Let V be a complex vector space of dimension $m + 1$. Define $V_* = V - \{0\}$. The *dual vector space* V^* of V consists of all linear functions on V. For $\xi \in V_*$ let $\mathbb{P}(\xi) = \mathbb{C}\xi$ be the complex line spanned by ξ. For $E \subseteq V$ define

$$\mathbb{P}(E) = \{ \, \mathbb{P}(\xi) \mid 0 \neq \xi \in E \, \}.$$

Then $\mathbb{P}(V)$ is a compact connected complex manifold of dimension m called the complex *projective space* of V. Observe that

$$\mathbb{P} : V_* \longrightarrow \mathbb{P}(V)$$

is a surjective holomorphic mapping. If E is a complex linear subspace of dimension $p + 1$ of V, then $\mathbb{P}(E)$ is a smoothly embedded complex submanifold of $\mathbb{P}(V)$ and $\mathbb{P}(E)$ is called a p-dimensional *projective plane*. They are parameterized by a Grassmann manifold.

The *Grassmann cone* of order p is the analytic subset

$$\tilde{G}_p(V) = \{ \, \xi_0 \wedge ... \wedge \xi_p \mid \xi_j \in V \text{ for } j = 0, ..., p \, \}$$

of $\bigwedge_{p+1} V$. The *Grassmann manifold* $G_p(V) = \mathbb{P}(\tilde{G}_p(V))$ of order p is a connected compact smoothly embedded complex submanifold of dimension

$$\dim G_p(V) = (m - p)(p + 1)$$

of $\mathbb{P}(\bigwedge_{p+1} V)$. Observe $G_0(V) = \mathbb{P}(V)$ and that $G_m(V)$ consists of one and only one point.

Take $x \in G_p(V)$. Then $x = \mathbb{P}(\xi)$ with $\xi = \xi_0 \wedge ... \wedge \xi_p \in \tilde{G}_p(V)$. Then

$$E(x) = \{ \, \eta \in V \mid \eta \wedge \xi = 0 \, \} = \mathbb{C}\xi_0 + ... + \mathbb{C}\xi_p$$

is a complex linear subspace of dimension $p + 1$ of V with $\xi_0, ..., \xi_p$ as a base, and $\ddot{E}(x) = \mathbb{P}(E(x))$ is a p-dimensional projective plane in $\mathbb{P}(V)$. The mapping $x \longmapsto \ddot{E}(x)$ provides a bijective parameterization of the set of p-dimensional projective planes in $\mathbb{P}(V)$.

Take $a = \mathbb{P}(\alpha) \in G_q(V^*)$ and define

$$E_p[a] = \{ \, \xi \in \tilde{G}_p(V) \mid \xi \angle \alpha = 0 \, \}.$$

Then $\ddot{E}_p[x] = \mathbb{P}(E_p[x])$ is a thin analytic subset of $G_p(V)$. Obviously the mapping $a \longmapsto \ddot{E}_0[a] := \ddot{E}[a]$ also provides a bijective parameterization of the set of $(m-q-1)$-dimensional projective planes in $\mathbb{P}(V)$.

A positive definite Hermitian structure $(,) : V \times V \longrightarrow \mathbb{C}$ is called a *Hermitian product* or a *Hermitian metric* on V. It defines a norm

$$\|\xi\| = (\xi, \xi)^{1/2}, \quad \text{for all } \xi \in V.$$

A complex vector space together with a Hermitian product is called a *Hermitian vector space*. The Hermitian product on V induces natural Hermitian products on V^* and $\bigwedge_{p+1} V$. Take $x = \mathbb{P}(\xi) \in \mathbb{P}(\bigwedge_{p+1} V)$ and $a = \mathbb{P}(\alpha) \in \mathbb{P}(\bigwedge_{q+1} V^*)$. Define the *projective distance* from x to a by

$$\|x, a\| = \frac{\|\xi \angle \alpha\|}{\|\xi\| \|\alpha\|}. \tag{B.64}$$

If $x = \mathbb{P}(\xi) \in \mathbb{P}(\bigwedge_{p+1} V)$ and $y = \mathbb{P}(\eta) \in \mathbb{P}(\bigwedge_{q+1} V)$, define

$$|x, y| = \frac{\|\xi \wedge \eta\|}{\|\xi\| \|\eta\|}. \tag{B.65}$$

Define a function by

$$\tau : V \longrightarrow \mathbb{R}_+, \quad \tau(\xi) = \|\xi\|^2,$$

and set

$$\hat{v} = dd^c \log \tau \quad \left(d^c = \frac{i}{4\pi}(\bar{\partial} - \partial) \right).$$

Then a closed positive form Ω of bidegree $(1, 1)$ on $\mathbb{P}(V)$ exists such that

$$\mathbb{P}^*(\Omega) = \hat{v}.$$

The form Ω is called the *Fubini-Study form* . It determines the Fubini-Study Käehler metric on $\mathbb{P}(V)$. Obviously

$$\hat{v}^{m+1} = \mathbb{P}^*(\Omega^{m+1}) = 0.$$

For $a \in \mathbb{P}(V^*)$, on $\mathbb{P}(V) - \ddot{E}[a]$ one has

$$\Omega(x) = -dd^c \log \|x, a\|^2.$$

Take $e \in (\bigwedge_{m+1} V)^*_*$ and $p \in \mathbb{Z}[0, m]$. A linear isomorphism

$$D_e : \bigwedge_{p+1} V \longrightarrow \bigwedge_{m-p} V^*$$

is defined by

$$\langle \eta, D_e \xi \rangle = \langle \xi \wedge \eta, e \rangle, \tag{B.66}$$

for $\xi \in \bigwedge_{p+1} V$, $\eta \in \bigwedge_{m-p} V$. Then a biholomorphic map

$$\delta : \mathbb{P}\left(\bigwedge_{p+1} V \right) \longrightarrow \mathbb{P}\left(\bigwedge_{m-p} V^* \right) \tag{B.67}$$

is uniquely defined such that

$$\delta \circ \mathbb{P} = \mathbb{P} \circ D_e.$$

The mapping δ does not depend on e. The mapping $\delta \circ \delta$ is the identity. We have the following identities: if $x \in \mathbb{P}(\overset{\wedge}{_{p+1}} V)$ and $y \in \mathbb{P}(\overset{\wedge}{_{q+1}} V)$ with $0 \le p + q + 1 \le m$, then

$$|x, y| = \|x, \delta y\| = \|y, \delta x\|. \tag{B.68}$$

The projective space $\mathbb{P}(\mathbb{C} \oplus V)$ is called the *projective closure* of V. We identify

$$V = \mathbb{P}(\{1\} \times V), \quad \xi = \mathbb{P}((1, \xi)) \text{ for all } \xi \in V, \tag{B.69}$$

$$\mathbb{P}(V) = \mathbb{P}(\{0\} \times V), \quad \mathbb{P}(\xi) = \mathbb{P}((0, \xi)) \text{ for all } \xi \in V_*. \tag{B.70}$$

Then $\mathbb{P}(\mathbb{C} \oplus V) = V \cup \mathbb{P}(V)$ is the disjoint union, in which $\mathbb{P}(V)$ is called the *infinite plane*. In particular, we have

$$\mathbb{P}^m = \mathbb{P}(\mathbb{C}^{m+1}) = \mathbb{C}^m \cup \mathbb{P}^{m-1}. \tag{B.71}$$

Here $\mathbb{P}^0 = \{\infty\}$ and $\mathbb{P}^1 = \mathbb{C} \cup \{\infty\}$, which we realize as a sphere of diameter 1 in $\mathbb{R}^3$. Take $x \in \mathbb{P}^m$. Then there is a $\xi = (z_0, z_1, ..., z_m) \in \mathbb{C}_*^{m+1}$ with

$$x = \mathbb{P}(\xi) := [z_0 : z_1 : ... : z_m], \tag{B.72}$$

which called the homogeneous coordinate of x. According to the identifying $(B.69)$, we have

$$\mathbb{C}^m = \mathbb{P}(\{1\} \times \mathbb{C}^m), \quad (z_1, ..., z_m) = [1 : z_1 : ... : z_m]. \tag{B.73}$$

For $x \in \mathbb{C}$ and $y \in \mathbb{C}$ we have $\mathbb{P}((1, x)) = x$ and $\mathbb{P}((1, y)) = y$. The chordal distance is given by

$$|x, y| = \frac{|x - y|}{(1 + |x|^2)^{1/2}(1 + |y|^2)^{1/2}}. \tag{B.74}$$

Thus on $\mathbb{C}$, one has

$$\Omega(z) = -dd^c \log |z, \infty|^2 = \frac{1}{(1 + |z|^2)^2} \frac{i}{2\pi} dz \wedge d\bar{z}.$$

Let $\{\omega_1, ..., \omega_{2m}\} \subset \mathbb{C}^m$ be linearly independent over $\mathbb{R}$. Let Γ be the subgroup of $\mathbb{C}^m$ generated by $\{\omega_1, ..., \omega_{2m}\}$

$$\Gamma = \left\{ \sum_{j=1}^{2m} n_j \omega_j \mid n_j \in \mathbb{Z}, \quad j = 1, ..., 2m \right\}.$$

The action of Γ on $\mathbb{C}^m$ is properly discontinuous and the quotient manifold $\mathbb{C}^m/\Gamma$ is called an m-dimensional complex *torus*. If $z_1, ..., z_m$ is the natural coordinate system in $\mathbb{C}^m$, then the holomorphic 1-form $dz_1, ..., dz_m$ can be considered as forms on $\mathbb{C}^m/\Gamma$. Every holomorphic 1-form on $\mathbb{C}^m/\Gamma$ is a linear combination of $dz_1, ..., dz_m$ with constant coefficients. In fact, every holomorphic 1-form on $\mathbb{C}^m/\Gamma$ is a linear combination of $dz_1, ..., dz_m$ with holomorphic functions as coefficients and since $\mathbb{C}^m/\Gamma$ is a compact complex manifold, these coefficient functions are constant.

Let $\mathbb{C}^n/\Gamma'$ be another complex torus and $u_1, ..., u_n$ the natural coordinate system in $\mathbb{C}^n$. A homomorphism $\mathbb{C}^m/\Gamma \to \mathbb{C}^n/\Gamma'$ is induced by a complex linear transformation of $\mathbb{C}^m$ into $\mathbb{C}^n$ which sends Γ into Γ'. If $f : \mathbb{C}^m/\Gamma \to \mathbb{C}^n/\Gamma'$ is a holomorphic mapping, then

$$f^*(du_k) = \sum_{j=1}^{m} a_k^j dz_j, \quad a_k^j \in \mathbb{C}, \quad k = 1, ..., n,$$

showing that f is induced by a mapping

$$\tilde{f} = (f_1, ..., f_n) : \mathbb{C}^m \longrightarrow \mathbb{C}^n$$

of the form

$$f_k = \sum_{j=1}^{m} a_k^j z_j + b_k, \quad b_k \in \mathbb{C}, \quad k = 1, ..., n,$$

such that $\pi' \circ \tilde{f} = f \circ \pi$, where $\pi : \mathbb{C}^m \to \mathbb{C}^m/\Gamma$ and $\pi' : \mathbb{C}^n \to \mathbb{C}^n/\Gamma'$ are the projections. Thus every holomorphic mapping of $\mathbb{C}^m/\Gamma$ into $\mathbb{C}^n/\Gamma'$ is a homomorphism modulo a translation in $\mathbb{C}^n/\Gamma'$.

B.7   Meromorphic mappings and order functions

Here we introduce a class of mapping which will be studied later as well as the method of value distribution which is often used in the study of complex dynamics of transcendental meromorphic mappings.

Let M be a connected complex manifold of dimension m. It is well-known that M is a Stein manifold if and only if there exists a strictly plurisubharmonic function $\tau_0 \in C^\infty(M, \mathbb{R})$, i.e., $dd^c\tau_0 > 0$, such that $M_{\tau_0}(r) = \{z \in M \mid \tau_0 < r\} \subset\subset M$ for every real number r. Recall that a continuous mapping is said to be proper if the inverse image of any compact set is compact. Thus if M is a Stein manifold, there exists a one-to-one proper mapping

$$\beta : M \longrightarrow \mathbb{C}^{2m+1}$$

with $\mathrm{rank}(f) = m$ at every point in M (see [115]). In this section, we mainly consider a class of non-negative functions on M. A real function

$$\tau : M \longrightarrow \mathbb{R}[0, b) \quad (0 < b \le \infty) \tag{B.75}$$

of class C^∞ is said to be an *exhaustion* of M if it is proper. If τ is an exhaustion of M, define

$$\upsilon = dd^c\tau, \quad \hat{\upsilon} = dd^c \log \tau, \quad \sigma_p = d^c \log \tau \wedge \hat{\upsilon}^p. \tag{B.76}$$

Then

$$\tau^{p+1}\hat{\upsilon}^p = \tau\upsilon^p \quad pd\tau \wedge d^c\tau \wedge \upsilon^{p-1}, \quad \sigma_p = \tau^{-p-1}d^c\tau \wedge \upsilon^p. \tag{B.77}$$

An initial remark is that every open manifold, real or complex, always admits an exhaustion function τ such that τ has only isolated critical points in $M - M_{\sqrt{\tau}}(r(\tau))$ for some $r(\tau)$, $0 \le r(\tau) < b$.

An exhaustion function τ is called *eventually concave* (resp., *concave*) iff $v \leq 0$ on $M - M_{\sqrt{\tau}}(r(\tau))$ for some $r(\tau)$, i.e., iff the eigenvalues of the Levi form v are nonpositive from $r(\tau)$ upward (resp., $v \leq 0$ on M). It is called *eventually convex* (resp., *convex*) iff $v \geq 0$ on $M - M_{\sqrt{\tau}}(r(\tau))$ for some $r(\tau)$, i.e., iff the eigenvalues of the Levi form v are nonnegative from $r(\tau)$ upward (resp., $v \geq 0$ on M). It is called *logarithmic concave or logarithmic convex* if $\log \tau$ is concave or convex. It is called *eventually parabolic* if

$$\hat{v} \geq 0, \quad \hat{v}^m \equiv 0, \quad v^m \not\equiv 0 \tag{B.78}$$

on $M - M_{\sqrt{\tau}}[r(\tau)]$ for some $r(\tau)$. A eventually parabolic exhaustion is called *parabolic* if $r(\tau) = 0$.

Griffiths and King [100] considered a special parabolic exhaustion which has only finitely many critical values and such that $\log \tau$ has only finitely many logarithmic singularities. The properties of parabolic exhaustions were discussed by Stoll [245]. If $\beta : M \longrightarrow \mathbb{C}^m$ is a surjective, proper, holomorphic mapping, then $\tau = \|\beta\|$ is a parabolic exhaustion of M called a *covering parabolic exhaustion*. Concave and convex exhaustions were used by Bott-Chern [49] and Wu [273].

If τ is eventually parabolic, then on $M - M_{\sqrt{\tau}}[r(\tau)]$,

$$\tau v^m = m d\tau \wedge d^c \tau \wedge v^{m-1} = m \tau^m d\tau \wedge \sigma_{m-1}, \quad d\sigma_{m-1} = \hat{v}^m = 0, \tag{B.79}$$

so Stokes Theorem implies that

$$\int_{M_{\sqrt{\tau}}\langle r \rangle} \sigma_{m-1} = \int_{M_{\sqrt{\tau}}\langle s \rangle} \sigma_{m-1}$$

for $r > s > r(\tau)$ with $r^2, s^2 \notin \tau(C_\tau)$. Denote this common number by ς. Obviously,

$$\varsigma = r^{-2m} \int_{M_{\sqrt{\tau}}[r]} v^m > 0 \quad (r > r(\tau)). \tag{B.80}$$

A complex manifold M is said to be parabolic if there exists an unbounded (i.e., $b = \infty$), parabolic exhaustion on M. A parabolic exhaustion τ is said to be *strict* if $v > 0$ on M which can be completely determined by the following result:

Theorem B.15 (Stoll [246], [247]) *If a parabolic manifold M of dimension m has a strict parabolic exhaustion τ, then there exists a biholomorphic mapping $h : \mathbb{C}^m \longrightarrow M$ such that*

$$\tau(h(z)) = \|z\|^2.$$

Alternative proofs were given by Burns [57] and Wong [268].

Theorem B.16 (Remmert [208]) *Let M and N be connected complex manifolds of dimensions m and n respectively. Let $f : M \longrightarrow N$ be a mapping and let*

$$G(f) = \{(x, f(x)) \mid x \in M\} \subset M \times N$$

be the graph of f. Then f is holomorphic iff $G(f)$ is an analytic subset of pure dimension m of $M \times N$.

Let S be an analytic subset of M such that $A = M - S \neq \emptyset$. Then A is dense in M. Let $f_A : A \longrightarrow N$ be a holomorphic mapping. The mapping f_A is said to be *meromorphic* on M and is denoted by $f : M \longrightarrow N$ if the closure $\overline{G(f_A)}$ of the graph $G(f_A)$ is analytic in $M \times N$ and if the projection

$$\pi_M : \overline{G(f_A)} \longrightarrow M$$

is proper. We set $G(f) = \overline{G(f_A)}$ which is called the graph of the meromorphic mapping $f : M \longrightarrow N$ determined by f_A. We let $\mathrm{Mer}(M, N)$ be the set of the meromorphic mappings between complex manifolds M and N. The next theorem shows that "meromorphic mapping" is an extended notion of "holomorphic mapping".

Theorem B.17 ([190]) *Take $f \in \mathrm{Mer}(M, N)$. Then f is represented by a holomorphic mapping $f_M : M \longrightarrow N$ iff $f(x)$ consists of one point for any $x \in M$.*

Assume that $f \in \mathrm{Mer}(M, N)$ and that $\pi_N : G(f) \longrightarrow N$ is the projection. For each $x \in M$, the set

$$f_0(x) = \pi_N \circ \pi_M^{-1}(x) = \{y \in N \mid (x, y) \in G(f)\}$$

is analytic and non-empty. Also the *indeterminacy*

$$I_f = \{x \in M \mid \#f_0(x) > 1\} = \pi_M(\{P \in G(f) \mid \mathrm{rank}_P(\pi_M) < \mathrm{rank}(\pi_M)\})$$

is analytic and contained in S with

$$\dim I_f \leq m - 2.$$

If $x \in I_f, y \in f_0(x)$, then $\dim_y f_0(x) > 0$. Also $f_0(x)$ is compact connected analytic subset of N. Thus the holomorphic mapping $f_A : A \longrightarrow N$ extends to a holomorphic mapping $f_{M-I_f} : M - I_f \longrightarrow N$. We also write $f(x) = f_0(x)$ for all $x \in M$. We define the *rank* of the meromorphic mapping $f : M \longrightarrow N$ by

$$\mathrm{rank}(f) = \mathrm{rank}(\pi_N).$$

Noting that $G(f)$ is irreducible and hence of pure dimension, i.e.,

$$\dim_P G(f) = \dim G(f) \quad \text{for all } P \in G(f),$$

we have

$$\mathrm{rank}(f) = \mathrm{rank}(\pi_N|_{G(f) - \pi_M^{-1}(I_f)}) = \mathrm{rank}(f_{M-I_f}).$$

Let $N = \mathbb{P}(V)$, where V is a Hermitian vector space of dimension $n + 1 > 1$, another equivalent definition of a meromorphic mapping exists. Assume M, S, A as above and let $f_A : A \longrightarrow \mathbb{P}(V)$ be a holomorphic mapping. Let $U \neq \emptyset$ be an open connected subset of M. A holomorphic vector function $\tilde{f} : U \longrightarrow V$ is said to be a *representation* of f_A on U if $\tilde{f} \not\equiv 0$ and if

$$\mathbb{P}(\tilde{f}(x)) = f_A(x) \quad \text{for all } x \in U - \tilde{f}^{-1}(0).$$

The representation is said to be *reduced* if $\dim \tilde{f}^{-1}(0) \leq m - 2$. If $x \in U$, we may say that $\tilde{f}$ is a representation of f_A at x. Then f_A is meromorphic on M iff there is a representation of f_A at every point of M. A (reduced) representation of f_A is also called a (reduced) representation of the meromorphic mapping $f : M \longrightarrow \mathbb{P}(V)$ determined by f_A. If $M = \mathbb{C}^m$ and if $f : \mathbb{C}^m \longrightarrow \mathbb{P}(V)$ is meromorphic, there exists a reduced representation $\tilde{f} : \mathbb{C}^m \longrightarrow V$ of f (cf. [190]).

Let M and N be complex manifolds of dimensions m and n respectively, and

$$f : M \longrightarrow N$$

a meromorphic mapping. To measure the *growth* of f, we assume as given an Hermitian metric on N with associated $(1,1)$-form ω and logarithmic convex exhaustion function $\tau : M \longrightarrow \mathbb{R}[0, \infty)$ with Levi form $v = dd^c \tau$. For $p \in \mathbb{Z}[0, m]$, the quantities

$$t_p(r) = t_{p,f}(r, \omega) = r^{2p-2m} \int_{M_{\sqrt{\tau}}[r]} f^*(\omega^p) \wedge v^{m-p},$$

$$T_p(r) = T_{p,f}(r, \omega) = \int_{r_0}^{r} t_p(x) \frac{dx}{x} \quad (r > r_0 \geq r(\tau))$$

will be called the *spherical image* (or *unintegrated order function*) and *order function* of order p respectively for the meromorphic mapping f. Then $t_p(r)$ increases with

$$t_p(0) = \lim_{r \to 0} t_p(r)$$

and with

$$t_p(r) = \int_{M_{\sqrt{\tau}}(0,r]} f^*(\omega^p) \wedge \hat{v}^{m-p} + t_p(0) \tag{B.81}$$

if $r > 0$. If $p \in \mathbb{Z}[0, m - 1]$ and if r^2 is a regular value of τ, then (cf. [245])

$$t_p(r) = \int_{M_{\sqrt{\tau}}\langle r \rangle} f^*(\omega^p) \wedge \sigma_{m-p-1}. \tag{B.82}$$

Obviously, T_p is increasing and $T_p(r) \to \infty$ as $r \to \infty$, unless of course $T_p(r) \equiv 0$. Also

$$t_p(\infty) = \lim_{r \to \infty} t_p(r) = \lim_{r \to \infty} \frac{T_p(r)}{\log r},$$

so that the conditions

$$t_p(r) = O(1), \quad T_p(r) = O(\log r)$$

are evidently equivalent.

Proposition B.9 *Take $p \in \mathbb{Z}[1, m]$. Assume that N is a compact Kähler manifold. Suppose that ω and $\tilde{\omega}$ are Kähler metrics in the same cohomology class on N. Then*

$$T_{p,f}(r, \tilde{\omega}) = T_{p,f}(r, \omega) + O(t_{p-1}(r)).$$

Proof. At first, assume that f is holomorphic. A C^∞ function u on N exists such that

$$\omega - \tilde{\omega} = dd^c u,$$

and consequently

$$\omega^p - \tilde{\omega}^p = dd^c u \wedge \chi_p,$$

where

$$\chi_1 = 1, \quad \chi_p = \sum_{j=0}^{p-1} \omega^j \wedge \tilde{\omega}^{p-1-j} \quad (p > 1).$$

Define $q = m - p$. Take $r > r_0 > 0$ with $r^2, r_0^2 \notin \tau(C_r)$. Then

$$
\begin{aligned}
&T_{p,f}(r, \omega) - T_{p,f}(r, \tilde{\omega}) \\
&= \int_{r_0}^{r} \int_{M_{\sqrt{\tau}}[t]} f^*(dd^c u) \wedge f^*(\chi_p) \wedge v^q t^{-2q-1} dt \\
&= \int_{r_0}^{r} \int_{M_{\sqrt{\tau}}<t>} f^*(d^c u) \wedge f^*(\chi_p) \wedge v^q t^{-2q-1} dt \\
&= \frac{1}{2} \int_{M_{\sqrt{\tau}}(r_0, r]} \tau^{-q-1} d\tau \wedge d^c(u \circ f) \wedge f^*(\chi_p) \wedge v^q \\
&= \frac{1}{2} \int_{M_{\sqrt{\tau}}(r_0, r]} \tau^{-q-1} d(u \circ f) \wedge f^*(\chi_p) \wedge d^c \tau \wedge v^q \\
&= \frac{1}{2} \int_{M_{\sqrt{\tau}}(r_0, r]} d(u \circ f) \wedge f^*(\chi_p) \wedge \sigma_q \\
&= \frac{1}{2} \int_{M_{\sqrt{\tau}}(r_0, r]} d((u \circ f) f^*(\chi_p) \wedge \sigma_q) - \frac{1}{2} \int_{M_{\sqrt{\tau}}(r_0, r]} (u \circ f) f^*(\chi_p) \wedge \hat{v}^{q+1} \\
&= \frac{1}{2} \int_{M_{\sqrt{\tau}}(r)} (u \circ f) f^*(\chi_p) \wedge \sigma_q - \frac{1}{2} \int_{M_{\sqrt{\tau}}(r_0)} (u \circ f) f^*(\chi_p) \wedge \sigma_q \\
&\quad - \frac{1}{2} \int_{M_{\sqrt{\tau}}(r_0, r]} (u \circ f) f^*(\chi_p) \wedge \hat{v}^{q+1}.
\end{aligned}
$$

Since N is compact, there are positive constants c_1, c_2, c_3 such that

$$|u| \leq c_1, \quad c_2 \omega \leq \tilde{\omega} \leq c_3 \omega.$$

Thus we have

$$\left(\sum_{j=0}^{p-1} c_2^{p-1-j} \right) \omega^{p-1} \leq \chi_p \leq \left(\sum_{j=0}^{p-1} c_3^{p-1-j} \right) \omega^{p-1}$$

so that the proposition follows from (B.81) and (B.82). If f is meromorphic, the same conclusion holds by going over the graph. □

Proposition B.10 ([58]) *Let $f : \mathbb{C}^m \longrightarrow N$ be a holomorphic mapping into a compact Kähler manifold N. Then f can be extended to a meromorphic mapping $f : \mathbb{P}^m \longrightarrow N$ iff*

$$T_p(r) = O(\log r) \quad (p = 1, ..., m).$$

Usually we abbreviate as

$$T_f(r,\omega) = T_{1,f}(r,\omega),$$

which is called the *characteristic function* of f with respect to ω. Assume that N is compact and that M is parabolic. Take another Hermitian metric form ω' on N. Then there are positive constants c and c' such that

$$c\omega' \leq \omega \leq c'\omega'.$$

Thus we have

$$cT_f(r,\omega') \leq T_f(r,\omega) \leq c'T_f(r,\omega'). \tag{B.83}$$

We define the *order* of f by

$$\mathrm{ord}(f) = \varlimsup_{r\to\infty} \frac{\log T_f(r,\omega)}{\log r}, \tag{B.84}$$

which is independent of the choice of Hermitian metric form ω by (B.83).

If ω and ω' are Hermitian metric forms on N, which are mutually cohomologous, by Proposition B.9, we have

$$T_f(r,\omega') = T_f(r,\omega) + O(1).$$

Hence we may define the characteristic function of f with respect to the cohomology class $[\omega] \in H^{1,1}(N,\mathbb{R})$, up to a bounded term, by

$$T_f(r,[\omega]) = T_f(r,\omega).$$

If $N = \mathbb{P}^n$, and if ω is the Fubini-Study form, we abbreviate as

$$T_f(r) = T_f(r,[\omega]).$$

Furthermore if $\tilde{f} : M \longrightarrow \mathbb{C}^{n+1}$ is a global reduced representation of f, then (see [248])

$$T_f(r) = \int_{M_{\sqrt{\tau}<r>}} \log\|\tilde{f}\|\sigma_{m-1} - \int_{M_{\sqrt{\tau}<r_0>}} \log\|\tilde{f}\|\sigma_{m-1}. \tag{B.85}$$

If W is a complex vector space, then $\mathbb{P}(W \oplus \mathbb{C})$ is called the *projective closure* of W. An analytic subset M of W is said to be an *affine algebraic variety embedded into* W if the closure $\overline{M}$ of M in $\mathbb{P}(W \oplus \mathbb{C})$ is analytic, which is the case if and only if M is the common zero set of polynomials on W (Chow [62]). A complex space M is said to be an *affine algebraic variety* if M is biholomorphically equivalent to an embedded affine algebraic variety. An analytic subset N of $\mathbb{P}(W)$ is said to be a *projective algebraic variety embedded into* $\mathbb{P}(W)$. A complex space N is said to be a *projective algebraic variety* if there exists a biholomorphic mapping $h : N \longrightarrow N'$ onto a projective algebraic variety N' embedded into some projective space. Let M be an affine algebraic variety embedded into W. Let N be a projective algebraic variety. Let $f : M \longrightarrow N$ be a meromorphic mapping. Then f is said to be *rational* if f extends to a meromorphic mapping $\tilde{f} : \overline{M} \longrightarrow N$. If N is embedded, then $\tilde{f}$ is given by rational functions. Note that affine algebraic varieties are parabolic (Griffiths and King [100],Stoll [245]).

Proposition B.11 ([100], [245]) *Let M be an affine algebraic variety. Let N be a projective algebraic variety with a Kähler form ω. Then a holomorphic mapping $f : M \longrightarrow N$ is rational iff $T_f(r, \omega) = O(\log r)$.*

Bibliography

[1] Abate, M., A characterization of hyperbolic manifolds, Proc. Amer. Math. Soc. 117(1993), 789-793.

[2] Abate, M., Iteration theory of holomorphic maps on taut manifolds, Mediterranean Press, Cosenza, 1989.

[3] Abate, M., Horospheres and iterates of holomorphic maps, Math. Z. 198(1988), 225-238.

[4] Abate, M. & Vigué, J. P., Common fixed points in hyperbolic Riemann surfaces and convex domain, Proc. Amer. Math. Soc. 112(1991),503-512.

[5] Abel, N. H., Détermination d'une fonction au moyen d'une èquation qui ne contient qu'une seule variable, Oeuvres complètes II, Christiania (1881), 36-39.

[6] Abercrombie, A. G. & Nair, R., An exceptional set in the ergodic theory of rational maps of the Riemann sphere, Ergod. Th. & Dynam. Sys. 17(1997), 253-267.

[7] Abraham, R. & Smale, S., Nongenericity of Ω-stability, Global Analysis (Proc. Sympos. Pure Math., Vol. 14, Berkeley, Calif., 1968), pp.5-8, Amer. Math. Soc., Providence, R. I., 1970.

[8] Ahern, P. & Forstneric, F., One parameter automorphism groups on $\mathbb{C}^2$, Complex Variables 27(1995), 245-268.

[9] Ahlfors, L. V., An extension of Schwarz's lemma, Trans. Amer. Math. Soc. 43(1938), 359-364.

[10] Aïzenberg, L. A. & Yuzhakov, A. P., Integral representations and residues in multidimensional complex analysis, Trans. Math. Monographs 58(1983), Amer. Math. Soc., Providence, Rhode Island.

[11] Aladro, G., Application of the Kobayashi metric to normal functions of several complex variables, Utilitas Math. 31(1987), 13-24.

[12] Aladro, G. & Krantz, S. G., A criterion for normality in $\mathbb{C}^n$, J. Math. Analysis and Appl. 161(1991), 1-8.

[13] Andersén, E., Volume preserving automorphisms of $\mathbb{C}^n$, Complex Variables 14(1990), 223-235.

[14] Andersén, E. & Lempert, L., On the group of holomorphic automorphisms of $\mathbb{C}^n$, Invent. Math. 110(1992), 371-388.

[15] Andreotti, A. & Stoll, W., Analytic and algebraic dependence of meromorphic functions, Lecture Notes in Math. 234(1971), Springer-Verlag.

[16] Anosov, D. V., Roughness of geodesic flows on compact Riemannian manifolds of negative curvature, Dokl. Akad. Nauk SSSR 145(4)(1962), 707-709. English transl.: Sov. Math. Dokl. 3(1962), 1068-1070.

[17] Anosov, D. V., Geodesic flows on closed Riemannian manifolds of negative curvature, Tr. Mat. Inst. Steklova 90. English transl.: Proc. Steklov Inst. Math. 90(1969).

[18] Arnold, V. I., Geometrical methods in the theory of ordinary differential equations, Grund. der Math. Wiss. 250, Springer-Verlag, 1988.

[19] Arnold, V. I., Sur une propriété topologique des applications globalement canoniques e à mécanique classique, C. R. Acad. Sci. Paris 261(1965), 3719-3722.

[20] Arnold, V. I., Mathematical methods in classical mechanics, Appendix 9, Springer-Verlag, 1978.

[21] Arrowsmith, D. K. & Place, C. M., An introduction to dynamical systems, Cambridge Univ. Press, Cambridge, 1990.

[22] Baker, I. N., Fixpoints of polynomials and rational functions, J. London Math. Soc. 39(1964), 615-622.

[23] Baker, I. N., The existence of fix-points of entire functions, Math. Z. 73(1960), 280-284.

[24] Baker, I. N. & Eremenko, A., A problem on Julia sets, Ann. Acad. Sci. Fenn., Series A, 12(1987), 229-236.

[25] Baker, I. N. & Singh, A., Wandering domains in the iteration of compositions of entire functions, Ann. Acad. Sci. Fenn. Ser. A I 20(1995), 149-153.

[26] Barnsley, M., Fractals everywhere, Academic Press, INC, 1988.

[27] Barth, T. J., Taut and tight complex manifolds, Proc. Amer. Math. Soc. 24(1970), 429-431.

[28] Barth, T. J., The Kobayashi distance induces the standard topology, Proc. Amer. Math. Soc. 35(1972), 439-441.

[29] Bass, H., Connell, E. H. & Wright, D., The Jacobian conjecture: reduction of degree and formal expansion of the inverse, Bull. Amer. Math. Soc. 7(1982), 287-330.

[30] Bass, H. & Meisters, G., Polynomial flows in the plane, Adv. in Math. 55(1985), 173-208.

[31] Beardon, A. F., Iteration of Rational Functions, Graduate Texts in Math. 132(1991), Springer-Verlag.

[32] Beardon, A. F., Symmetries of Julia sets, Bull. London Math. Soc. 22(1990), 576-582.

[33] Bedford, E., Lyubich, M. & Smillie, J., Polynomial diffeomorphisms of $\mathbb{C}^2$. IV: the measure of maximal entropy and Laminar currents, Stony Brook Preprint No.8(1992).

[34] Bedford, E. & Smillie, J., Polynomial diffeomorphisms of $\mathbb{C}^2$: currents, equilibrium measure and hyperbolicity, Invent. Math. 103(1991), 69-99.

[35] Bedford, E. & Smillie, J., Polynomial diffeomorphisms of $\mathbb{C}^2$. II: stable manifolds and recurrence, J. Amer. Math. Soc. 4(1991), 657-679.

[36] Bedford, E. & Smillie, J., Fatou-Bieberbach domains arising from polynomial automorphisms, Indiana Univ. Math. J. 40(1991), 789-792.

[37] Bedford, E. & Smillie, J., Polynomial diffeomorphisms of $\mathbb{C}^2$: ergodicity, exponents and entropy of the equilibrium measure, preprint, 1991.

[38] Berenstein, C., Chang, D. C. & Li, B. Q., Exponential sums and shared values for meromorphic functions in several complex variables, to appear in Advances of Math..

[39] Bergweiler, W., Iteration of meromorphic functions, Bull. Amer. Math. Soc. (2) 29 (1993), 151-188.

[40] Bergweiler, W. & Wang, Y., On the dynamics of composite entire functions, preprint.

[41] Biancofiore, A. & Stoll, W., Another proof of the lemma of the logarithmic derivative in several complex variables, Annals of Math. Studies 100(1981), 29-45, Princeton Univ. Press, Princeton, N.J..

[42] Bieberbach, L., Beispiel zweier ganzer Funktionen zweier komplexer Variablen, welche eine schlichte volumtreue Abbildung des $\mathbb{R}^4$ auf einen Teil seiner selbst vermitteln, S. B. Preuss Akad. Wiss. 14/15(1933), 476-479.

[43] Birkhoff, G. D., Proof of the ergodic theorem, Proc. Nat. Acad. Sci. USA 17(1931), 656-660.

[44] Birkhoff, G. D., Proof of the ergodic theorem, Nat. Acad. USA 18(1932).

[45] Birkhoff, G. D., Dynamical systems, Amer. Math. Soc., 1927, revised edition 1966.

[46] Bochner, S. & Martin, W., Several complex variables, Princeton Univ. Press, Princeton, N.J., 1948.

[47] Bochner, S. & Montgomery, D., Locally compact groups of differentiable transformations, Ann. Math. 47(1946), 639-654.

[48] Borel, E., Sur les Zéros des fonctions entières, Acta Math. 20(1897), 357-396.

[49] Bott, R. & Chern, S. S., Hermitian vector bundles and the equidistribution of the zeros of their holomorphic sections, Acta Math. 114(1965), 71-112.

[50] Bowen, R., Equilibrium states and the ergodic theory of Anosov diffeomorphisms, LNM 470(1975), Springer, Berlin.

[51] Bowen, R., On Axiom A diffeomorphisms, C. B. M. S. Regional Conf. Series in Math., no. 38, Amer. Math. Soc., Providence, R. I., 1978.

[52] Bowen, R., Methods of symbolic dynamics, collected papers, Mir, Moscow, 1979(Russian).

[53] Bremermann, H. J., Die Characterisierung von Regularitätsgebieten durch pseudokonvexe Funktionen, Schriftenreihe Math. Inst. Univ. Münster, No. 5(1951), pp. 1-92.

[54] Brody, R., Compact manifolds and hyperbolicity, Trans. Amer. Math. Soc. 235(1978), 213-219.

[55] Brown, J. R., Ergodic theory and topological dynamics, Pure and Applied Math. 70(1976), Academic Press, New York, London.

[56] Büger, M., On the Julia set of the composition of meromorphic functions, Analysis 16(1996), 385-397.

[57] Burns, D., Curvature of the Monge-Ampère foliations and parabolic manifolds, Ann. of Math. 115(1982), 349-373.

[58] Carlson, J. & Griffiths, Ph., The order function for entire holomorphic mappings, Value-Distribution Theory Part A, Pure and Appl. Math. 25(1974), 225-248, Marcel Dekker, New York.

[59] Cartan, H., Sur les fonctions de deux variables complexes: les transformations d'un domaine borné D en un domaine intérieur á D, Bull. Soc. Math. France 58(1930), 199-219.

[60] Cegrell, U., Removable singularities for plurisubharmonic functions and related problems, Proc. London Math. Soc. 36(1978), 310-336.

[61] Chern, S. S., On holomorphic mappings of Hermitian manifolds of the same dimension, Proc. Symposia in Pure Math. 11(1968), 157-170, Publ. Amer. Math. Soc..

[62] Chow, W. L., On compact analytic varieties, Amer. J. Math. 71(1949), 893-914.

[63] Chow, Mallet-Paret & Yorke, Springer Lecture Notes 1007(1983), 109-131.

[64] Conley, C., Isolated invariant sets and the Morse index, C. B. M. S. Regional Conf. Series in Math., no. 38, Amer. Math. Soc., Providence, R. I., 1978.

[65] Cornalba, M. & Shiffman, B., A counterexample to the "transcendental Bézout problem", Ann. of Math. 96(1972), 402-406.

[66] Cornfeld, I. P., Fomin, S. V. & Sinai, Ya. G., Ergodic theory, Springer-Verlag, 1982.

[67] Coven, E. M., Madden, J. & Nitecki, Z., A note on generic properties of continuous maps, Ergodic Theory and Dynamical systems II (1982), pp.97-101, Birkhäuser, Boston.

[68] Dektyarev, I. M., Multidimensional value distribution theory, Several Complex Variables III (ed. G. M. Khenkin), Springer-Verlag, 1989.

[69] Denjoy, A., Sur l'itération des fonctions analytiques, C. R. Acad. Sci. Paris, Sér I 182(1926), 255-257.

[70] Dixon, P. G. & Esterle, J., Michael's problem and the Poincaré-Fatou-Bieberbach phenomenon, Bull. Amer. Math. Soc. 15(1986), 127-187.

[71] Dieudonné, J., Grundzüge der modernen Analysis I (Logik und Grundlagen der Mathematik 8), Vieweg, 1985.

[72] Douady, A., Julia sets of polynomials, Preprint, 1994.

[73] Dunford, N. & Schwartz, J. T., Linear operators I, Interscience Publishers, London-New York, 1958.

[74] Edelstein, M., On fixed and periodic points under contractive mappings, J. London Math. Soc. 37(1962), 74-79.

[75] Eremenko, A. E. & Lyubich, M. Y., The dynamics of analytic transformations, Leningrad Math. J. 1 (3) (1990), 563-634.

[76] Eustice, D. J., Holomorphic idempotents and common fixed points on the 2-disk, Michigan Math. J. 19(1972), 347-352.

[77] Fabritiis, C. D., Commuting holomorphic functions and hyperbolic automorphisms, Proc. Amer. MAth. Soc. 124(1996), 3027-3037.

[78] Fabritiis, C. D. & Gentili, G., On holomorphic maps which commute with hyperbolic automorphisms, to appear in Advances in Math..

[79] Fatou, P., Sur certaines fonctions uniformes de deux variables, C. R. Acad. Sci. Paris 175(1922), 1030-1033.

[80] Floer, A., Morse theory for Lagrangian intersections, J. Diff. Geom. 28(1988), 513-547.

[81] Floer, A., Symplectic fixed points and holomorphic spheres, preprint, Courant Institute, New York University, 1988.

[82] Fornaess, J. E. & Sibony, N., Complex dynamics in higher dimension. I, S. M. F. Astérisque 222(1994), 201-231.

[83] Fornaess, J. E. & Sibony, N., Classification of recurrent domains for some holomorphic maps, Math. Ann. 301(1995), 813-820.

[84] Fornaess, J. E. & Sibony, N., Critically finite rational maps on $\mathbb{P}^2$, Preprint.

[85] Fornaess, J. E. & Sibony, N., Complex dynamics in higher dimension. II, Annals of Math. Studies 137, Princeton Univ. Press, 1995.

[86] Fornaess, J. E. & Sibony, N., Complex Hénon mappings in $\mathbb{C}^2$ and Fatou-Bieberbach domains, Duke Math. J. 65(1992), 345-380.

[87] Fornaess, J. E. & Sibony, N., Oka's inequality for currents and applications, Math. Ann. 301(1995), 399-419.

[88] Fornaess, J. E. & Sibony, N., The closing lemma for holomorphic maps, Ergod. Th. & Dynam. Sys. 17(1997), 821-837.

[89] Forstneric, F., Actions of $(\mathbb{R}, +)$ and $(\mathbb{C}, +)$ on complex manifolds, Math. Z., to appear.

[90] Franks, J. M., Homology and dynamical systems, C. B. M. S. Regional Conf. Series in Math., no. 49, Amer. Math. Soc., Providence, R. I., 1982.

[91] Friedland, S. & Milnor, J., Dynamical properties of plane polynomial automorphisms, Ergod. Th. & Dynam. Sys. 9(1989), 67-99.

[92] Friedman, J., On the convergence of Newton's method, Journal of Complexity 5(1989), 12-33.

[93] Gauthier, P. & Ngo Van Que, Problème de surjectivité des applications holomorphes, Ann. Sci. Norm. Sup. Pisa (2) 27 (1973), 555-559.

[94] Gavosto, E. A., Attracting basins in $\mathbb{P}^2$, Journal of Geometric Analysis, to appear.

[95] Ghys, E., Holomorphic Anosov systems, Invent. Math. 119(1995), 585-614.

[96] Grauert, H. & Reckxiegel, H., Hermiteoche Metriken und normale Famiben holomorpher Abbildungen, Math. Z. 89(1965), 108-125.

[97] Grauert, H. & Remmert, R., Coherent analytic sheaves, Springer, 1984.

[98] Green, M., Holomorphic maps into complex projective spaces omitting hyperplanes, Trans. Amer. Math. Soc. 169(1972), 89-103.

[99] Griffiths, P., Holomorphic mapping into canonical algebraic varieties, Ann. of Math. 93(1971), 439-458.

[100] Griffiths, P. & King, J., Nevanlinna theory and holomorphic mappings between algebraic varieties, Acta Math. 130(1973), 145-220.

[101] Griffiths, P. & Harris, J., Principles of algebraic geometry, John Wiley & Sons, New York, 1978.

[102] Gunning, R. C., Introduction to holomorphic functions of several variables, Wadsworth (Belmont), 1990.

[103] Gutwirth, A., An inequality for certain pencils of plane curves, Proc. Amer. Math. Soc. 12(1961), 631-638.

[104] Heath, L. F. & Suffridge, T. J., Holomorphic retracts in complex n-space, Illinois J. Math. 25(1981), 125-135.

[105] Heinemann, S. M., Julia sets for holomorphic endomorphisms of $\mathbb{C}^n$, Ergod. Th. & Dynam. Sys. 16(1996), 1275-1296.

[106] Heins, M. H., On the iteration of functions which are analytic and single-valued in a given multiply-connected region, Amer. J. Math. 63(1941), 461-480.

[107] Heins, M. H., A generalization of the Aumann-Carathéodory "Starrheitssatz", Duke Math. J. 8(1941), 312-316.

[108] Hénon, M., A two-dimensional mapping with a strange attractor, Comm. Math. Phys. 50(1976), 69-77.

[109] Hill, R. & Velani, S., The ergodic theory of shrinking targets, Invent. Math., To appear.

[110] Hille, E., Methods in classical and functional analysis, Addison-Wesley, 1972.

[111] Hirsch, M., Differential topology, Springer-Verlag, 1976.

[112] Hirsch, M., Palis, J., Pugh, C. & Shub, M., Neighborhoods of hyperbolic sets, Invent. Math. 9(1970), 121-134.

[113] Hofer, H., Ljusternik-Schnirelman-theory for Lagrangian intersections, Ann. Henri Poincaré-analyse nonlinéaire 5(1988), 465-499.

[114] Hopf, E., The general temporally discrete Markov process, J. Rat. Mech. Anal. 3(1954), 13-45.

[115] Hörmander, L., An introduction to complex analysis in several variables, North-Holland, Amsterdam, New York, Oxford, Tokyo, 1990.

[116] Horst, C., Compact varieties of surjective holomorphic mappings, Math. Z. 196(1987), 259-269.

[117] Horst, C., A finiteness criterion for compact varieties of surjective holomorphic mappings, preprint.

[118] Hu, P. C., Value distribution theory and complex dynamics, PhD thesis, The Hong Kong University of Science and Technology, 1996.

[119] Hu, P. C. & Yang, C. C., Uniqueness of meromorphic functions on $\mathbb{C}^n$, Complex Variables 30(1996), 235-270.

[120] Hu, P. C. & Yang, C. C., Fixed points and factorization of meromorphic mappings, Complex Variables (to appear).

[121] Hu, P. C. & Yang, C. C., Dynamics in high dimensional spaces, Preprint.

[122] Hu, P. C. & Yang, C. C., Invariant sets in dynamics, Bulletin of the Hong Kong Math. Soc. 2(1998), 75-88.

[123] Hu, P. C. & Yang, C. C., An invariant property of Kobayashi metrics, Preprint.

[124] Hu, P. C. & Yang, C. C., Fatou-Julia theory in high dimensional spaces, International Workshop on The Value Distribution Theory and Its Applications, Hong Kong, 1996; Bull. of Hong Kong Math. Soc. 1 (2) (1997), 273-287.

[125] Hu, P. C. & Yang, C. C., Fatou-Julia theory in differentiable dynamics, Proc. of ISAAC'97 Congress, University of Delaware, U. S. A.

[126] Hu, P. C. & Yang, C. C., Dynamics of composite mappings, Proc. Japan Acad. 74(1998), 146-148.

[127] Hubbard, J. H. & Papadopol, P., Superattractive fixed points in $\mathbb{C}^n$, Indiana Univ. Math. J. 43(1994), 321-365.

[128] Hurewicz, W. & Wallman, H., Dimension theory, Princeton Univ. Press, Princeton, N. J., 1942.

[129] Hurley, M., On proofs of the C^0 general density theorem, Proc. Amer. Math. Soc. 124(1996), 1305-1309.

[130] Hurley, M., Properties of attractors of generic homeomorphisms, Ergod. Th. & Dynam. Sys. 16(1996), 1297-1310.

[131] Jung, H. W. E., Über ganze birationale transformationen der Ebene, J. Reine Angew. Math. 184(1942), 161-174.

[132] Karlin, S. & McGregor, J., Iteration of analytic functions of several variables, Problems in analysis (Edited by R. C. Gunning), Princeton Math. Ser. 31(1970), 81-92, Princeton Univ. Press.

[133] Katok, A., Lyapunov exponents, entropy and periodic orbits for diffeomorphisms, Pub. Math. (IHES) 51(1980), 137-173.

[134] Katok, A. & Hasselblatt, B., Introduction to the modern theory of dynamical systems, Encyclopedia of Math. & its Appl. 54(1995), Cambridge Univ. Press.

[135] Kelley, J. L., General Topology, Graduate Texts in Math. 27(1975), Springer-Verlag.

[136] Kennedy, J. A., The topology of attractors, Ergod. Th. & Dynam. Sys. 16(1996), 1311-1322.

[137] Khintchine, A., Zur Birkhoffs Lösung des Ergodenproblems, Math. Ann. 107(1932).

[138] Kiernan, P., On the relations between taut, tight and hyperbolic manifolds, Bull. Amer. Math. Soc. 76(1970), 49-51.

[139] Klingenberg, W., Riemannian geometry, de Gruyter Studies in Mathematics 1(1982), Walter de Gruyter, Berlin, New York.

[140] Klimek, M., Metrics associated with extremal plurisubharmonic functions, Proc. Amer. Math. Soc. 123(9)(1995), 2763-2770.

[141] Klimek, M., Pluripotential theory, Oxford University Press, 1993.

[142] Kneser, H., Math. Ann. 91(1924), 135-155.

[143] Kobayashi, S., Hyperbolic manifolds and holomorphic mappings, New York, Marcel Dekker, 1970.

[144] Kobayashi, S., Volume elements, holomorphic mappings and Schwarz lemma, Proc. Symposia in Pure Math. 11(1968), 253-260, Publ. Amer. Math. Soc..

[145] Kobayashi, S. & Nomizu, K., Foundations of differential geometry, I, Interscience Publishers, New York · London, 1963.

[146] Kobayashi, S. & Ochiai, T., Meromorphic mappings onto compact complex spaces of general type, Invent. Math. 31(1975), 7-16.

[147] Kodaira, K., Holomorphic mappings of polydiscs into compact complex manifold, J. Differential Geom. 6(1971), 33-46.

[148] Kosek, M., Hölder continuity property of filled-in Julia sets in $\mathbb{C}^n$, Proc. Amer. Math. Soc. 125(7)(1997), 2029-2032.

[149] Krantz, S. G., Function theory of severy complex variables, Wyley, New York, 1982.

[150] Krengel, U., Ergodic theorems, de Gruyter Studies in Math. 6(1985), Waiter de Gruyter, Berlin, New York.

[151] Kriete, H., On the dynamics of the relaxed Newton's method, Math. Institute der Georg-August-Universiät, 1997.

[152] Kubota, Y., Iteration of holomorphic maps of the unit ball into itself, Proc. Amer. Math. Soc. 88(1983), 476-480.

[153] Lang, S., Introduction to Complex Hyperbolic Spaces, Springer-Verlag, 1987.

[154] Lang, S., Higher dimensional Diophantine problems, Bull. Amer. Math. Soc. 80(1974), 779-787.

[155] Lang, S., Hyperbolic and Diophantine analysis, Bull. Amer. Math. Soc. 14(1986), 159-205.

[156] Ledrappier, F., Some relation between dimension and Lyapunov exponents, Commun. Math. Phys. 81(1981), 229-238.

[157] Ledrappier, F. & Young, L. S., The metric entropy of diffeomorphisms, Ann. Math. 122(1985), 509-574.

[158] Lelong, P., Domaines convexes par rapport aux fonctions plurisousharmoniques, J. Anal. Math. 2(1952), 178-208.

[159] Li, T. Y. & Yorke, J. A., Period three implies chaos, Amer. Math. Monthly 82(1975), 985-992.

[160] Liapounoff, A., Problème général de la stabilite du mouvement, Ann. Fac. Sci. Toulouse (2)(1907), 203-474, reprinted, Ann. Math. Studies no. 17.

[161] Ma, D., On iterates of holomorphic maps, Math. Z. 207(1991), 417-428.

[162] MacCluer, B. D., Iterates of holomorphic self-maps of the unit ball in $\mathbb{C}^N$, Michigan Math. J. 30(1983), 97-106.

[163] Mañé, R., A proof of the C^1-stability conjecture, IHES Publ. Math. 66(1988), 161-210.

[164] Mañé, R., Ergodic theory and differentiable dynamics, Springer-Verlag, 1987.

[165] Manning, A., A relation between Lyapunov exponents, Hausdorff dimension and entropy, Ergodic Theory Dyn. Systems 1(1981), 451-459.

[166] Marotto, F. R., Snap-back repellers imply chaos in $\mathbb{R}^n$, J. Math. Anal. Appl. 63(1978), 199-223.

[167] Marty, F., Reserches sur la répartition des valeurs d'une fonction méromorphe, Ann. Fac. Sci. Univ. Toulouse 23(1931), 183-196.

[168] Mather, J., Anosov diffeomorphism, Bull. Amer. Math. Soc. 73(1967), 747-817.

[169] Mazur, T., Recurrence theorem for biholomorphic mappings, Complex Variables 35(1998), 209-212.

[170] Mazur, T. & Skwarczyński, M., Spectral properties of holomorphic automorphism with fixed point, Glasgow Math. J. 28(1986), 25-30.

[171] McKay, J. H. & Wang, S. S., An elementary proof of the automorphism theorem for the polynomial ring in two variables; to appear.

[172] Milnor, J., Dynamics in one complex variable: Introductory Lectures, Suny Stony Brook, Institute for Mathematical Sciences, Preprint #1990/5.

[173] Milnor, J., Morse theory, Princeton Univ. Press, Princeton, N.J., 1963.

[174] Milnor, J., On the concept of attractor, Commun. Math. Phys. 99(1985), 177-195.

[175] Misiurewicz, M., Chaos almost everywhere, Lecture Notes in Math. 1163(1985), 125-130.

[176] Misiurewicz, M. & Smital, J., Smooth chaotic maps with zero topological entropy, Ergod. Th. & Dynam. Sys. 8(1988), 421-424.

[177] Morosawa, S., On the residual Julia sets of rational functions, Ergod. Th. & Dynam. Sys. 17(1997), 205-210.

[178] Moser, J., Lecture on Hamiltonian systems, New York, Courant Inst. Math. Sci., 1968.

[179] Myers, S. B. & Steenrod, N., The group of isometries of a Riemannian manifold, Ann. of Math. 40(1939), 400-416.

[180] Narasimhan, R., Analysis on real and complex manifolds, Masson & Cie, Paris, 1968.

[181] Narasimhan, R., Several complex variables, The Univ. of Chicago Press, Chicago and London, 1971.

[182] Narasimhan, R., Un Analogue holomorphe du théorème de Lindemann, Ann. Inst. Fourier, Grenoble 21(1971), 271-278.

[183] Nevanlinna, R., Einige Eindentigkeitssätze in der Theorie der meromorphe Functionen, Acta Math. 48(1926), 367-391.

[184] Nevanlinna, R., Le théorème de Picard-Borel et la théorie des functions méromorphes, Paris, 1929.

[185] Newhouse, S. E., Nondensity of axiom A(a) on S^2, Global Analysis (Proc. Sympos. Pure Math., Vol. 14, Berkeley, Calif., 1968), pp. 191-202, Amer. Math. Soc., Providence, R. I., 1970.

[186] Newhouse, S. E., Diffeomorphisms with infinitely many sinks, Topology 12(1974), 9-18.

[187] Newlander, A. & Nirenberg, L., Complex analytic coordinate in almost complex manifolds, Ann. of Math. 65(1957), 391-404.

[188] Noguchi, J., Moduli spaces of holomorphic mappings into hyperbolically imbedded complex spaces and locally symmetric spaces, Invent. Math. 93(1988), 15-34.

[189] Noguchi, J., Hyperbolic fibre spaces and Mordell's conjecture over function fields, Publ. RIMS, Kyto University 21(1985), 27-46.

[190] Noguchi, J. & Ochiai, T., Geometric Function Theory in Several Complex Variables, Transl. Math. Monographs 80(1990), Amer. Math. Soc., Providence, Rhode Island.

[191] Oka, K., Domaines pseudoconvexes, Tokoku Math. J. 49(1942), 15-52.

[192] Palis, J., Pugh, C., Shub, M. & Sullivan, D., Genericity theorem in topological dynamics, Dynamical Systems–Warwick 1974, Lecture Notes in Math. 468(1975), 241-250, Springer.

[193] Pesin, Ya. B. & Sinai, Ya. G., Hyperbolicity and stochasticity of dynamical systems, Mathematical Physics Reviews, Gordon and Breach Press, Harwood Acad. Publ., USA, pp.53-115, 1981.

[194] Petersen, K., Ergodic theory, Cambridge studies in advanced Math. 2(1983), Cambridge Univ. Press.

[195] Pilyugin, S. Yu., The space of dynamical systems with the C^0-topology, Lecture Notes in Mathematics 1571(1994), Springer-Verlag.

[196] Pliss, V. A., Integral sets of periodic systems of differential equations (in Russian). Moscow. 1977.

[197] Poincaré, H., Les méthodes nouvelles de la mécanique céleste, I(1892), II(1893), III(1899), Gauthiers-Villars, Paris. Also Dover, New York, 1957 and NASA TTF 450-2, Washington, D. C., 1967.

[198] Poincaré, H., Sur une classe nouvelle de transcendantes uniformes, J. de Math. 6(1890), 313-365.

[199] Pollicott, M., Lectures on ergodic theory and Pesin theory on compact manifolds, London Math. Soc. Lecture Note series 180(1993), Cambridge Univ. Press.

[200] Poon, K. K. & Yang, C. C., Dynamics of composite functions, preprint.

[201] Pugh, Ch., The closing lemma, Amer. J. Math. 89(1967), 956-1009; 1010-1021.

[202] Qiao, J. Y., Buried points of Julia sets for rational or entire functions, Sci. in China Ser. A 25 (11) (1995), 1139-1146.

[203] Qiao, J. Y., Topological complexity of Julia sets, Sci. in China Ser. A 27(1997), 775-781.

[204] Qiu, W. Y., Ren, F. Y. & Yin, Y. C., Iterate for small random perturbations of rational functions and polynomials, Preprint.

[205] Raikov, D. A., On some arithmetical properties of summable functions, Mat. Sb. 1(43) (1936), 377-384.

[206] Reich, L., Das Typenproblem bei formal holomorphen Abbildungen mit anziehendem Fixpunkt, Math. Ann. 179(1969), 227-250.

[207] Reich, L., Normalformen biholomorpher Abbildungen mit anziehendem Fixpunkt, Math. Ann. 180(1969), 233-255.

[208] Remmert, R., Holomorphe und meromorphe Abbildungen komplexer Räume, Math. Annalen 133(1957), 328-370.

[209]* Ren, F. Y., Complex analytic dynamics, Fudan Univ. Press, 1997.

[210] Rentschler, R., Opérations du groupe additif sur lc plan affine, C. R. Acad. Sc. Paris 267(1968), 384-387.

[211] Reshetnyak, Yu. G., Space mappings with bounded distortion, Translations of Math. Monographs, Vol. 73, Amer. Math. Soc., 1989.

[212] Riesz, F., Sur la théorie ergodique, Comment. Math. Helv. 17 (1945), 221-239.

[213] Robbin, J., A structural stability theorem, Ann. Math. 94(1971), 447-493.

[214] Robinson, C., Structural stability for C^1-diffeomorphisms, J. Diff. Equat. 22(1976), 28-73.

[215] Rosay, J. P., Un exemple d'ouvert borné de $\mathbb{C}^3$ "taut" mais non hyperbolique complet, Pacific J. Math. 98(1982), 153-156.

[216] Rosay, J. P. & Rudin, W., Holomorphic maps from $\mathbb{C}^n$ to $\mathbb{C}^n$, Trans. Amer. Math. Soc. 310(1988), 47-86.

[217] Rosenbloom, P. C., The fix-points of entire functions, Medd. Lunds Univ. Mat. Sem. Suppl. Bd. M. Riesz (1952), 186-192.

[218] Rudin, W., Function theory in the unit ball of $\mathbb{C}^n$, Grundlehren der Math., 241, Springer, New York, 1980.

[219] Ruelle, D., An inequality for the entropy of differentiable maps, Boletim da Sociedade Brasileira Matemática 9(1978), 83-87.

[220] Ruelle, D., Elements of differentiable dynamics and bifurcation theory, Acad. Press, 1989.

[221] Rund, H., The differential geometry of Finsler spaces, Springer-Verlag, 1959.

[222] Rüssmann, H., Über das verhalten analytischer Hamiltonscher Differentiagleichungen in der Nähe einer Gleichgewichtslösung, Math. Ann. 154(1964), 285-300.

[223] Sadullaev, A., On Fatou's example, Mat. Zametki 6(1969), 437-441 (Russian), translated in Math. Notes 6(1969), 717-719.

[224] Schröder, E., Ueber unendlich viele Algorithmen zur Auflösung der Gleichungen, Math. Ann. 2(1870), 317-365.

[225] Schröder, E., Uber iterierte Funktionen, Math. Ann. 3(1871), 296-322.

[226] Schwartz, A. J., Amer. J. Math. 85(1963), 453-458.

[227] Shields, A. L., On fixed points of commuting analytic functions, Proc. Amer. Math. Soc. 15(1964), 703-706.

[228] Shub, M., Global stability of dynamical systems, Springer-Verlag, 1987.

[229] Shub, M. & Sullivan, D., Topology 13(1974), p.189.

[230] Sinai, Ya. G., Dynamical systems II, Encyclopaedia of Mathematical Sciences, Springer-Verlag, 1989.

[231] Smale, S., Dynamical systems on n-dimensional manifolds, Differential Equations and Dynamical Systems, Proc. Int. Puerto Rico 1965, 483-486.

[232] Smale, S., Differentiable dynamical systems, Bull. Amer. Math. Soc. 73(1967), 747-817.

[233] Smale, S., On the efficiency of algorithms of analysis, Bull. Amer. Math. Soc. 13(1985), 87-121.

[234] Smale, S., Complexity and real computation, City Univ. of Hong Kong, 1996.

[235] Smale, S., Mathematical problems for the next century, preprint.

[236] Smale, S., Dynamical systems and the topological conjugacy problem for diffeomorphisms, Proc. Int. Congress Math. (Stockholm, 1962), pp. 490-496; Inst. Mittag-Leffler, Djursholm(1963).

[237] Smillie, J., The entropy of polynomial diffeomorphisms of $\mathbb{C}^2$, Ergodic Theory & Dynamical Systems 10(1990), 823-827.

[238] Smital, J., A chaotic function with some extremal properties, Proc. Amer. Math. Soc. 87(1)(1983), 54-56.

[239] Smital, J., A chaotic function with scrambled set of positive Lebesgue measure, Proc. Amer. Math. Soc. 92(1)(1984), 50-54.

[240] Steenrod, N., The topology of fibre bundles, Princeton Univ. Press, Princeton, 1951.

[241] Stehlé, J. L., Plongements du disque dans $\mathbb{C}^2$, Séminaire P. Lelong (Analysis) Lecture Notes in Math. 275(1970), 119-130, Springer-Verlag.

[242] Sternberg, S., Local contractions and a theorem of Poincaré, Amer. J. Math. 79(1957), 809-823.

[243] Stoll, W., Algebroid reduction of Nevanlinna theory, Lecture Notes in Mathematics 1277(1987), 131-241, Springer-Verlag.

[244] Stoll, W., The multiplicity of a holomorphic map, Invent. Math. 2(1966), 15-58.

[245] Stoll, W., Value distribution on parabolic spaces, Lecture Notes in Mathematics 600(1977), Springer-Verlag.

[246] Stoll, W., The characterization of strictly parabolic manifolds, Ann. Scuola. Norm. Sup. Pisa 7(1980), 87-154.

[247] Stoll, W., The characterization of strictly parabolic manifolds, Compositio Math. 44(1981), 305-373.

[248] Stoll, W., The Ahlfors-Weyl theory of meromorphic maps on parabolic manifolds, Lecture Notes in Mathematics 981(1983), 101-219, Springer-Verlag.

[249] Stoll, W., A Bézout estimate for complete intersections, Ann. of Math. 96(1972), 361-401.

[250] Stolzenberg, G., Volumes, limits, and extensions of analytic varieties, Lecture Notes in Mathematics 19(1966), Springer.

[251] Suffridge, T. J., Common fixed points of commuting holomorphic maps of the hyperball, Michigan Math. J. 21(1974), 309-314.

[252] Suzuki, M., Sur les opérations holomorphes du groupe additif complexe sur l'espace de deux variables complexes (I), Ann. sc. Éc. Norm. Sup., 4 Ser., 10(1977), 517-546.

[253] Suzuki, M., Propriétés topologiques des polynômes de deux variables complexes, et automorphismes algébriques de l'espace $\mathbb{C}^2$, J. Math. Soc. Japan 26(1973), 241-257.

[254] Szlenk, W., An introduction to the theory of smooth dynamical systems, PWN-Polish Scientific Publishers, Warszawa, 1984.

[255] Targonski, G., Topics in Iteration Theory, Studia Mathematica Skript 6, Vandenhoeck & Ruprecht in Göttingen, 1981.

[256] Tsuji, M., Potential theory in modern function theory, Maruzen Co., LTD., Tokyo, 1959.

[257] Ueda, T., Fatou set in complex dynamics in projective spaces, J. Math. Soc. Japan 46(1994), 545-555.

[258] Ueda, T., Critical orbits of holomorphic maps on projective spaces, to appear in Journal of Geometric Analysis.

[259] Ura, T., On the flow outside a closed invariant set Contr. Diff. Eq. 3(1964), 249-294.

[260] van Dantzig, D. & van der Waerden, B. L., Über metrisch homogenen Räume, Abhandl. Math. Sem. Univ. Hamburg 6(1928), 374-376.

[261] Van der Kulk, W., On polynomial rings in two variables, Nieuw Arch. Wisk (3)1(1953), 33-41.

[262] Vitter, A., The lemma of the logarithmic derivative in several complex variables, Duke Math. J. 44(1977), 89-104.

[263] von Neumann, J., Proof of the quasi-ergodic hypothesis, Proc. Nat. Acad. Sci. USA 18 (1932), 70-82.

[264] Walters, P., An introduction to ergodic theory, GTM 79(1982), Springer, Berlin.

[265] Wiener, N., The ergodic theorem, Duke Math. J. 5(1939), 1-18.

[266] Wilson, W., Smoothing derivatives of functions and applications, Trans. Amer. Math. Soc. 139(1969), 413-428.

[267] Wolff, J., Sur l'itération des fonctions bornées, C. R. Acad. Sci. Paris, Sér I 182(1926), 200-201.

[268] Wong, P. M., Geometry of the homogeneous complex Monge-Ampère equation, Invent. Math. 67(1982), 261-274.

[269] Wu, He, Stable manifolds of automorphisms of $\mathbb{C}^n$ at fixed points, Complex Variables 20(1992), 139-144.

[270] Wu, He, Complex Pesin's stable manifold theorem and "generalized Hénon" mappings, preprint, SISSA 123/91/M(1991).

[271] Wu, He, Complex stable manifold of holomorphic diffeomorphisms, Indiana Univ. Math. J. 42(1993), 1349-1358.

[272] Wu, H., Normal families of holomorphic mappings, Acta Math. 119(1967), 193-233.

[273] Wu, H., Remarks on the first main theorem in equidistribution theory, I, II, III, IV, J. Diff. Geom. 2(1968), 197-202; 369-384; ibid 3(1969), 83-94; 433-446.

[274] Wu, H., The equidistribution theory of holomorphic curves, Annals of Math. Studies 64, Princeton Univ. Press, Princeton, NJ, 1970.

[275] Yang, C. C. & Zheng, J. H., On the fix-points of composite meromorphic functions and generalizations, J. d'Analyse Math. 68(1996), 59-93.

[276] Yau, S. T., A general Schwarz lemma for Kähler manifolds, Amer. J. of Math. 100(1978), 197-203.

[277] Yin, Y. C., Discontinuity of Julia sets for polynomials, Acta Math. Sinica 38(1995), 99-102.

[278] Yoccoz, J. C., Travaux de Herman sur les tores invariants, Seminaire Bourbaki, no. 154, Asterisque 206(1992), 311-344.

[279] Yomdin, Y., Nonautonomous linearization, Lecture Notes in Mathematics 1342(1988), 718-726, Springer-Verlag.

[280] Yosida, K. & Kakutani, S., Birkhoff's ergodic theorem and the maximal ergodic theorem, Proc. Japan Acad. 15(1939), 165-168.

[281] Young, L. S., Dimension, entropy and Lyapunov exponents, Ergod. Th. & Dynam. Sys. 2(1982), 109-124.

[282] Zalcman, L., A heuristic principle in complex function theory, Amer. Math. Monthly 82(1975), 813-817.

[283] Zehnder, E., The Arnold conjecture for fixed points of symplectic mappings and periodic solutions of Hamiltonian systems, Proceedings of the International Congress of Mathematicians, Berkeley 1986, pp. 1237-1246.

[284] Zhang, W. J., Periodic points of polynomial self-mappings on $\mathbb{C}^n$, Preprint.

[285] Zhang, W. J. & Ren, F. Y., Iterations of holomorphic self-maps of $\mathbb{C}^N$, J. of Fudan University(Natural Science) 33(1994), 452-462.

Index

335

Other *Mathematics and Its Applications* titles of interest:

V.I. Istratescu: *Fixed Point Theory. An Introduction.* 1981, 488 pp.
out of print, ISBN 90-277-1224-7

A. Wawrynczyk: *Group Representations and Special Functions.* 1984, 704 pp.
ISBN 90-277-2294-3 (pb), ISBN 90-277-1269-7 (hb)

R.A. Askey, T.H. Koornwinder and W. Schempp (eds.): *Special Functions: Group Theoretical Aspects and Applications.* 1984, 352 pp.   ISBN 90-277-1822-9

A.V. Arkhangelskii and V.I. Ponomarev: *Fundamentals of General Topology. Problems and Exercises.* 1984, 432 pp.   ISBN 90-277-1355-3

J.D. Louck and N. Metropolis: *Symbolic Dynamics of Trapezoidal Maps.* 1986, 320 pp.
ISBN 90-277-2197-1

A. Bejancu: *Geometry of CR-Submanifolds.* 1986, 184 pp.   ISBN 90-277-2194-7

R.P. Holzapfel: *Geometry and Arithmetic Around Euler Partial Differential Equations.* 1986, 184 pp.   ISBN 90-277-1827-X

P. Libermann and Ch.M. Marle: *Sympletic Geometry and Analytical Mechanics.* 1987, 544 pp.   ISBN 90-277-2438-5 (hb), ISBN 90-277-2439-3 (pb)

D. Krupka and A. Svec (eds.): *Differential Geometry and its Applications.* 1987, 400 pp.
ISBN 90-277-2487-3

Shang-Ching Chou: *Mechanical Geometry Theorem Proving.* 1987, 376 pp.
ISBN 90-277-2650-7

G. Preuss: *Theory of Topological Structures. An Approach to Categorical Topology.* 1987, 318 pp.   ISBN 90-277-2627-2

V.V. Goldberg: *Theory of Multicodimensional (n+1)-Webs.* 1988, 488 pp.
ISBN 90-277-2756-2

C.T.J. Dodson: *Categories, Bundles and Spacetime Topology.* 1988, 264 pp.
ISBN 90-277-2771-6

A.T. Fomenko: *Integrability and Nonintegrability in Geometry and Mechanics.* 1988, 360 pp.   ISBN 90-277-2818-6

L.A. Cordero, C.T.J. Dodson and M. de Leon: *Differential Geometry of Frame Bundles.* 1988, 244 pp.   *out of print,* ISBN 0-7923-0012-2

E. Kratzel: *Lattice Points.* 1989, 322 pp.   ISBN 90-277-2733-3

E.M. Chirka: *Complex Analytic Sets.* 1989, 396 pp.   ISBN 0-7923-0234-6

Kichoon Yang: *Complete and Compact Minimal Surfaces.* 1989, 192 pp.
ISBN 0-7923-0399-7

A.D. Alexandrov and Yu.G. Reshetnyak: *General Theory of Irregular Curves.* 1989, 300 pp.
ISBN 90-277-2811-9

B.A. Plamenevskii: *Algebras of Pseudodifferential Operators.* 1989, 304 pp.
ISBN 0-7923-0231-1

Other *Mathematics and Its Applications* titles of interest:

Ya.I. Belopolskaya and Yu.L. Dalecky: *Stochastic Equations and Differential Geometry.* 1990, 288 pp. ISBN 90-277-2807-0

V. Goldshtein and Yu. Reshetnyak: *Quasiconformal Mappings and Sobolev Spaces.* 1990, 392 pp. ISBN 0-7923-0543-4

A.T. Fomenko: *Variational Principles in Topology. Multidimensional Minimal Surface Theory.* 1990, 388 pp. ISBN 0-7923-0230-3

S.P. Novikov and A.T. Fomenko: *Basic Elements of Differential Geometry and Topology.* 1990, 500 pp. ISBN 0-7923-1009-8

B.N. Apanasov: *The Geometry of Discrete Groups in Space and Uniformization Problems.* 1991, 500 pp. ISBN 0-7923-0216-8

C. Bartocci, U. Bruzzo and D. Hernandez-Ruiperez: *The Geometry of Supermanifolds.* 1991, 242 pp. ISBN 0-7923-1440-9

N.J. Vilenkin and A.U. Klimyk: *Representation of Lie Groups and Special Functions. Volume 1: Simplest Lie Groups, Special Functions, and Integral Transforms.* 1991, 608 pp. ISBN 0-7923-1466-2

A.V. Arkhangelskii: *Topological Function Spaces.* 1992, 206 pp.   ISBN 0-7923-1531-6

Kichoon Yang: *Exterior Differential Systems and Equivalence Problems.* 1992, 196 pp. ISBN 0-7923-1593-6

M.A. Akivis and A.M. Shelekhov: *Geometry and Algebra of Multidimensional Three-Webs.* 1992, 358 pp. ISBN 0-7923-1684-3

A. Tempelman: *Ergodic Theorems for Group Actions.* 1992, 400 pp.
ISBN 0-7923-1717-3

N.Ja. Vilenkin and A.U. Klimyk: *Representation of Lie Groups and Special Functions, Volume 3. Classical and Quantum Groups and Special Functions.* 1992, 630 pp.
ISBN 0-7923-1493-X

N.Ja. Vilenkin and A.U. Klimyk: *Representation of Lie Groups and Special Functions, Volume 2. Class I Representations, Special Functions, and Integral Transforms.* 1993, 612 pp. ISBN 0-7923-1492-1

I.A. Faradzev, A.A. Ivanov, M.M. Klin and A.J. Woldar: *Investigations in Algebraic Theory of Combinatorial Objects.* 1993, 516 pp. ISBN 0-7923-1927-3

M. Puta: *Hamiltonian Mechanical Systems and Geometric Quantization.* 1993, 286 pp.
ISBN 0-7923-2306-8

V.V. Trofimov: *Introduction to Geometry of Manifolds with Symmetry.* 1994, 326 pp.
ISBN 0-7923-2561-3

J.-F. Pommaret: *Partial Differential Equations and Group Theory. New Perspectives for Applications.* 1994, 473 pp. ISBN 0-7923-2966-X

Other *Mathematics and Its Applications* titles of interest:

Kichoon Yang: *Complete Minimal Surfaces of Finite Total Curvature.* 1994, 157 pp.
ISBN 0-7923-3012-9

N.N. Tarkhanov: *Complexes of Differential Operators.* 1995, 414 pp.
ISBN 0-7923-3706-9

L. Tamássy and J. Szenthe (eds.): *New Developments in Differential Geometry.* 1996, 444 pp.
ISBN 0-7923-3822-7

W.C. Holland (ed.): *Ordered Groups and Infinite Permutation Groups.* 1996, 255 pp.
ISBN 0-7923-3853-7

K.L. Duggal and A. Bejancu: *Lightlike Submanifolds of Semi-Riemannian Manifolds and Applications.* 1996, 308 pp.
ISBN 0-7923-3957-6

D.N. Kupeli: *Singular Semi-Riemannian Geometry.* 1996, 187 pp.   ISBN 0-7923-3996-7

L.N. Shevrin and A.J. Ovsyannikov: *Semigroups and Their Subsemigroup Lattices.* 1996, 390 pp.
ISBN 0-7923-4221-6

C.T.J. Dodson and P.E. Parker: *A User's Guide to Algebraic Topology.* 1997, 418 pp.
ISBN 0-7923-4292-5

B. Rosenfeld: *Geometry of Lie Groups.* 1997, 412 pp.   ISBN 0-7923-4390-5

A. Banyaga: *The Structure of Classical Diffeomorphism Groups.* 1997, 208 pp.
ISBN 0-7923-4475-8

A.N. Sharkovsky, S.F. Kolyada, A.G. Sivak and V.V. Fedorenko: *Dynamics of One-Dimensional Maps.* 1997, 272 pp.   ISBN 0-7923-4532-0

A.T. Fomenko and S.V. Matveev: *Algorithmic and Computer Methods for Three-Manifolds.* 1997, 346 pp.   ISBN 0-7923-4770-6

A. Mallios: *Geometry of Vector Sheaves.* An Axiomatic Approach to Differential Geometry. Volume I: Vector Sheaves. General Theory. 1998, 462 pp.   ISBN 0-7923-5004-9
Volume II: Geometry. Examples and Applications. 1998, 462 pp.   ISBN 0-7923-5005-7
(Set of 2 volumes: 0-7923-5006-5)

G. Luke and A.S. Mishchenko: *Vector Bundles and Their Applications.* 1998, 262 pp.
ISBN 0-7923-5154-1

D. Repovš and P.V. Semenov: *Continuous Selections of Multivalued Mappings.* 1998, 364 pp.   ISBN 0-7923-5277-7

E. García-Río and D.N. Kupeli: *Semi-Riemannian Maps and Their Applications.* 1999, 208 pp.   ISBN 0-7923-5661-6

K.L. Duggal: *Symmetries of Spacetimes and Riemannian Manifolds.* 1999, 224 pp.
ISBN 0-7923-5793-0

P.-C. Hu and C.-C. Yang: *Differentiable and Complex Dynamics of Several Variables.* 1999, 354 pp.   ISBN 0-7923-5771-X